Iosif Ilyich Gikhman was born on the 26th of May 1918 in the city of Uman, Ukraine. He studied in Kiev, graduating in 1939, then remained there to teach and do research under the supervision of N. Bogolyubov, defending a "candidate" thesis on the influence of random processes on dynamical systems in 1942 and a doctoral dissertation on Markov processes and mathematical statistics in 1955.
I.I. Gikhman is one of the founders of the theory of stochastic differential equations and also contributed significantly to mathematical statistics, limit theorems,multidimensional martingales, and stochastic control. He died in 1985, in Donetsk.

Anatoli Vladimirovich Skorokhod was born on September 10th, 1930 in the city Nikopol, Ukraine. He graduated from Kiev University in 1953, after which his graduate studies at Moscow University were directed by E.B.Dynkin.
From 1956 to 1964 Anatoli Skorokhod was a professor of Kiev University. Thereafter he worked at the Institute of Mathematics of the Ukrainian Academy of Sciences, but he has also, since 1993, been professor of Statistics and Probability at Michigan State University. Skorokhod was elected to the Ukrainian Academy of Sciences in 1985 and became a Fellow of American Academy of Arts and Sciences in 2000.
His mathematical research interests are the theory of stochastic processes, stochastic differential equations,Markov processes, randomly perturbed dynamical systems.

Classics in Mathematics

Iosif I. Gikhman · Anatoli V. Skorokhod

The Theory of Stochastic Processes III

Iosif I. Gikhman · Anatoli V. Skorokhod

The Theory of Stochastic Processes III

Reprint of the 1974 Edition

Authors

Iosif Ilyich Gikhman

Anatoli Vladimirovich Skorokhod

Translator

Samuel Kotz

Originally published as Vol. 218 of the
Grundlehren der mathematischen Wissenschaften

Mathematics Subject Classification (2000): 34F05, 60Hxx

Library of Congress Control Number: 2007900646

ISSN 1431-0821
ISBN 978-3-540-49940-4 Springer Berlin Heidelberg New York

Springer is a part of Springer Science+Business Media

springer.com

Production: LE-TEX Jelonek, Schmidt & Vöckler GbR, Leipzig
Cover-design: WMX Design GmbH, Heidelberg

SPIN 11920502 41/3100YL - 5 4 3 2 1 0 Printed on acid-free paper

Grundlehren der mathematischen Wissenschaften 232

A Series of Comprehensive Studies in Mathematics

I. I. Gihman   A. V. Skorohod

The Theory of Stochastic Processes III

Translated from the Russian
by S. Kotz

Springer-Verlag
Berlin Heidelberg New York

Iosif Il'ich Gihman

Academy of Sciences of the Ukranian SSR
Institute of Applied Mathematics and Mechanics
Donetsk
USSR

Anatolii Vladimirovich Skorohod

Academy of Sciences of the Ukranian SSR
Institute of Mathematics
Kiev
USSR

Translator:

Samuel Kotz

Department of Mathematics
Temple University
Philadelphia, PA 19122
USA

AMS Subject Classification: 34F05, 60Hxx

Library of Congress, Cataloging in Publication Data

Gikhman, Iosif Il'ich.
The theory of stochastic processes.
(Die Grundlehren der mathematischen Wissenschaften in Einzeldarstellungen mit besonderer Berücksichtigung der Anwendungsgeblete; Bd. 210, 218, 232)
Translation of Teoriya sluchainyh protsessov.
Includes bibliographies and indexes.
1. Stochastic processes. I. Skorokhod, Anatolli Vladimirovich, joint author. II. Title. III. Series: Die Grundlehren der mathematischen Wissenschaften in Einzeldarstellungen; Bd. 210, [etc.].
QA274.G5513 519.2 74-2552
ISBN 0-387-06573-3 (v. I)
ISBN 0-387-07247-0 (v. II)

Title of the Russian Original Edition: Teoriya sluchainyh protsessov, Tom III. Publisher: Nauka, Moscow, 1975.

Printed in the United States of America.

9 8 7 6 5 4 3 2 1

ISBN 0-387-90375-5 New York Heidelberg Berlin
ISBN 3-540-90375-5 Berlin Heidelberg New York

Preface

It was originally planned that the *Theory of Stochastic Processes* would consist of two volumes: the first to be devoted to general problems and the second to specific classes of random processes. It became apparent, however, that the amount of material related to specific problems of the theory could not possibly be included in one volume. This is how the present third volume came into being.

This volume contains the theory of martingales, stochastic integrals, stochastic differential equations, diffusion, and continuous Markov processes.

The theory of stochastic processes is an actively developing branch of mathematics, and it would be an unreasonable and impossible task to attempt to encompass it in a single treatise (even a multivolume one). Therefore, the authors, guided by their own considerations concerning the relative importance of various results, naturally had to be selective in their choice of material. The authors are fully aware that such a selective process is not perfect. Even a number of topics that are, in the authors' opinion, of great importance could not be included, for example, limit theorems for particular classes of random processes, the theory of random fields, conditional Markov processes, and information and statistics of random processes.

With the publication of this last volume, we recall with gratitude our associates who assisted us in this endeavor, and express our sincere thanks to G. N. Sytaya, L. V. Lobanova, P. V. Boĭko, N. F. Ryabova, N. A. Skorohod, V. V. Skorohod, N. I. Portenko, and L. I. Gab.

I. I. Gihman and A. V. Skorohod

Table of Contents

Chapter I

Martingales and Stochastic Integrals

§1. Martingales and Their Generalizations

Survey of preceding results. We start by recalling and making more precise the definitions and previously obtained results pertaining to martingales and semimartingales (cf. Volume I, Chapter II, Section 2 and Chapter III, Section 4).

Let $\{\Omega, \mathfrak{S}, \mathsf{P}\}$ be a probability space, let T be an arbitrary ordered set (in what follows only those cases where T is a subset of the extended real line $[-\infty, +\infty]$ will be discussed) and let $\{\mathfrak{F}_t, t \in T\}$ be a current of σ-algebras ($\mathfrak{F}_t \subset \mathfrak{S}$): if $t_1 < t_2$ then $\mathfrak{F}_{t_1} \subset \mathfrak{F}_{t_2}$. The symbol $\{\xi(t), \mathfrak{F}_t, t \in T\}$ or simply $\{\xi(t), \mathfrak{F}_t\}$ denotes an object consisting of a current of σ-algebras $\{\mathfrak{F}_t, t \in T\}$ on the measurable space $\{\Omega, \mathfrak{S}\}$ and a random process $\xi(t)$, $t \in T$, adopted to $\{\mathfrak{F}_t, t \in T\}$ (i.e., $\xi(t)$ if $\mathfrak{F}_t$-measurable for each $t \in T$). This object will also be referred to in what follows as a *random process*.

A random process $\{\xi(t), \mathfrak{F}_t, t \in T\}$ is called an *$\mathfrak{F}_t$-martingale* (or *martingale* if there is no ambiguity concerning the current of σ-algebras $\mathfrak{F}_t$ under consideration) provided

$$\mathsf{E}|\xi(t)| < \infty \qquad \forall t \in T \tag{1}$$

and

$$\mathsf{E}\{\xi(t) \mid \mathfrak{F}_s\} = \xi(s) \quad \text{for } s < t,\ s, t \in T;$$

it is called a *supermartingale* (*submartingale*) if it satisfies condition (1) and moreover

$$\begin{aligned} &\mathsf{E}\{\xi(t) \mid \mathfrak{F}_s\} \leqslant \xi(s), \qquad s < t, \quad s, t \in T \\ &(\mathsf{E}(\xi(t) \mid \mathfrak{F}_s\} \geqslant \xi(s), \qquad s < t). \end{aligned} \tag{2}$$

Observe that the above definition differs from that presented in Volume I since we now require finiteness of the mathematical expectation of the quantity $\xi(t)$ in all cases. Previously, in the case of the supermartingale, for example, only the finiteness of the expectation $\mathsf{E}\xi^-(t)$ was assumed.

The definition presented herein is equivalent to the following: $\{\xi(t), \mathfrak{F}_t, t \in T\}$ is a martingale (supermartingale) if for any set $B_s \in \mathfrak{F}_s$ and for any s and t belonging to T such that $s < t$,

$$\int_{B_s} \xi(t) d\mathsf{P} = \int_{B_s} \xi(s)\, d\mathsf{P} \qquad \left(\int_{B_s} \xi(t)\, d\mathsf{P} \leq \int_{B_s} \xi(s)\, d\mathsf{P}\right).$$

Supermartingales and submartingales are also called *semimartingales*.

In this section we shall consider mainly semimartingales of a continuous argument.

The space of all real-valued functions on the interval $[0, T]$ which possess the left-hand limit for each $t \in (0, T]$ and which are continuous from the right on $[0, T)$ will be denoted by $\mathscr{D}$ or by $\mathscr{D}[0, T]$.

Analogous meaning is attached to the notation $\mathscr{D}[0, T)$, $\mathscr{D}[0, \infty)$, and $\mathscr{D}[0, \infty]$.

A number of inequalities and theorems concerning the existence of limits plays an important role in the martingale theory. The following relationships were established in Volume I, Chapter II, Section 2:

If $\xi(t)$, $t \in T$, is a separable submartingale, then

$$\mathsf{P}\left\{\sup_{t\in T} \xi^+(t) \geq C\right\} \leq \frac{\sup_{t\in T} \mathsf{E}\xi^+(t)}{C},$$

$$\mathsf{E}\left[\sup_{t\in T} \xi^+(t)\right]^p \leq q^p \sup_{t\in T} \mathsf{E}[\xi^+(t)]^p, \qquad q = \frac{p}{p-1}, \quad p > 1, \tag{4}$$

$$\mathsf{E}\nu[a, b) \leq \sup_{t\in T} \frac{\mathsf{E}(\xi(t) - b)^+}{b - a}; \tag{5}$$

here $a^+ = a$ for $a \geq 0$ and $a^+ = 0$ for $a < 0$ and $\nu[a, b)$ denotes the number of crossings downward of the half-interval $[a, b)$ by the sample function of the process $\xi(t)$ (a more precise definition is given in Volume I, Chapter II, Section 2).

We now recall the definition of a closure of a semimartingale.

Let $\{\xi(t), \mathfrak{F}_t, t \in T\}$ be a semimartingale and let the set T possess no largest (smallest) element. The random variable η is called a *closure from the right* (*left*) of *the semimartingale* $\xi(t)$ if one can extend the set T by adding one new element b (a) which satisfies

$$t < b \ (t > a) \qquad \forall t \in T$$

and complete the current of σ-algebras $\{\mathfrak{F}_t, t \in T\}$ by adding the corresponding σ-algebra $\mathfrak{F}_b$ ($\mathfrak{F}_a$) so that the extended family of the random variables $\xi(t)$, $t \in T'$, $T' = T \cup \{b\}$ ($T' = T \cup \{a\}$) also forms an $\mathfrak{F}_t$-semimartingale.

Theorem 1. *Let $\xi(t)$, $t \in T$, be a separable submartingale, $T \subset (a, b)$, and the points a and b be the limit points for the set T ($-\infty \leq a < b \leq \infty$). Then a set Λ of probability 0 exists such that for $\omega \notin \Lambda$:*

a) *in every interior point t of the set T the limits $\xi(t-)$ and $\xi(t+)$ exist;*

b) *if* $\sup\{\mathsf{E}\xi^+(t), t\in T\}<\infty$, *then the limit $\xi(b-)$ exists; moreover, if for some t_0 the family of random variables $\{\xi(t), t\in[t_0, b)\}$ is uniformly integrable, then the limit $\xi(b-)$ exists in L_1 as well and $\xi(b-)$ is a closure from the right of the submartingale;*

c) *if* $\lim_{t\to a}\mathsf{E}\xi(t)>-\infty$ *then the family of random variables $\{\xi(t), t\in(a, t_0]\}$ is uniformly integrable, the limit $\xi(a+)$ exists for every $\omega\notin\Lambda$ also in the sense of convergence in L_1, and $\xi(a+)$ is a closure from the left of the submartingale.*

Proof. The existence with probability 1 of the one-sided limits $\xi(t-)$ and $\xi(t+)$ for each $t\in[a, b]$ under the condition that $\sup\{\mathsf{E}\xi^+(t), t\in(a, b)\}<\infty$ was established in Volume I, Chapter III, Section 4.

Furthermore, if the family $\{\xi(t), t\in[t_0, b)\}$ is uniformly integrable then the convergence of $\xi(t)$ to $\xi(b-)$ with probability 1 as $t\uparrow b$ implies that this convergence is also valid in L_1 and that $\xi(b-)$ is a closure from the right of the submartingale $\xi(t)$, $t\in(a, b)$. This is proved in the same manner as in the case of submartingales of a discrete argument (Volume I, Chapter II, Section 2).

We need now to verify assertion c).

Let $l=\lim_{t\downarrow a}\mathsf{E}\xi(t)$. Since $\mathsf{E}\xi(t)$ is a monotonically nondecreasing function the existence of the limit is assured. Moreover, the equality

$$|\xi(t)|=2\xi^+(t)-\xi(t)$$

implies that

$$\sup_{t\in(a,t_0]}\mathsf{E}|\xi(t)|\leq 2\mathsf{E}\xi^+(t_0)-l=C<\infty.$$

In view of Chebyshev's inequality $\mathsf{P}(B_t)\leq C/N$, where $B_t=\{|\xi(t)|>N\}$, i.e., $\mathsf{P}(B_t)\to 0$ as $N\to\infty$ uniformly in t. Let $\varepsilon>0$ be arbitrary and t_1 be such that $\mathsf{E}\xi(t)-l<\varepsilon/2$ for all $t<t_1$. Then for $t\in(a, t_1]$

$$\begin{aligned}\int_{B_t}|\xi(t)|\,d\mathsf{P}&=\int_{\{\xi(t)>N\}}\xi(t)\,d\mathsf{P}+\int_{\{\xi(t)>-N\}}\xi(t)\,d\mathsf{P}-\mathsf{E}\xi(t)\\&\leq\int_{\{\xi(t)>N}\xi(t_1)\,d\mathsf{P}+\int_{\{\xi(t)>-N\}}\xi(t_1)\,d\mathsf{P}-\mathsf{E}\xi(t)\\&\leq\int_{B_t}|\xi(t_1)|\,d\mathsf{P}+\frac{\varepsilon}{2},\end{aligned}$$

so that $\int_{B_t}|\xi(t)|\,d\mathsf{P}<\varepsilon$ for all $t\in(a, t_1]$ and N sufficiently large. Thus the family $\{\xi(t), t\in(a, t_0]\}$ is uniformly integrable. The limit $\lim_{t\downarrow a}\xi(t)$ exists with probability 1; therefore it also exists in the sense of convergence in L_1.

Now $\xi(a+)$ is indeed a closure from the left of the submartingale $\{\xi(t), t\in T\}$. This follows from the fact that it is permissible to approach the limit as $s\downarrow a$ under the integral sign in the inequality

$$\int_B\xi(s)\,d\mathsf{P}\leq\int_B\xi(t)\,d\mathsf{P},\qquad s<t,\quad B\in\bigcap_{t\in T}\mathfrak{F}_t.\quad\square$$

Remark. It is evident that assertion c) of the theorem is valid for sequences also. In this case the assertion can be stated as follows:

If $\{\cdots \xi(-n), \xi(-n+1), \ldots, \xi(0)\}$ *is a submartingale and* $\lim_n \mathsf{E}\xi(-n) > -\infty$, *then the sequence* $\{\xi(-n)\}$ *is uniformly integrable, the limit* $\xi_\infty = \lim \xi(-n)$ *exists with probability* 1 *in* L_1 *as well and is a closure from the left of the submartingale* $\{\xi(n), n = \cdots -k, -k+1, \ldots, 0\}$.

In what follows we shall call a semimartingale *uniformly integrable* if a corresponding family of random variables $\xi(t), t \in T$, is uniformly integrable. We shall refer to a martingale as *integrable* if $\sup\{\mathsf{E}|\xi(t)|, t \in T\} < \infty$.

Theorem 2. *Let* $T \subset (a, b)$ *and let a and b be limit points of the set* T $(-\infty \leq a < b \leq \infty)$. *In order that martingale* $\{\xi(t), \mathfrak{F}_t, t \in T\}$ *be uniformly integrable, it is necessary and sufficient that a random variable* η *exist such that*

$$\mathsf{E}|\eta| < \infty, \qquad \xi(t) = \mathsf{E}\{\eta \mid \mathfrak{F}_t\}, \qquad t \in T. \tag{6}$$

If this condition is satisfied, we can set $\eta = \lim_{t \uparrow b} \xi(t)$ *and the variable* η *will be uniquely determined* (mod P) *in the class of all* $\sigma\{\mathfrak{F}_t, t \in T\}$*-measurable random variables.*

Proof. By Theorem 1, if the martingale $\{\xi(t), \mathfrak{F}_t, t \in T\}$ is uniformly integrable, then it possesses a closure from the right and therefore admits representation (6).

Now let martingale $\xi(t)$ admit the representation given by formula (6). Then

$$\int_A \xi(t)\, d\mathsf{P} = \int_A \eta\, d\mathsf{P} \quad \forall A \in \mathfrak{F}_t,$$

which implies that

$$\int_B |\xi(t)|\, d\mathsf{P} \leq \int_B |\eta|\, d\mathsf{P} \qquad \forall B \in \mathfrak{F}_t. \tag{7}$$

In particular, $\mathsf{E}|\xi(t)| \leq \mathsf{E}|\eta|$. Therefore Chebyshev's inequality implies that $\mathsf{P}\{|\xi(t)| > N\} \to 0$ as $N \to \infty$ uniformly in t. Applying inequality (7) to the set $B = B_t = \{|\xi(t)| > N\}$ we verify that the family $\{\xi(t), t \in T\}$ is uniformly integrable.

We need now to prove the uniqueness of representation (6) in the class of all $\sigma\{\mathfrak{F}_t, t \in T\}$-measurable random variables. Suppose that two such representations exist in terms of the random variables η_i, $i = 1, 2$. Then

$$\mathsf{E}\{\zeta \mid \mathfrak{F}_t\} = 0 \qquad \forall t \in T,$$

where $\zeta = \eta_1 - \eta_2$.

Thus

$$\int_A \zeta\, d\mathsf{P} = 0$$

for all A belonging to $\mathfrak{F}_t$ and all $t \in T$, and consequently for all A belonging to

$\sigma\{\mathfrak{F}_t, t\in T\}$. Since the variable ζ is $\sigma\{\mathfrak{F}_t, t\in T\}$-measurable, it follows that $\zeta=0$ (mod P). □

Remark. If $\{\xi(t), t\in T\}$ is a martingale and T contains a maximal element then the family of random variables $\xi(t)$, $t\in T$, is uniformly integrable.

A $\sigma\{\mathfrak{F}_t, t\in T\}$-measurable random variable η appearing in representation (6) is called *the boundary value of the martingale* $\xi(t)$, $t\in T$.

It was shown in Volume I, Chapter III, Section 4 that under very general assumptions there exists—for a given semimartingale—a stochastically equivalent process $\{\zeta(t), \mathfrak{F}_t, t\geq 0\}$ with sample functions belonging to $\mathscr{D}[0,\infty)$, and, moreover, the current of σ-algebras $\mathfrak{F}_t$ is continuous from the right, i.e.,

$$\mathfrak{F}_{t+}=\mathfrak{F}_t \qquad \forall t>0.$$

In this section we shall assume, unless stated otherwise, that the semimartingales under consideration possess these properties.

Quasi-martingales. Let $\{\mathfrak{F}_t, t\geq 0\}$ be a current of σ-algebras continuous from the right ($\mathfrak{F}_t=\mathfrak{F}_{t+}$).

Definition. A process $\{\xi(t), t\geq 0\}$ adopted to $\mathfrak{F}_t$ is called a *quasi-martingale* (*$\mathfrak{F}_t$-quasi-martingale*) if

$$\mathsf{E}|\xi(t)|<\infty \qquad \forall t\geq 0$$

and

$$\sup\sum_{k=0}^{n-1}\mathsf{E}|\xi(t_k)-\mathsf{E}\{\xi(t_{k+1})\,|\,\mathfrak{F}_{t_k}\}|=V<\infty,$$

where the supremum is taken over arbitrary values of n and $t_1, t_2, \ldots, t_n$; $0\leq t_0<t_1<t_2<\cdots<t_n<\infty$.

It will be shown below that a study of quasi-martingales can be reduced to a study of semimartingales.

As examples of quasi-martingales, one may note martingales, super-(sub)martingales for which $\inf \mathsf{E}\xi(t)>-\infty$ ($\sup \mathsf{E}\xi(t)<\infty$) and also processes which are differences of two supermartingales. It turns out that these examples exhaust all the possible quasi-martingales.

Set

$$\delta(s,t)=\xi(s)-\mathsf{E}\{\xi(t)\,|\,\mathfrak{F}_s\} \qquad (s<t), \quad a(t)=\mathsf{E}\xi(t).$$

Then

$$\sum_{k=0}^{n-1}|a(t_k)-a(t_{k+1})|=\sum_{k=0}^{n-1}|\mathsf{E}\delta(t_k, t_{k+1})|\leq V,$$

i.e., $a(t)$ is a function of bounded variation. In particular, for any $t>0$ there exist limits $a(t-)$ and $a(t+)$ and, moreover, $a(\infty)=\lim_{t\to\infty} a(t)$.

Inequalities (3)–(5) can be generalized for the case of quasi-martingales. For this purpose observe that inequalities (21) and (23) derived in Volume I, Chapter II, Section 2 for countable sequences can easily be adapted for separable quasi-martingales and can be written in this case in the form

(8) $$\mathsf{P}\{\sup \xi(t) \geqslant C\} \leqslant \frac{\sup E\xi^+(t)+V}{C},$$

(9) $$\mathsf{E}\nu[a, b) \leqslant \frac{\sup_t \mathsf{E}(\xi(t)-b)^+ + V}{b-a},$$

where $\nu[a, b)$ is the number of downward crossings of the interval $[a, b)$. Utilizing inequality (9) one can prove the following theorem (analogously to the proofs of Theorems 6 and 7 for semimartingales in Volume I, Chapter III, Section 4).

Theorem 3. *A separable quasi-martingale $\xi(t)$, $t>0$, possesses with probability* 1 *for each t the left-hand and right-hand limits. Moreover, $\{\xi(t+), \mathfrak{F}_{t+}, t\geqslant 0\}$ is also a quasi-martingale with sample functions which are continuous from the right and* $\mathsf{P}\{\xi(t)=\xi(t+)\}=1$ *at each point t such that $\mathfrak{F}_t=\mathfrak{F}_{t+}$ and $\mathsf{E}\xi(t)$ is continuous.*

In view of this theorem, we can, without loss of generality, concentrate, in what follows, only on those quasi-martingales with sample functions belonging to $\mathscr{D}$ with probability 1 and for which $\mathfrak{F}_t=\mathfrak{F}_{t+}$ for all $t\geqslant 0$. In this subsection we shall assume that the stipulated conditions are satisfied.

Theorem 4. *An arbitrary quasi-martingale admits representation*

$$\xi(t)=\mu(t)+\zeta(t),$$

where $\mu(t)$ is a martingale and $\mathsf{E}|\zeta(t)|\to 0$ as $t\to\infty$.

This decomposition is unique.

If $\xi(t)$ is a supermartingale satisfying condition $\inf \mathsf{E}\xi(t)>-\infty$, *then $\zeta(t)$ is a nonnegative supermartingale.*

Proof. For each $s>0$ and $t\geqslant 0$ we set

$$\xi(s, t)=\mathsf{E}\{\xi(s+t)|\mathfrak{F}_t\},$$

and consider a separable modification of the process $\xi(s, t)$. We show that for a fixed t, $\xi(s, t)$ as a function of s is of bounded variation with probability 1.

Indeed,

$$\begin{aligned}\sum_{k=0}^{n-1} |\xi(s_k, t)-\xi(s_{k+1}, t)| &= \sum_{k=0}^{n-1} |\mathsf{E}\{\xi(s_k+t)-\mathsf{E}\{\xi(s_{k+1}+t)|\mathfrak{F}_{s_k+t}\}|\mathfrak{F}_t\}| \\ &= \sum_{k=0}^{n-1} \mathsf{E}\{|\delta(s_k+t, s_{k+1}, +t)||\mathfrak{F}_t\}\end{aligned}$$

and

$$\mathsf{E}\sum_{k=0}^{n-1}|\xi(s_k,t)-\xi(s_{k+1},t)|\leq V.$$

It may be assumed that the set of separability points of function $\xi(s,t)$ in variables s and t is of the form $I\times I$. For each $t\in I$ we choose sequences $\{s_0,s_1,\dots,s_n\}$, $s_k\in I$, such that these sequences—viewed as sets—increase monotonically as n increases and in the limit exhaust the whole set I. Moreover, the sums $\sum_{k=0}^{n-1}|\xi(s_k,t)-\xi(s_{k+1},t)|$ are monotonically nondecreasing and approach their upper bound $V(t)$. Thus $\mathsf{E}\,V(t)\leq V$ and $V(t)<\infty$ with probability 1 for each $t\in I$. This implies the existence of a set $N\in\mathfrak{S}$ with $\mathsf{P}(N)=0$, such that if $\omega\notin N$ then $V(t)<\infty$ for *any* t. Thus, there exists with probability 1 the limit

$$\mu(t)=\lim_{s\uparrow\infty}\xi(s,t).$$

Let $s_n\uparrow\infty$. Since

$$|\mu(t)-\xi(s_n,t)|\leq\sum_{n}^{\infty}|\xi(s_k,t)-\xi(s_{k+1},t)|\leq V(t),$$

$\mu(t)$ is an integrable random variable and the sequence $\xi(s_n,t)$ possesses an integrable majorant. Hence we have for $t_1<t_2$

$$\mathsf{E}\{\mu(t_2)|\mathfrak{F}_{t_1}\}=\mathsf{E}\left\{\lim_{s\to\infty}\mathsf{E}\{\xi(s+t_2)|\mathfrak{F}_{t_2}\}\mathfrak{F}_{t_1}\right\}$$

$$=\lim_{s\to\infty}\mathsf{E}\{\xi(s+t_2)|\mathfrak{F}_{t_1}\}=\mu(t_1).$$

Thus $\mu(t)$ is an $\mathfrak{F}_t$-martingale.

Now set $\zeta(t)=\xi(t)-\mu(t)$. Then $\mathsf{E}|\zeta(t)|\to 0$ as $t\to\infty$.

Indeed, assume the contrary. Then an $\varepsilon>0$ exists and for each $N>0$ a $t=t_N$ satisfying $t_N>N$ such that $\mathsf{E}|\zeta(t_N)|>\varepsilon$. Now choose some t_1. Since

$$\mathsf{E}|\zeta(t_1)|=\mathsf{E}|\xi(t_1)-\lim_{s\to\infty}\xi(t_1,s)|=\lim_{s\to\infty}\mathsf{E}|\xi(t_1)-\xi(t_1,s)|,$$

one can find s_1 such that $\mathsf{E}|\xi(t_1)-\xi(t_1,s_1)|>\varepsilon$. Set $t_2=t_1+s_1$ and choose $t_3>t_2$ such that $\mathsf{E}|\zeta(t_3)|>\varepsilon$. We continue this process indefinitely. Then

$$\mathsf{E}\sum_{1}^{2n-1}|\delta(t_k,t_{k+1})|\geq\mathsf{E}\sum_{1}^{n}|\delta(t_{2k-1},t_{2k})|$$

$$=\mathsf{E}\sum_{1}^{n}|\xi(t_{2k})-\mathsf{E}\{\xi(t_{2k})|\mathfrak{F}_{t_{2k-1}}\}|\geq n\varepsilon\to\infty,$$

which contradicts the definition of a quasi-martingale.

Thus the existence of a decomposition satisfying the conditions of the theorem has been established. We shall now prove its uniqueness.

Let there be two decompositions $\xi(t)=\mu_1(t)-\zeta_1(t)=\mu_2(t)-\zeta_2(t)$. Then $\mu_1(t)-\mu_2(t)=\zeta_2(t)-\zeta_1(t)$, while $\mathsf{E}|\zeta_1(t)-\zeta_2(t)|\to 0$ as $t\to\infty$. On the other hand, $|\mu_1(t)-\mu_2(t)|$ is a submartingale and $\mathsf{E}|\mu_1(t)-\mu_2(t)|$ is a monotonically nondecreasing function of t. Consequently, $\mathsf{E}|\mu_1(t)-\mu_2(t)|\equiv 0$ and $\mu_1(t)=\mu_2(t)$ (mod P). Finally if $\xi(t)$ is a supermartingale, then

$$\mu(t)=\lim_{s\uparrow\infty}\xi(s,t)=\lim_{s\to\infty}\mathsf{E}\{\xi(s+t)|\mathfrak{F}_t\}\leqslant\xi(t),$$

which implies that $\zeta(t)=\xi(t)-\mu(t)\geqslant 0$ in this case. □

Definition. A nonnegative supermartingale satisfying condition $\mathsf{E}\xi(t)\to 0$ as $t\to\infty$ is called a *potential.*

Observe that for a potential the limit $\xi_\infty=\lim_{t\to\infty}\xi(t)$ exists and $\xi_\infty=0$ with probability 1.

Corollary. *Supermartingale $\xi(t)$ satisfying the condition* $\inf \mathsf{E}\xi(t)>-\infty$ *admits the decomposition $\xi(t)=\mu(t)+\pi(t)$, where $\mu(t)$ is a martingale and $\pi(t)$ is a potential. This decomposition is unique.*

By analogy with the classical theory of superharmonic functions this decomposition is called the Riesz decomposition. We shall agree to call the decomposition which was established in Theorem 4 Riesz's decomposition as well, and a quasi-martingale $\zeta(t)$ satisfying condition $\mathsf{E}|\zeta(t)|\to 0$ as $t\to\infty$ will be called *quasi-potential.*

We now show that an arbitrary quasi-potential can be represented as a difference of two potentials.

Let $\zeta(t)$ be an arbitrary quasi-potential. Set

$$\delta_{k,n}=\delta\left(\frac{k}{2^n},\frac{k+1}{2^n}\right),$$

$$\delta^+_{k,n}=\max(\delta_{k,n},0),\qquad \delta^-_{k,n}=\delta^+_{k,n}-\delta_{k,n},$$

$$\pi^n_+(t)=\mathsf{E}\left\{\sum_{k=j(t)}^{\infty}\delta^+_{k,n}\,\middle|\,\mathfrak{F}_t\right\},\qquad \pi^n_-(t)=\mathsf{E}\left\{\sum_{k=j(t)}^{\infty}\delta^-_{k,n}\,\middle|\,\mathfrak{F}_t\right\},$$

where $j(t)$ is an integer defined by conditions $(j(t)-1)/2^n<t\leqslant j(t)/2^n$.

Note that for $t=j/2^n$

$$\zeta(t)=\mathsf{E}\left\{\sum_{k=j}^{\infty}\delta_{k,n}\,\middle|\,\mathfrak{F}_t\right\}=\pi^n_+(t)-\pi^n_-(t),$$

and that the absolute convergence (mod P) of the series $\sum_{k=0}^{\infty}\delta_{k,n}$ and its integrability follow from the definition of a quasi-martingale. Clearly $\mathsf{E}\{\pi_+^n(t)\,|\,\mathfrak{F}_s\}\leqslant \pi_+^n(s)$ for $s<t$ so that $\pi_+^n(t)$ is a potential.

We show that $\pi_+^n(t)\leqslant \pi_+^{n+1}(t)$, $n=1, 2, \ldots$. Consider a summand appearing in the expression for $\pi_+^n(t)$, for example $\mathsf{E}\{\delta_{k,n}^+\,|\,\mathfrak{F}_t\}$ $(k/2^n\geqslant t)$. We have

$$\mathsf{E}\{\delta_{k,n}^+\,|\,\mathfrak{F}_t\}=\mathsf{E}\Big\{\Big[\zeta\Big(\frac{2k}{2^{n+1}}\Big)-\mathsf{E}\Big\{\zeta\Big(\frac{2k+1}{2^{n+1}}\Big)\Big|\,\mathfrak{F}_{\frac{2k}{2^{n+1}}}\Big\}$$
$$+\zeta\Big(\frac{2k+1}{2^{n+1}}\Big)-\mathsf{E}\Big\{\zeta\Big(\frac{2k+2}{2^{n+1}}\Big)\Big|\,\mathfrak{F}_{\frac{2k+1}{2^{n+1}}}\Big\}\Big|\,\mathfrak{F}_{\frac{k}{2^n}}\Big]^+\Big|\,\mathfrak{F}_t\Big\}$$
$$\leqslant \mathsf{E}\{[\delta_{2k,n+1}^+ +\delta_{2k+1,n+1}^-]\,|\,\mathfrak{F}_t\}.$$

This implies the monotonicity of the sequences $\pi_+^n(t)$. Moreover, it follows from that proven above that $\pi_-^n(t)$ is also a potential and that $\pi_-^n(t)\leqslant \pi_-^{n+1}(t)$.

Now set

$$\pi_+(t)=\lim \pi_+^n(t), \qquad \pi_-(t)=\lim \pi_-^n(t).$$

These limits exist for each t with probability 1. Since $\mathsf{E}(\pi_+^n(t)+\pi_-^n(t))=\mathsf{E}\sum_{k=0}^{\infty}|\delta_{k,n}|\leqslant V$, it follows that $\mathsf{E}\pi_\pm(t)<\infty$. Clearly $\pi_\pm(t)$ are supermartingales.

It is easy to verify that $\mathsf{E}\pi_\pm^n(t)\to 0$ as $t\to\infty$ uniformly in n. Hence $\pi_\pm(t)$ are potentials.

We now define processes $\pi_\pm(t)$ for all $t\geqslant 0$ in such a manner that their sample functions will be continuous from the right with probability 1. Taking into account the fact that the process $\zeta(t)$ is also continuous from the right, we observe that equality $\zeta(t)=\pi_+(t)-\pi_-(t)$ is valid for all $t\geqslant 0$ with probability 1.

Theorem 5. *If $\zeta(t)$ is a quasipotential with sample functions belonging to $\mathscr{D}$, then potentials $\pi_+(t)$ and $\pi_-(t)$ exist such that with probability* 1

$$\zeta(t)=\pi_+(t)+\pi_-(t) \qquad \forall t\geqslant 0.$$

Stopping and random time substitution. Now we shall consider submartingales

$$\{\xi(t), \mathfrak{F}_t, t\in T\},$$

where

$$T=N=\{0, 1, \ldots, n, \ldots\} \quad \text{or} \quad T=[0, \infty).$$

If $T=[0,\infty)$ we shall assume that the sample functions of the process $\xi(t)$ belong to $\mathscr{D}[0,\infty)$, $\mathfrak{F}_t=\mathfrak{F}_{t+}$, $t\in[0,\infty)$, and $\mathfrak{F}_0$ contains all the subsets of probability 0.

Recall the definition of random time (Volume I, Chapter I, Section 1, Definition 14). Let $\{\mathfrak{F}_t, t\in T\}$ be a current of σ-algebras. A function $\tau=f(\omega)$, $\omega\in\Omega_\tau\subset\Omega$, with values in T is called a *random time on* $\{\mathfrak{F}_t, t\in T\}$ or an $\mathfrak{F}_t$*-random time* if $\{\tau\leqslant t\}\in\mathfrak{F}_t$ for all $t\in T$.

In what follows we shall discuss only those random times which are defined on the whole space Ω ($\Omega_\tau = \Omega$).

A σ-algebra of events $\mathfrak{F}_\tau$ generated by events up to time τ which is called a σ-algebra corresponds to each random time τ. This σ-algebra consists of events $B \in \mathfrak{S}$ such that

$$B \cap \{\tau \leq t\} \in \mathfrak{F}_t \qquad \forall t \in T.$$

It is easy to verify that if $\tau_1 \leq \tau_2$ then $\mathfrak{F}_{\tau_1} \subset \mathfrak{F}_{\tau_2}$ (Volume I, Chapter I, Section 1). In Volume I, Chapter II, Section 2 (p. 55) the following result was proved.

Lemma 1. *Let T be a finite set, τ_k, $k = 1, \ldots, s$, be a sequence of random times on $\{\mathfrak{F}_t, t \in T\}$ defined on the whole space Ω and such that $\tau_1 \leq \tau_2 \leq \cdots \leq \tau_s$, and let $\mathfrak{F}_k^* = \mathfrak{F}_{\tau_k}$ be a σ-algebra generated by the random time τ_k ($k = 1, \ldots, s$). If $\{\xi(t), \mathfrak{F}_t, t \in T\}$ is a supermartingale (martingale) then $\{\xi(\tau_k), \mathfrak{F}_k^*, k = 1, \ldots, s\}$ is also a supermartingale (martingale).*

We now generalize this result to the case of semimartingales considered in this section.

Let $\{\xi(t), \mathfrak{F}_t, t \in T\}$ be a supermartingale satisfying the condition: an integrable random variable η exists such that

$$\xi(t) \geq E\{\eta \mid \mathfrak{F}_t\}. \tag{10}$$

Consider a random time τ taking on values in T and also possibly the value $t = \infty$. Set

$$\mathfrak{F}_\infty = \sigma\{\sigma(\eta), \mathfrak{F}_t, t \in [0, \infty)\},$$

$$\xi_\tau = \begin{cases} \xi(t) & \text{for } \tau = t, t \in T, \\ \eta & \text{for } \tau = \infty. \end{cases}$$

The random variable ξ_τ is $\mathfrak{F}_\tau$-measurable. Indeed, in the case of a discrete time, this assertion was proved earlier, in Volume I, Chapter I, Section 1, Lemma 5. For $T = [0, \infty)$ the proof is as follows.

Proof. Introduce discrete approximations of the random time τ by setting $\tau^{(n)} = (k+1)/2^n$ if $\tau \in (k/2^n, (k+1)/2^n]$; $\tau^{(n)} = 0$ if $\tau = 0$. The continuity from the right of the sample functions of the process $\xi(t)$ implies that $\lim \xi(\tau_n) = \xi(\tau)$. On the other hand

$$\{\xi(\tau) < a\} \cap \{\tau < t\} = \lim \{\xi(\tau^{(n)}) < a\} \cap \{\tau^{(n)} < t\}.$$

For $t \in (k/2^n, (k+1)/2^n]$ we have

$$\{\xi(\tau^{(n)}) < a\} \cap \{\tau^{(n)} < t\} \in \mathfrak{F}_{(k+1)/2^n}.$$

Thus

$$\{\xi(\tau)<a\}\cap\{\tau<t\}\in\mathfrak{F}_{t+}=\mathfrak{F}_t.$$

Utilizing once again equality $\mathfrak{F}_{t+}=\mathfrak{F}_t$ we obtain

$$\{\xi(\tau)<a\}\cap\{\tau\leq t\}\in\mathfrak{F}_t$$

which implies that $\xi(\tau)$ is $\mathfrak{F}_t$-measurable. □

Theorem 6. *Let supermartingale $\xi(t)$ satisfy condition* (10); *let σ and τ be random times with $\sigma\leq\tau$. Then the variables ξ_σ and ξ_τ are integrable and*

$$\mathsf{E}\{\xi_\tau\,|\,\mathfrak{F}_\sigma\}\leq\xi_\sigma. \tag{11}$$

Proof. First consider the case $T=N$. Let $\sigma_k=\sigma\wedge k$ and $\tau_k=\tau\wedge k$. Set $\xi(t)=\zeta(t)+\eta(t)$, where $\eta(t)=\mathsf{E}\{\eta\,|\,\mathfrak{F}_t\}$, $\zeta(t)=\xi(t)-\eta(t)$. Conditions of the theorem imply that $\zeta(t)\geq 0$ and, moreover, $\zeta(t)$ is a supermartingale.

We shall start with the process $\zeta(t)$. Lemma 1 implies that $\mathsf{E}\zeta_{\tau_k}\leq\mathsf{E}\zeta_0$. Approaching the limit as $k\to\infty$ and utilizing Fatou's lemma we obtain

$$\mathsf{E}\zeta_\tau=\mathsf{E}\lim_{k\to\infty}\zeta_{\tau_k}\leq\mathsf{E}\zeta_0$$

so that $\mathsf{E}\zeta_\tau<\infty$.

Now let $B=\mathfrak{F}_\sigma$. Then in view of Lemma 1

$$\int_{B\cap\{\tau\leq k\}}\zeta_\tau\,d\mathsf{P}\leq\int_{B\cap\{\sigma\leq k\}}\zeta_\tau\,d\mathsf{P}\leq\int_{B\cap\{\sigma\leq k\}}\zeta_{\sigma_k}\,d\mathsf{P}=\int_{B\cap\{\sigma\leq k\}}\zeta_\sigma\,d\mathsf{P}.$$

Taking into account that $\zeta_\tau=\zeta_\sigma=0$ for $\sigma=\infty$ and approaching the limit as $k\to\infty$ in the obtained relationships, we have

$$\int_B\zeta_\tau\,d\mathsf{P}\leq\int_B\zeta_\sigma\,d\mathsf{P}; \tag{12}$$

this yields the assertion of the theorem for the process $\zeta(t)$.

We now consider process $\eta(t)$. This process is a uniformly integrable martingale. Observe that

$$\eta_\tau=\mathsf{E}\{\eta\,|\,\mathfrak{F}_t\}|_{t=\tau}=\mathsf{E}\{\eta\,|\,\mathfrak{F}_\tau\}. \tag{13}$$

Indeed, if $A\in\mathfrak{F}_\tau$, $A_k=A\cap\{\tau=k\}$, $k=0,1,\dots,n,\dots,\infty$, then

$$\int_{A_k}\eta_\tau\,d\mathsf{P}=\int_{A_k}\mathsf{E}\{\eta\,|\,\mathfrak{F}_k\}\,d\mathsf{P}=\int_{A_k}\eta\,d\mathsf{P}.$$

Summing up these equalities over all the values of k, we obtain

$$\int_A\eta_\tau\,d\mathsf{P}=\int_A\eta\,d\mathsf{P}.$$

Since η_τ is an $\mathfrak{F}_\tau$-measurable random variable, the last equality implies (13) and also that the expectation $\mathsf{E}\eta_\tau$ is finite. Moreover, it implies that for any $B=\mathfrak{F}_\sigma$ $(\mathfrak{F}_\sigma\subset\mathfrak{F}_\tau)$

$$\int_B \eta_\tau\, d\mathsf{P}=\int_B \eta_\sigma\, d\mathsf{P}.$$

Adding the last equality to inequality (12) we obtain the required inequality (11).

Now consider the case $T=[0,\infty)$. We introduce discrete approximations $\tau^{(n)}$ and $\sigma^{(n)}$ of the quantities τ and σ by setting

$$\tau^{(n)}=\frac{k}{2^n}\quad \text{if } \tau\in\left(\frac{k-1}{2^n},\frac{k}{2^n}\right],$$

$$\tau^{(n)}=\infty\quad \text{if } \tau=\infty.$$

The quantity $\sigma^{(n)}$ is defined analogously. From the above we have

$$\int_A \xi_{\tau^{(n)}}\, d\mathsf{P}\geqslant\int_A \xi_{\sigma^{(n)}}\, d\mathsf{P}\qquad \forall A\in\mathfrak{F}_{\sigma^{(n)}}.$$

Moreover,

$$\sigma\leqslant\sigma^{(n+1)}\leqslant\sigma^{(n)},\qquad \mathfrak{F}_\sigma\subset\mathfrak{F}_{\sigma^{(n)}}\quad \forall n.$$

Hence the last relationship is valid for all $A\in\mathfrak{F}_\sigma$. Since the process is continuous from the right it follows that $\xi_{\tau^{(n)}}\to\xi_\tau$ and $\xi_{\sigma^{(n)}}\to\xi$ with probability 1. Therefore to prove the theorem it is sufficient to show that the sequences of random variables $\{\xi_{\tau^{(n)}}, n=1,2,\ldots\}$ and $\{\xi_{\sigma^{(n)}}, n=1,2,\ldots\}$ are uniformly integrable.

Observe that $\sigma^{(n)}\leqslant\sigma^{(n-1)}$. Therefore

$$\xi_{\sigma^{(n)}}\geqslant\mathsf{E}\{\xi_{\sigma^{(n-1)}}\mid\mathfrak{F}_{\sigma^{(n)}}\}.$$

If we set $\xi_{\sigma^{(n)}}=\eta_{-n}$ and $\mathfrak{F}_{\sigma^{(n)}}=\mathfrak{G}_{-n}$, then the inequality

$$\eta_{-n}\geqslant\mathsf{E}\{\eta_{-n+1}\mid\mathfrak{G}_{-n}\}$$

implies that $\{\eta_k,\mathfrak{G}_k, k=\ldots,-n,-n+1,\ldots,-1\}$ forms a supermartingale and since $\mathsf{E}\eta_{-n}\leqslant\mathsf{E}\xi_{\sigma^{(n)}}\leqslant\mathsf{E}\xi_0$ we have $\mathsf{E}\eta_{-n}\leqslant C$. Thus the family of random variables η_{-n} is uniformly integrable. The same argument is also applicable to the sequence $\xi_{\tau^{(n)}}$. □

Corollary 1. *If $\{\xi_t,\mathfrak{F}_t, t\geqslant 0\}$ is a uniformly integrable martingale and σ and τ are random times such that $\sigma\leqslant\tau$, then*

$$\xi_\sigma=\mathsf{E}\{\xi_\tau\mid\mathfrak{F}_\sigma\}$$

and the variable ξ_τ is integrable.

Corollary 2. *If* $\eta(t)=\mathsf{E}\{\eta\,|\,\mathfrak{F}_t\}$ *and* τ *is a random time then*

$$\eta(\tau)=\mathsf{E}\{\eta\,|\,\mathfrak{F}_\tau\}.$$

Corollary 3. *If* $\{\xi(t), \mathfrak{F}_t, t\geq 0\}$ *is a uniformly integrable supermartingale (martingale) then the process*

$$\{\eta(t), \mathfrak{F}_t^*, t\geq 0\} \quad \text{where } \eta(t)=\xi(\tau\wedge t), \quad \mathfrak{F}_t^*=\mathfrak{F}_{\tau\wedge t},$$

is also a supermartingale (martingale).

This process is called *τ-stopping* of the process $\xi(t)$.

Theorem 6 will be utilized repeatedly later on. As an application of this theorem we shall prove an assertion which will also be utilized in what follows.

Theorem 7. *Let* $\{\xi(t), t\geq 0\}$ *be a nonnegative supermartingale continuous from the right. Set*

$$\tau=\inf\{t\colon \xi(t)=0 \quad \text{or} \quad \xi(t-)=0\}$$

if the corresponding set of values of t *is nonempty and* $\tau=\infty$ *otherwise. Then we have with probability* 1 $\xi(t)=0$ *for all* $t\geq\tau$ $(\tau<\infty)$.

Proof. Let $\tau_n=\inf(t\colon \xi(t)<1/n\}$ (it is assumed that $\inf\varnothing=\infty$) and χ_n be the indicator of the event $\tau_n<\infty$. Clearly $\tau_n\leq\tau_{n+1}\leq\tau$. Let $\sigma=\sup\tau_n$ and χ be the indicator of the event $\sigma<\infty$. Then $\sigma\leq\tau$. It follows from Theorem 6 that

$$\mathsf{E}\xi_{\tau_n}\chi_n\geq\mathsf{E}\xi_{\sigma\vee t}\chi_n\geq\mathsf{E}\xi_{\sigma\vee t}\chi.$$

Since $\mathsf{E}\xi_{\tau_n}\chi_n\leq 1/n$ we have $\mathsf{E}\xi_{\sigma\vee t}\chi=0$, i.e., $\xi(t)=0$ with probability 1 on the set $t>\sigma$, $\sigma<\infty$ for each and every t. The sample functions of the process $\xi(t)$ are continuous from the right (mod P); this yields that $\xi(t)=0$ with probability 1 for all $t>\sigma$ provided $\sigma<\infty$. □

Denote by $\mathcal{T}$ or $\mathcal{T}(T)$ the family of all random times of $(\mathfrak{F}_t, t\in T\}$.

Definition. The family of random variables $\{\xi(t), t\in T\}$ adopted to $\{\mathfrak{F}_t, t\in T\}$ is called a *completely uniformly integrable family* or a *process of class D* provided the family $\{\xi_\tau, \tau\in T\}$ is uniformly integrable.

It is called a *process of class DL* if for any $a>0$ the family $\{\xi_t, t\in\mathcal{T}([0, a])\}$ is uniformly integrable.

Theorem 8. a) *A uniformly integrable martingale* $\{\xi(t), t\in T\}$ *is completely uniformly integrable.*

b) *If $\{\xi(t), t \in T\}$ is a nonnegative submartingale and a random variable η exists such that*

$$\eta \geqslant 0, \qquad \xi(t) \leqslant \mathsf{E}\{\eta \mid \mathfrak{F}_t\} \qquad \forall t \in T,$$

then the family $\{\xi(t), t \in T\}$ is completely uniformly integrable.

c) *If $T = N$ and the submartingale $\{\xi(t), t \in N\}$ is uniformly integrable, then it is completely uniformly integrable.*

Proof. a) Let $\xi(t) = \mathsf{E}\{\eta \mid \mathfrak{F}_t\}$, where $\eta = \lim_{t \to \infty} \xi(t)$. Set $B = \{|\xi_\tau| > C\}$. Since

$$\mathsf{P}\{|\xi_\tau| > C\} \leqslant \mathsf{P}\{\sup |\xi(t)| > C\} \leqslant \frac{\sup \mathsf{E}|\xi(t)|}{C},$$

it follows that $\mathsf{P}(B) \to 0$ as $C \to \infty$ uniformly in all $\tau \in \mathcal{T}$. Since $|\xi(t)|$ is a submartingale, we have (from Theorem 6) that

$$\int_B |\xi(t)| \, d\mathsf{P} \leqslant \int_B |\eta| \, d\mathsf{P} \qquad \forall B \in \mathfrak{F}_\tau,$$

which proves assertion a) of the theorem.

Assertion b) is proved analogously. Firstly,

$$\int_{\{\xi_\tau \geqslant C\}} \xi_\tau \, d\mathsf{P} \leqslant \int_{\{\xi_\tau \geqslant C\}} \eta \, d\mathsf{P},$$

and furthermore

$$\mathsf{P}\{\xi_\tau \geqslant C\} \leqslant \frac{\mathsf{E}\xi_\tau}{C} \leqslant \frac{\mathsf{E}\eta}{C} \to 0$$

uniformly in τ as $C \to \infty$.

c) Since a uniformly integrable submartingale can be represented as a difference of a uniformly integrable martingale and a potential, it is sufficient, in view of assertion a), to consider only the case of a potential $\pi(t)$. We have

$$\int_{\{\pi_\tau \geqslant C\}} \pi_\tau \, d\mathsf{P} = \sum_{j=1}^{k} \int_{\{\pi_\tau \geqslant C\} \cap \{\tau = j\}} \pi_\tau \, d\mathsf{P} + \int_{\{\pi_\tau \geqslant C\} \cap \{\tau > k\}} \pi_\tau \, d\mathsf{P}.$$

Taking into account that $\{\pi_\tau \geqslant C\} \cap \{\tau > k\} \in \mathfrak{F}_\tau$ and applying Theorem 6 to random times τ and k we obtain

$$\int_{\{\pi_\tau \geqslant C\} \cap \{\tau > k\}} \pi_\tau \, d\mathsf{P} \leqslant \int_{\{\pi_\tau \geqslant C\} \cap \{\tau > k\}} \pi(k) \, d\mathsf{P}.$$

Now choose an arbitrary $\varepsilon > 0$ and select k from the condition $\mathsf{E}\pi(k) < \varepsilon/2$, which is possible since $\mathsf{E}\pi(n) \to 0$. Next choose C in such a manner that

$$\sum_{j=1}^{k} \int_{\{\pi(j) \geqslant C\} \cap \{\tau = j\}} \pi(j) \, d\mathsf{P} < \varepsilon/2.$$

We have thus established the existence of a constant C such that independently of τ

$$\int_{\{\pi_\tau \geq C\}} \pi_\tau \, d\mathsf{P} < \varepsilon. \quad \square$$

Remark 1. The argument presented in the proof of assertion c) of the theorem can be carried over to the case of submartingales of a continuous argument.

Denote by $\mathcal{T}_a$ the class of all random times τ such that $\tau \leqslant a \pmod{\mathsf{P}}$. Then if the submartingale $\{\xi(t), t \geqslant 0\}$ is uniformly integrable and for each $a > 0$ the family of random variables $\{\xi_\tau, \tau \in \mathcal{T}_a\}$ is uniformly integrable, it follows that the family $\{\xi(t), t \geqslant 0\}$ is completely uniformly integrable.

Indeed we have, from Riesz's representation that it is sufficient to consider a potential $\{\pi(t), t \geqslant 0\}$. Now choose a such that $\mathsf{E}\pi(t) < \varepsilon/2$ for $t \geqslant a$, where ε is an arbitrary positive number chosen in advance. Then

$$\int_{\{\pi_\tau \geq C\}} \pi_\tau \, d\mathsf{P} \leqslant \int_{\{\pi_\tau \geq C\} \cap \{\tau \leqslant a\}} \pi_\tau \, d\mathsf{P} + \frac{\varepsilon}{2} \leqslant \int_{\{\pi_{\tau \wedge a} \geq C\}} \pi_\tau \, d\mathsf{P} + \frac{\varepsilon}{2}$$

On the other hand, the quantity $\int_{\{\pi_{\tau \wedge a} \geq C\}} \pi_\tau \, d\mathsf{P} \to 0$ as $C \to \infty$ uniformly in τ. This implies the uniform integrability of the family $\{\pi_\tau, \tau \in \mathcal{T}\}$.

Remark 2. Let semimartingale $\{\xi(t), t \in T\}$ belong to class D. Then the family $\{\xi(t), t \in T\}$ is uniformly integrable and the limit $\lim_{t \to \infty} \xi(t) = \xi_\infty$ exists. Therefore ξ_τ can be defined for random times admitting infinite values as well. Since a closure of a family of uniformly integrable random functions (with respect to a.s. convergence) is also uniformly integrable, it follows that the family $\{\xi_\tau, \tau \in \mathcal{T}_\infty\}$, where $\mathcal{T}_\infty$ is a class of random times admitting values $\leqslant \infty$, is also uniformly integrable.

When discussing completely uniformly integrable semimartingales $\xi(t)$, $t \geqslant 0$, we shall, in what follows, always assume that ξ_τ are defined for $\tau \in \mathcal{T}_\infty$ in the manner just indicated.

A theorem of decomposition of supermartingales. We begin with the case of discrete time. We show that the supermartingale $\{\xi_n, n = 0, 1, \ldots\}$ can be represented as a difference of two sequences—a martingale and a nondecreasing sequence of random variables.

Set

$$\Delta\xi_n = \xi_n - \xi_{n-1}, \qquad n = 1, 2, \ldots$$

$$\begin{aligned}
\eta_0 &= \xi_0, & \alpha_0 &= 0,\\
\eta_1 &= \eta_0 + (\Delta\xi_1 - \mathsf{E}\{\Delta\xi_1 \mid \mathfrak{F}_0\}), & \alpha_1 &= -\mathsf{E}\{\Delta\xi_1 \mid \mathfrak{F}_0\},\\
&\vdots & &\vdots\\
\eta_n &= \eta_{n-1} + (\Delta\xi_n - \mathsf{E}\{\Delta\xi_n \mid \mathfrak{F}_{n-1}\}), & \alpha_n &= \alpha_{n-1} - \mathsf{E}\{\Delta\xi_n \mid \mathfrak{F}_{n-1}\},\\
&\vdots & &\vdots
\end{aligned}$$

Then

(14) $$\xi_n = \eta_n - \alpha_n,$$

where $\alpha_n \geqslant \alpha_{n-1}$ (since for a supermartingale $\mathsf{E}\{\Delta\xi_n \mid \mathfrak{F}_{n-1}\} \leqslant 0$), α_n is an $\mathfrak{F}_{n-1}$-measurable random variable, and $\{\eta_n, n=0,1,2,\ldots\}$ is a martingale. Relation (14) is called *Doob's decomposition* for the supermartingale ξ_n.

It is easy to verify that representation (14) with η_n a martingale and α_n adopted to a current of σ-algebras $\{\mathfrak{F}_{n-1}, n=1,2,\ldots\}$ and $\alpha_0=0$ is unique.

Indeed, assume that the quantities α_k and η_k, $k=0,1,2,\ldots,n$, are uniquely defined (by definition, this is valid for $n=0$). Since α_{n+1} is $\mathfrak{F}_n$-measurable, we have $\mathsf{E}\{\xi_{n+1} \mid \mathfrak{F}_n\} = \mathsf{E}\{\eta_{n+1} - \alpha_{n+1} \mid \mathfrak{F}_n\} = \eta_n - \alpha_{n+1}$, so that α_{n+1} and hence also η_{n+1} are uniquely defined. Observe that if instead of the $\mathfrak{F}_n$-measurability of the quantity α_{n+1} one requires only that α_{n+1} be $\mathfrak{F}_{n+1}$-measurable, then the uniqueness of representation (14) will not be valid in general.

If ξ_n is a potential then the sequence ξ_n, $n=1,2,\ldots$, is uniformly integrable (since it is convergent to zero in L_1). The corresponding process α_n in relation (13) is also uniformly integrable since $0 \leqslant \alpha_n \leqslant \alpha_{n+1} \leqslant \alpha_\infty$, where $\alpha_\infty = \lim \alpha_n$ and, moreover, $\mathsf{E}\alpha_\infty = \lim \mathsf{E}\eta_n = \mathsf{E}\eta_0$. Therefore martingale η_n is uniformly integrable so that $\eta_n = \mathsf{E}\{\eta_\infty \mid \mathfrak{F}_n\}$, where $\eta_\infty = \lim \eta_n = \lim \alpha_n = \alpha_\infty$. We have thus obtained the following result.

Lemma 2. *A potential* $\{\xi_n, \mathfrak{F}_n, n=0,1,\ldots\}$ *admits decomposition*

$$\xi_n = \mathsf{E}\{\alpha_\infty \mid \mathfrak{F}_n\} - \alpha_n,$$

where:

a) α_n *is an* $\mathfrak{F}_{n-1}$*-measurable random variable*, $n=1,2,\ldots$, $\alpha_\infty \geqslant 0$, *and* $\mathsf{E}\alpha_\infty < \infty$, $\alpha_0 = 0$;

b) *the sequence* α_n *is nonnegative, monotonically nondecreasing, and* $\alpha_\infty = \lim \alpha_n$.

Under condition a) the sequence α_n *is uniquely determined* (mod P).

A process $\{\alpha_n, n=0,1,2,\ldots\}$ satisfying the conditions of Lemma 2 is called a *process associated with the potential (supermartingale)* $\xi(t)$.

We now reformulate condition a) in a form more convenient for an extension to the case of continuous time.

We shall refer to $\{\alpha_n, \mathfrak{F}_n, n \in N\}$ as an *integrable increasing process* if $\alpha_0 = 0$, $\alpha_n \leqslant \alpha_{n+1}$ and $\mathsf{E}\alpha_\infty < \infty$, where $\alpha_\infty = \lim \alpha_n$.

Lemma 3. *In order that an integrable increasing process* $\{\alpha_n, \mathfrak{F}_n, n \in N\}$ *be adopted to the current* $\{\mathfrak{F}'_n, n=1,2,\ldots\}$, *where* $\mathfrak{F}'_n = \mathfrak{F}_{n-1}$, $n=1,2,\ldots$, *it is necessary and sufficient that for any* $\mathfrak{F}_n$*-martingale* $\{\eta_n, n=0,1,2,\ldots\}$ *bounded with probability* 1 *equality*

(15) $$\mathsf{E}\sum_{n=1}^{\infty} \eta_{n-1}\Delta\alpha_n = \mathsf{E}\eta_\infty\alpha_\infty,$$

be satisfied, where $\Delta\alpha_n = \alpha_n - \alpha_{n-1}$ *and* $\eta_\infty = \lim \eta_n$.

Proof. Necessity. Note that $\sum_{n=1}^{N} \eta_{n-1}\, \Delta\alpha_n = \sum_{n=1}^{N-1} \alpha_n(\eta_{n-1}-\eta_n)+\alpha_N\eta_{N-1}$. Applying the Lebesgue dominated convergence theorem and the fact that sequence η_n is bounded, we obtain

$$\begin{aligned}\mathsf{E}\sum_{n=1}^{\infty} \eta_{n-1}\,\Delta\alpha_n &= \lim \mathsf{E}\sum_{1}^{N} \eta_{n-1}\,\Delta\alpha_n \\ &= \lim \sum_{1}^{N} \mathsf{E}\{\alpha_n\, \mathsf{E}(\eta_{n-1}-\eta_n)\,|\,\mathfrak{F}_{n-1}\} + \lim \mathsf{E}\alpha_N\eta_{N-1} \\ &= \lim \mathsf{E}\alpha_N\eta_{N-1} = \mathsf{E}\alpha_\infty\eta_\infty.\end{aligned}$$

Sufficiency. Set $\eta_0=\eta_1=\cdots=\eta_{n-1}=0$, $\eta_n=\eta_{n+1}=\cdots=\zeta$, where $\zeta=\eta-\mathsf{E}\{\eta\,|\,\mathfrak{F}_{n-1}\}$ and η is an arbitrary $\mathfrak{F}_n$-measurable bounded variable. Clearly $\mathsf{E}\{\zeta\,|\,\mathfrak{F}_{n-1}\}=0$. It follows from (15) that $\mathsf{E}\zeta(\alpha_\infty-\alpha_n)=\mathsf{E}\zeta\alpha_\infty$ or $\mathsf{E}\zeta\alpha_n=0$. Since for any $\mathfrak{F}_{n-1}$-measurable random variable ψ the expectation $\mathsf{E}\zeta\psi=0$, we have

$$\mathsf{E}\zeta(\alpha_n-\mathsf{E}\{\alpha_n\,|\,\mathfrak{F}_{n-1}\})=0,$$

which implies that

$$\mathsf{E}\eta(\alpha_n-\mathsf{E}\{\alpha_n\,|\,\mathfrak{F}_{n-1}\})=0.$$

Let ψ^c denote the "truncated" variable ψ, i.e. $\psi^c=\psi$ for $|\psi|\leqslant c$ and $\psi^c=0$ for $|\psi|\geqslant c$. The previous equality yields that

$$\mathsf{E}(\alpha_n-\mathsf{E}\{\alpha_n\,|\,\mathfrak{F}_{n-1}\})^c(\alpha_n-\mathsf{E}\{\alpha_n\,|\,\mathfrak{F}_{n-1}\})=0$$

or

$$\mathsf{E}[(\alpha_n-\mathsf{E}\{\alpha_n\,|\,\mathfrak{F}_{n-1}\})^c]^2=0$$

for any $c>0$. Thus $\alpha_n=\mathsf{E}\{\alpha_n\,|\,\mathfrak{F}_{n-1}\}(\text{mod }\mathsf{P})$. □

In this item, Lemma 2 will be generalized for the case of supermartingales with a continuous argument. Unlike the situation in the discrete case, the corresponding proof is not simple and the result we are going to establish is quite profound.

In what follows we shall assume the existence of a fixed current of σ-algebras $\{\mathfrak{F}_t,\, t\geqslant 0\}$, $\mathfrak{F}_t=\mathfrak{F}_{t+}$, and that all the semimartingales under consideration (unless explicitly specified) are $\mathfrak{F}_t$-semimartingales. We call $\alpha(t)$ an *increasing process* provided it is adopted to $\{\mathfrak{F}_t,\, t\geqslant 0\}$, its sample functions $\alpha(t)$ are monotonically nondecreasing with probability 1, it is continuous from the right, and $\alpha(0)=0$.

An increasing process is called *integrable* if $\sup \mathsf{E}\alpha(t)<\infty$.

Definition. An integrable increasing process $\alpha(t)$, $t\geqslant 0$, is called *natural* if for an arbitrary nonnegative martingale $\eta(t)$, bounded with probability 1,

(16) $$\mathsf{E}\int_0^\infty \eta(t-)\,d\alpha(t)=\mathsf{E}\alpha(\infty)\eta(\infty).$$

A process $\alpha(t)$ satisfying condition (16) is called *natural* also in the case when it is a difference of two integrable increasing processes.

Relation (16) is a continuous analog of equality (15). The integral in the left-hand side of formula (16) is the ordinary Lebesgue–Stieltjes integral and exists with probability 1 for each and every sample function of the process $\alpha(t)$. Indeed, a monotone function $\alpha(t)=\alpha(t,\omega)$ generates for a fixed ω a measure $\alpha(A)$ on $[0,\infty)$ such that $\alpha(c,d]=\alpha(d)-\alpha(c)$ and the integral

$$\int_0^\infty \varphi(t)\,d\alpha(t)$$

is defined with probability 1 for an arbitrary random process $\varphi(t)$ with sample functions being (with probability 1) Borel functions and integrable with respect to measure $d\alpha$ (or are nonnegative).

We note the following formula of integration by parts.

Let $\alpha(t)$, $\beta(t)$ be two increasing processes,

$$\alpha(0)=\beta(0)=0, \qquad \alpha(\infty)<\infty, \qquad \beta(\infty)<\infty.$$

Then

$$\int_0^\infty \beta(s)\,d\alpha(s)+\int_0^\infty \alpha(s-)\,d\beta(s)=\alpha(\infty)\beta(\infty). \tag{17}$$

This formula can be easily obtained if one represents the double integral $\int_0^\infty\int_0^\infty d\alpha(s)\,d\beta(u)$—which equals $\alpha(\infty)\beta(\infty)$—as the sum of integrals over the regions $\{(s,u): s>u, u\in[0,\infty)\}$ and $\{(s,u): u\in[0,\infty), u\le s\}$. Applying Fubini's theorem we thus obtain

$$\alpha(\infty)\beta(\infty)=\int_0^\infty [\beta(\infty)-\beta(s)]\,d\alpha(s)+\int_0^\infty [\alpha(\infty)-\alpha(u-)]\,d\beta(u),$$

which yields (17).

A few remarks concerning equality (16) are in order:

a) If $\eta(t)$ is an arbitrary nonnegative bounded martingale and $\alpha(t)$ is an integrable increasing process, then

$$\mathsf{E}\int_0^\infty \eta(t)\,d\alpha(t)=\mathsf{E}\eta(\infty)\alpha(\infty). \tag{18}$$

Indeed, taking into account that $\eta(t)$ is continuous from the right we obtain the following equalities:

$$\begin{aligned}\mathsf{E}\int_0^\infty \eta(t)\,d\alpha(t)&=\mathsf{E}\lim\sum_{k=0}^\infty \eta\left(\frac{k+1}{2^n}\right)\left[\alpha\left(\frac{k+1}{2^n}\right)-\alpha\left(\frac{k}{2^n}\right)\right]\\&=\lim\mathsf{E}\left\{\eta(\infty)\alpha(\infty)-\sum_{k=0}^\infty\left[\eta\left(\frac{k+1}{2^n}\right)-\eta\left(\frac{k}{2^n}\right)\right]\alpha\left(\frac{k}{2^n}\right)\right\}\\&=\mathsf{E}\eta(\infty)\alpha(\infty).\end{aligned}$$

Here we have utilized the fact that

$$\sum_{k=0}^{\infty} \eta\left(\frac{k+1}{2^n}\right)\left[\alpha\left(\frac{k+1}{2^n}\right)-\alpha\left(\frac{k}{2^n}\right)\right] \leqslant \sup \eta(t)\alpha(\infty).$$

b) Let $\alpha(t)$ be a natural process and $\eta(t)$ be a positive uniformly integrable martingale. Then

$$\mathsf{E}\int_0^\infty \eta(t-)\,d\alpha(t)=\mathsf{E}\int_0^\infty \eta(t)\,d\alpha(t). \tag{19}$$

To prove this assertion note that equality (19) is valid for the martingale $\eta^{(n)}(t)=\mathsf{E}\{\eta(\infty)\wedge n \mid \mathfrak{F}_t\}$. On the other hand the sequence $\eta^{(n)}(t)$ is monotonically nondecreasing and

$$\mathsf{P}(\sup_t (\eta(t)=\eta^{(n)}(t))>\varepsilon)\leqslant\frac{1}{\varepsilon}\,\mathsf{F}|\eta(\infty)\quad \eta^{(n)}(\infty)| \xrightarrow[n\to\infty]{} 0.$$

Therefore a sequence n_j exists such that $\eta^{(n_j)}(t)\to\eta(t)$ with probability 1 uniformly in t. Approaching the limit as $n_j\to\infty$ in relation (19) with $\eta(t)=\eta^{(n_j)}(t)$ we obtain formula (19) in the general case.

c) A continuous integrable increasing process is a natural process.

d) If $\alpha(t)$ is a natural process, $\eta(t)$ is a nonnegative bounded martingale and τ a random time, then

$$\mathsf{E}\int_{[0,\tau]} \eta(t-)\,d\alpha=\mathsf{E}\eta(\tau)\alpha(\tau). \tag{20}$$

Indeed, let $\bar\eta(t)=\eta(t)\chi(t\leqslant\tau)+\eta(\tau)\chi(t>\tau)$. Then $\bar\eta(t)$ is also a bounded martingale and

$$\begin{aligned}\mathsf{E}\bar\eta(\infty)\alpha(\infty)&=\mathsf{E}\int_0^\infty \bar\eta(t-)\,d\alpha\\ &=\mathsf{E}\int_{[0,\tau]}\eta(t-)\,d\alpha+\mathsf{E}\eta(\tau)[\alpha(\infty)-\alpha(\tau)],\end{aligned}$$

which yields equality (20).

e) In particular, under the preceding assumptions

$$\mathsf{E}\int_0^t \eta(s-)\,d\alpha(s)=\mathsf{E}\eta(t)\alpha(t). \tag{21}$$

f) If $\alpha(t)$ is a natural increasing process then $\bar\alpha(t)=\alpha(t)\chi(t\leqslant\tau)+\alpha(t)\chi(t>\tau)$ is also a natural increasing process.

Indeed, the requirement that $\bar\alpha(t)$ is a natural process is equivalent to the condition

$$\mathsf{E}\eta(\infty)\bar\alpha(\infty)=\mathsf{E}\int_0^\infty \eta(t-)\,d\bar\alpha=\mathsf{E}\int_{[0,\tau]}\eta(t-)\,d\alpha \tag{22}$$

for an arbitrary bounded nonnegative martingale $\eta(t)$. On the other hand,

$$\mathsf{E}\eta(\infty)\bar{\alpha}(\infty)=\mathsf{E}\eta(\infty)\alpha(\tau)=\mathsf{E}\{\alpha(\tau)\mathsf{E}\{\eta(\infty)|\mathfrak{F}_\tau\}\}=\mathsf{E}\alpha(\tau)\eta(\tau),$$

so that condition (22) coincides with condition (20).

We also note the following alternative version of equality (19). It follows from (19) and arguments analogous to those leading to relation (22) that

g)

$$\mathsf{E}\int_{[0,\tau]}\eta(t-)\,d\alpha=\mathsf{E}\int_{[0,\tau]}\eta(t)\,d\alpha. \tag{23}$$

h) If $\alpha(t)$ is a natural process then $\beta(t)=\alpha(t)\wedge n$ is also a natural process. Indeed, if $\tau=\inf\{t,\alpha(t)\geq n\}$, then

$$\begin{aligned}\mathsf{E}\int_0^\infty\eta(t-)\,d\beta&=\mathsf{E}\int_{[0,\tau]}\eta(t-)\,d\alpha\\&=\mathsf{E}\eta(\tau)\alpha(\tau)=\mathsf{E}\eta(\infty)\alpha(\tau)=\mathsf{E}\eta(\infty)\beta(\infty),\end{aligned}$$

since $\beta(\infty)=\alpha(\tau)$.

To prove the important theorem of Mayer (Theorem 9), which is the basic result in this subsection, the following lemma will be required. This lemma will also be utilized later on for other purposes.

Lemma 4. *Let* $\{\xi^{(z)}(n),\mathfrak{F}_n, n=0,1,\ldots\}$, $z\in Z$, *be a set of potentials,* $\alpha^{(z)}(n)$ *be a process associated with* $\xi^{(z)}(n)$, *and* $\mathcal{T}$ *be a set of all finite* $\mathfrak{F}_n$-*random times on N. Assume that*

$$\sup_z\sup_{\tau\in\mathcal{T}}\int_{\{\xi_\tau\geq C\}}\xi_\tau^{(z)}\,d\mathsf{P}=\rho(C)<\infty$$

and $\rho(C)\to 0$ *as* $C\to\infty$. *Then the family of random variables* $\{\alpha^{(z)}(\infty), z\in Z\}$ *is uniformly integrable.*

Proof. For convenience we shall omit index z in the designation of $\xi^{(z)}(n)$ and $\alpha^{(z)}(n)$. Set $\tau_N=\inf\{n,\alpha_{n+1}>N\}$. Since $\alpha(n)$ is an $\mathfrak{F}_{n-1}$-measurable random variable, τ_N is a random time on $\{\mathfrak{F}_n, n=0,1,\ldots\}$. It follows from the relation $\xi(n)=\mathsf{E}\{\alpha(\infty)|\mathfrak{F}_n\}-\alpha(n)$ that

$$\mathsf{E}\{\alpha(\infty)|\mathfrak{F}_{\tau_N}\}=\xi(\tau_N)+\alpha(\tau_N)$$

and

$$\int_{\{\alpha(\infty)>N\}}\alpha(\infty)\,d\mathsf{P}\leq\int_{\{\alpha(\infty)>N\}}\xi(\tau_N)\,d\mathsf{P}+N\mathsf{P}\{\alpha(\infty)>N\}. \tag{24}$$

Applying this inequality we obtain

$$
\begin{aligned}
N\mathsf{P}\{\alpha(\infty)>2N\} &\leqslant \int_{\{\alpha(\infty)>2N\}} (\alpha(\infty)-N)\,d\mathsf{P} \\
&\leqslant \int_{\{\alpha(\infty)>N\}} (\alpha(\infty)-N)\,d\mathsf{P} \leqslant \int_{\{\alpha(\infty)>N\}} \xi(\tau_N)\,d\mathsf{P}.
\end{aligned}
$$

Substituting $2N$ in place of N in relation (24) and utilizing the last inequality we observe that

$$
\int_{\{\alpha(\infty)>2N\}} \alpha(\infty)\,d\mathsf{P} \leqslant \int_{\{\alpha(\infty)>2N\}} \xi(\tau_{2N})\,d\mathsf{P} + 2\int_{\{\alpha(\infty)>N\}} \xi(\tau_N)\,d\mathsf{P}.
$$

Since

$$
\begin{aligned}
\int_{\{\alpha(\infty)>N\}} \xi(\tau_N)\,d\mathsf{P} &\leqslant \rho(C) + \int_{\{\alpha(\infty)>N\}\cap\{\xi(\tau_N)\leqslant C\}} \xi(\tau_N)\,d\mathsf{P} \\
&\leqslant \rho(C) + C\mathsf{P}\{\alpha(\infty)>N\}, \quad N\mathsf{P}\{\alpha(\infty)>N\} \leqslant \mathsf{E}\alpha(\infty) = \mathsf{E}\xi(0),
\end{aligned}
$$

it follows that

$$
\int_{\{\alpha(\infty)>2N\}} \alpha(\infty)\,d\mathsf{P} \leqslant 3\rho(C) + \frac{5C\mathsf{E}\xi(0)}{2N}. \tag{25}
$$

Setting, for example, $C = N^{1/2}$ in the last inequality, we obtain the required result. □

Theorem 9 (Meyer's theorem). *In order that a supermartingale $\xi(t)$, $t \geqslant 0$, admit a representation in the form*

$$
\xi(t) = \mu(t) - \alpha(t), \tag{26}
$$

where $\mu(t)$ is a uniformly integrable martingale and $\alpha(t)$ is an integrable increasing process, it is necessary and sufficient that process $\xi(t)$ belong to class D. If this condition is satisfied then the process $\alpha(t)$ can be chosen to be natural; moreover, in the class of decompositions with natural processes, decomposition (26) *is unique.*

In the case where $\mu(t)$ is a martingale and $\alpha(t)$ is a natural process, decomposition (26) is called *Doob's decomposition* of supermartingale $\xi(t)$.

Proof. Necessity of the condition of the theorem is almost self-evident. If a martingale $\mu(t)$ is uniformly integrable then it belongs to class D (Theorem 5). On the other hand, the family $\{\alpha(t), t \geqslant 0\}$ is completely uniformly integrable since $\alpha(\tau) \leqslant \alpha(\infty)$ for any $\tau \in \mathcal{T}$ and $\mathsf{E}\alpha(\infty) < \infty$. Therefore the process $\{\xi(t), t \geqslant 0\}$ is completely uniformly integrable.

Now let $\{\xi(t), t \geqslant 0\}$ be a supermartingale in class D. As such it admits a unique representation of the form $\xi(t) = \eta(t) - \pi(t)$, where $\eta(t)$ is a martingale and $\pi(t)$ is a potential in class D. Therefore it is sufficient to prove Theorem 9 for potentials. Thus, in what follows, let $\xi(t)$ represent a potential in class D.

For every integer n the sequence $\xi_k^{(n)}=\xi(k/2^n)$, $k=0,1,2,\ldots$, is a potential with respect to the current of σ-algebras $\{\mathfrak{F}_k^{(n)}=\mathfrak{F}_{k/2^n}, k=0,1,\ldots\}$. In view of Lemma 2 a sequence $\alpha_n(k)$ exists which can be more suitably denoted by $\alpha_n(k/2^n)$, $k=0,1,\ldots$, $n=1,2,\ldots$, such that

$$\xi\left(\frac{k}{2^n}\right)=\mathsf{E}\{\alpha_n(\infty)|\mathfrak{F}_{k/2^n}\}-\alpha_n\left(\frac{k}{2^n}\right), \tag{27}$$

where

$$\alpha_n(\infty)=\lim_{k\to\infty}\alpha_n\left(\frac{k}{2^n}\right),\qquad \alpha_n\left(\frac{k+1}{2^n}\right)\geqslant\alpha_n\left(\frac{k}{2^n}\right),$$

and $\alpha_n(k/2^n)$ is an $\mathfrak{F}_{(k-1)/2^n}$-measurable random variable. Since $\xi(t)$ is a potential in class D the quantity

$$\rho(C)=\sup_{\tau\in\mathcal{T}}\int_{\{\xi(\tau)\geqslant C\}}\xi(\tau)\,d\mathsf{P}<\infty$$

and $\rho(C)\to 0$ as $C\to\infty$. Lemma 4 is applicable to the family of potentials $\{\xi(k/2^n), \mathfrak{F}_{k/2^n}, k=0,1,\ldots\}$, $n=1,2,\ldots$. Consequently, the sequence of variables $\{\alpha_n(\infty), n=1,2,\ldots\}$ is uniformly integrable. This implies the existence of a sequence n_j such that $\alpha_{n_j}(\infty)$ is weakly convergent to a limit α_∞, i.e., for an arbitrary bounded random variable η

$$\mathsf{E}\alpha_{n_j}(\infty)\eta\to\mathsf{E}\alpha_\infty\eta.$$

Observe that the sequence of random variables $\mu_n(r)=\mathsf{E}\{\alpha_n(\infty)|\mathfrak{F}_r\}$, $n=1,2,\ldots$, is uniformly integrable. Indeed,

$$N\mathsf{P}\{\mu_n(r)>N\}\leqslant\int_{\{\mu_n(r)>N\}}\mu_n(r)\,d\mathsf{P}\leqslant\int\alpha_n(\infty)\,d\mathsf{P}.$$

Therefore $\mathsf{P}\{\mu_n(r)>N\}\to 0$ uniformly in N and it follows from the equality

$$\int_{\{\mu_n(r)>N\}}\mu_n(r)\,d\mathsf{P}=\int_{\{\mu_n(r)>N\}}\alpha_n(\infty)\,d\mathsf{P} \tag{28}$$

that the sequence $\mu_n(r)$, $n=1,2,\ldots$, is uniformly integrable. Thus the sequence $\mu_n(r)$ is also weakly compact. By means of a diagonal process one can choose a subsequence of indices k_j such that $\alpha_{k_j}(\infty)=\alpha_\infty$ and $\mu_{k_j}(r)\to\mu_\infty(r)$ for every dyadic rational number r in the sense of weak convergence. Since for any $B_r\in\mathfrak{F}_r$

$$\int_{B_r}\mu_\infty(r)\,d\mathsf{P}=\lim_{j\to\infty}\int_{B_r}\mathsf{E}\{\alpha_{k_j}(\infty)|\mathfrak{F}_r\}\,d\mathsf{P}$$
$$=\lim\int_{B_r}\alpha_{k_j}(\infty)\,d\mathsf{P}=\int_{B_r}\alpha_\infty\,d\mathsf{P},$$

it follows that $\mu_\infty(r)=\mathsf{E}\{\alpha_\infty|\mathfrak{F}_r\}$.

Let $s<r$ and s and r be dyadic rational numbers. From the relation

$$\alpha_n(s)=\mu_n(s)-\xi(s)\leq\mu_n(r)-\xi(r),$$

passing to the limit we obtain that for any measurable B.

$$\int_B(\mu_\infty(s)-\xi(s))\,d\mathsf{P}\leq\int_B(\mu_\infty(r)-\xi(r))\,d\mathsf{P}.$$

Thus, $\mu_\infty(s)-\xi(s)\leq\mu_\infty(r)-\xi(r)$. Set

$$\alpha(t)=\mathsf{E}\{\alpha_\infty\,|\,\mathfrak{F}_t\}-\xi(t),\qquad t>0. \tag{29}$$

Here $\mathsf{E}\{\alpha_\infty\,|\,\mathfrak{F}_t\}$ represents a martingale with sample functions belonging to $\mathcal{D}$. Therefore sample functions of process $\alpha(t)$ also belong to $\mathcal{D}$ and are monotonically nondecreasing. We now show that the process $\alpha(t)$ is natural.

Let $\eta(t)$ be an arbitrary bounded martingale, with sample functions belonging to $\mathcal{D}$. Then we have with probability 1

$$\int_0^\infty\eta(s-)\,d\alpha(s)=\lim_{n\to\infty}\sum_{k=0}^\infty\eta\left(\frac{k}{2^n}\right)\left[\alpha\left(\frac{k+1}{2^n}\right)-\alpha\left(\frac{k}{2^n}\right)\right],$$

and in view of Lebesgue's dominated convergence theorem we have

$$\mathsf{E}\int_0^\infty\mu(s-)\,d\alpha(s)=\lim_{n\to\infty}\sum_{k=0}^\infty\mathsf{E}\eta\left(\frac{k}{2^n}\right)\left[\alpha\left(\frac{k+1}{2^n}\right)-\alpha\left(\frac{k}{2^n}\right)\right].$$

It follows from the definition of $\alpha(t)$ and Lemma 3 that

$$\begin{aligned}\sum_{k=0}^\infty\mathsf{E}\eta\left(\frac{k}{2^n}\right)\left[\alpha\left(\frac{k+1}{2^n}\right)-\alpha\left(\frac{k}{2^n}\right)\right]&=\sum_{k=0}^\infty\mathsf{E}\eta\left(\frac{k}{2^n}\right)\left[\xi\left(\frac{k}{2^n}\right)-\xi\left(\frac{k+1}{2^n}\right)\right]\\&=\sum_{k=0}^\infty\mathsf{E}\eta\left(\frac{k}{2^n}\right)\left[\alpha_n\left(\frac{k+1}{2^n}\right)-\alpha_n\left(\frac{k}{2^n}\right)\right]\\&=\mathsf{E}\eta(\infty)\alpha_n(\infty).\end{aligned}\tag{29'}$$

Now from the definition of α_∞ we have

$$\mathsf{E}\eta(\infty)\alpha_{k_j}(\infty)\to\mathsf{E}\eta(\infty)\alpha_\infty\quad\text{as } k_j\to\infty.$$

Whence, taking the preceding equalities into account, we obtain

$$\mathsf{E}\int_0^\infty\eta(s-)\,d\alpha(s)=\mathsf{E}\eta(\infty)\alpha_\infty,$$

which proves that the process $\alpha(t)$ is natural.

We need only show that the process $\alpha(t)$ is unique (mod P).

It follows from the preceding deliberations that the sequence $\alpha_n(\infty)$ converges weakly to $\alpha(\infty)$. Indeed, first of all

$$\mathsf{E}\eta(\infty)\alpha_\infty = \lim \mathsf{E}\eta(\infty)\alpha_n(\infty).$$

Let η be an arbitrary $\mathfrak{S}$-measurable bounded variable and $\eta_\infty = \mathsf{E}\{\eta \mid \mathfrak{F}_\infty\}$; then

$$\begin{aligned}\mathsf{E}\eta\alpha_\infty &= \mathsf{E}\mathsf{E}\{\eta\alpha_\infty \mid \mathfrak{F}_\infty\} = \mathsf{E}\eta_\infty\alpha_\infty \\ &= \lim_{n\to\infty} \mathsf{E}\eta_\infty\alpha_n(\infty) = \lim_{n\to\infty} \mathsf{E}\eta\alpha_n(\infty),\end{aligned}$$

which proves that $\alpha_n(\infty)$ converges weakly to α_∞.

Assume now that a representation for potential $\xi(t)$,

$$\xi(t) = \mathsf{E}\{\beta_\infty \mid \mathfrak{F}_t\} - \beta(t)$$

exists where $\beta(t)$ is a natural process. It follows from equation (29′) that

$$\begin{aligned}\mathsf{E}\int_0^\infty \eta(s-)\,d\beta(s) &= \lim_{n\to\infty} \sum_{k=0}^{\infty} \mathsf{E}\eta\left(\frac{k}{2^n}\right)\left[\alpha_n\left(\frac{k+1}{2^n}\right) - \alpha_n\left(\frac{k}{2^n}\right)\right] \\ &= \lim_{n\to\infty} \mathsf{E}\eta(\infty)\alpha_n(\infty) = \mathsf{E}\eta(\infty)\alpha_\infty,\end{aligned}$$

and since $\beta(t)$ is natural,

$$\mathsf{E}\eta(\infty)\alpha_\infty = \mathsf{E}\eta(\infty)\beta(\infty),$$

which implies that $\alpha_\infty = \beta(\infty)$ and $\beta(t) = \alpha(t) (\text{mod } \mathsf{P})$. □

Remark 1. If $\xi(t) = \mu(t) - \alpha(t)$ is Doob's decomposition for supermartingale $\xi(t)$ in class D, then the equality $\xi(t \wedge \tau) = \mu(t \wedge \tau) - \alpha(t \wedge \tau)$ is Doob's decomposition of the supermartingale $\xi(t \wedge \tau)$.

Remark 2. In the proof of the uniqueness of Doob's decomposition the assumption of the monotonicity of the process $\beta(t)$ was not actually utilized. It is therefore sufficient to assume that the process $\beta(t)$ is representable as a difference of two natural integrable increasing processes, in which case the proof presented above yields $\beta(t) = \alpha(t)$.

A generalization of Meyer's theorem. We extend the definition of a natural process to the case of arbitrary (i.e., generally nonintegrable) increasing processes. Namely, an increasing process $\alpha(t)$ will be called *natural* if for an arbitrary bounded nonnegative martingale $\eta(t)$ and any $a > 0$

$$\mathsf{E}\int_0^a \eta(t-)\,d\alpha(t) = \mathsf{E}\int_0^a \eta(t)\,d\alpha(t).$$

As it follows from (21) an integrable natural process is natural also in the sense just described.

Theorem 10. *A supermartingale $\xi(t)$ admits the decomposition*

$$\xi(t)=\mu(t)-\alpha(t), \tag{32}$$

where $\mu(t)$ is a martingale and $\alpha(t)$ is an increasing process, if and only if $\xi(t)$ belongs to the class DL.

A decomposition in which the process $\alpha(t)$ is natural is unique.

Proof. Let $\xi(t)$ be a supermartingale of class DL. Then $\xi_a(t)=\xi(a\wedge t)$, $a>0$, is a supermartingale of class D, and in view of Theorem 9

$$\xi_a(t)=\mu_a(t)-\alpha_a(t),$$

where $\mu_a(t)$ is a uniformly integrable martingale, and $\alpha_a(t)$ is an integrable natural process. Let $b>a$. Then $\xi_a(t)=\xi_b(t\wedge a)=\mu_b(t\wedge a)-\alpha_b(t\wedge a)$, and it follows from the uniqueness of Doob's decomposition that $\mu_b(t)=\mu_a(t)$ and $\alpha_b(t)=\alpha_a(t)$ for $t\leqslant a$. Hence, limits $\mu(t)=\lim\mu_a(t)$ and $\alpha(t)=\lim\alpha_a(t)$ exist with probability 1 and, moreover, $\mu(t)$ is clearly a martingale, while $\alpha(t)$ is a natural process and $\xi(t)=\mu(t)-\alpha(t)$. The existence of decomposition (32) is thus verified.

Assume now that the process $\xi(t)$ is given by formula (32). For each $a>0$ the equation $\xi(a\wedge t)=\mu(a\wedge t)-\alpha(a\wedge t)$ represents Doob's decomposition in which $\mu(a\wedge t)$ is a uniformly integrable martingale, $\alpha(a\wedge t)$ is an integrable natural process and in view of Theorem 9 $\xi(a\wedge t)$ is a supermartingale in class D and hence $\xi(t)$ is a supermartingale in class DL.

The uniqueness of decomposition (32) with a natural $\alpha(t)$ can also be easily deduced from the uniqueness of the decomposition proved in Theorem 9. □

A generalization will now be required of the notion of a martingale.

Definition. A process $\{\xi(t), \mathfrak{F}_t, t\geqslant 0\}$ is called a *local martingale* if its sample functions belong to $\mathscr{D}$ and a monotonically increasing sequence of $\mathfrak{F}_t$-random times τ_n, $n=1,2,\ldots$ exists such that

i) $\lim\tau_n=\infty$ (mod P),
ii) $\xi(\tau_n\wedge t)$ is a uniformly integrable martingale with respect to $\{\mathfrak{F}_{t\wedge\tau_n}, t\geqslant 0\}$, $n=1,2,\ldots$.

A sequence τ_n, $n=1,2,\ldots$, satisfying the conditions of the definition is called *completely reducing local martingale* $\xi(t)$ and the random time τ for which $\xi(\tau\wedge t)$ is a uniformly integrable martingale is called *reducing martingale* $\xi(t)$.

Theorem 11. *Let $\xi(t)$ be a nonnegative supermartingale. Then it admits decomposition* (32) *in which $\mu(t)$ is a local martingale and $\alpha(t)$ is an increasing integrable natural process. Such a decomposition is unique.*

Proof. The proof is analogous to the proof of the preceding theorem. Introduce a sequence of random times $\tau_n = \inf\{t: \xi(t) \geq n\}$, $n = 1, 2, \ldots$, and a sequence of stopping supermartingales $\xi_n(t) = \xi(\tau_n \wedge t)$. Evidently $\xi_n(t)$ belongs to class D, since $\xi_n(t) \leq \max(\xi(\tau_n), n)$. Hence in view of Theorem 9 there exists decomposition $\xi_n(t) = \mu_n(t) - \alpha_n(t)$, where $\alpha_n(t)$ is an integrable natural process. In the same manner as in the proof of Theorem 10 we verify that $\mu_n(t) = \mu_{n+1}(t)$ and $\alpha_n(t) = \alpha_{n+1}(t)$ for $t \leq \tau_n$ for each n and that with probability 1 there exist limits $\mu(t) = \lim \mu_n(t)$ and $\alpha(t) = \lim \alpha_n(t)$. Since $\sup \xi(t) < \infty (\text{mod } \mathsf{P})$, it follows that $\xi(t) = \lim \xi_n(t) = \mu(t) - \alpha(t)$. However, in the present case one can assert about the process $\mu(t)$ only that it is a local martingale. On the other hand $\alpha(t)$ is an integrable process. Indeed, in view of the theorem on monotone convergence

$$\begin{aligned} \mathsf{E}\alpha(t) &= \lim \mathsf{E}\alpha_n(t) = \lim \mathsf{E}(\mu_n(t) - \xi_n(t)) \\ &= \lim \mathsf{E}(\xi_n(0) - \xi_n(t)) \leq \underline{\lim}\, \mathsf{E}\xi_n(0) = \mathsf{E}\xi(0). \end{aligned}$$

The uniqueness of this decomposition easily follows from the uniqueness of the decomposition in Theorem 9. □

Utilizing the theorem just proved and Theorem 5 we easily obtain the following result:

Theorem 12. *An arbitrary quasi-martingale admits decomposition*

$$\xi(t) = \mu(t) + \nu(t) + \beta(t), \tag{33}$$

where $\mu(t)$ is a martingale, $\nu(t)$ is the difference between two integrable nonnegative local martingales, and $\beta(t)$ is the difference between two natural increasing integrable processes.

Clearly the converse is also true. If a process $\xi(t)$ admits decomposition (33) it is a quasi-martingale.

To prove this it is sufficient to verify that process $\nu(t)$ is a quasi-martingale. This follows from the fact that a nonnegative local martingale is a supermartingale. The latter, however, follows from the inequalities

$$\begin{aligned} \mathsf{E}\{\nu(t) \mid \mathfrak{F}_s\} &= \mathsf{E}\{\lim_{n\to\infty} \nu(\tau_n \wedge t) \mid \mathfrak{F}_s\} \\ &\leq \lim \mathsf{E}\{\nu(\tau_n \wedge t) \mid \mathfrak{F}_s\} \leq \lim \nu(\tau_n \wedge s) = \nu(s). \end{aligned}$$

(Here $s < t$, $\nu \geq 0$, and ν is an integrable local martingale.)

Regular supermartingales

Definition. A natural process $\alpha(t)$ which appears in the decomposition of supermartingale $\xi(t)$ is called a *process associated with* $\xi(t)$.

In turn a unique (mod P) *associated potential*

$$\pi(t)=\mathsf{E}\{\alpha(\infty)|\mathfrak{F}_t\}-\alpha(t)$$

corresponds to each integrable natural process $\alpha(t)$.

In a number of cases there are important and sufficiently simple connections among properties of associated processes. Some of these are discussed in this section.

Definition. A semimartingale $\{\xi(t), t\geqslant 0\}$ is called *regular* if for an arbitrary increasing sequence of random times τ_n which converges to a bounded random time with probability 1

$$\lim \mathsf{E}\xi(\tau_n)=\mathsf{E}\xi(\tau). \tag{34}$$

Clearly every martingale is regular. Observe that if a supermartingale is regular and is completely uniformly integrable then relation (34) is fulfilled for an arbitrary nondecreasing sequence of random times taking on possibly infinite values also. Here we assume that the value of $\xi(\tau)$ at $\tau=\infty$ is defined in the manner indicated above.

The main result of the present subsection is as follows:

Theorem 13. *Let $\xi(t)$ be a supermartingale of class D. The associated process $\alpha(t)$ is continuous if and only if the process $\xi(t)$ is regular.*

To prove this theorem, a number of results will be needed which are of interest in their own right. They are connected with the possibility of approximating potentials by means of bounded and continuous potentials.

Theorem 14. *If $\xi(t)$ is a completely uniformly integrable potential and $\alpha(t)$ is the associated process, then*

$$\mathsf{E}\alpha^2(\infty)=\mathsf{E}\int_0^\infty [\xi(t)+\xi(t-)]\,d\alpha(t). \tag{35}$$

Proof. Let $\alpha_n(t)=\alpha(t)\wedge n$ and $\xi_n(t)$ be the potential associated with $\alpha_n(t)$, $\eta_n(t)=\mathsf{E}\{\alpha_n(\infty)|\mathfrak{F}_n\}$. Since $\alpha(t)$ is a natural process, we have

$$\mathsf{E}\alpha(\infty)\alpha_n(\infty)=\mathsf{E}\int_0^\infty \eta_n(t-)\,d\alpha(t).$$

On the other hand (cf. (18))

$$\mathsf{E}\alpha(\infty)\alpha_n(\infty)=\mathsf{E}\int_0^\infty \eta_n(t)\,d\alpha(t),$$

so that

$$\begin{aligned}2\mathsf{E}\alpha(\infty)\alpha_n(\infty)&=\mathsf{E}\int_0^\infty (\eta_n(t-)+\eta_n(t))\,d\alpha(t)\\&=\mathsf{E}\int_0^\infty (\xi_n(t)+\xi_n(t-))\,d\alpha(t)\\&\quad+\mathsf{E}\int_0^\infty (\alpha_n(t)+\alpha_n(t-))\,d\alpha(t).\end{aligned}$$

It follows from the formula of integration by parts (17) that

$$(36)\quad \int_0^\infty (\alpha_n(t)+\alpha_n(t-))\, d\alpha(t) = 2\alpha(\infty)\alpha_n(\infty) - \int_0^\infty [\alpha(t)+\alpha(t-)]\, d\alpha_n(t).$$

Hence,

$$\mathsf{E}\int_0^\infty (\xi_n(t)+\xi_n(t-))\, d\alpha(t) = \mathsf{E}\int_0^\infty [\alpha(t)+\alpha(t-)]\, d\alpha_n(t).$$

By the theorem on monotone convergence

$$\lim_{n\to\infty} \mathsf{E}\int_0^\infty [\alpha(t)+\alpha(t-)]\, d\alpha_n(t) = \mathsf{E}\int_0^\infty [\alpha(t)+\alpha(t-)]\, d\alpha(t),$$

which in view of (36) equals $\mathsf{E}\alpha^2(\infty)$.

On the other hand, $\xi_n(t) = \mathsf{E}\{\alpha_n(\infty)-\alpha_n(t) \mid \mathfrak{F}_t\}$, $n = 1, 2, \ldots$, form a monotonically nondecreasing sequence of nonnegative random variables, and to establish formula (35) it is sufficient to verify that a sequence of indices n_j exists such that $\xi_{n_j}(t)$ with probability 1 converges uniformly to $\xi(t)$ (in this case $\xi_n(t-)\to \xi(t-)$ also, with probability 1). Since $\xi_n(t) = \eta_n(t) - \alpha_n(t)$ and $\sup(\alpha(t)-\alpha_n(t)) = \alpha(\infty) - n \wedge \alpha(\infty) \to 0$, it is sufficient to verify the required property of the sequence $\xi_n(t)$ for the sequence $\eta_n(t)$.

However,

$$\mathsf{P}(\sup|\eta(t)-\eta_n(t)|>\varepsilon) \leq \frac{1}{\varepsilon}\sup \mathsf{E}|\eta(t)-\eta_n(t)| = \frac{1}{\varepsilon}\mathsf{E}|\alpha(\infty)-\alpha_n(\infty)| \to 0.$$

In view of Riesz's lemma a subsequence of indices n_j exists such that

$$\sup_t |\eta(t)-\eta_{n_j}(t)| \to 0$$

with probability 1. □

Corollary 1. *Under the conditions of the preceding theorem,* $\mathsf{E}\alpha^2(\infty) \leq 2c^2$ *provided* $|\xi(t)| \leq c$.

Indeed,

$$\mathsf{E}\alpha^2(\infty) \leq 2c\,\mathsf{E}\alpha(\infty) = 2c\,\mathsf{E}\xi(0) \leq 2c^2.$$

Corollary 2. *Since a sum of two potentials in class D also belongs to D,* (35) *implies equality*

$$2\mathsf{E}\alpha_1(\infty)\alpha_2(\infty) = \mathsf{E}\int_0^\infty [\xi_1(t)+\xi_1(t-)]\, d\alpha_2(t) + \mathsf{E}\int_0^\infty [\xi_2(t)+\xi_2(t-)]\, d\alpha_1(t),$$

where $\alpha_i(t)$ *are processes associated with potentials* $\xi_i(t)$, $i = 1, 2$. *This implies that if*

$\eta(t)=\xi_1(t)-\xi_2(t)$, $\beta(t)=\alpha_1(t)-\alpha_2(t)$, *then*

$$\mathsf{E}\beta^2(\infty)=\mathsf{E}\int_0^\infty[\eta(t)+\eta(t-)]\,d\beta(t). \tag{37}$$

Let $\{\xi(t), \mathfrak{F}_t, t>0\}$ be a supermartingale. Set

$$\eta_h(t)=\mathsf{E}\{\xi(t+h)|\mathfrak{F}_t\}, \qquad h>0.$$

We now establish several properties of the process $\eta_h(t)$.

a) *The process $\eta_h(t)$ is a supermartingale.*

Proof. Indeed, let $s<t$; then

$$\begin{aligned}\mathsf{E}\{\eta_h(t)|\mathfrak{F}_s\}&=\mathsf{E}\{\xi(t+h)|\mathfrak{F}_s\}\\&=\mathsf{E}\{\mathsf{E}\{\xi(t+h)|\mathfrak{F}_{s+h}\}|\mathfrak{F}_s\}\\&\leq\mathsf{E}\{\xi(s+h)|\mathfrak{F}_s\}=\eta_h(s).\quad\square\end{aligned}$$

Since the function $\mathsf{E}\eta_h(t)=\mathsf{E}\xi(t+h)$ is continuous from the right a modification of the process $\eta_h(t)$ exists with sample functions continuous from the right. Such a modification will be considered in what follows.

b) *If $\xi(t)$ is a potential in class D, then $\eta_h(t)$ is also a potential in the same class.*

Proof. The fact that $\eta_h(t)$ is a potential is obvious while its complete uniform integrability follows from the inequality $\eta_h(t)\leq\xi(t)$. $\square$

c) *Let τ be a random time finite or infinite and let $\xi(t)$ be a potential in class D. Then*

$$\eta_h(\tau)=\mathsf{E}\{\xi(\tau+h)|\mathfrak{F}_\tau\}. \tag{38}$$

Proof. Set $\tau_n=k/2^n$ if $\tau\in[(k-1)/2^n, k/2^n)$ and let $A\in\mathfrak{F}_\tau$.

Then,

$$\begin{aligned}\int_A\xi(\tau_n+h)\,d\mathsf{P}&=\sum_{k=0}^\infty\int_{A\cap\{\tau_n=k/2^n\}}\xi\left(\frac{k}{2^n}+h\right)d\mathsf{P}\\&=\sum_k\int_{A\cap\{\tau_n=k/2^n\}}\mathsf{E}\left\{\xi\left(\frac{k}{2^n}+h\right)\Big|\,\mathfrak{F}_{k/2^n}\right\}d\mathsf{P}=\int_A\eta_h(\tau_n)\,d\mathsf{P},\end{aligned}\tag{39}$$

which yields formula (38) for $\tau=\tau_n$.

Now let $n\to\infty$. Since the process $\eta_h(t)$ is continuous from the right it follows that $\eta_h(\tau_n)\to\eta_h(\tau)$ with probability 1. Also, since the family of random variables $\xi(\tau_n+h)$ and $\eta_h(\tau)$ is uniformly integrable one can approach the limit as $n\to\infty$ at the extreme sides of equality (39). $\square$

d) *If $\xi(t)$ is a regular potential in class D, so is $\eta_h(t)$.*

Proof. Let $\tau_n \uparrow \tau$ where τ_n is a random time. In this case

$$\mathsf{E}\eta_h(\tau_n) = \mathsf{E}\mathsf{E}\{\xi(\tau_n+h)|\mathfrak{F}_{\tau_n}\} = \mathsf{E}\xi(\tau_n+h) \to \mathsf{E}\xi(\tau+h) = \mathsf{E}\eta_h(\tau). \quad \square$$

Let $\xi(t)$ be a potential in class D. Set

$$\alpha_h(t) = \int_0^t \frac{\xi(s)-\eta_h(s)}{h}\, ds. \tag{40}$$

Clearly, $\alpha_h(t)$ is a monotonically nondecreasing continuous process. We show that this process is integrable. We have

$$\begin{aligned} h\mathsf{E}\alpha_h(t) &= \int_0^t (\mathsf{E}\xi(s) - \mathsf{E}\xi(s+h))\, ds \\ &= \int_0^t \mathsf{E}\xi(s)\, ds - \int_h^{t+h} \mathsf{E}\xi(s)\, ds \leqslant \int_0^h \mathsf{E}\xi(s)\, ds, \end{aligned}$$

which implies that $\mathsf{E}\alpha_h(\infty) < \infty$.

Denote by $\xi_h(t)$ a potential associated with the increasing process $\alpha_h(t)$. Then

$$\begin{aligned} \xi_h(t) &= \mathsf{E}\{\alpha_h(\infty) - \alpha_h(t)|\mathfrak{F}_t\} = \mathsf{E}\left\{\int_t^\infty \frac{\xi(s)-\eta_h(s)}{h}\, ds \,\middle|\, \mathfrak{F}_t\right\} \\ &= \lim_{N\to\infty} \mathsf{E}\frac{1}{h}\left\{\int_t^N (\xi(s) - \mathsf{E}\{\xi(s+h)|\mathfrak{F}_s\})\, ds \,\middle|\, \mathfrak{F}_t\right\} \\ &= \lim_{N\to\infty} \frac{1}{h}\int_t^N (\mathsf{E}\{\xi(s)|\mathfrak{F}_t\} - \mathsf{E}\{\xi(s+h)|\mathfrak{F}_t\})\, ds \\ &= \lim_{N\to\infty} \frac{1}{h}\left[\int_t^{t+h} \mathsf{E}(\xi(s)|\mathfrak{F}_t)\, ds - \int_N^{N+h} \mathsf{E}(\xi(s)|\mathfrak{F}_t\}\, ds\right] \\ &= \frac{1}{h}\int_t^{t+h} \mathsf{E}\{\xi(s)|\mathfrak{F}_t\}\, ds = \mathsf{E}\left\{\frac{1}{h}\int_t^{t+h} \xi(s)\, ds \,\middle|\, \mathfrak{F}_t\right\}. \end{aligned}$$

Since for $s > t$, $h > 0$,

$$\mathsf{E}\{\xi(s+h)|\mathfrak{F}_t\} = \mathsf{E}\{\mathsf{E}\{\xi(s+h)|\mathfrak{F}_s\}|\mathfrak{F}_t\} \leqslant \mathsf{E}\{\xi(s)|\mathfrak{F}_t\},$$

i.e., the function $\mathsf{E}\{\xi(s), |\mathfrak{F}_t\}$, $s > t$, is monotonically nonincreasing and continuous from the right, we deduce the following result from the formula obtained:

Theorem 15. *Let $\xi(t)$ be a potential in class D. The process*

$$\xi_h(t) = \mathsf{E}\left\{\frac{1}{h}\int_t^{t+h} \xi(s)\, ds \,\middle|\, \mathfrak{F}_t\right\} \tag{41}$$

is a potential associated with the continuous increasing process

$$\alpha_h(t) = \frac{1}{h}\int_0^t (\xi(s) - \eta_h(s))\, ds.$$

As $h \downarrow 0$ the variables $\xi_h(t)$ are monotonically nondecreasing and $\xi_h(t) \to \xi(t)$ as $h \to 0$ with probability 1 for each t. Moreover, $\xi_h(t) \in D$.

Proof. The last assertion is a result of the following considerations. It is easy to verify that the calculations leading to formula (41) are valid also in the case when t is replaced by an arbitrary random time τ. Hence

$$\xi_h(\tau) = \mathsf{E}\left\{\frac{1}{h}\int_\tau^{\tau+h} \xi(s)\, ds \,\middle|\, \mathfrak{F}_\tau\right\},$$

which implies the following equality for any $A \in \mathfrak{F}_\tau$:

$$\int_A \xi_h(\tau)\, d\mathsf{P} = \frac{1}{h}\int_0^h \left(\int_A \xi(s+\tau)\, d\mathsf{P}\right) ds.$$

Since $\xi(t)$ is uniformly integrable we can find for each $s > 0$ and any $\varepsilon > 0$ a $\delta > 0$ such that $\int_A \xi(s)\, d\mathsf{P} < \varepsilon$ provided $\mathsf{P}(A) < \delta$. This implies the inequality $\int_A \xi_h(\tau)\, d\mathsf{P} < \varepsilon$ for any random time τ, which proves the uniform integrability of the family of random variables $\{\xi_h(\tau), \tau \in \mathcal{T}\}$. □

Lemma 5. *Every potential $\xi(t)$ in class D can be represented in the form*

$$\xi(t) = \sum_{n=1}^{\infty} \xi_n(t),$$

where $\xi_n(t)$ is a bounded potential. If potential $\xi(t)$ is regular, then $\xi_n(t)$ can also be chosen as such.

Proof. Let

$$\xi(t) = \mathsf{E}\{\alpha(\infty) | \mathfrak{F}_t\} - \alpha(t), \qquad \alpha_n(t) = \alpha(t) \wedge n,$$

$$\beta_n(t) = \alpha_{n+1}(t) - \alpha_n(t), \qquad \xi_n(t) = \mathsf{E}\{\beta_n(\infty) | \mathfrak{F}_t\} - \beta_n(t).$$

Then $\xi_n(t)$ is a bounded potential and

$$\sum_{n=1}^{\infty} \xi_n(t) = \mathsf{E}\left\{\sum_{n=0}^{\infty} \beta_n(\infty) \,\middle|\, \mathfrak{F}_t\right\} - \left(\sum_{n=0}^{\infty} \beta_n(t)\right) = \xi(t).$$

Observe that the difference $\xi(t) - \xi_n(t) = \zeta(t)$ is a completely uniformly integrable potential. For notational convenience set $\xi_n(t) = \eta(t)$.

Let τ_n be an arbitrary bounded nondecreasing sequence of random times, $\tau_n \uparrow \tau$. Since $\mathsf{E}\eta(\tau_n) \geqslant \mathsf{E}\eta(\tau)$ and $\mathsf{E}\zeta(\tau_n) \geqslant \mathsf{E}\zeta(\tau)$, relation

$$\mathsf{E}\eta(\tau)+\mathsf{E}\zeta(\tau)=\mathsf{E}\xi(\tau)=\lim \mathsf{E}\xi(\tau_n)=\lim (\mathsf{E}\eta(\tau_n)+\mathsf{E}\zeta(\tau_n))$$

follows from the regularity of the process $\xi(t)$; this is possible, however, if and only if $\lim \mathsf{E}\eta(\tau_n)=\mathsf{E}\eta(\tau)$ and $\lim \mathsf{E}\zeta(\tau_n)=\mathsf{E}\zeta(\tau)$, i.e. the process $\eta(t)=\xi_n(t)$ is regular. □

Lemma 6. *Let $\xi(t)$ be a regular potential in class D, $\xi_n(t)$ be an arbitrary nondecreasing sequence of potentials converging to $\xi(t)$ with probability 1 and let $\varepsilon>0$. Set*

$$\tau_{n\varepsilon}=\inf\{t\colon \xi(t)-\xi_n(t)>\varepsilon\}.$$

Then

$$\lim_{n\to\infty} \mathsf{P}\{\tau_{n\varepsilon}<\infty\}=0.$$

Proof. Clearly, $\tau_{n\varepsilon}$ is a nondecreasing sequence of random times. Set $\tau=\lim \tau_{n\varepsilon}$. Then $\lim_{n\to\infty} \mathsf{E}\xi_p(\tau_{n\varepsilon})\geqslant \mathsf{E}\xi_p(\tau)$. Since $\xi(t)$ is regular, $\mathsf{E}(\xi(\tau)-\xi_p(\tau))\geqslant \lim \mathsf{E}(\xi(\tau_{n\varepsilon})-\xi_p(\tau_{n\varepsilon}))$. Also the fact that $\xi(\tau_{n\varepsilon})-\xi_p(\tau_{n\varepsilon})\geqslant \xi(\tau_{n\varepsilon})-\xi_n(\tau_{n\varepsilon})\geqslant \varepsilon\chi(\tau_{n\varepsilon}<\infty)$ for $p\leqslant n$, yields that

$$\mathsf{E}(\xi(\tau)-\xi_p(\tau))\geqslant \varepsilon \lim_{n\to\infty} \mathsf{P}\{\tau_{n\varepsilon}<\infty\} \quad \text{for all } p>0.$$

Approaching the limit in the last inequality as $p\to\infty$ we obtain the assertion of the lemma. □

Lemma 7. *Let the conditions of the preceding lemma be satisfied, and the potential $\xi(t)$ be bounded. Then*

$$\mathsf{E}[\alpha(\infty)-\alpha_n(\infty)]^2\to 0 \quad \text{as } n\to\infty,$$

where $\alpha(t)$ and $\alpha_n(t)$ are natural processes associated with $\xi(t)$ and $\xi_n(t)$, respectively.

Proof. Let $\xi(t)\leqslant c$, then $\xi_n(t)\leqslant c$ also, and in view of Corollary 1 to Theorem 14, $\mathsf{E}\alpha^2(t)<\infty$ and $\mathsf{E}\alpha_n^2(\infty)<\infty$.

Set $\eta(t)=\xi(t)-\xi_n(t)$ and $\beta(t)=\alpha(t)-\alpha_n(t)$. Utilizing Lemma 14, we have

$$\begin{aligned}
\mathsf{E}(\alpha(\infty)-\alpha_n(\infty))^2&=\mathsf{E}\int_0^\infty (\eta(t)+\eta(t-))\,d\beta(t)=\mathsf{E}\Big(\int_{[0,\tau_{n\varepsilon})}+\int_{[\tau_{n\varepsilon},\infty)}\Big)\\
&\leqslant 2\varepsilon\mathsf{E}(\alpha(\infty)+\alpha_n(\infty))+\mathsf{E}\Big\{\chi(\tau_{n\varepsilon}<\infty)\int_0^\infty 4c\,d(\alpha+\alpha_n)\Big\}\\
&\leqslant 4\varepsilon\mathsf{E}\xi(0)+8c\mathsf{E}\chi(\tau_{n\varepsilon}<\infty)\alpha(\infty)\\
&\leqslant 4\varepsilon\mathsf{E}\xi(0)+8c(\mathsf{P}(\tau_{n\varepsilon}<\infty))^{1/2}.
\end{aligned}$$

Applying Lemma 6 we obtain the required assertion. □

Proof of Theorem 13. We can assume without loss of generality that $\xi(t)$ is a potential. We will show that if this process is regular, then the associated process is continuous. First we shall assume that the potential $\xi(t)$ is regular and bounded. Let

$$\mu(t)=\xi(t)+\alpha(t)=\mathsf{E}\{\alpha(\infty)|\mathfrak{F}_t\},$$
$$\mu_h(t)=\xi_h(t)+\alpha_h(t)=\mathsf{E}\{\alpha_h(\infty)|\mathfrak{F}_t\},$$

where $\alpha(t)$ and $\xi_h(t)$ are processes defined by formulas (40) and (41). It follows from Lemma 7 that

$$\mathsf{E}[\alpha(\infty)-\alpha_h(\infty)]^2\to 0 \quad \text{as } h\to 0.$$

Consequently, $\sup_t|\mu(t)-\mu_h(t)|^2\to 0$ in probability and in view of Lemma 6 $\sup_t|\xi(t)-\xi_h(t)|\to 0$ also in probability. Hence $\sup_t|\alpha_h(t)-\alpha(t)|\to 0$ as $h\to 0$ in probability. In view of Riesz's lemma a subsequence h_j exists such that $\sup_t(\alpha_{h_j}(t)-\alpha(t))\to 0$ as $j\to\infty$ with probability 1. Therefore $\alpha(t)$ is a continuous function with probability 1.

Now assume that $\xi(t)$ is not necessarily a bounded potential. In view of Lemma 5, $\xi(t)=\sum_{n=1}^{\infty}\xi_n(t)$, where $\xi_n(t)$ are regular bounded potentials in class D. Let $\alpha_n(t)$ be a process associated with $\xi_n(t)$. This process is continuous. Since $\mathsf{E}\sum\alpha_n(\infty)=\mathsf{E}\sum\xi_n(0)=\mathsf{E}\xi(0)<\infty$, the series $\sum\alpha_n(\infty)$ converges with probability 1. In this case, however, the series $\sum\alpha_n(t)$ converges uniformly in t with probability 1 and is a continuous function. Clearly the process $\alpha(t)$ is natural and is associated with potential $\xi(t)$.

We now prove the converse. If the process $\alpha(t)$ is continuous and is associated with potential $\xi(t)$, then $\xi(t)$ is regular.

Let τ_n be an increasing sequence of random times convergent to τ. In this case $\lim\mathsf{E}\alpha(\tau_n)=\mathsf{E}\alpha(\tau)$. Hence

$$\lim\mathsf{E}\xi(\tau_n)=\lim\mathsf{E}(\alpha(\infty)-\alpha(\tau_n))=\mathsf{E}[\alpha(\infty)-\alpha(\tau)]=\mathsf{E}\xi(\tau).\quad\square$$

Square integrable martingales. Let $\{\mu(t),\mathfrak{F}_t, t\geq 0\}$ be a martingale with sample functions in $\mathcal{D}$, $\mathfrak{F}_t=\mathfrak{F}_{t+}$ and

$$\sup_{t>0}\mathsf{E}|\mu(t)|^2<\infty.$$

A martingale possessing this property is called *square integrable*, and the class of all square integrable martingales with respect to a given current of σ-algebras $\{\mathfrak{F}_t, t\geq 0\}$ and a probability measure P will be denoted by $\mathcal{M}_2=\mathcal{M}_2\{\mathfrak{F}_t,\mathsf{P}\}$. The subset of $\mathcal{M}_2$ with sample functions which are continuous with probability 1 (we shall call these martingales *continuous square integrable*) will be denoted by $\mathcal{M}_2^c=\mathcal{M}_2^c\{\mathfrak{F}_t,\mathsf{P}\}$.

Since (cf. (4))

$$\mathsf{E}\sup_{t>0}|\mu(t)|^2\leq 4\sup_{t>0}\mathsf{E}|\mu(t)|^2<\infty,$$

the family of random variables $\{\mu(t), t \geq 0\}$ is uniformly integrable and the limit

$$\mu(\infty) = \lim_{t\to\infty} \mu(t)$$

exists with probability 1 as well as in the mean square, and, moreover, $\mu(t) = \mathsf{E}\{\mu(\infty) | \mathfrak{F}_t\}$.

We introduce in $\mathcal{M}_2$ the scalar product by setting

$$(\mu, \nu) = \mathsf{E}\mu(\infty)\nu(\infty), \qquad \mu, \nu \in \mathcal{M}_2.$$

It is easy to verify that the bilinear form introduced herein possesses all the properties of a scalar product. Moreover, an isometric correspondence exists between the class $\mathcal{M}_2$ and the class $L_2 = L_2\{\mathfrak{F}_\infty, \mathsf{P}\}$ of all the $\mathfrak{F}_\infty$-measurable random variables η such that $\mathsf{E}\eta^2 < \infty$. This correspondence is given by

$$\eta \to \mu(t) = \mathsf{E}\{\eta \,|\, \mathfrak{F}_t\}, \qquad \mu(t) \to \eta = \lim_{t\to\infty} \mu(t),$$

and, moreover, if $\eta \leftrightarrow \mu(t)$, then

$$\mathsf{E}\mu^2(t) \leq \mathsf{E}\eta^2.$$

Theorem 16. *$\mathcal{M}_2$ is a Hilbert space and $\mathcal{M}_2^c$ is closed in $\mathcal{M}_2$.*

Proof. Let $\{\mu_n(t), n = 1, 2, \ldots\}$ be a Cauchy sequence in $\mathcal{M}_2$. Then the limit $\lim \mu_n(\infty) = \mu_\infty$ exists. We construct the martingale $\mu(t) = \mathsf{E}\{\mu_\infty | \mathfrak{F}_t\}$ with sample functions in $\mathcal{D}$. This martingale belongs to $\mathcal{M}_2$,

$$\mathsf{E}\mu^2(t) = \mathsf{E}(\mathsf{E}\{\mu_\infty | \mathfrak{F}_t\})^2 \leq \mathsf{E}\mathsf{E}\{\mu_\infty^2 | \mathfrak{F}_t\} = \mathsf{E}\mu_\infty^2 < \infty,$$

and is the limit in $\mathcal{M}_2$ of the sequence $\mu_n(t)$. Since

$$\mathsf{E}|\sup_t |\mu_n(t) - \mu(t)|^2 \leq 4\mathsf{E}|\mu_n(\infty) - \mu_\infty|^2 \to 0,$$

a subsequence $\mu_{n_k}(t)$, $k = 1, 2, \ldots$, exists such that $\sup_t |\mu_{n_k}(t) - \mu(t)| \to 0$ as $k \to \infty$ with probability 1. In particular, if processes $\mu_n(t)$ are continuous, then the martingale $\mu(t)$ is also continuous. □

Remark. We have also shown that if $\mu_n(t)$ approaches $\mu(t)$ in $\mathcal{M}_2$, then a subsequence of martingales $\mu_{n_k}(t)$ exists with sample functions which with probability 1 converge to the sample functions of $\mu(t)$ uniformly in $t \geq 0$.

Let $\mu \in \mathcal{M}_2$. Then $\mu^2(t)$ is a submartingale and

$$\mu^2(t) \leq \mathsf{E}\{\mu^2(\infty) | \mathfrak{F}_t\} = \xi(t),$$

where $\xi(t)$ is a uniformly integrable martingale. Since $\xi(t)$ is a completely uniformly integrable process, $\mu^2(t)$ is a submartingale in class $\mathscr{D}$. In view of Meyer's theorem, a uniquc integrable natural process $\alpha(t)$ and a martingale $\nu(t)$ exist such that

(42) $$\mu^2(t) - \nu(t) + \alpha(t).$$

Definition. A natural process $\alpha(t)$ in decomposition (42) is called a *characteristic* of martingale $\mu(t)$ ($\mu(t) \in \mathcal{M}_2$) and is denoted by $\langle\mu, \mu\rangle_t$.

Since for $t > s$

$$\mathsf{E}\{(\mu(t) - \mu(s))^2 \mid \mathfrak{F}_s\} = \mathsf{E}\{\mu^2(t) - \mu^2(s) \mid \mathfrak{F}_s\},$$

we obtain

(43) $$\mathsf{E}\{(\mu(t) - \mu(s))^2 \mid \mathfrak{F}_s\} = \mathsf{E}\{\langle\mu, \mu\rangle_t - \langle\mu, \mu\rangle_s \mid \mathfrak{F}_s\}.$$

The converse is also obvious: if $\langle\mu, \mu\rangle_t$ is a natural process and if for arbitrary t and s (with $s < t$) equality (43) is satisfied, then $\langle\mu, \mu\rangle_t$ is a characteristic of martingale $\mu(t)$.

Example. Let $\mu(t)$ be a process with independent increments and $\mu \in \mathcal{M}_2$. Then

$$\mathsf{E}\mu(t) = \alpha = \text{const.}, \qquad \mathsf{E}(\mu(t) - \alpha)^2 = \sigma^2(t) < \infty,$$

and

$$\mathsf{E}\{(\mu(t) - \mu(s))^2 \mid \mathfrak{F}_s\} = \mathsf{E}(\mu(t) - \mu(s))^2 = \sigma^2(t) - \sigma^2(s).$$

Thus a characteristic of a process with independent increments $\langle\mu, \mu\rangle_t = \sigma^2(t)$ does not depend on chance.

Let σ and τ be two random times with respect to a current of σ-algebras $\{\mathfrak{F}_t, t \geq 0\}$ and let $\sigma \leq \tau (\text{mod } \mathsf{P})$. In view of Corollary 1 to Theorem 6 we obtain

$$\begin{aligned}\mathsf{E}\{(\mu(\tau) - \mu(\sigma))^2 \mid \mathfrak{F}_\sigma\} &= \mathsf{E}\{\mu^2(\tau) - 2\mu(\sigma)\mathsf{E}\{\mu(\tau) \mid \mathfrak{F}_\sigma\} + \mu^2(\sigma) \mid \mathfrak{F}_\sigma\} \\ &= \mathsf{E}\{\mu^2(\tau) - \mu^2(\sigma) \mid \mathfrak{F}_\sigma\},\end{aligned}$$

which implies the following equality generalizing formula (43):

(44) $$\mathsf{E}\{(\mu(\tau) - \mu(\sigma))^2 \mid \mathfrak{F}_\sigma\} = \mathsf{E}\{\langle\mu, \mu\rangle_\tau - \langle\mu, \mu\rangle_\sigma \mid \mathfrak{F}_\sigma\}.$$

In general the product of two square integrable martingales is not a martingale.

Theorem 17. *The product* $\mu_1(t)\mu_2(t)$ $(\mu_i(t)\in\mathcal{M}_2,\ i=1,2)$ *is a martingale if and only if*

$$\langle\mu_1+\mu_2,\mu_1+\mu_2\rangle_t=\langle\mu_1,\mu_1\rangle_t+\langle\mu_2,\mu_2\rangle_t.$$

Proof. Necessity follows from the uniqueness of the characteristic and equality

$$(\mu_1(t)+\mu_2(t))^2=\nu_1(t)+\nu_2(t)+2\mu_1(t)\mu_2(t)+(\alpha_1(t)+\alpha_2(t)),$$

where $\mu_i^2(t)=\nu_i(t)+\alpha_i(t)$ is Doob's decomposition of submartingale $\mu_i^2(t)$. The same equality also implies that if $\alpha_1(t)+\alpha_2(t)$ is a characteristic of martingale $\mu_1(t)+\mu_2(t)$, then $\mu_1(t)\mu_2(t)$ is a martingale. □

Definition. The random process

$$\langle\mu_1,\mu_2\rangle_t=\tfrac{1}{2}[\langle\mu_1+\mu_2,\mu_1+\mu_2\rangle_t-\langle\mu_1,\mu_1\rangle_t-\langle\mu_2,\mu_2\rangle_t]$$

is called the *joint characteristic* of martingales $\mu_1(t)$ and $\mu_2(t)$ $(\mu_i(t)\in\mathcal{M}_2,\ i=1,2)$.

Clearly, the joint characteristic of martingales $\mu_1(t)$ and $\mu_2(t)$ possesses the following properties: it is adopted to the current of σ-algebras $\{\mathfrak{F}_t,t\geq 0\}$, $\langle\mu_1,\mu_2\rangle_0=0$, and the process $\langle\mu_1,\mu_2\rangle_t$ can be represented as the difference of two natural processes.

The usefulness of the notion of a joint characteristic of two martingales is due to the fact that the process

$$\mu_1(t)\mu_2(t)-\langle\mu_1,\mu_2\rangle_t$$

is a martingale. This follows immediately from the equality (which can easily be verified):

$$\mu_1(t)\mu_2(t)-\langle\mu_1,\mu_2\rangle_t=\tfrac{1}{2}(\nu_3(t)-\nu_1(t)-\nu_2(t)),$$

where $\nu_3(t)=(\mu_1(t)+\mu_2(t))^2-\langle\mu_1+\mu_2,\mu_1+\mu_2\rangle_t$ and $\nu_i(t)(i=1,2)$ are as defined above. This implies in particular that if σ and τ are two random times and $\sigma\leq\tau$, then

$$\mathsf{E}\{\mu_1(\tau)\mu_2(\tau)-\mu_1(\sigma)\mu_2(\sigma)\,|\,\mathfrak{F}_\sigma\}=\mathsf{E}\{\langle\mu_1,\mu_2\rangle_\tau-\langle\mu_1,\mu_2\rangle_\sigma\,|\,\mathfrak{F}_\sigma\}.$$

This last relation can also be written in the form

$$(45)\qquad \mathsf{E}\{(\mu_1(\tau)-\mu_1(\sigma))(\mu_2(\tau)-\mu_2(\sigma))\,|\,\mathfrak{F}_\sigma\}=\mathsf{E}\{\langle\mu_1,\mu_2\rangle_\tau-\langle\mu_1,\mu_2\rangle_\sigma\,|\,\mathfrak{F}_\sigma\}.$$

This implies, for example, that

$$(46)\qquad |\mathsf{E}\{\Delta\langle\mu_1,\mu_2\rangle\,|\,\mathfrak{F}_\sigma\}|\leq\sqrt{\mathsf{E}\{\Delta\langle\mu_1,\mu_1\rangle\,|\,\mathfrak{F}_\sigma\}\mathsf{E}\{\Delta\langle\mu_2,\mu_2\rangle\,|\,\mathfrak{F}_\sigma\}},$$

where $\Delta\langle\mu_i,\mu_i\rangle=\langle\mu_i,\mu_i\rangle_\tau-\langle\mu_i,\mu_i\rangle_\sigma$ and $\Delta\langle\mu_1,\mu_2\rangle$ is defined analogously.

Theorem 18. *For two arbitrary martingales $\mu_1(t)$ and $\mu_2(t)(\mu_i(t) \in \mathcal{M}_2, i=1,2)$ a process $\langle\mu_1, \mu_2\rangle_t$ exists with the following properties: $\langle\mu_1, \mu_2\rangle_t$ is the difference of two natural processes and the process $\mu_1(t)\mu_2(t) - \langle\mu_1, \mu_2\rangle_t$ is a martingale.*

This process is unique and satisfies inequality (46).

The existence of process $\langle\mu_1, \mu_2\rangle$ follows from the preceding considerations and constructions. We now prove its uniqueness.

Proof. Let $\alpha(t)$ denote a process satisfying the conditions of the theorem. Then $\beta(t) = \langle\mu_1, \mu_1\rangle_t + 2\alpha(t) + \langle\mu_2, \mu_2\rangle_t$ possesses the following properties: $\beta(t)$ is the difference of two natural processes and $(\mu_1(t)+\mu_2(t))^2 - \beta(t)$ is a martingale. In view of Meyer's theorem $\beta(t) = \langle\mu_1+\mu_2, \mu_1+\mu_2\rangle_t$ so that $\alpha(t) = \frac{1}{2}[\langle\mu_1+\mu_2, \mu_1+\mu_2\rangle_t - \langle\mu_1, \mu_1\rangle_t - \langle\mu_2, \mu_2\rangle_t]$. Hence the process $\alpha(t)$ is uniquely defined (mod P). □

Local square integrable martingales. We now introduce the notion of a local square integrable martingale which is quite analogous to the notion of a local martingale.

Definition. A process $\{\mu(t), \mathfrak{F}_t, t \geq 0\}$ is called a *local square integrable martingale* if its sample functions belong to $\mathcal{D}$ and if a monotone nondecreasing sequence of $\mathfrak{F}_t$-random times τ_n, $n = 1, 2, \ldots$, exists such that

1) $\lim \tau_n = \infty$ (mod P),
2) process $\mu(t \wedge \tau_n)$ is a square integrable martingale with respect to the current $\{\mathfrak{F}_{t\wedge\tau_n}, t \geq 0\}$.

In this connection the sequence τ_n is said to be *completely reducing a local square integrable martingale* $\mu(t)$, and an arbitrary $\mathfrak{F}_t$-random time τ for which the remaining process $\mu(t \wedge \tau)$ is a square integrable martingale is said to be *reducing* $\mu(t)$.

The class of all local integrable martingales with respect to a given current of σ-algebras $\{\mathfrak{F}_t, t \geq 0\}$ is denoted by $l\mathcal{M}$ and the class of local square integrable martingales by $l\mathcal{M}_2$. The subclass of $l\mathcal{M}(l\mathcal{M}_2)$ consisting of processes with the sample functions being continuous with probability 1 is denoted by $l\mathcal{M}^c(l\mathcal{M}_2^c)$.

If $\mu(t) \in l\mathcal{M}_2$ and τ_n is a sequence of random times completely reducing $\mu(t)$, then a sequence of natural increasing processes $\alpha_n(t)$, $t \geq 0$, $n = 1, 2, \ldots$, exists such that the process $\zeta_n(t) = \mu^2(t \wedge \tau_n) - \alpha_n(t)$ is a martingale for any $n = 1, 2, \ldots$. Since $\mu^2((t \wedge \tau_{n'}) \wedge \tau_n) = \mu^2(t \wedge \tau_n)$ for $n < n'$, the uniqueness of the characteristic implies that $\alpha_{n'}(t \wedge \tau_n) = \alpha_n(t)$. Thus for $t < \tau_n$, $\alpha_n(t) = \alpha_{n+1}(t) = \cdots$, i.e., for any $t > 0$ the variables $\alpha_n(t)$ will be identical starting with some number $n_0 = n_0(\omega, t)$.

Set $\alpha(t) = \lim \alpha_n(t)$. It is easy to verify that the process $\alpha(t)$ does not depend (mod P) on the choice of sequence τ_n. Indeed, let τ_n' be another sequence of random times completely reducing process $\mu(t)$, and let $\alpha_n'(t)$ be the characteristic of the martingale $\mu(t \wedge \tau_n')$, $\alpha'(t) = \lim \alpha_n'(t)$. It then follows from the relation $\mu((t \wedge \tau_n') \wedge \tau_m) = \mu((\tau \wedge \tau_m) \wedge \tau_n')$ and the uniqueness of the characteristic that

$\alpha_m(t)=\alpha'_n(t)$ for $t<\tau_m\wedge\tau'_n$. As m and $n\to 0$ we obtain

$$\alpha(t)=\alpha'(t)(\text{mod P}).$$

The same argument shows that if τ is an arbitrary random time reducing $\mu(t)$, then $\mu^2(t\wedge\tau)-\alpha(t\wedge\tau)$ is a martingale and the process $\alpha(t\wedge\tau)$ is a natural process (with respect to the current of σ-algebras $\{\mathfrak{F}_{t\wedge\tau}, t\geqslant 0\}$; cf. equation (20)). Moreover, the process possessing these properties is unique.

The above considerations are also applicable to the case of a product of two local square integrable martingales. Thus we have obtained the following assertion:

Theorem 19. 1) *If $\mu(t)\in l\mathcal{M}_2$, then a nonnegative increasing process $\alpha(t)$, $t\geqslant 0$, exists such that $\mu^2(t\wedge\tau)-\alpha(t\wedge\tau)$ is an $\mathfrak{F}_{t\wedge\tau}$-martingale for any random time τ reducing martingale $\mu(t)$ and $\alpha(t\wedge\tau)$ is a natural process. A process $\alpha(t)$ possessing this property is unique.*

2) *If $\mu_i(t)\in l\mathcal{M}_2$, $i=1,2$, then a process $\beta(t)$ representable in the form $\beta(t)=\alpha_1(t)-\alpha_2(t)$ exists (where $\alpha_i(t)$ are increasing processes) such that the process $\gamma(t\wedge\tau)=\mu_1(t\wedge\tau)\mu_2(t\wedge\tau)-\beta(t\wedge\tau)$ is an $\mathfrak{F}_{t\wedge\tau}$-martingale for an arbitrary random time τ reducing martingales $\mu_1(t)$ and $\mu_2(t)$ and $\alpha_i(t\wedge\tau)$ are natural processes $(i=1,2)$. Moreover, a process $\beta(t)$ possessing these properties is also unique.*

As above, we shall refer to $\alpha(t)$ as the characteristic of a local square integrable process $\mu(t)$ and to $\beta(t)$ as the joint characteristic of processes $\mu_1(t)$ and $\mu_2(t)$. We shall also continue to use the notation

$$\alpha(t)=\langle\mu,\mu\rangle_t,\qquad \beta(t)=\langle\mu_1,\mu_2\rangle_t.$$

Here

$$\langle\mu_1,\mu_2\rangle_t=\tfrac{1}{2}[\langle\mu_1+\mu_2,\mu_1+\mu_2\rangle_t-\langle\mu_1,\mu_1\rangle_t-\langle\mu_2,\mu_2\rangle_t].$$

Remark. If τ is an arbitrary random time then the function $\langle\mu,\mu\rangle_{t\wedge\tau}$ is the characteristic of the process $\mu(t\wedge\tau)$ $(\mu(\cdot)\in l\mathcal{M}_2)$.

Martingales with continuous characteristics. Let $\mu(t)\in\mathcal{M}_2$ and $\langle\mu,\mu\rangle_t$ be characteristic of $\mu(t)$. In view of Meyer's theorem (Theorem 13) the process $\langle\mu,\mu\rangle_t$ is continuous if and only if for an arbitrary monotonically nondecreasing sequence of random times τ_n converging to a finite random time τ

$$\mathsf{E}\mu^2(\tau_n)\to\mathsf{E}\mu^2(\tau). \tag{47}$$

On the other hand, if this condition is satisfied then

$$\mathsf{E}\mu(\tau)\mu(\tau_n)=\mathsf{E}\{\mu(\tau_n)\mathsf{E}\{\mu(\tau)|\mathfrak{F}_{\tau_n}\}\}=\mathsf{E}\mu^2(\tau_n)\to\mathsf{E}\mu^2(\tau),$$
$$\mathsf{E}(\mu(\tau)-\mu(\tau_n))^2=\mathsf{E}\{\mu^2(\tau)-2\mu(\tau)\mu(\tau_n)+\mu^2(\tau_n)\}\to 0.$$

Since $\mu^2 \in D$, the last relation is fulfilled if and only if

$$\text{P-lim}\, \mu(\tau_n) = \mu(\tau). \tag{48}$$

Conversely, if (48) is valid so is (47).

Definition. A martingale $\{\mu(t), \mathfrak{F}_t, t \geq 0\}$ is called *quasi-continuous from the left* if it satisfies condition (48) for any monotonically nondecreasing sequence of random times τ_n such that $\tau = \lim \tau_n < \infty$ (mod P). A local martingale $\mu(t)$ is called *quasi-continuous from the left* if (48) is satisfied for monotonically nondecreasing sequences τ_n such that τ reduces $\mu(t)$.

Theorem 20. *In order that characteristic $\langle \mu, \mu \rangle_t$ of process $\mu(t) \in l\mathcal{M}_2$ be continuous, it is necessary and sufficient that the process $\mu(t)$ be quasi-continuous from the left.*

The proof easily follows from the preceding argument.

Corollary. *If the characteristics of processes $\mu_1(t)$ and $\mu_2(t) (\mu_i(t) \in l\mathcal{M}_2, i = 1, 2)$ are continuous, so is their joint characteristic $\langle \mu_1, \mu_2 \rangle_t$.*

Indeed, quasi-continuity from the left of processes $\mu_1(t)$ and $\mu_2(t)$ implies the quasi-continuity from the left of their sum.

We shall now verify several propositions which will allow us in certain cases to obtain characteristics of martingales and local martingales.

Let $\xi(t) = \mu(t) + \alpha(t)$, $t \in [0, T]$, where $\mu(t)$ is a $\{\mathfrak{F}_t, t \in [0, T]\}$-martingale and $\alpha(t)$ is an arbitrary continuous increasing process integrable on $[0, T]$ adopted to a current of σ-algebras $\{\mathfrak{F}_t, t \in [0, T]\}$ and $\alpha(0) = 0$. Consider an arbitrary subdivision λ on the interval $[0, T]$, $\lambda = \{0 = t_0, t_1, \ldots, t_n = T\}$. We shall write $\lambda \to 0$ if $\max_{k=1,\ldots,n} (t_k - t_{k-1}) \to 0$. Set

$$\alpha_\lambda = \mathsf{E}\{\xi(t_1) - \xi(t_0) | \mathfrak{F}_0\} + \mathsf{E}\{\xi(t_2) - \xi(t_1) | \mathfrak{F}_{t_1}\} + \cdots + \mathsf{E}\{\xi(t_n) - \xi(t_{n-1}) | \mathfrak{F}_{t_{n-1}}\}.$$

Then

$$\alpha_\lambda = \mathsf{E}\{\alpha(t_1) | \mathfrak{F}_0\} + \mathsf{E}\{\alpha(t_2) - \alpha(t_1) | \mathfrak{F}_{t_1}\} + \cdots + \mathsf{E}\{\alpha(t_n) - \alpha(t_{n-1}) | \mathfrak{F}_{t_{n-1}}\}.$$

We thus see that α_λ does not depend on $\mu(t)$ and is completely determined by the process $\alpha(t)$.

Theorem 21. *If $\alpha(t)$ is an increasing process, continuous and integrable on $[0, T]$, then*

$$\lim \alpha_\lambda = \alpha(T) \tag{49}$$

in the sense of convergence in L_1.

Proof. First we shall assume that $\mathsf{E}\alpha^2(T)<\infty$. Then

$$\mathsf{E}(\alpha_\lambda-\alpha(T))^2=\mathsf{E}\Big[\sum_{k=1}^{n}(\mathsf{E}\{\Delta\alpha_k\,|\,\mathfrak{F}_{t_{k-1}}\}-\Delta\alpha_k)\Big]^2$$

when $\Delta\alpha_k=\alpha(t_k)-\alpha(t_{k-1})$. Since the different summands in the sum on the right-hand side of the equality are orthogonal, we have

$$\begin{aligned}\mathsf{E}(\alpha_\lambda-\alpha(T))^2&=\sum_{k=1}^{n}\mathsf{E}[\mathsf{E}\{\Delta\alpha_k\,|\,\mathfrak{F}_{t_{k-1}}\}-\Delta\alpha_k]^2\\&=\sum_{k=1}^{n}\{\mathsf{E}\Delta\alpha_k^2-\mathsf{E}[\mathsf{E}\{\Delta\alpha_k\,|\,\mathfrak{F}_{t_{k-1}}\}]^2\}\\&\leq\mathsf{E}\sum_{k=1}^{n}\Delta\alpha_k^2\leq\mathsf{E}(\sup_k\Delta\alpha_k\cdot\alpha(T)).\end{aligned}$$

Since $\sup_k\Delta\alpha_k\cdot\alpha(T)\leq\alpha^2(T)$, $\mathsf{E}\alpha^2(T)<\infty$, and $\sup\Delta\alpha_k\to 0$ with probability 1, Lebesgue's dominated convergence theorem implies that

$$\mathsf{E}(\alpha_\lambda-\alpha(T))^2\to 0.$$

We now proceed to the general case. Set $\beta(t)=\alpha(t)\wedge n$, $\gamma(t)=\alpha(t)-\beta(t)$. Processes $\beta(t)$ and $\gamma(t)$ are continuous and monotonically nondecreasing and also $\beta(t)\leq n$. Moreover,

$$\mathsf{E}|\alpha(T)-\alpha_\lambda|\leq\mathsf{E}[|\beta(T)-\beta_\lambda|+\gamma(T)+(\alpha_\lambda-\beta_\lambda)].$$

It follows from relations

$$0\leq\mathsf{E}(\alpha_\lambda-\beta_\lambda)=\mathsf{E}\gamma(T)\leq\mathsf{E}\alpha(T)\chi\{\alpha(T)>n\}$$

that for any $\varepsilon>0$ we have for n sufficiently large $\mathsf{E}\gamma(T)<\varepsilon/3$ and $\mathsf{E}(\alpha_\lambda-\beta_\lambda)<\varepsilon/3$ independently of λ. Since $\mathsf{E}\beta^2(T)<n^2$ we have for n chosen in this manner that $\mathsf{E}|\beta(T)-\beta_\lambda|\to 0$ as $\lambda\to 0$. This implies that

$$\mathsf{E}|\alpha(T)-\alpha_\lambda|\to 0\quad\text{as }\lambda\to 0.$$

Remark. If $\mathsf{E}\alpha^2(T)<\infty$, then relationship (49) holds also in the sense of convergence in L_2.

Corollary 1. *If the characteristic of a martingale is continuous, then*

$$\langle\mu,\mu\rangle_t=\lim_{\lambda\to 0}\sum_{0<t_0<t_1<\cdots<t_n=t}\mathsf{E}\{(\mu(t_k)-\mu(t_{k-1}))^2\,|\,\mathfrak{F}_{t_{k-1}}\}.$$

Corollary 2. *If the characteristics of martingales* $\mu_i(t)$ $(\mu_i(t) \in \mathcal{M}_2, i=1, 2)$ *are continuous, then*

$$(50)\quad \langle \mu_1, \mu_2 \rangle = \lim_{\lambda \to 0} \sum_{\substack{k=1 \\ 0=t_0<t_1<\cdots<t_n=t}}^{n} \mathsf{E}\{(\mu_1(t_k)-\mu_1(t_{k-1}))(\mu_2(t_k)-\mu_2(t_{k-1})) \mid \mathfrak{F}_{t_{k-1}}\}.$$

Also one should bear in mind that the continuity of processes $\langle \mu_i, \mu_i \rangle_t$ implies the continuity of the process $\langle \mu_1+\mu_2, \mu_1+\mu_2 \rangle_t$ (cf. the corollary to Theorem 20).

Corollary 3. *If the characteristics of martingales* $\mu_i(t)$, $i=1, 2$, *are continuous, then*

$$(51)\quad |\Delta\langle \mu_1, \mu_2 \rangle_t|^2 \leqslant \Delta\langle \mu_1, \mu_1 \rangle_t \Delta\langle \mu_2, \mu_2 \rangle_t.$$

This inequality is a substantial improvement (for the case under consideration) over inequality (46).

It is also obvious that this inequality is valid as well for the case of local square integrable martingales with continuous characteristics.

In the case when sample functions of a martingale or a local martingale are continuous (mod P) one can prove another limit theorem which allows us to compute the characteristic of a process.

First let us observe that a martingale (local martingale) which is continuous is quasi-continuous from the left. Next, in this case a local martingale is locally square integrable. Indeed, let

$$\tau_n = \inf\{t : |\mu(t)| \geqslant n\},$$

where if $A_n = \{t : |\mu(t)| \geqslant n\} = \varnothing$, we define $\tau_n = \infty$. Set $\mu_n(t) = \mu(t \wedge \tau_n)$; then for $\omega \in A_n$ and $t \leqslant \tau_n$, $|\mu_n(t)| \leqslant n$ and since $\mu(t)$ is continuous, $|\mu_n(t)| = n$ for $t \geqslant \tau_n$. If, however, $\omega \notin A_n$, then $|\mu_n(t)| < n$ for each t. Thus $|\mu_n(t)| \leqslant n$ with probability 1. Consequently if $\mu(t)$ is a continuous local martingale, then a sequence of random times τ_n exists completely reducing $\mu(t)$ such that $\mu(t \wedge \tau_n)$ is a bounded martingale with probability 1.

We now introduce the notion of a square variation of a process.

As before let λ denote a subdivision of the interval $[0, T]$ with subdividing points $0 = t_0 < t_1 < \cdots < t_n = T$, and $\zeta(t)$, $t \in [0, T]$, be an arbitrary random process.

Definition. *The quadratic variation of the process* $\zeta(t)$ on the interval $[0, T]$ is the limit (as $\lambda \to 0$) in the sense of convergence in probability of the sums

$$\sigma_\lambda^2 = \sigma_\lambda^2(T) = \sum_{k=1}^{n} [\zeta(t_k) - \zeta(t_{k-1})]^2,$$

provided this limit exists.

The quadratic variation of process $\zeta(t)$ on the interval $[0, T]$ is denoted by $[\zeta, \zeta]_t$, $[\zeta, \zeta]_t = \mathsf{P}\text{-}\lim_{\lambda \to 0} \sigma_\lambda^2(t)$.

Remark. If the sample functions of a process $\zeta(t)$ are continuous and of bounded variation on $[0, T]$ with probability 1, then $[\mu, \mu]_T = 0$.

Indeed, $\sigma_\lambda^2 \leq \max_k |\zeta(t_k) - \zeta(t_{k-1})| V_T(\zeta)$, where

$$V_T(\zeta) = \sup_\lambda \sum_{k=1}^n |\zeta(t_k) - \zeta(t_{k-1})|$$

is the variation of $\zeta(t)$ on $[0, T]$. Since $\max |\zeta(t_k) - \zeta(t_{k-1})| \to 0$ as $\lambda \to 0$, it follows that $\sigma_\lambda^2 \to 0$ with probability 1. □

Lemma 8. *If $\mu(t)$ is a square integrable martingale on $[0, T]$ then the family of variables $\{\sigma_\lambda^2(T)\}$ is uniformly integrable.*

Proof. Define for each λ the potential $\xi(n) = \xi_\lambda(n)$ by setting

$$\xi(0) = \mathsf{E}\{\mu^2(T) | \mathfrak{F}_0\},$$
$$\xi(k) = \mathsf{E}\{\mu^2(T) | \mathfrak{F}_{t_k}\} - \mu^2(t_k) + (\mu(t_k) - \mu(t_{k-1}))^2, \qquad k = 1, 2, \ldots, n.$$

Moreover,

$$\mathsf{E}\{\xi(k) | \mathfrak{F}_{t_{k-1}}\} = \mathsf{E}\{\mu^2(T) | \mathfrak{F}_{t_{k-1}}\} - \mu^2(t_{k-1}) \leq \xi(k-1)$$

and

$$\Delta\alpha_k = (\mu(t_{k-1}) - \mu(t_{k-2}))^2,$$

where

$$\Delta\alpha_k = \mathsf{E}\{\xi(t_{k-1}) - \xi(t_k) | \mathfrak{F}_{t_{k-1}}\}, \qquad k = 1, 2, \ldots, n$$

($\mu(t_{-1})$ is defined to be 0) and

$$\alpha(\infty) = \sum_{k=1}^n \Delta\alpha_k = \sigma_\lambda^2(T).$$

On the other hand, we have

$$\xi(k) \leq \mathsf{E}\{\mu^2(T) | \mathfrak{F}_{t_k}\} + 4 \sup_{0 \leq t \leq T} \mu^2(t)$$

independently of the choice of the subdivision λ, and since the martingale $\mathsf{E}\{\mu^2(T) | \mathfrak{F}_t\}$ is uniformly integrable, $\rho(c) = \sup_{\lambda,\tau} \int_{\{\xi_\tau > c\}} \xi_\tau \, d\mathsf{P} \to 0$ as $c \to \infty$, where τ is an arbitrary random time on $\{\mathfrak{F}_{t_k}; k = 0, \ldots, n\}$. Consequently, Lemma 4 is applicable in this case, which implies that the variables $\sigma_\lambda^2(t)$ form a uniformly integrable family. □

Theorem 22. *Let $\mu(t) \in l\mathcal{M}^c$. The square variation of the process $\mu(t)$, $t \in [0, T]$, exists for each $T > 0$ and coincides with its characteristic, i.e.,*

$$[\mu, \mu]_t = \langle \mu, \mu \rangle_t \pmod{\mathsf{P}}. \tag{52}$$

Proof. First let $\mu(t)$ be a square integrable martingale. Define the random time $\tau = \tau_c$ by setting

$$\tau = \tau_c = \inf \{s: (|\mu(s)| \geqslant c) \cup (\langle \mu, \mu \rangle_s \geqslant c^2), s \in [0, t]\},$$

provided the set appearing in the braces is nonvoid, and setting $\tau = t$ otherwise. Let

$$\mu'(t) = \mu(t \wedge \tau), \qquad \alpha'(t) = \langle \mu, \mu \rangle_{t \wedge \tau}, \qquad \alpha(t) = \langle \mu, \mu \rangle_t,$$

$$\bar{\sigma}_\lambda^2 = \sum_{k=1}^n [\mu'(t_k) - \mu'(t_{k-1})]^2, \qquad \sigma_\lambda^2 = \sum_{k=1}^n [\mu(t_k) - \mu(t_{k-1})]^2.$$

Moreover, $|\mu'(t)| \leqslant c$, $\alpha'(t) \leqslant c^2$.

It follows from the inequality

$$\mathsf{P}\{|\sigma_\lambda^2 - \langle \mu, \mu \rangle_t| < \varepsilon\} \leqslant \mathsf{P}\{|\bar{\sigma}_\lambda^2 - \alpha'(t)| > \varepsilon\} + \mathsf{P}\{\tau < t\}$$

and the relation

$$\mathsf{P}\{\tau < t\} = \mathsf{P}\{\sup |\mu(s)| \geqslant c) \cup (\alpha(t) > c^2)\} \to 0 \quad \text{as } c \to \infty$$

that to prove the theorem in the case under consideration it is sufficient to verify that $\bar{\sigma}_\lambda^2 \to \alpha'(t)$ in probability as $\lambda \to 0$.

For this purpose, observe that

$$\mathsf{E}(\bar{\sigma}_\lambda^2 - \alpha'(t))^2 = \mathsf{E}\left[\sum_{k=1}^n (\Delta\mu_k')^2 - \Delta\alpha_k'\right]^2 \leqslant 2 \sum_{k=1}^n \mathsf{E}[(\Delta\mu_k')^4 + (\Delta\alpha_k')^2],$$

where

$$\Delta\mu_k' = \mu'(t_k) - \mu'(t_{k-1}), \qquad \Delta\alpha_k' = \alpha'(t_k) - \alpha'(t_{k-1}).$$

Moreover,

$$\sum_{k=1}^n \mathsf{E}(\Delta\mu_k')^4 \leqslant 4c^2 \mathsf{E}(\max_k |\Delta\mu_k'|^2),$$

$$\sum_{k=1}^n \mathsf{E}(\Delta\alpha_k')^2 \leqslant c^2 \mathsf{E}(\max_k |\Delta\alpha_k'|).$$

Since $\max (\Delta\mu_k')^2 \to 0$, $\max \Delta\alpha_k' \to 0$ with probability 1 and these quantities are bounded above by an absolute constant, it follows that $\mathsf{E}(\bar{\sigma}_\lambda^2-\alpha'(t))^2\to 0$ and the theorem is proved for the case of square integrable martingales.

Now let $\mu(t)\in l\mathcal{M}^c$ and $\{\tau_n\}$ be a sequence of random times reducing $\mu(t)$. Then

$$\begin{aligned}\mathsf{P}\{|\sigma_\lambda^2-\alpha(t)|>\varepsilon\}&\le \mathsf{P}\left\{\left|\sum_{k=1}^n(\Delta\mu_k^{(r)})^2-\alpha(t\wedge\tau_r)\right|>\frac{\varepsilon}{3}\right\}\\&\quad+\mathsf{P}\left\{|\alpha(t\wedge\tau_r)-\alpha(t)|>\frac{\varepsilon}{3}\right\}\\&\quad+\mathsf{P}\left\{\left|\sum_{k=1}^n(\Delta\mu_k^{(r)})^2-(\Delta\mu_k)^2\right|>\frac{\varepsilon}{3}\right\}\\&\le 2\mathsf{P}\{\tau_r<t\}+\mathsf{P}\left\{\left|\sum_{k=1}^n(\Delta\mu_k^{(r)})^2-\alpha(t\wedge\tau_r)\right|>\frac{\varepsilon}{3}\right\},\end{aligned}$$

where $\mu^{(r)}(t)=\mu(t\wedge\tau_r)$. The assertion of the theorem easily follows from the last inequality. □

Corollary. *Let* $\mu_i(t)\in l\mathcal{M}^c$. *Set*

$$[\mu_1,\mu_2]_t=\mathsf{P}\text{-}\lim_{\lambda\to\infty}\sum_{k=1}^n(\mu_1(t_k)-\mu_1(t_{k-1}))(\mu_2(t_k)-\mu_2(t_{k-1})). \tag{53}$$

Then

$$[\mu_1,\mu_2]_t=\langle\mu_1,\mu_2\rangle_t.$$

The following lemma is utilized quite often in what follows.

Lemma 9. *Let* $\mu(t)\in l\mathcal{M}_2$ *and the characteristic* $\langle\mu,\mu\rangle_t$ *be continuous. Then*

$$\mathsf{P}\{\sup_{0\le t\le T}|\mu(t)|>\varepsilon\}\le\frac{N}{\varepsilon^2}+\mathsf{P}\{\langle\mu,\mu\rangle_T\ge N\} \tag{54}$$

for any positive ε *and* N.

Proof. It is sufficient to prove inequality (54) for martingales in $\mathcal{M}_2$ since it can obviously be generalized to the case of local square integrable martingales by means of a limit transition. Thus, let $\mu(t)\in\mathcal{M}_2$. Set $\tau_N=\inf\{t\colon\langle\mu,\mu\rangle_t\ge N,\ t\le T\}$ provided that if the set of values of t in the braces { } is void, we then set $\tau_N=T$. In view of the continuity of the process $\langle\mu,\mu\rangle_t$ inequality $\langle\mu,\mu\rangle_\tau\le N$ is fulfilled with probability 1.

On the other hand,

$$\begin{aligned}\mathsf{P}\{\sup_{0\le t\le T}|\mu(t)|>\varepsilon\}&=\mathsf{P}\{(\sup_{0\le t\le T}|\mu(t)|>\varepsilon)\cap\{\tau<T\}\}\\&\quad+\mathsf{P}\{(\sup_{0\le t\le T}|\mu(t)|>\varepsilon)\cap(\tau=T)\}\\&\le\mathsf{P}\{\tau<T\}+\mathsf{P}\{\sup_{0\le t\le\tau}|\mu(t)|>\varepsilon\}.\end{aligned}$$

Utilizing the fact that

$$\mathsf{P}\{\tau < T\} \leq \mathsf{P}\{\langle \mu, \mu \rangle_T \geq N\},$$

$$\mathsf{P}\{\sup_{0 \leq t \leq \tau} |\mu(t)| > \varepsilon\} \leq \frac{\mathsf{E}|\mu(\tau)|^2}{\varepsilon^2} \leq \frac{N}{\varepsilon^2},$$

we obtain inequality (54). □

Corollary. *If $\mu_n(t) \in l\mathcal{M}_2$ and characteristics $\langle \mu, \mu \rangle_n$, $n = 1, 2, \ldots$ are continuous then the relation*

$$\langle \mu_{n+p} - \mu_n, \mu_{n+p} - \mu_n \rangle_t \xrightarrow{\mathsf{P}} 0 \qquad \forall t > 0$$

yields the existence of the process $\mu(t) \subset l\mathcal{M}_2$ such that

$$\mathsf{P}\{\sup_{0 \leq t \leq T} |\mu(t) - \mu_n(t)| > \varepsilon\} \to 0 \quad \text{as } n \to \infty \quad \forall T > 0$$

The arguments utilized in the course of the proof of Lemma 9 are applicable also to a martingale with a discrete time. We shall note those points in the proof which require additional amplification. Let $\{\mu_k, \mathfrak{F}_k, k = 1, 2, \ldots, n\}$ be a square integrable martingale. Set

$$\Delta\mu_k = \mu_k - \mu_{k-1}, \qquad \Delta\alpha_k = \mathsf{E}\{\Delta\mu_k^2 \mid \mathfrak{F}_{k-1}\}, \qquad \alpha_k = \sum_{j=1}^{k} \Delta\alpha_j,$$

and let $\tau = k$ provided a k $(1 \leq k \leq n-1)$ exists such that

$$\alpha_1 \leq N, \qquad \alpha_2 \leq N, \ldots, \alpha_k \leq N, \qquad \alpha_{k+1} > N,$$

and $\tau = n$ provided no such k exists. Since α_{k+1} is an $\mathfrak{F}_k$-measurable random variable, τ is a random time. Set $\tilde{\mu}_j = \mu_{j \wedge \tau}$, $i = 1, \ldots, n$. Then $\mathsf{E}\tilde{\mu}_j^2 \leq \mathsf{E}\alpha_{j \wedge \tau} \leq N$. Repeating the argument presented in the course of the proof of Lemma 9 we obtain the following result:

Lemma 10. *Let $\{\mu_k, \mathfrak{F}_k, k = 1, \ldots, n\}$ be a square integrable martingale; then for any $\varepsilon > 0$, $N > 0$,*

$$\mathsf{P}\{\max_{1 \leq k \leq n} |\mu_k| > \varepsilon\} \leq \frac{N}{\varepsilon^2} + \mathsf{P}\{\alpha_n \geq N\}. \tag{55}$$

Remark. If $\mu(t) = (\mu^1(t), \ldots, \mu^s(t))$ is a vector-valued martingale, inequality (54) can be replaced by the following:

$$\mathsf{P}\{\sup_{0 \leq t \leq T} |\mu(t)| > \varepsilon\} \leq \frac{N}{\varepsilon^2} + \mathsf{P}\left\{\sum_{j=1}^{s} \langle \mu^j, \mu^j \rangle_T \geq N\right\}.$$

Inequality (55) is generalized analogously. The proof is identical to the proof of Lemma 9.

§2. Stochastic Integrals

Integration of piecewise constant functions. The definition of a stochastic integral

$$\int_a^b f(t)\, d\zeta(t), \tag{1}$$

where $\zeta(t)$ is a process with orthogonal increments and $f(t)$ is a nonrandom function was given in Volume I, Chapter IV, Section 4. It is easy to see that the construction of the integral described therein is in general not applicable for the case of *random* functions $f(t)$. For a nonrandom function $f(t)$ the integral (1) is an element of the closure of a linear hull of values of the random variables $\zeta(t)-\zeta(a)$; however, if $f(t)$ is a random process, then this is generally not the case. Nevertheless, being sufficiently cautious with the definitions and utilizing additional assumptions concerning the process $\zeta(t)$ it is possible to develop a general theory (which is convenient for applications and yet is sufficiently extensive) of integration of random functions over processes $\zeta(t)$ with sample functions which are in general of an unbounded variation with probability 1.

The following remark points up the difficulties which may arise in this connection.

Let $f(t), t \in (0, 1]$, be a random function and $f_n(t)=\sum \beta_n^k \chi_{(t_{k-1},t_k]}(t)$, $n = 1, 2, \ldots$, be a sequence of simple functions ($0=t_0<t_1<\cdots<t_n=1$) convergent to $f(t)$. Then the sequence of "integral sums"

$$\sum_{k=1}^{n} \beta_n^k[\zeta(t_k)-\zeta(t_{k-1})],$$

which is natural to consider as the values of integral (1) for the functions $f_n(t)$ in general does not approach any limit even under very strong assumptions on the convergence of $f_n(t)$ to $f(t)$. We can illustrate this phenomenon by the following simple example.

Assume that we are to define the integral

$$\int_0^1 w(t)\, dw(t),$$

where $w(t)$ is a Wiener process. Observe first that if we set $f_n(t)=w(\theta_k)$, where $\theta_k \in (t_{k-1}, t_k]$, $0=t_0<t_1<\cdots<t_n=1$, then $f_n(t)\to w(t)$ in any one of the usual topologies.

Thus this convergence is uniform in t with probability 1. Moreover, $\mathsf{E}\sup_t |f_n(t)-w(t)|^2 \to 0$ and

$$\int_0^1 |f_n(t)-w(t)|^2\, dt \to 0$$

with probability 1. On the other hand, the integral under consideration cannot be defined as a mean square limit of the quantities

$$\int_0^1 f_n(t)\,dw(t) = \sum_{k=1}^{n} w(\theta_k)[w(t_k) - w(t_{k-1})],$$

since should such a limit exist, so would the limit of the sequence

$$\mathsf{E}\int_0^1 f_n(t)\,dw(t) = \sum_{k=1}^{n} \mathsf{E}w(\theta_k)[w(t_k) - w(t_{k-1})] = \sum (\theta_k - t_{k-1}),$$

while the set of limit points of this last sum coincides with the interval [0, 1].

We shall subdivide the definition of a stochastic integral into several stages of increasing generality.

Firstly we shall consider the definition of the integral in the case when $\zeta(t)$ is a square integrable martingale. As far as the class of integrable processes $f(t)$ is concerned, we shall start with functions which are piecewise constant and bounded with probability 1. We then proceed to the integration of functions with finite moments of the second order and next to classes of random processes which possess no finite moments of any order. Further extension of the notion of a stochastic integral is connected with the extension of the class of integrating processes. In place of square integrable martingales $\zeta(t)$ we shall consider local square integrable martingales. Finally, we shall consider integrals over martingale measures.

At the same time, not striving for the greatest generality, we shall in most cases confine ourselves to the study of martingales (local martingales) with continuous characteristics.

Thus, let $\{\mathfrak{F}_t, t \geq 0\}$ be a fixed current of σ-algebras, $\mathfrak{F}_t = \mathfrak{F}_{t^+}$, where $\mathfrak{F}_0$ contains all the subsets of P-measure 0 (we shall assume in this section that this assumption is always satisfied), let $\mu(t)$ be a square integrable $\mathfrak{F}_t$-martingale with sample functions belonging to $\mathscr{D}$ with probability 1 and with characteristic $\langle\mu, \mu\rangle_t$:

$$\sup_{0 \leq t < \infty} \mathsf{E}|\mu(t)|^2 < \infty,$$

$$\mathsf{E}\{(\mu(t) - \mu(s))^2 \mid \mathfrak{F}_s\} = \mathsf{E}\{\langle\mu, \mu\rangle_t - \langle\mu, \mu\rangle_s \mid \mathfrak{F}_s\}, \qquad s < t.$$

Observe that $\mu(t)$ is a process with orthogonal increments. Indeed, for $t_1 < t_2 < t_3$

(2) $$\mathsf{E}\{(\mu(t_3) - \mu(t_2))(\mu(t_2) - \mu(t_1)) \mid \mathfrak{F}_{t_2}\} = (\mu(t_2) - \mu(t_1))\mathsf{E}\{\mu(t_3) - \mu(t_2) \mid \mathfrak{F}_{t_2}\} = 0,$$

whence, in particular,

$$\mathsf{E}(\mu(t_3) - \mu(t_2))(\mu(t_2) - \mu(t_1)) = 0.$$

Proceeding now to describe the class of integrable functions we shall first agree to consider those processes which are adopted to $\{\mathfrak{F}_t, t \geq 0\}$.

Let $\mathfrak{L}_0 = \mathfrak{L}_0\{\mathfrak{F}_t, t \geq 0\}$ denote the class of processes adopted to $\{\mathfrak{F}_t, t \geq 0\}$ and taking on constant bounded (mod P) values on a finite number of semiintervals of the form $(s_{k-1}, s_k]$ and vanish outside these intervals. In other words, $\eta(t) \in \mathfrak{L}_0$ if and only if

$$\eta(t) = \sum_{k=1}^{n} \eta_{k-1} \chi_{\Delta_k}(t),$$

where $\Delta_k = (s_{k-1}, s_k]$, $0 \leq s_0 < s_1 < \cdots < s_n$, $\chi_\Delta(t)$ is the indicator of the set Δ, and η_k is an $\mathfrak{F}_k$-measurable random variable bounded by a constant ($k = 0, 1, \ldots, n-1$). Observe that the function $\eta(t)$ is continuous from the left.

First define the integral

$$I(\eta) = \int_0^\infty \eta(s) \mu(ds) = \int_0^\infty \eta \, d\mu$$

for processes $\eta(t) \in \mathfrak{L}_0$. Namely, we set

$$I(\eta) = \int_0^\infty \eta(s) \mu(ds) = \sum_{k=1}^{n} \eta_{k-1}[\mu(s_k) - \mu(s_{k-1})]. \tag{3}$$

Although the representation of the function $\eta(t)$ as a linear combination of indicators of semiintervals is not unique, it is easy to verify that the value of $I(\eta)$ is uniquely determined. It is also clear that the operator $I(\eta)$ is linear, i.e.,

$$I(\gamma_1 \eta^{(1)} + \gamma_2 \eta^{(2)}) = \gamma_1 I(\eta^{(1)}) + \gamma_2 I(\eta^{(2)}), \tag{4}$$

where γ_i are $\mathfrak{F}_0$-measurable bounded random variables, $\eta^{(i)} = \eta^{(i)}(t) \in \mathfrak{L}_0$, $i = 1, 2$. Since

$$\mathsf{E}\left\{\sum_{k=1}^{n} \eta_{k-1}[\mu(s_k) - \mu(s_{k-1})] \,\middle|\, \mathfrak{F}_0\right\} = \mathsf{E}\left\{\sum_{k=1}^{n} \eta_{k-1} \mathsf{E}\{\mu(s_k) - \mu(s_{k-1}) \mid \mathfrak{F}_{s_{k-1}}\} \,\middle|\, \mathfrak{F}_0\right\} = 0,$$

it follows that

$$\mathsf{E}\{I(\eta) \mid \mathfrak{F}_0\} = 0. \tag{5}$$

Let $\eta_i(t) \in \mathfrak{L}_0$, $i = 1, 2$. Without loss of generality, we may assume that

$$\eta_i(t) = \sum_{k=1}^{n} \eta_{k-1}^{(i)} \chi_{\Delta_k}(t).$$

For abbreviation, set $\Delta\mu_k = \mu(s_k) - \mu(s_{k-1})$. Taking into account (2) we have

$$\mathsf{E}\{I(\eta_1) I(\eta_2) \mid \mathfrak{F}_0\} = \mathsf{E}\left\{\sum_{j,k=1}^{n} \eta_{k-1}^{(1)} \eta_{j-1}^{(2)} \, \Delta\mu_k \Delta\mu_j \,\middle|\, \mathfrak{F}_0\right\}$$

$$= \mathsf{E}\left\{\sum_{k=1}^{n} \eta_{k-1}^{(1)} \eta_{k-1}^{(2)}, \Delta\langle \mu, \mu\rangle_{s_k} \,\middle|\, \mathfrak{F}_0\right\},$$

where $\Delta\langle\mu,\mu\rangle_{s_k}=\langle\mu,\mu\rangle_{s_k}-\langle\mu,\mu\rangle_{s_{k-1}}$. Thus

$$\mathsf{E}\left\{\int_0^\infty \eta_1\,d\mu\int_0^\infty \eta_2\,d\mu\,|\,\mathfrak{F}_0\right\}=\mathsf{E}\left\{\int_0^\infty \eta_1(s)\eta_2(s)\,d\langle\mu,\mu\rangle_s\,|\,\mathfrak{F}_0\right\}. \tag{6}$$

In particular,

$$\mathsf{E}\left|\int_0^\infty \eta_1\,d\mu-\int_0^\infty \eta_2\,d\mu\right|^2=\mathsf{E}\int_0^\infty (\eta_1(s)-\eta_2(s))^2\,d\langle\mu,\mu\rangle_s. \tag{7}$$

Observe that $\langle\mu,\mu\rangle_t$ is a monotonically nondecreasing function and the integrals appearing in the right-hand side of formulas (6) and (7) are the ordinary Lebesgue–Stieltjes integrals which exist with probability 1.

In addition to the integral defined above we shall consider the indefinite integral

$$I_t=I_t(\eta)\underset{\mathrm{Def}}{=}\int_0^\infty \chi_{(0,t]}(s)\eta(s)\mu(ds)\underset{\mathrm{Def}}{=}\int_0^t \eta(s)\mu(ds).$$

Clearly

$$I_t=\sum_{j=1}^{k}\eta_{j-1}\Delta\mu_j+\eta_k[\mu(t)-\mu(s_k)],\qquad t\in(s_k,s_{k+1}]. \tag{8}$$

We note several properties of the process I_t:

a) *The process I_t is adopted to the current of σ-algebras $\{\mathfrak{F}_t,t\geq 0\}$; the sample functions of the process are continuous from the right and possess limits from the left for all $t\geq 0$ (mod P) and*

$$\delta I_t=\eta(t)\,\delta\mu(t),$$

where $\delta\mu(t)$ is the jump of the process $\mu(t)$ at point t, i.e.,

$$\delta\mu(t)=\mu(t)-\mu(t-).$$

b) *The process I_t is an $\mathfrak{F}_t$-martingale.*

Proof. Indeed, let $0\leq t_1<t_2$. Without loss of generality, we can assume that $t_1=s_k$ and $t_2=s_l$ for some k and l.

Then in view of (2)

$$I_{t_2}-I_{t_1}=\sum_{j=k+1}^{l}\eta_{j-1}[\mu(s_j)-\mu(s_{j-1})]$$

and

$$\mathsf{E}\{I(t_2)-I(t_1)\,|\,\mathfrak{F}_{t_1}\}=0.\quad\square \tag{9}$$

c) $I_t \in \mathcal{M}_2$; *moreover,*

(10) $$\mathsf{E}\{(I_{t_2}-I_{t_1})^2 \mid \mathfrak{F}_{t_1}\} = \mathsf{E}\{\int_{t_1}^{t_2} \eta^2(s)\, d\langle\mu,\mu\rangle_s \mid \mathfrak{F}_{t_1}\}.$$

Proof. Formula (10) can be directly deduced from (6). □

One can also easily verify that the process

$$\beta(t) = \int_0^t \eta^2(s)\, d\langle\mu,\mu\rangle_s$$

is a natural process. Indeed, if the process $\alpha(t) = \langle\mu,\mu\rangle_t$ is natural, then (cf. Section 1)

$$\mathsf{E}\int_\Delta \xi(s-)\, d\alpha(s) = \mathsf{E}\int_\Delta \xi(s)\, d\alpha(s)$$

for any semiinterval $\Delta = (a, b]$ and an arbitrary nonnegative martingale $\xi(t)$ bounded with probability 1.

Let η be an arbitrary bounded $\mathfrak{F}_a$-measurable random variable. Set $\xi_1(t) = \xi(b)\eta^2$ for $t \geq b$, $\xi_1(t) = \xi(t)\eta^2$ for $t \in [a, b]$ and $\xi_1(t) = \mathsf{E}\{\xi(a)\eta^2 \mid \mathfrak{F}_t\}$ for $t \in [0, a)$. This definition assures us that $\xi_1(t)$ is a bounded martingale. Applying the preceding relation to martingale $\xi_1(b)$ we obtain

$$\mathsf{E}\int_\Delta \xi(s-)\eta^2\, d\alpha(s) = \mathsf{E}\int_\Delta \xi(s)\eta^2\, d\alpha(s).$$

Setting in the last equality $\Delta = \Delta_k = (s_{k-1}, s_k]$, $\eta = \eta_{k-1}$ and summing over all $k = 1, 2, \ldots, n$ we have

$$\mathsf{E}\int_0^\infty \xi(s-)\eta^2(s)\, d\alpha(s) = \mathsf{E}\int_0^\infty \xi(s)\eta^2(s)\, d\alpha(s),$$

which implies (cf. Section 1, (18)) that

$$\mathsf{E}\int_0^\infty \xi(s-)\, d\beta = \mathsf{E}\xi(+\infty)\beta(+\infty).$$

This shows that $\beta(t)$ is a natural process. □

Thus, equality (10) can be refined in the following manner:

d) *The characteristic of a stochastic integral I_t is equal to*

$$\langle I, I\rangle_t = \int_0^t \eta^2(s)\, d\langle\mu,\mu\rangle_s.$$

From this formula one can easily deduce the following expression for the joint characteristic of stochastic integral $I'_t = I_t(\eta')$ and $I''_t = I_t(\eta'')$ $(\eta', \eta'' \in \mathfrak{L}_0)$:

(11) $$\langle I', I''\rangle_t = \int_0^t \eta'(s)\eta''(s)\, d\langle\mu,\mu\rangle_s.$$

Analogously, one can obtain an expression for the joint characteristic of $I_t(\eta)$ and

$\nu(t)$, where $\nu(t)$ is an arbitrary martingale in $\mathcal{M}_2$. Let $0 \leq t_1 \leq t_2$. Assume as above that without loss of generality $t_1 = s_k$ and $t_2 = s_l$. Then

$$\begin{aligned} \mathsf{E}\{I_{t_2}\nu(t_2) - I_{t_1}\nu(t_1) \,|\, \mathfrak{F}_{t_1}\} &= \mathsf{E}\left\{ \sum_{j=k+1}^{l} \mathsf{E}\{\Delta I_{s_j} \Delta \nu_{s_j} \,|\, \mathfrak{F}_{s_{j-1}}\} \,|\, \mathfrak{F}_{t_1} \right\} \\ &= \mathsf{E}\left\{ \sum_{j=k+1}^{l} \eta_{j-1} \mathsf{E}\{\Delta \mu_{s_j} \Delta \nu_{s_j} \,|\, \mathfrak{F}_{s_{j-1}}\} \,|\, \mathfrak{F}_{t_1} \right\} \\ &= \mathsf{E}\{\textstyle\int_{t_1}^{t_2} \eta(s)\, d\langle \mu, \nu\rangle_s \,|\, \mathfrak{F}_{t_1}\}. \end{aligned}$$

Thus,

$$\mathsf{E}\{(I_{t_2} - I_{t_1})(\nu(t_2) - \nu(t_1)) \,|\, \mathfrak{F}_{t_1}\} = \mathsf{E}\{\textstyle\int_{t_1}^{t_2} \eta(s)\, d\langle \mu, \nu\rangle_s \,|\, \mathfrak{F}_{t_1}\}. \tag{12}$$

Assertion c) above implies that the process

$$\int_0^t \eta(s)\, d\langle \mu, \nu\rangle_s$$

is the difference of two natural processes and in view of the uniqueness of the joint characteristic of the two martingales

$$\langle I, \nu\rangle_t = \int_0^t \eta(s)\, d\langle \mu, \nu\rangle_s.$$

Also note the following inequality:

$$\mathsf{E} \int_0^\infty |\eta(s)|\, d\|\langle \mu, \nu\rangle\|_s \leq (\mathsf{E}\langle \nu, \nu\rangle_\infty)^{1/2} \left(\mathsf{E} \int_0^\infty \eta^2(s)\, d\langle \mu, \mu\rangle_s\right)^{1/2}. \tag{13}$$

Here μ and $\nu \in \mathcal{M}_2$, $\eta \in \mathfrak{L}_0$, and $\|\langle \mu, \nu\rangle\|_s$ denotes the total variation of the function $\langle \mu, \nu\rangle_t$ on the interval $[0, s]$.

Proof. To prove this we utilize inequality (46) in Section 1, which implies that

$$\begin{aligned} &\mathsf{E}\{|\eta_{k-1}|\, \Delta\langle \mu, \mu\rangle_{s_k} \,|\, \mathfrak{F}_{s_{k-1}}\} \\ &\quad \leq |\eta_{k-1}| (\mathsf{E}\{\Delta\langle \mu, \mu\rangle_{s_k} \,|\, \mathfrak{F}_{s_{k-1}}\})^{1/2} (\mathsf{E}\{\Delta\langle \nu, \nu\rangle_{s_k} \,|\, \mathfrak{F}_{s_{k-1}}\})^{1/2} \\ &\quad = (\mathsf{E}\{\eta_{k-1}^2\, \Delta\langle \mu, \mu\rangle_{s_k} \,|\, \mathfrak{F}_{s_{k-1}}\})^{1/2} (\mathsf{E}\{\Delta\langle \nu, \nu\rangle_{s_k} \,|\, \mathfrak{F}_{s_{k-1}}\})^{1/2}. \end{aligned}$$

Consider now an arbitrary subdivision $t_1, t_2, \ldots, t_n$ of the interval $[s_{k-1}, s_k]$. Applying the last inequality to each of the intervals $[t_{k-1}, t_k]$, using the Cauchy–Schwarz inequality, and approaching the limit we obtain

$$\begin{aligned} &\mathsf{E}\{|\eta_{k-1}| (\|\langle \mu, \nu\rangle\|_{s_k} - \|\langle \mu, \nu\rangle\|_{s_{k-1}}) \,|\, \mathfrak{F}_{s_{k-1}}\} \\ &\leq (\mathsf{E}\{\eta_{k-1}^2\, \Delta\langle \mu, \mu\rangle_{t_{k-1}} \,|\, \mathfrak{F}_{s_{k-1}}\})^{1/2} (\mathsf{E}\{\Delta\langle \nu, \nu\rangle_{t_{k-1}} \,|\, \mathfrak{F}_{s_{k-1}}\})^{1/2}. \end{aligned}$$

Summing up these inequalities and applying the Cauchy–Schwarz inequality once again we arrive at inequality (13). □

The stochastic integral in the sense of mean square convergence. We now extend the definition of a stochastic integral to a wider class of random processes $\eta(t)$ and show that the properties of a stochastic integral which were established above for functions $\eta(t)\in\mathfrak{L}_0$ are retained for this extension.

We introduce in $\mathfrak{L}_0$ a Hilbert metric generated by the norm

$$\|\eta(\cdot)\|=(\mathsf{E}\int_0^\infty \eta^2(s)\,d\langle\mu,\mu\rangle_s)^{1/2}. \tag{14}$$

The preceding results show that the correspondence $\eta(\cdot)\to I_t(\eta)$ is a single-valued linear and isometric mapping of $\mathfrak{L}_0$ into $\mathcal{M}_2$. Consider the completion of $\mathfrak{L}_0$ via the metric introduced above. The Hilbert space obtained will be denoted by $\mathfrak{L}_2=\mathfrak{L}_2\{\mathfrak{F}_t,\mu\}$. It can be described in the following manner. Consider on the space $[0,\infty)\times\Omega$ the class of sets of the form $[0]\times A_0$, where $[0]$ is the singleton consisting of point 0 and A_0 is an $\mathfrak{F}_0$-measurable set, and sets of the form $(a,b]\times A$, where A is an arbitrary $\mathfrak{F}_a$-measurable set in Ω, $0\leqslant a<b$. This class is a semiring $\mathcal{R}$. We define a measure P' on this semiring in the following manner. Let $\alpha(t,\omega)=\langle\mu,\mu\rangle_t$. For almost all ω one can define a measure $\alpha(B,\omega)$ on the Borel subsets of the half-line $[0,\infty)$. Set

$$\begin{aligned}\mathsf{P}'((a,b]\times A)&=\int_A\int_a^b \alpha(ds,\omega)\mathsf{P}(d\omega)\\&=\mathsf{E}\chi_A[\langle\mu,\mu\rangle_b-\langle\mu,\mu\rangle_a].\end{aligned}$$

Measure P' can be continued over the minimal σ-algebra $\mathfrak{T}$ containing the semiring $\mathcal{R}$ and its completion $\tilde{\mathfrak{T}}$. Moreover, for any nonnegative $\tilde{\mathfrak{T}}$-measurable function $\eta(t)=\eta(t,\omega)$

$$\int_0^\infty\int_\Omega \eta(t,\omega)\mathsf{P}'(dt,d\omega)=\mathsf{E}\int_0^\infty \eta(t)\,d\langle\mu,\mu\rangle_t.$$

The space $\mathfrak{L}_2$ is the space of *equivalence classes of* $\tilde{\mathfrak{T}}$*-measurable functions* $g(t,\omega)$: $g(t,\omega)$ and $g'(t,\omega)$ are equivalent if $\tilde{\mathsf{P}}'\{(t,\omega): g(t,\omega)\neq g'(t,\omega)\}=0$. For simplicity we shall not distinguish in what follows between the class of equivalent functions (in the sense just indicated) and representatives of this class. Furthermore, the process $\eta(t)=g(t,\omega)$ (more precisely the corresponding equivalence class) is an element of $\mathfrak{L}_2\{\mathfrak{F}_t,\mu\}$ if and only if a sequence of simple processes $\eta_n(t)\in\mathfrak{L}_0(\mathfrak{F}_t)$ exists such that

$$\mathsf{E}\int_0^\infty(\eta(t)-\eta_n(t))^2\,d\langle\mu,\mu\rangle_t\to 0$$

as $n\to 0$.

It follows from general theorems in $\mathfrak{L}_2$-theory that the space $\mathfrak{L}_2\{\mathfrak{F}_t,\mu\}$ consists of all the $\tilde{\mathfrak{T}}$-measurable functions $\eta(t)=\eta(t,\omega)$ such that

$$\int_0^\infty\int_\Omega \eta^2(t,\omega)\mathsf{P}'(dt,d\omega)=\mathsf{E}\int_0^\infty \eta^2(t)\,d\langle\mu,\mu\rangle_t<\infty.$$

We present examples of classes of $\tilde{\mathfrak{T}}$-measurable processes. Firstly, a process of the form $\xi(t)=\eta\chi_{(a,b]}(t)$, where η is an $\mathfrak{F}_a$-measurable random variable and $\chi_{(a,b]}$

is the indicator of the semiinterval $(a, b]$, is $\mathfrak{T}$-measurable. Consequently, the process $\xi(t)=\sum \eta_k\chi_{(s_k,s_{k+1}]}(t)$, where $s_1<s_2<\cdots$, and η_k are $\mathfrak{F}_{s_k}$-measurable random variables, is also $\mathfrak{T}$-measurable.

An arbitrary process $\xi(t)$, $t\geq 0$, with sample functions continuous from the left (with probability 1) in t is $\tilde{\mathfrak{T}}$-measurable. Indeed $\xi(t)=\lim_{n\to\infty}(\xi_0\chi_{[0]}(t)+\sum_{k=0}^{\infty}\xi(k/n)\chi_{(k/n,(k+1)/n]}(t))$ with probability 1 for all t.

The process $\xi(t)$ for which a sequence of $\tilde{\mathfrak{T}}$-measurable processes $\xi_n(t)$ exists such that

$$\mathsf{E}\int_0^\infty [\xi(t)-\xi_n(t)]^2\, d\langle\mu,\mu\rangle_t\to 0 \quad \text{as } n\to\infty$$

is also a $\tilde{\mathfrak{T}}$-measurable process.

We use this remark in the case when the characteristic $\langle\mu,\mu\rangle_t$ is absolutely continuous in t, i.e.,

$$\langle\mu,\mu\rangle_t=\int_0^t \psi(s)\,ds$$

where $\varphi(t)$ is a nonnegative integrable function. In this case an arbitrary process $\xi(t)$ adopted to the current $\{\mathfrak{F}_t, t\geq 0\}$ and satisfying condition

$$\mathsf{E}\int_0^\infty \xi^2(t)\varphi(t)\,dt<\infty$$

belongs to $\mathfrak{L}_2\{\mathfrak{F}_t,\mu\}$. To prove this assertion it is sufficient to consider the case when $|\xi(t)|\leq c$ and $\xi(t)=0$ for $t\geq T$.

Here the process

$$\zeta_h(t)=\xi(t)-\frac{1}{h}\int_{t-h}^t \xi(s)\,ds,\qquad h>0,$$

is adopted to the current $\{\mathfrak{F}_t, t\geq 0\}$, is bounded with probability 1 and tends to 0 as $h\downarrow 0$ for almost all t and hence P'-almost for all (t,ω). Furthermore, the process $(1/h)\int_{t-h}^t\xi(s)\,ds$ is continuous and belongs to $\mathfrak{L}_2\{\mathfrak{F}_t,\mu\}$. Since

$$\mathsf{E}\int_0^\infty |\zeta_h(t)|^2\varphi(t)\,dt\to 0 \quad \text{as } h\to 0,$$

we have $\xi(\cdot)\in\mathfrak{L}_2\{\mathfrak{F}_t,\mu\}$.

Since the mapping $I\colon \eta\to\{I_t(\eta)\}$ is a linear and an isometric mapping of $\mathfrak{L}_0$ into $\mathcal{M}_2$ it is uniquely extendable to a linear isometric mapping of $\mathfrak{L}_2\{\mathfrak{F}_t,\mu\}$ into $\mathcal{M}_2$. This extended mapping is referred to as a *stochastic integral* as before and is denoted by the symbol $I=I(\eta)$ and the value of the process $I(\eta)$ at time t is

$$I_t(\eta)=\int_0^t \eta(s)\mu(ds)=\int_0^t \eta\,d\mu.$$

In the following theorem the basic properties of a stochastic integral are enumerated. All these properties are almost a direct consequence of the discussion above.

Theorem 1. *There is a correspondence between process $\eta(t)\in\mathfrak{L}_2\{\mathfrak{F}_t,\mu\}$ and a process $\xi(t)\in\mathcal{M}_2$ which is called the stochastic integral of $\eta(t)$ over martingale $\mu(t)$ ($\mu(t)\in\mathcal{M}_2$). This correspondence is given by*

$$\xi(t)=I_t(\eta)=\int_0^t \eta(s)\mu(ds)$$

and satisfies:

a) $I_\infty[\chi_{(0,a]})=\mu(a)-\mu(0)$;
b) $I_\infty(\chi_{(0,a]}\eta)=I_a(\eta)$;
c) $I_t(c_1\eta_1+c_2\eta_2)=c_1I_t(\eta_1)+c_2I_t(\eta_2)$;
d) $\mathsf{E}I_\infty(\eta_1)I_\infty(\eta_2)=\mathsf{E}\int_0^\infty \eta_1(s)\eta_2(s)\,d\langle\mu,\mu\rangle_s$;
e) *for any martingale* $\nu(t)\in\mathcal{M}_2$

$$\mathsf{E}\{\nu(t)I_t(\eta)-\nu(s)I_s(\eta)\,|\,\mathfrak{F}_s\}=\mathsf{E}\{\int_s^t \eta(\theta)\,d\langle\mu,\nu\rangle_\theta\,|\,\mathfrak{F}_s\};$$

in particular,

$$\mathsf{E}\{I_t^2-I_s^2\,|\,\mathfrak{F}_s\}=\mathsf{E}\{\int_s^t \eta^2(\theta)\,d\langle\mu,\mu\rangle_\theta\,|\,\mathfrak{F}_s\};$$

f) $\mathsf{E}\int_0^\infty |\eta(s)|\,d\|\langle\mu,\nu\rangle\|_s\leq(\mathsf{E}\langle\nu,\nu\rangle_\infty)^{1/2}(\mathsf{E}\int_0^\infty \eta^2(s)\,d\langle\mu,\mu\rangle_s)^{1/2}$;
g) *if* $\mu_i\in\mathcal{M}_2$, $i=1,2$, *and* $\eta\in\mathfrak{L}_2\{\mathfrak{F}_t,\mu_1\}\cap\mathfrak{L}_2\{\mathfrak{F}_t,\mu_2\}$, then

$$\int_0^t \eta\,d(\mu_1+\mu_2)=\int_0^t \eta\,d\mu_1+\int_0^t \eta\,d\mu_2;$$

h) *if* $\mu(t)\in\mathcal{M}_2^c$, *then* $\xi(t)\in\mathcal{M}_2^c$.

Proof. The existence of a mapping with properties a)–d) and h) was established above. Moreover, the construction above shows that properties a)–d) uniquely define a stochastic integral.

We now prove property e). This relationship is valid for $\eta\in\mathfrak{L}_0$. Let $\eta^{(n)}(\cdot)\in\mathfrak{L}_0$ and $\eta^{(n)}(\cdot)\overset{L_2}{\to}\eta(\cdot)$. It follows from the inequality

$$\mathsf{E}|[\nu I(\eta^{(n)})-\nu I(\eta)]_s^t|\leq\{\mathsf{E}(\nu|_s^t)^2\}^{1/2}\{\mathsf{E}([I(\eta^{(n)})-I(\eta)]_s^t)^2\}^{1/2}$$

that $\nu I(\eta^{(n)})|_s^t$ approaches $\nu I(\eta)|_s^t$ in L_1. Therefore one can approach the limit as $n\to\infty$ in the relation

$$\int_{B_s}\nu I_s(\eta^{(n)})|_s^t\,d\mathsf{P}=\int_{B_s}(\int_s^t \eta^{(n)}\,d\langle\mu,\nu\rangle)\,d\mathsf{P},\qquad B_s\in\mathfrak{F}_s.$$

This yields an equality which is equivalent to e). Inequality f), which was previously established for functions $\eta\in\mathfrak{L}_0$, and relation g) are self-evident for these functions. In the general case, f) and g) are obtained by means of a limit transition. □

It follows from the theorem just proved that the characteristic of martingale $I_t=I_t(\eta)$ is

(15) $$\langle I,I\rangle_t=\int_0^t \eta^2\,d\langle\mu,\mu\rangle.$$

Note that the family of random variables I_t^2 is uniformly integrable. Indeed,

$$I_t^2 \leqslant \sup_t I_t^2, \qquad \mathsf{E}(\sup_t I_t^2) \leqslant 4\mathsf{E} I_\infty^2 < \infty.$$

Moreover, the set $\{I_\tau^2, \tau \in \mathcal{T}\}$, where $\mathcal{T}$ is a family of all random times on $\{\mathfrak{F}_t, t \in [0, \infty]\}$, is also uniformly integrable.

Let σ and $\tau \in \mathcal{T}$ with $\sigma \leqslant \tau$ (mod P). Set

$$\int_\sigma^\tau \eta \, d\mu \underset{\text{Def}}{=} I_\tau - I_\sigma.$$

It follows from the general theory of martingales (Section 1, Theorem 1, Corollary 1, and Section 1, equation (44)) that

$$\mathsf{E}\{\int_\sigma^\tau \eta \, d\mu \mid \mathfrak{F}_\sigma\} = 0,$$
$$\mathsf{E}\{(\int_\sigma^\tau \eta \, d\mu)^2 \mid \mathfrak{F}_\sigma\} = \mathsf{E}\{I_\tau^2 - I_\sigma^2 \mid \mathfrak{F}_\sigma\} - \mathsf{E}\{\langle I, I\rangle_\tau - \langle I, I\rangle_\sigma \mid \mathfrak{F}_\sigma\}.$$

The last equality can be written in the following form:

$$\mathsf{E}\{(\int_\sigma^\tau \eta \, d\mu)^2 \mid \mathfrak{F}_\sigma\} = \mathsf{E}\{\int_\sigma^\tau \eta^2 \, d\langle \mu, \mu\rangle \mid \mathfrak{F}_\sigma\}. \tag{16}$$

Let $\mu_\tau(t)$ be the τ-stopping of the martingale $\mu(t)$: $\mu_\tau(t) = \mu(t \wedge \tau)$. If $\eta(\cdot) \in \mathfrak{L}_2\{\mathfrak{F}_t, \mu\}$, then $\eta(t \wedge \tau) \in \mathfrak{L}_2\{\mathfrak{F}_{t\wedge\tau}, \mu_\tau\}$ (Section 1, Theorem 6, Corollary 3) and the integral $\int_0^t \eta(s \wedge \tau)\mu_\tau(ds)$ is well defined in accordance with the preceding definition.

Lemma 1.

$$\int_0^t \eta(s \wedge \tau)\mu_\tau(ds) = \int_0^{t\wedge\tau} \eta(s)\mu(ds) = \int_0^t \eta_\tau(s)\mu(ds), \tag{17}$$

where $\eta_\tau(t) = \eta(t)$ for $t < \tau$ and $\eta_\tau(t) = 0$ for $t \geqslant \tau$.

Proof. For function $\eta \in \mathfrak{L}_0$ equality (17) follows directly from the definition of a stochastic integral. For arbitrary processes $\eta(\cdot)$ belonging to $\mathfrak{L}_2\{\mathfrak{F}_t, \mu\}$ it follows from the fact that the class of functions $\eta(\cdot)$ in $\mathfrak{L}_2\{\mathfrak{F}_t, \mu\}$ for which this equality is valid is closed in $\mathfrak{L}_2\{\mathfrak{F}_t, \mu\}$. □

Lemma 2. *If characteristic $\langle \mu, \mu\rangle_t$ is continuous and $\eta \in \mathfrak{L}_2$, then for any $\varepsilon > 0$ and $N > 0$*

$$\mathsf{P}\left\{\sup_{0 \leqslant t < \infty} \left|\int_0^t \eta(s)\mu(ds)\right| \geqslant \varepsilon\right\} \leqslant \frac{N}{\varepsilon^2} + \mathsf{P}\{\int_0^\infty \eta^2(s) \, d\langle \mu, \mu\rangle_s > N\}. \tag{18}$$

Proof. This inequality follows directly from Lemma 9 in Section 1. □

A general definition of a stochastic integral over a martingale. We now extend the definition of a stochastic integral to a more general class of functions $\eta(t)$.

We call a sequence of functions $\eta_n(t) \in \mathfrak{L}_0$, $n = 1, 2, \ldots$, *H_2-fundamental* (or H_2-Cauchy) if

$$\text{(19)} \qquad \mathsf{P}\text{-}\lim_{n\to\infty,\, m>0} \int_0^\infty (\eta_n(t) - \eta_{n+m}(t))^2 \, d\langle \mu, \mu \rangle_t = 0.$$

If a sequence $\eta_n(t)$ is H_2-fundamental, then a function $g(t)$ exists defined on $[0, \infty) \times \Omega$ which is $\mathfrak{T} \times \mathfrak{S}$-measurable and such that

$$\text{(20)} \qquad \begin{gathered} \int_0^\infty g^2(t) \, d\langle \mu, \mu \rangle_t < \infty \ (\text{mod } \mathsf{P}), \\ \mathsf{P}\text{-}\lim_{n\to\infty} \int_0^\infty (g(t) - \eta_n(t))^2 \, d\langle \mu, \mu \rangle_t = 0. \end{gathered}$$

The class of functions $g(t)$ obtained in such a manner is denoted by H_2 or $H_2\{\mathfrak{F}_t, \mu\}$. We introduce a topology in H_2: a sequence of function $\eta_n(t)$ in H_2 is said to be *convergent* to the limit $g(t, \omega)$ provided relation (20) is satisfied. Clearly, H_2 is a linear and a complete space, i.e., an arbitrary H_2-fundamental sequence converges to a certain limit in H_2. Moreover, $H_2 \supset \mathfrak{L}_2$; therefore $\mathfrak{L}_2$ and hence $\mathfrak{L}_0$ is everywhere dense in H_2. It is easy to verify that H_2 consists of all $\mathfrak{T}$-measurable functions $g(t) = g(t, \omega)$ for which the first of the conditions in (20) is satisfied. For example, an arbitrary process $g(t)$, continuous from the left and adopted to the current of σ-algebras $\{\mathfrak{F}_t, t \geq 0\}$ satisfying the first of the conditions (20) belongs to H_2. The same argument as in the case of the space $\mathfrak{L}_2\{\mathfrak{F}_t, \mu\}$ shows that if $\langle \mu, \mu \rangle_t = \int_0^t \varphi(s) \, d(s)$, an arbitrary process $g(t)$ adopted to the current $\{\mathfrak{F}_t, t \geq 0\}$ and satisfying the first of the conditions (20) belongs to H_2.

In what follows we shall restrict our attention to integration over processes $\mu(t)$ with continuous characteristics.

The subspace of the space $\mathcal{M}_2$ consisting of martingales with continuous characteristics is denoted by $\mathcal{M}_2^r$ (if $\mu(\cdot) \in \mathcal{M}_2^r$ then $\mu^2(t)$ is a regular submartingale).

Let

$$\mu(\cdot) \in \mathcal{M}_2^r, \qquad \eta(\cdot) \in H_2\{\mathfrak{F}_t, \mu\}, \qquad \eta_n(\cdot) \in \mathfrak{L}_2, \qquad n = 1, 2, \ldots,$$

and $\eta_n(\cdot)$ be convergent to $\eta(\cdot)$ in H_2. Set

$$\text{(21)} \qquad I_t(\eta) \underset{\text{Def}}{=} \int_0^t \eta(s) \mu(ds) \underset{\text{Def}}{=} \mathsf{P}\text{-}\lim \int_0^t \eta_n(s) \mu(ds).$$

It follows from inequality (18) that the limit in the right-hand side of relation (21) exists for the sequence $\eta_n \in \mathfrak{L}_2$ convergent to $\eta(\cdot)$ in H_2, and hence this limit depends only on $\eta(\cdot)$.

Equation (21) defines the value of $I_t(\eta)$ for each t only with probability 1. This fact can be utilized for defining the $I_t(\eta)$ in such a manner that its realizations belong to $\mathcal{D}$ with probability 1.

Indeed, let $\eta_n(t)\in\mathfrak{L}_0$ and let processes $\eta_n(t)$ be convergent to $\eta(t)$ in H_2. Inequality (18) implies the existence of a sequence of integers n_k such that $\mathsf{P}(A_k)<2^{-k}$ where

$$A_k=\left\{\sup_t\left|\int_0^t\eta_{n_k}\,d\mu-\int_0^t\eta_{n_{k+1}}\,d\mu\right|>\frac{1}{2^k}\right\}.$$

It follows from Borel–Cantelli's theorem that the series

$$\sum_{k=1}^{\infty}\sup_t\left|\int_0^t\eta_{n_k}\,d\mu-\int_0^t\eta_{n_{k+1}}\,d\mu\right|$$

is convergent with probability 1. Therefore the series

$$\int_0^t\eta_{n_0}\,d\mu+\sum_{k=1}^{\infty}\left(\int_0^t\eta_{n_{k+1}}\,d\mu-\int_0^t\eta_{n_k}\,d\mu\right)$$

converges with probability 1 uniformly in t ($t\geqslant 0$) and its sum is with probability 1 continuous from the right and possesses left-hand limits since the stochastic integrals $I_t(\eta_{n_k})$ possess this property. In the case when the martingale $\mu(\cdot)$ is continuous the same argument shows that the sum of the series under consideration is continuous for all $t\geqslant 0$. Consequently, for an arbitrary process $\eta(\cdot)\in H_2\{\mathfrak{F}_t,\mu\}$ one can define the process $I_t(\eta)$, $t\geqslant 0$, in such a manner that its realizations belong to $\mathscr{D}$ with probability 1, and in the case when the martingale $\mu(\cdot)$ is continuous the realizations would be continuous with probability 1; moreover, for any sequence $\eta_n(\cdot)\in\mathfrak{L}_2\{\mathfrak{F}_t,\mu\}$ convergent in the H_2-topology to $\eta(\cdot)$ and for any fixed $t>0$, relation (21) will be satisfied.

Definition. *A stochastic integral*

$$I_t(\eta)=\int_0^t\eta(s)\mu(ds),$$

where $\mu(\cdot)\in\mathcal{M}_2^r$ *and* $\eta(\cdot)\in H_2\{\mathfrak{F}_t,\mu\}$ *is a random process satisfying for each* $t>0$ *relation (21) with sample functions belonging to* $\mathscr{D}$.

In the case when μ is a continuous martingale it is assumed that the realizations of the process $I_t(\eta)$ are continuous for all $t\geqslant 0$ with probability 1.

Theorem 2. *If* $\eta(\cdot)\in H_2$ *and* $\mu(\cdot)\in\mathcal{M}_2^r$ *then the stochastic integral* $I_t(\eta)$ *exists and possesses the following properties*:

1) *The process* $I_t(\eta)$ *is adopted to the current of* σ*-algebras* $\{\mathfrak{F}_t, t\geqslant 0\}$ *and if* $\eta\in\mathfrak{L}_2$ *then the new definition of the integral coincides with the one presented above.*

2) $I_t(\eta)$ *is a linear functional on* η.

3) *Inequality* (18) *is satisfied for arbitrary functions* $\eta(\cdot)\in H_2$.

4) *Sample functions of the process* $I_t(\eta)$ *are bounded on the half-line* $[0,\infty)$ *with probability* 1.

5) *If τ is an $\mathfrak{F}_t$-random time and $\eta_1 = \eta_2$ for $t < \tau$, then $I_t(\eta_1) = I_t(\eta_2)$ for all $t \leqslant \tau$ with probability* 1.

6) *The process $I_t(\eta)$ is a local square integrable martingale with respect to the current σ-algebras $\{\mathfrak{F}_t, t \geqslant 0\}$ with continuous characteristic*

$$\langle I(\eta), I(\eta)\rangle_t = \int_0^t \eta^2(s)\, d\langle\mu, \mu\rangle_s.$$

7) *If*

$$\eta \in H_2\{\mu_1, \mathfrak{F}_t\} \cap H_2\{\mu_2, \mathfrak{F}_t\}, \qquad \mu_i \in \mathcal{M}_2^c\{\mathfrak{F}_t\},$$

$i = 1, 2$, *then*

$$\int_0^t \eta\, d(\mu_1 + \mu_2) = \int_0^t \eta\, d\mu_1 + \int_0^t \eta\, d\mu_2.$$

8) *If τ is an $\mathfrak{F}_t$-random time, $\mu_\tau(t) = \mu(t \wedge \tau)$, then*

$$\int_0^t \eta_\tau(s)\mu_\tau(ds) = \int_0^{t\wedge\tau} \eta(s)\mu(ds),$$

where $\eta_\tau(t) = \eta(t)$ for $t < \tau$ and $\eta_\tau(t) = 0$ for $t \geqslant \tau$.

Proof. Assertion 1) follows directly from relation (21). Assertion 2) is obvious, assertion 3) follows easily from Lemma 2, and 8) follows from Lemma 1 by means of a limit transition.

Since

$$\mathsf{P}\Big\{\sup_{t\geqslant 0} |I_t(\eta)| > C\Big\} \leqslant \frac{N}{C^2} + \mathsf{P}\{\textstyle\int_0^\infty \eta^2\, d\langle\mu, \mu\rangle > N\},$$

$\mathsf{P}\{\sup_{t\geqslant 0} |I_t(\eta)| = \infty\} = 0$. Thus assertion 4) is verified.

We shall now prove 5). First let us assume that $\eta_i(\cdot) \in \mathfrak{L}_2$. Then

$$\mathsf{E}|I_{t\wedge\tau}(\eta_1) - I_{t\wedge\tau}(\eta_2)|^2 = \mathsf{E} \int_0^{t\wedge\tau} [\eta_1(s) - \eta_2(s)]^2\, d\langle\mu, \mu\rangle_s = 0,$$

i.e., $I_{t\wedge\tau}(\eta_1) - I_{t\wedge\tau}(\eta_2) = 0$ for *each* t with probability 1. However, from the continuity from the right of the sample functions of the process $I_t(\eta)$ it follows that this equality is satisfied with probability 1 for *all* $t \geqslant 0$ as well. Consider now the general case.

Let $\tau_N = \inf\{t: \min_{i=1,2} \int_0^t \eta_i^2\, d\langle\mu, \mu\rangle \geqslant N\}$ if the set in the braces is nonempty and $\tau_N = \infty$ otherwise; $\eta_i^N(t) = \eta_i(t)$ for $t < \tau$ and $\eta_i^N(t) = 0$ for $t \geqslant \tau$. Then $\eta_i^N(\cdot) \in \mathfrak{L}_2$ and $\tau_N \to \infty$ as $N \to \infty$ with probability 1. Moreover,

$$\int_0^\infty (\eta_i - \eta_i^N)^2\, d\langle\mu, \mu\rangle = \int_{\tau_N}^\infty \eta_i^2\, d\langle\mu, \mu\rangle \to 0$$

with probability 1. Therefore, as it follows from inequality (18), $\sup [I_t(\eta_i) - I_t(\eta_i^N)] \to 0$ in probability as $N \to \infty$. As it was shown above, $I_t(\eta_1^N) = I_t(\eta_2^N)$ for all $t \leqslant \tau$ (mod P). Approaching the limit in this equality as $N \to \infty$ we obtain the required result.

All that remains is to prove assertion 6). Let

$$\tau_N = \inf\{t: \int_0^t \eta^2\, d\langle\mu,\mu\rangle > N\},$$

and $\eta^N(t)$ be defined as above. Then $\eta^N(t) \in \mathfrak{L}_2$. Assertion 5) implies that $I_t(\eta) = I_t(\eta^N)$ for $t \leq \tau$ and it follows from equality

$$\mathsf{E}|I_{t\vee\tau_N}(\eta^N) - I_{\tau_N}(\eta^N)|^2 = \mathsf{E}\int_{\tau_N}^{t\vee\tau_N} (\eta^N)^2\, d\langle\mu,\mu\rangle = 0$$

that $I_t(\eta^N) = I_{\tau_N}(\eta^N)$ for $t > \tau_N$. Thus

$$I_{t\wedge\tau_N}(\eta) = I_t(\eta^N) \tag{22}$$

for all $t > 0$.

Since the integral $\int_0^\infty \eta^2\, d\langle\mu,\mu\rangle$ is finite (mod P), $I_t(\eta^N) \in \mathcal{M}_2$ and $\tau_N \to \infty$ as $N \to \infty$. This proves that $I_t(\eta)$ is a local square integrable martingale, that the above defined sequence τ_N reduces $I_t(\eta)$, and that

$$\langle I(\eta), I(\eta)\rangle_{t\wedge\tau_N} = \langle I(\eta^N), I(\eta^N)\rangle_t = \int_0^{t\wedge\tau_N} \eta^2\, d\langle\mu,\mu\rangle. \quad \square$$

Corollary 1. *If* $\eta_n \in H_2$ *and*

$$\text{P-lim} \int_0^\infty (\eta - \eta_n)^2\, d\langle\mu,\mu\rangle = 0,$$

then

$$\mathsf{P}\left\{\sup_t \left|\int_0^t \eta\, d\mu - \int_0^t \eta_n\, d\mu\right| > \varepsilon\right\} \to 0 \quad \text{as } n \to \infty. \tag{23}$$

Corollary 2. *If* $\eta_i(\cdot) \in H_2$, $i = 1, 2$, *then the joint characteristic of the processes* $I_t(\eta_i)$ *is given by*

$$\langle I(\eta_1), I(\eta_2)\rangle_t = \int_0^t \eta_1(s)\eta_2(s)\, d\langle\mu,\mu\rangle_s.$$

Corollary 3. *Assume that* $\mu_i \in \mathcal{M}_2^c$ $(i = 1, 2)$, $\eta_i \in H_2\{\mathfrak{F}_t, \mu_1\} \cap H_2\{\mathfrak{F}_t, \mu_2\}$, *and let* $I_t^{\mu_i}(\eta) = \int_0^t \eta\, d\mu_i$. *Then*

$$\langle I^{\mu_i}(\eta_1), I^{\mu_i}(\eta_2)\rangle_t = \int_0^t \eta_1(s)\eta_2(s)\, d\langle\mu_1,\mu_2\rangle_s. \tag{24}$$

Integration over local square integrable martingales. We now take yet another step on the road toward extending the notion of a stochastic integral. Namely, we shall assume that μ is a local square integrable martingale with a continuous characteristic. The class of these processes will be denoted by $l\mathcal{M}_2^c$ or $l\mathcal{M}_2^c\{\mathfrak{F}_t, t \geq 0\}$. The definition of the space $H_2\{\mathfrak{F}_t, \mu\}$ associated with the process $\mu \in l\mathcal{M}_2^c$, more precisely with its characteristic, does not require any modification. Let τ_n be a sequence of random times completely reducing μ and let $\mu_n(t) = \mu(t \wedge \tau_n)$.

As it is known from the above (Section 1, Theorem 19) $\langle\mu_n, \mu_n\rangle_t = \langle\mu, \mu\rangle_{t\wedge\tau_n}$. Set

$$I_n(t) = \int_0^t \eta(s \wedge \tau_n)\mu_n(ds).$$

Assertion 5) of Theorem 2 implies that for $n' > n$

$$I_{n'}(t) - I_n(t) = \int_{t\wedge\tau_n}^{t\wedge\tau_{n'}} \eta(s)\mu(ds) = 0 \quad \text{as } t < \tau_n.$$

Thus, the integrals $I_n(t)$ become identical with probability 1 starting with some $n = n_0 = n_0(\omega)$. Set

$$I(t) = \int_0^t \eta(s)\mu(ds) \underset{\text{Def}}{=} \lim \int_0^t \eta(s \wedge \tau_n)\mu_n(ds). \tag{25}$$

It is easy to verify that $I(t)$ does not depend on the choice of sequence τ_n.

Definition. The limit (25) is called a *stochastic integral* (over a local square integrable martingale μ).

Lemma 3. *A stochastic integral over process $\mu(\cdot) \in l\mathcal{M}_2^r$ possesses all the properties stipulated in Theorem* 2.

Proof. Indeed, properties 1), 2), 3), 5), 7), and 8) are valid for $I_n(t)$, and starting with some $n = n(\omega)$ $I(t)$ coincides with $I_n(t)$ for each ω. Therefore $I(t)$ possesses all these properties as well. □

Inequality (18) can be easily obtained if one applies it first to the integrable martingale $\mu_n(t) = \mu(t \wedge \tau_n)$ and then approaches the limit as $n \to \infty$. To prove that $I(t)$ is a local square integrable martingale, introduce random times $\sigma_n = \inf\{t: \int_0^t \eta^2 \, d\langle\mu, \mu\rangle \geq n\}$ and let τ_n be a sequence of random times as before, completely reducing μ. Set $\tau'_n = \sigma_n \wedge \tau_n$. It follows from Theorem 2 that $I(t \wedge \tau'_n)$ is a square integrable martingale and obviously $\tau'_n \to \infty$ with probability 1.

Theorem 3. *Let*

$$\mu(\cdot) \in l\mathcal{M}_2^r, \qquad \eta(\cdot) \in H_2(\mathfrak{F}_t, \mu), \qquad \lambda(t) = \int_0^t \eta(s)\mu(ds).$$

Then $\lambda(\cdot) \in l\mathcal{M}_2^r$ and

$$\int_0^t \zeta(s)\lambda(ds) = \int_0^t \eta(s)\zeta(s)\mu(ds), \tag{26}$$

provided $\zeta(\cdot) \in H_2(\mathfrak{F}_t, \lambda)$.

Proof. For the case of piecewise constant functions $\zeta(\cdot)$ formula (26) is trivially verified to be valid. If, however, $\zeta(\cdot)$ is an arbitrary process in $H_2(\mathfrak{F}_t, \lambda)$ and

$$\text{P-lim} \int_0^\infty (\zeta - \zeta_n)^2 \, d\langle\lambda, \lambda\rangle = 0,$$

where $\zeta_n(\cdot)\in\mathfrak{L}_0$, then

$$\int_0^t \zeta\, d\lambda = \text{P-lim} \int_0^t \zeta_n\, d\lambda = \text{P-lim} \int_0^t \eta(s)\zeta_n(s)\mu(ds).$$

On the other hand,

$$\int_0^\infty (\eta(s)\zeta_n(s)-\eta(s)\zeta(s))^2\, d\langle\mu,\mu\rangle_s = \int_0^\infty (\zeta_n(s)-\zeta(s))^2\eta(s)^2\, d\langle\mu,\mu\rangle_s$$
$$= \int_0^\infty (\zeta_n(s)-\zeta(s))^2\, d\langle\lambda,\lambda\rangle_s \to 0$$

in probability. Hence,

$$\text{P-lim} \int_0^t \eta(s)\zeta_n(s)\mu(ds) = \int_0^t \eta(s)\zeta(s)\mu(ds).$$

Thus it is proved that equality (26) is valid for each t with probability 1. Since the processes appearing in both sides of this equality are continuous from the right, it follows that this equality is valid with probability 1 for all values of t. □

Vector-valued stochastic integrals. Consider a vector-valued process $\mu(t) = (\mu^1(t), \mu^2(t), \ldots, \mu^m(t))$ with components $\mu^k(t)$ belonging to $\mathcal{M}_2$ ($l\mathcal{M}_2, \ldots$) ($k = 1, \ldots, m$). In this case we shall agree to write $\mu(t)\in\mathcal{M}_2$ ($l\mathcal{M}_2, \ldots$), and refer to $\mu(t)$ as a *vector-valued square integrable (local square integrable) martingale.*

Let $\eta(t)$ be a scalar process and $\eta(\cdot)\in\bigcap_{k=1}^m H_2(\mathfrak{F}_t, \mu_k)$. The integral

$$I_t = \int_0^t \eta(t)\mu(dt)$$

represents a vector-valued process with components $\int_0^t \eta(t)\mu^k(t)$.

We introduce matrix $\langle\mu,\mu\rangle_t$ with elements $\langle\mu^k,\mu^j\rangle$ and call it the *matrix characteristic of a vector-valued process* $\mu(t)$. Observe that matrix $\Delta\langle\mu,\mu\rangle_t = \langle\mu,\mu\rangle_{t+\Delta t} - \langle\mu,\mu\rangle_t$ is nonnegatively definite.

Indeed for any numbers $z_1, \ldots, z_m$ we have

$$\sum_{k,j=1}^m z_k z_j \Delta\langle\mu^k,\mu^j\rangle_t = \Delta\left\langle \sum_{k=1}^m z^k\mu^k, \sum_{k=1}^m z_k\mu^k\right\rangle_t \geq 0.$$

The class of processes $\mu(t)\in\mathcal{M}_2$ ($l\mathcal{M}_2$) with continuous characteristics for their components is denoted by $\mathcal{M}_2^c$ ($l\mathcal{M}_2^c$). Since the continuity of characteristics of two local martingales implies continuity of their joint characteristic (Section 1, corollary to Theorem 20) the functions $\langle\mu^k,\mu^j\rangle_t$ are continuous with probability 1 provided $\mu(t)\in l\mathcal{M}_2^c$.

We note that the matrix characteristic of a vector-valued square integrable martingale $\mu(t)$ with independent components is of the form

$$\langle\mu,\mu\rangle_t = \mathsf{E}\mu(t)\mu^*(t)$$

and is a nonrandom function. In particular, an m-dimensional Wiener process $w(t)$ is a vector-valued square integrable martingale with independent increments, continuous sample functions (with probability 1), and independent components. Its matrix characteristic is $\langle w, w\rangle_t = \mathbf{I}t$, where $\mathbf{I}$ is the unit matrix.

Stochastic integrals over martingale measures. In the preceding item, stochastic martingales were discussed in which integration was carried out over a real variable t. In what follows integrals over random measures in multidimensional spaces will be required. Here one should distinguish between the time and space variables which play different roles in the definition of an integral. Otherwise, the construction below is analogous to the preceding ones.

Let $\{\mathfrak{F}_t, t \geq 0\}$ be a fixed current of σ-algebras in the basic probability space $\{\Omega, \mathfrak{S}, \mathsf{P}\}$, $\{U, \mathfrak{U}\}$ be a measurable space and $\mathfrak{U}_0$ be a semiring of sets generating σ-algebra $\mathfrak{U}$ ($\mathfrak{U} = \sigma\{\mathfrak{U}_0\}$).

Definition. *A martingale measure* $\mu(t, A)$, $t \in [0, \infty)$, $A \in \mathfrak{U}_0$, is a random function possessing the following properties:

1) For a fixed $A \in \mathfrak{U}_0$, $\mu(t, A)$ is a square integrable $\mathfrak{F}_t$-martingale with sample functions belonging to $\mathscr{D}$ and for a fixed t, $\mu(t, \cdot)$ is an additive function on $\mathfrak{U}_0$:

$$\mu(t, A \cup B) = \mu(t, A) + \mu(t, B), \qquad (A \cap B = \varnothing, A, B \in \mathfrak{U}_0);$$

2) If $A \cap B = 0$, then the product $\mu(t, A)\mu(t, B)$ is a martingale, i.e.,

$$\mathsf{E}\{\mu(\Delta t, A)\mu(\Delta t, B) | \mathfrak{F}_t\} = 0,$$

where

$$\mu(\Delta t, C) = \mu(t + \Delta t, C) - \mu(t, C).$$

Denote by $\pi(t, A)$ the characteristic of the martingale $\mu(t, A)$. It follows from 2) that if $A \cap B = \varnothing$, then the characteristic of martingale $\mu(t, A \cup B) = \mu(t, A) + \mu(t, B)$ is equal to $\pi(t, A) + \pi(t, B)$. Thus $\pi(t, A \cup B) = \pi(t, A) + \pi(t, B)$ if $A \cap B = \varnothing$, i.e., $\pi(t, A)$ is an additive function on $\mathfrak{U}_0$ for each t. We shall also assume that the following condition is satisfied:

3) the characteristic $\pi(t, A)$ of martingale $\mu(t, A)$ can be defined in such a manner that it would be a measure on $\mathfrak{U}$ (for each t with probability 1), and for fixed $A \in \mathfrak{U}_0$ it would be a continuous monotone nondecreasing function of the argument t.

It follows from 2) that the joint characteristic of martingales $\mu(t, A)$ and $\mu(t, B)$ for any A and B belonging to $\mathfrak{U}_0$ is equal to $\pi(t, A \cap B)$ and

$$\mathsf{E}\{\mu(\Delta t, A)\mu(\Delta t, B) | \mathfrak{F}_t\} = \mathsf{E}\{\pi(\Delta t, A \cap B) | \mathfrak{F}_t\}.$$

Definition. A random function $\mu(t, A)$ is called a *local martingale measure* if a monotone nondecreasing sequence of $\mathfrak{F}_t$-random times τ_n exists such that $\lim \tau_n = \infty$ and $\mu(t \wedge \tau_n, A)$, $n = 1, 2, \ldots$, is a martingale measure (with respect to the current $\{\mathfrak{F}_{t \wedge \tau_n}, t \geq 0\}$).

It is easy to verify (in the same manner as was done in Theorem 19 in Section 1) that a local martingale measure possesses a unique characteristic $\pi(t, A)$ which is (with probability 1) a continuous monotonically nondecreasing function of argument t and which is a measure on $\mathfrak{U}$ for a fixed t.

Denote by $\mathfrak{L}_0\{\mathfrak{T}_0\times\mathfrak{U}_0\}$ the class of all simple and bounded functions with probability 1 on the semiring of set of the form $\Delta\times A$, $\Delta=(a, b]$, $A\in\mathfrak{U}_0$, adopted to the current $\{\mathfrak{F}_t, t\geq 0\}$. Thus $\varphi\in\mathfrak{L}_0\{\mathfrak{T}_0\times\mathfrak{U}_0\}$ if and only if

$$\varphi(t, u)=\sum_{k=1}^{n}\gamma_k\chi_{\Delta_k\times A_k}(t, u),$$

where

$$\Delta_k=(t_{k-1}, t_k],\qquad 0\leq t_1<\cdots<t_n,\qquad A_k\in\mathfrak{U}_0,$$

and γ_k is an $\mathfrak{F}_{t_{k-1}}$-measurable random variable bounded with probability 1, $|\gamma_k|\leq C$, $k=1,\ldots,n$ (mod P), and C is an absolute (nonrandom) constant.

Set

$$I(\varphi)=\int_0^\infty\int_U\varphi(s, u)\mu(ds, du)\underset{\text{Def}}{=}\sum_{k=1}^{n}\gamma_k\mu(\Delta_k, A_k),$$

$$I_t(\varphi)=\int_0^t\int_U\varphi(s, u)\mu(ds, du)$$

$$\underset{\text{Def}}{=}\int_0^\infty\int_U\chi_{(0,t]}(s)\varphi(s, u)\mu(ds, du),\qquad t>0,$$

where $\chi_{(0,t]}(s)$ is the indicator of the half-interval $(0, t]$ and call $I(\varphi)$ and $I_t(\varphi)$ *stochastic integrals with respect to a martingale (local martingale) measure.*

We note several properties of the integrals just defined. Let μ be a martingale measure, φ, φ_1, and $\varphi_2\in\mathfrak{L}_0(\mathfrak{T}_0\times\mathfrak{U}_0\}$.

a) *If γ_i $(i=1, 2)$ are $\mathfrak{F}_0$-measurable random variables bounded with probability 1, then*

$$I(\gamma_1\varphi_1+\gamma_2\varphi_2)=\gamma_1 I(\varphi_1)+\gamma_2 I(\varphi_2).$$

b) $$\mathsf{E}\{I(\varphi)\mid\mathfrak{F}_0\}=0,$$

$$\mathsf{E}\{I(\varphi_1)I(\varphi_2)\mid\mathfrak{F}_0\}=\mathsf{E}\{\int_0^\infty\int_U\varphi_1(s, u)\varphi_2(s, u)\pi(ds, du)\mid\mathfrak{F}_0\}.$$

In particular,

$$\mathsf{E}\{[I(\varphi_1)-I(\varphi_2)]^2\mid\mathfrak{F}_0\}=\mathsf{E}\{\int_0^\infty\int_U(\varphi_1(s, u)-\varphi_2(s, u))^2\pi(ds, du)\mid\mathfrak{F}_0\}.$$

c) *Sample functions of the process $I_t(\varphi)$ are continuous from the right and possess left-hand limits for each $t>0$.*

d) *The process $I_t(\varphi)$ is a square integrable martingale with respect to the current of σ-algebras $\{\mathfrak{F}_t, t \geq 0\}$ and*

$$\mathsf{E}\{(\Delta I_t(\varphi))^2 \mid \mathfrak{F}_t\} = \mathsf{E}\{\int_t^{t+\Delta t} \int_U \varphi^2(s, u)\pi(ds, du) \mid \mathfrak{F}_t\},$$

where $\Delta I_t(\varphi) = I_{t+\Delta t}(\varphi) - I_t(\varphi)$. The joint characteristic of martingales $I_t(\varphi_1)$ and $I_t(\varphi_2)$ is continuous and is equal to

$$(27) \qquad \langle I(\varphi_1), I(\varphi_2)\rangle_t = \int_0^t \int_U \varphi_1(s, u)\varphi_2(s, u)\pi(ds, du).$$

e) *Let $\mu_i(t, A)$, $i = 1, 2$, be two martingale measures (with respect to the same current of σ-algebras $\{\mathfrak{F}_t, t \geq 0\}$) and let $\varphi_2 \in \mathfrak{L}_0$, $i = 1, 2$. Set*

$$\pi^*(t, A, B) = \langle \mu_1(\cdot, A), \mu_2(\cdot, B)\rangle_t,$$

$$I_t^i(\varphi_i) = \int_0^t \int_U \varphi_i(s, u)\mu_i(ds, du).$$

Then $\pi^*(t, A, B)$ admits representation as a difference of two measures on $\mathfrak{U} \times \mathfrak{U}$ (mod P), finite for $A \in \mathfrak{U}_0$, $B \in \mathfrak{U}_0$, and

$$(28) \qquad \langle I^1(\varphi_1), I^2(\varphi_2)\rangle_t = \int_0^t \int_U \int_U \varphi_1(s, u)\varphi_2(s, v)\pi^*(dt, du, dv).$$

The function $\pi^*(t, A, B)$ is called the *joint characteristic of two martingale measures.*

f) *For any $N > 0$, $\varepsilon > 0$,*

$$(29) \qquad \mathsf{P}\left\{\sup_{0 \leq t < \infty} \left|\int_0^t \int_U \varphi(s, u)\mu(ds, du)\right| > N\right\} \leq \frac{N}{\varepsilon^2} + \mathsf{P}\{\int_0^\infty \int_U \varphi^2(s, u)\pi(ds, du) > N\}.$$

These assertions are proved in the same manner as in the case of integration of piecewise constant functions over a martingale.

We now extend the definition of a stochastic integral to a wider class of random functions $\varphi(t, u)$.

First we introduce a semiring $\mathcal{R}^*$ of all the sets of the form $(a, b] \times A \times S$ on the space $[0, \infty) \times U \times \Omega$, where $0 \leq a < b \leq \infty$, $A \in \mathfrak{U}_0$, $S \in \mathfrak{F}_a$. On this semiring we define the measure P^* by setting

$$\mathsf{P}^*((a, b] \times A \times S) = \mathsf{E}\chi_S[\pi(b, A) - \pi(a, A)],$$

where χ_S is the indicator of the set S. This measure admits a unique continuation to the σ-algebra $\mathfrak{T}^*$ generated by the semiring $\mathcal{R}^*$ as well as to its completion $\bar{\mathfrak{T}}^*$. Moreover, for an arbitrary nonnegative $\bar{\mathfrak{T}}^*$-measurable function $\varphi(t, u) = \varphi(t, u, \omega)$

$$\int_0^\infty \int_U \int_\Omega \varphi(t, u, w)\mathsf{P}^*(dt, du, dw) = \mathsf{E}\int_0^\infty \int_U \varphi(t, u)\pi(dt, du).$$

Now introduce the space $H_2^\pi = H_2(\mathfrak{F}_t, \mu)$ of random functions $\varphi(t, u) = \varphi(t, u, \omega)$ in the same manner as the space H_2 was introduced above. Namely, H_2^π consists of all $\mathfrak{T}^*$-measurable functions $\varphi(t, u)$ such that

$$(30) \qquad \int_0^\infty \int_U |\varphi(t, u)|^2 \pi(ds, du) < \infty \quad (\text{mod } \mathsf{P}).$$

For each function $\varphi(t, u) \in H_2^\pi$ there exists a sequence $\varphi_n(t, u) \in \mathfrak{L}_0$ such that

$$(31) \qquad \mathsf{P}\text{-lim} \int_0^\infty \int_U |\varphi(t, u) - \varphi_n(t, u)|^2 \pi(dt, du) = 0.$$

It is easy to verify, as was done above, that an arbitrary function $\varphi(t, u)$ continuous from the left in t for all u and for P-almost all ω adopted to the current $\{\mathfrak{F}_t, t \geq 0\}$ is $\mathfrak{T}^*$-measurable. If, however, the function $\pi(t, A)$ admits representation

$$\pi(t, A) = \int_0^t \int_A \varphi(s, u)\, ds\, m(du),$$

where $m(\cdot)$ is a measure on $\mathfrak{U}$, then an arbitrary random function $\varphi(t, u)$ adopted to the current $\{\mathfrak{F}_t, t \geq 0\}$ and satisfying condition (30) belongs to H_2^π. An arbitrary sequence of random functions $\varphi(t, u) \in H_2^\pi$ is called convergent in H_2^π to the function $\varphi(t, u)$ if relation (31) is fulfilled.

We also introduce the subspace $\mathfrak{L}_2^\pi$ of the space H consisting of functions $\varphi(t, u) \in H_2^\pi$ satisfying the additional condition

$$\mathsf{E} \int_0^\infty \int_U |\varphi(s, u)|^2 \pi(ds, du) < \infty.$$

We say that a sequence φ_n, $n = 1, 2, \ldots$, $\varphi_n \in \mathfrak{L}_2^\pi$ converges in $\mathfrak{L}_2^\pi$ to the limit φ if for all $t > 0$

$$\mathsf{E} \int_0^t \int_U |\varphi(s, u) - \varphi_n(s, u)|^2 \pi(ds, du) \to 0 \quad \text{as } n \to \infty.$$

Let $\varphi(t, u) \in H_2^\pi$ and $\varphi_n(t, u)$ converge in H_2^π to $\varphi(t, u)$. Then the sequence $\varphi_n(t, u)$ is a Cauchy sequence in H_2^π, i.e.,

$$\mathsf{P}\text{-}\lim_{n_1, n_2 \to \infty} \int_0^\infty \int_U |\varphi_{n_1}(s, u) - \varphi_{n_2}(s, u)|^2 \pi(ds, du) = 0.$$

It follows from the property f) of a stochastic integral that a sequence of random variables $I_t(\varphi)$ is a Cauchy sequence in the sense of convergence in probability for each $t > 0$. Moreover, since $\mathcal{M}_2$ is closed one can define P-lim $I_t(\varphi_n) = I_t(\varphi)$ in such a manner that the sample functions of process $I_t(\varphi)$ will belong to $\mathcal{D}[0, \infty)$ with probability 1, and if martingales $\mu(t, A)$ are continuous with probability 1 for each A, then the realizations of the process $I_t(\varphi)$ will be continuous for all $t > 0$ (mod P).

Definition. A random process

$$I_t(\varphi) \underset{\text{Def}}{=} \int_0^t \int_U \varphi(s, u) \mu(ds, du) \underset{\text{Def}}{=} \mathsf{P}\text{-lim} \int_0^t \int_U \varphi_n(s, u) \mu(ds, du),$$

with sample functions belonging to $\mathscr{D}[0, \infty)$ with probability 1 is called *the stochastic integral of a function* $\varphi(t, u) \in H_2^\pi$ *over a martingale measure* $\mu(t, A)$.

Here $\varphi_n(t, u)$ is an arbitrary sequence of functions belonging to $\mathfrak{L}_0(\mathfrak{T}_0 \times \mathfrak{U}_0)$ satisfying relation (30).

It follows from the discussion above that if $\mu(t, A) \in \mathcal{M}_2^c$ for each $A \in \mathfrak{U}_0$, then the sample functions of the process $I_t(\varphi)$ belong to $\mathscr{C}[0, \infty)$ with probability 1.

The following theorem is proved in the same manner as analogous theorems in the preceding subsections.

Theorem 4. *If* $\mu(t, A)$ *is a martingale measure with characteristic* $\pi(t, A)$ *and* $\varphi(t, u) \in H_2^\pi$, *then the following hold*:

a) *The stochastic integral* $I_t(\varphi)$ *exists and is a local square integrable martingale with continuous characteristic*

$$\langle I(\varphi), I(\varphi)\rangle_t = \int_0^t \int_U \varphi^2(s, u)\pi(ds, du).$$

b) *If* c_i $(i = 1, 2)$ *are* $\mathfrak{F}_0$*-measurable random variables,* $\varphi_i \in H_2^\pi$, *then*

$$I_t(c_1\varphi_1 + c_2\varphi_2) = c_1 I_t(\varphi_1) + c_2 I_t(\varphi_2).$$

c) *If* $\mu_i(t, A)$ $(i = 1, 2)$ *are two martingale measures with respect to the same current of* σ*-algebras,* $I_t^i(\varphi)$ *is an integral over measure* μ_i *and* $\varphi_i \in H_2^\pi$, *then the joint characteristic of processes* $I^1(\varphi_1)$ *and* $I^2(\varphi_2)$ *is given by formula* (28), *where* $\pi^*(t, A, B)$ *is the joint characteristic of martingales* $\mu_1(t, A)$ *and* $\mu_2(t, B)$.

d) *The inequality* (29) *remains valid for an arbitrary function* $\varphi \in H_2^\pi$.

e) *For any* $\mathfrak{F}_t$*-random time* τ

$$\int_0^t \int_U \varphi(s, u)\mu_\tau(ds, du) = \int_0^t \int_U \varphi_\tau(s, u)\mu(ds, du), \tag{32}$$

where $\mu_\tau(t, A) = \mu(t \wedge \tau, A)$, $\varphi_\tau(t, u) = \varphi(t, u)$ *for* $t < \tau$ *and* $\varphi_\tau(t, u) = 0$ *for* $t \geqslant \tau$.

Now let $\mu(t, A)$ be a local martingale measure, τ_n be a monotonically nondecreasing sequence of random times such that $\lim \tau_n = \infty$ (mod P), and measures $\mu_n(t, A) = \mu(t \wedge \tau_n, A)$ be martingale measures $(n = 1, 2, \ldots)$. Denote by $\pi(t, A)$ the characteristic of the measure $\mu(t, A)$ and set $\pi_n(t, A) = \pi(t \wedge \tau_n, A)$.

Definition. *A stochastic integral* $I_t(\varphi)$ *over a local martingale measure* $\mu(t, A)$ is a process with sample functions belonging with probability 1 to $\mathscr{D}[0, \infty)$ defined by the relationship

$$I_t(\varphi) \underset{\text{Def}}{=} \int_0^t \int_U \varphi(s, u)\mu(ds, du) \underset{\text{Def}}{=} \text{P-lim} \int_0^t \int_U \varphi(s, u)\mu_n(ds, du).$$

Theorem 5. *All the assertions of Theorem 4 are valid for stochastic integrals over a local martingale measure.*

The proof of this theorem is omitted since it is analogous to the proof of Lemma 3 above.

§3. Itô's Formula

In this section analogs and corollaries of the chain rule of differentiation of composite functions are studied in the case when differentiation is interpreted as the inverse operation of stochastic integration. The results obtained are of importance in what follows.

Unless stated otherwise, the stochastic processes are assumed here to be defined on a fixed probability space $\{\Omega, \mathfrak{S}, \mathsf{P}\}$ and are adopted to a given current of σ-algebras $\{\mathfrak{F}_t, t \geq 0\}$, $\mathfrak{F}_t \subset \mathfrak{S}$.

Itô's formula for continuous processes. Denote by $\mathscr{V}$ or $\mathscr{V}(\mathfrak{F}_t)$ the class of random processes $\alpha(t)$, $t \geq 0$, adopted to $\{\mathfrak{F}_t, t \geq 0\}$ representable as a difference of two monotonically nondecreasing continuous from the right processes and by $\mathscr{V}^c = \mathscr{V}^c(\mathfrak{F}_t)$ the subclass of $\mathscr{V}$ consisting of processes representable as a difference of two monotonically nondecreasing processes with sample functions continuous for each $t \geq 0$ with probability 1.

If $\gamma(t)$, $t > 0$, is an arbitrary process progressively measurable with respect to $\{\mathfrak{F}_t, t \geq 0\}$ with sample functions which are bounded with probability 1 on every finite interval, and $\alpha(t) \in \mathscr{V}$, then the integral $\int_0^t (\gamma(s)\, d\alpha(s)$ is defined with probability 1 for all $t > 0$ as an ordinary Lebesgue–Stieltjes integral of sample functions of processes $\gamma(t)$ and $\alpha(t)$. Moreover, the process $\zeta(t) = \int_0^t \gamma(s)\, d\alpha(s)$ is adopted to the current of σ-algebras $\mathfrak{F}_t$ and the sample functions of this process are continuous with probability 1 at each point t at which the functions $\alpha(t)$ is continuous, are continuous from the right for any t, and are of the bounded variation on an arbitrary interval $[0, t]$. Let

$$\xi(t) = \xi_0 + \alpha(t) + \mu(t),$$

where ξ_0 is an $\mathfrak{F}_0$-measurable random variable, $\alpha(t) \in \mathscr{V}$, and $\mu(t) \in l\mathscr{M}_2$. Set, by definition,

$$\int_0^t \gamma(s)\, d\xi(s) \underset{\text{Def}}{=} \int_0^t \gamma(s)\, d\alpha(s) + \int_0^t \gamma(s)\, d\mu(s),$$

provided both integrals on the right-hand side of this equality exist.

If $\xi(t)$ is a process with values in R^m, $\xi(t) = \{\xi^1(t), \ldots, \xi^m(t)\}$, $\gamma(t)$ is a scalar random process, then the integral $\int_0^t \gamma(s)\, d\xi(s)$ is interpreted as a vector process with components $\int_0^t \gamma(s)\, d\xi^k(s)$, $k = 1, \ldots, m$. Set

$$\xi^k(t) = \alpha^k(t) + \mu^k(t), \qquad \alpha(t) = (\alpha^1(t), \ldots, \alpha^m(t)),$$
$$\mu(t) = (\mu^1(t), \ldots, \mu^m(t)).$$

Denote by λ a subdivision of a fixed interval $[0, t]$ with subdividing points $t_0 = 0 < t_1 < \cdots < t_n = t$ and let

$$|\lambda| = \max_{1 \leq r \leq n} (t_r - t_{r-1}),$$

$$\Delta\xi_r = \xi(t_r) - \xi(t_{r-1}), \qquad \Delta\langle\mu^k, \mu^j\rangle_r = \langle\mu^k, \mu^j\rangle_{t_r} - \langle\mu^k, \mu^j\rangle_{t_{r-1}}.$$

Lemma 1. *Let $\varphi(t)$ be a continuous process adopted to $\{\mathfrak{F}_t, t \geq 0\}$ and let $\mu^k(t) \in l\mathcal{M}^c$. Then*

(1)
$$\text{P-lim} \sum_{r=1}^{n} \varphi(t_{r-1}) \Delta\xi_r = \int_0^t \varphi(s)\, d\xi(s),$$

(2)
$$\text{P-lim} \sum_{r=1}^{n} \varphi(t_{r-1}) \Delta\xi_r^k \Delta\xi_r^j = \int_0^t \varphi(s)\, d\langle\mu^k, \mu^j\rangle_s.$$

Proof. It is sufficient to prove formula (1) for a one-dimensional process $\xi(t) = \mu(t) \in l\mathcal{M}^c$. Recall that $l\mathcal{M}^c = l\mathcal{M}_2^c$. Then

$$\int_0^t [\varphi(s) - \varphi_\lambda(s)]^2\, d\langle\mu, \mu\rangle_s \to 0 \quad \text{as } |\lambda| \to 0.$$

Since $\sum_1^n \varphi(t_{r-1}) \Delta\xi_r = \int_0^t \varphi_\lambda(s)\, d\xi(s)$, equation (1) follows from Lemma 2 in Section 2. We now proceed to prove formula (2). It is sufficient to consider the one-dimensional case

$$\xi^k(t) = \xi^j(t) = \xi(t) = \alpha(t) + \mu(t).$$

First,

$$\sum_{r=1}^{n} \varphi(t_{r-1})(\Delta\xi_r)^2 = \sum_{r=1}^{n} \varphi(t_{r-1})(\Delta\alpha_r)^2 + 2\sum_{r=1}^{n} \varphi(t_{r-1}) \Delta\alpha_r \Delta\mu_r + \sum_{r=1}^{n} \varphi(t_{r-1})(\Delta\mu_r)^2$$
$$= S_1 + 2S_2 + S_3,$$

and, moreover, with probability 1

$$|S_1| \leq \max_{0 \leq s \leq t} |\varphi(s)|\, V_t^0(\alpha) \max_r |\Delta\alpha_r| \to 0.$$

Here $V_t^0(\alpha)$ is the total variation of the function $\alpha(s)$ on the interval $[0, t]$. Next,

$$|S_2| \leq \max_{0 \leq s \leq t} |\varphi(s)| \left\{ \sum_{r=1}^{n} (\Delta\alpha_r)^2 \sum_{r=1}^{n} (\Delta\mu_r)^2 \right\}^{1/2}$$
$$\leq \max_{0 \leq s \leq t} |\varphi(s)| (V_0^t(\alpha) \max_r |\Delta\alpha_r|)^{1/2} \left(\sum_{r=1}^{n} (\Delta\mu_r)^2 \right)^{1/2}.$$

Since $\sum_{r=1}^{n}(\Delta\mu_r)^2 \to \langle\mu,\mu\rangle$ in probability (Theorem 22 in Section 1) it follows that $|S_2| \to 0$ in probability.

It remains to show that

$$\text{(2')} \qquad \text{P-}\lim_{|\lambda|\to 0} \sum_{r=1}^{n} \varphi(t_{r-1})(\Delta\mu_r)^2 = \int_0^t \varphi(s)\, d\langle\mu,\mu\rangle_s.$$

Choose an arbitrary $\varepsilon > 0$ and fix a subdivision λ_0 of the interval $[0, t]$ such that

$$\mathsf{P}\left\{\int_0^t |\varphi_{\lambda_0}(s) - \varphi(s)|\, d\langle\mu,\mu\rangle_s > \frac{\varepsilon}{4}\right\} < \frac{\varepsilon}{4}.$$

Let λ' denote the subdivision obtained when an arbitrary subdivision λ is superimposed on λ_0, and let S, S' and S_0 be the sums appearing in the left-hand side of equality (2') corresponding to subdivision λ, λ' and λ_0. Clearly, the sums S and S' for a fixed λ_0 differ from each other by at most a finite number of summands each one of which tends to zero with probability 1. Hence

$$\mathsf{P}\left\{|S' - S| > \frac{\varepsilon}{4}\right\} < \frac{\varepsilon}{4}$$

for $|\lambda|$ sufficiently small.

Let $t_1, t_2, \ldots, t_N$ be subdividing points of λ' $(t_N = t)$, t_{j_k}, $k = 1, 2, \ldots, m$ be the corresponding points of λ_0 $(t_{j_m} = t_N)$ and $S^* = \sum \varphi(t_{j_{k-1}}) \sum_{r=j_{k-1}+1}^{j_k} (\Delta\mu_r)^2$. Then

$$\begin{aligned}\mathsf{P}\{|S - \textstyle\int_0^t \varphi(s)\, d\langle\mu,\mu\rangle_s| > \varepsilon\} &\le \mathsf{P}\left\{|S - S'| > \frac{\varepsilon}{4}\right\} \\ &\quad + \mathsf{P}\left\{|S' - S^*| > \frac{\varepsilon}{4}\right\} + \mathsf{P}\left\{|S^* - \textstyle\int_0^t \varphi_{\lambda_0}(s)\, d\langle\mu,\mu\rangle_s| > \frac{\varepsilon}{4}\right\} \\ &\quad + \mathsf{P}\left\{|\textstyle\int_0^T [\varphi_{\lambda_0}(s) - \varphi(s)]\, d\langle\mu,\mu\rangle_s| > \frac{\varepsilon}{4}\right\}.\end{aligned}$$

Furthermore, we have

$$|S' - S^*| \le \sum_{k=1}^{m} \sum_{r=j_{k-1}+1}^{j_k} |\varphi(t_{j_{k-1}}) - \varphi(t_{r-1})|\, |\Delta\mu(t_r)|^2 \le \delta \sum_{r=1}^{N} (\Delta\mu(t_r))^2,$$

where $\delta = \max_{|t-s|\le|\lambda_0|} |\varphi(t) - \varphi(s)|$ and $\delta \to 0$ as $|\lambda_0| \to 0$ with probability 1. Since the sums $\sum_{r=1}^{N} (\Delta\mu(t_r))^2$ converge in probability, it follows that for a $|\lambda_0|$ sufficiently small and for any λ, $\mathsf{P}\{|S' - S^*| > \varepsilon/4\} < \varepsilon/4$. Finally, in view of Theorem 22 in Section 1 we have for a fixed λ_0 and for $|\lambda| \to 0$

$$S^* - \int_0^t \varphi_{\lambda_0}(s)\, d\langle\mu,\mu\rangle_s = \sum_{k=1}^{m} \varphi(t_{j_{k-1}})\left\{\sum_{r=j_{k-1}}^{j_k} [\Delta\mu(t_r)]^2 - [\langle\mu,\mu\rangle_{t_{j_k}} - \langle\mu,\mu\rangle_{t_{j_{k-1}}}]\right\} \to 0$$

in probability. Thus choosing first a suitable λ_0 one can find ε_0 such that for $|\lambda|<\varepsilon_0$

$$\mathsf{P}\{|S-\int_0^t \varphi(s)\,d\langle\mu,\mu\rangle_s|>\varepsilon\}<\varepsilon. \quad \square$$

Theorem 1 (Itô's formula). *Let $f(x)$, $x\in\mathcal{R}^m$ be a twice continuously differentiable function, $\alpha^k(t)\in\mathcal{V}^c$, $\mu^k(t)\in l\mathcal{M}^c$, $k=1,\ldots,m$, $\xi(t)=\xi_0+\alpha(t)+\mu(t)$, and $\alpha(0)=\mu(0)=0$. Then*

$$f(\xi(t))=f(\xi_0)+\int_0^t\sum_{k=1}^m \nabla^k f(\xi(s))\,d\xi^k(s)+\frac{1}{2}\int_0^t\sum_{k,j=1}^m \nabla^k\nabla^j f(\xi(s))\,d\langle\mu^k,\mu^j\rangle_s. \tag{3}$$

Here

$$\nabla^k f(x)=\frac{\partial f}{\partial x^k},\qquad \nabla^k\nabla^j f(x)=\frac{\partial^2 f}{\partial x^k\,\partial x^j}.$$

Proof. Assume first that function $f(x)$ is thrice continuously differentiable. Then

$$\begin{aligned} f(\xi(t))-f(\xi(0)) &= \sum_{r=1}^n f(\xi(t_r))-f(\xi(t_{r-1})) \\ &= \sum_{r=1}^n\sum_{k=1}^m \nabla^k f(\xi(t_{r-1}))(\xi^k(t_r)-\xi^k(t_{r-1})) \\ &\quad +\tfrac{1}{2}\sum_{r=1}^n\sum_{k,j=1}^m \nabla^k\nabla^j f(\xi(t_{r-1}))(\xi^k(t_r)-\xi^k(t_{r-1})) \\ &\quad \times(\xi^j(t_r)-\xi^j(t_{r-1}))+S, \end{aligned}$$

where

$$\begin{aligned} S=\frac{1}{3!}\sum_{r=1}^n\sum_{k,j,i=1}^m &\nabla^k\nabla^j\nabla^i f(\widetilde{\xi(t_r)})(\xi^k(t_r)-\xi^k(t_{r-1})) \\ &\times(\xi^j(t_r)-\xi^j(t_{r-1}))(\xi^i(t_r)-\xi^i(t_{r-1})), \end{aligned}$$

and $\widetilde{\xi(t_r)}$ is a point located on the interval joining $\xi(t_{r-1})$ and $\xi(t_r)$. Since the third partial derivatives of the function $f(x)$ are uniformly bounded, a constant $C=C(\omega)$ exists such that

$$S\le C\max_r|\xi(t_r)-\xi(t_{r-1})|\cdot\sum_{r=1}^n|\xi(t_r)-\xi(t_{r-1})|^2.$$

Moreover, the continuity of process $\xi(t)$ and Theorem 22 in Section 1 imply that $S\to 0$ in probability. Now taking into account the fact that the functions $\nabla^k f(\xi(t))$ are continuous and utilizing Lemma 1 we verify formula (3) for the case under consideration.

Now let $f(x)$ be an arbitrary twice continuously differentiable function. One can then construct a sequence of thrice continuously differentiable functions $f_n(x)$ vanishing outside a compact set and uniformly convergent (together with their first- and second-order partial derivatives) to $f(x)$ (and the corresponding derivatives) on an arbitrary compact set in $\mathcal{R}^m$. Applying formula (3) to $f_n(x)$ we observe that it is permissible to approach the limit under the integral sign as $n \to \infty$ in the relations obtained. □

Corollary 1. *If $f(x, t)$, $t \geq 0$, $x \in \mathcal{R}^m$, is continuously differentiable with respect to t and twice continuously differentiable with respect to x, then*

$$f(t, \xi(t)) - f(t, \xi_0) = \alpha_1(t) + \beta(t), \tag{4}$$

where

$$\alpha_1(t) = \int_0^t \frac{\partial f}{\partial s}(s, \xi(s))\, ds + \sum_{k=1}^m \int_0^t \nabla^k f(s, \xi(s))\, d\alpha^{(k)}(s) + \frac{1}{2} \sum_{k,j=1}^m \int_0^t \nabla^k \nabla^j f(s, \xi(s))\, d\langle \mu^{(k)}, \mu^{(j)} \rangle_s, \tag{5}$$

$$\beta(t) = \int_0^t \sum_{k=1}^m \nabla^k f(s, \xi(s))\, d\mu^k(s); \tag{6}$$

moreover, $\alpha_1(t) \in \mathcal{V}^c$ and $\beta(t) \in \mathcal{LM}^c$.

Proof. In the case when $f(t, x)$ is twice continuously differentiable with respect to (t, x) the formula obtained follows directly from (3) if one considers an $(m+1)$-dimensional process $\eta(t) = (\xi(t), t)$ with the last component being $\eta^{(m+1)} = t \in \mathcal{V}^c$ and $\mu^{(m+1)}(t) \equiv 0$. Utilizing the structure of formulas (4)–(6) and the fact that an arbitrary function $f(t, x)$ which satisfies the conditions of the corollary can be approximated in terms of functions $f_n(t, x)$ twice continuously differentiable with respect to (t, x) and converging to $f(t, x)$ uniformly on an arbitrary compact set in $[0, \infty) \times \mathcal{R}^m$, we conclude that formulas (4)–(6) are also valid under the conditions stipulated in the corollary. □

Corollary 2. *Let $w(t) = \{w^1(t), \ldots, w^m(t)\}$ be an m-dimensional Wiener process and $f(x)$ be a twice continuously differentiable function; then*

$$f(w(t)) = f(0) + \frac{1}{2} \int_0^t \Delta f(\xi(s))\, ds + \int_0^t (\nabla f(\xi(s)), dw(s)), \tag{7}$$

where

$$\Delta f = \sum_{k=1}^m \frac{\partial^2 f}{(\partial x^k)^2}, \qquad \nabla f = (\nabla^1 f, \nabla^2 f, \ldots, \nabla^m f).$$

Stochastic differentials. Let

$$\alpha(t)=(\alpha^{(1)}(t),\dots,\alpha^{(m)}(t)),$$
$$\mu(t)=(\mu^{(1)}(t),\dots,\mu^{(m)}(t)),$$
$$\alpha^{(k)}(t)\in\mathcal{V}^c,\qquad \mu^{(k)}(t)\in l\mathcal{M}^c,\qquad k=1,\dots,m.$$

We say that a process $\eta(t)$ *possesses a stochastic differential* (*of a continuous type*)

$$d\eta=(\varphi,d\alpha)+(\psi,d\mu)=\sum_{k=1}^m\varphi^{(k)}(t)\,d\alpha^{(k)}(t)+\sum_{k=1}^m\psi^{(k)}(t)\,d\mu^{(k)}(t)$$

for $t\in[0,T]$, where $\psi^{(k)}(t)\in H_2(\mathfrak{F}_t,\mu^{(k)})$ and the process $\varphi^{(k)}(t)$ is progressively measurable with respect to $\{\mathfrak{F}_t, t\geq 0\}$ whose realizations are (with probability 1) bounded functions if

$$\eta(t)=\eta(0)+\int_0^t(\varphi,d\alpha)+\int_0^t(\psi,d\mu)$$
$$=\eta(0)+\sum_{k=1}^m\int_0^t\varphi^{(k)}(s)\,d\alpha^{(k)}(s)+\sum_{k=1}^m\int_0^t\psi^{(k)}(s)\,d\mu^{(k)}(s)$$

(with probability 1) for each and every $t\in[0,T]$.

Clearly a process possessing a stochastic differential of a continuous type has a continuous modification. In what follows, we shall consider just this kind of modification of the process $\eta(t)$. Theorem 1 can be formulated using the concept of a stochastic differential in the following manner.

If processes $\xi^k(t)$, $k=1,2,\dots,m$, possess stochastic differentials $d\xi^k=d\alpha^k+d\mu^k$ and the function $f(t,x)=f(t,x^1,\dots,x^m)$ is continuously differentiable with respect to t and is twice continuously differentiable with respect to x, then the process $\eta(t)=f(t,\xi(t))$ also possesses a stochastic differential and, moreover,

$$d\eta=f'_t(t,\xi(t))\,dt+\sum_{k=1}^m\nabla^k f(t,\xi(t))\,d\alpha^k$$
$$+\tfrac{1}{2}\sum_{k,j=1}^m\nabla^k\nabla^j f(t,\xi(t))\,d\langle\mu^k,\mu^j\rangle_t+\sum_{k=1}^m\nabla^k f(t,\xi(t))\,d\mu^k$$
$$=f'_t(t,\xi(t))\,dt+\tfrac{1}{2}\sum_{k,j=1}^m\nabla^k\nabla^j f(t,\xi(t))\,d\langle\mu^k,\mu^j\rangle_t+\sum_{k=1}^m\nabla^k f(t,\xi(t))\,d\xi^k.$$

This implies

Theorem 2. *If* $\xi^k(t)$, $k=1,2,\dots,m$ *possess stochastic differential* $d\xi^k=d\alpha^k+d\mu^k$ *and* $\xi(t)=(\xi^1(t),\dots,\xi^m(t))$, *then*

$$d(\xi^1+\xi^2)=d\xi^1+d\xi^2,$$
$$d(\xi^1\xi^2)=\xi^1\,d\xi^2+\xi^2\,d\xi^1+d\langle\mu^1,\mu^2\rangle_t,$$
$$d\left(\frac{\xi_1}{\xi_2}\right)=\frac{\xi_2\,d\xi_1-\xi_1\,d\xi_2}{\xi_2^2}+\frac{\xi_1 d\langle\mu^2,\mu^2\rangle_t-\xi_2\,d\langle\mu^1,\mu^2\rangle_t}{\xi_2^3},$$
$$de^{(\xi,u)}=e^{(\xi,u)}\sum_{k=1}^m u^k\,d\xi^k+\tfrac{1}{2}e^{(\xi,u)}\sum_{k,j=1}^m\mu^k\mu^j\,d\langle\mu^k,\mu^j\rangle_t,$$

and the formula for $d(\xi_1/\xi_2)$ *is applicable provided* $\xi_2(t)\geq\delta>0$.

The formulas presented in Theorem 2 follow directly from Theorem 1 when applied successively to functions $x_1+x_2, x_1x_2, x_1/x_2$ and $e^{(x,u)}$.

Some applications of Itô's formula

Theorem 3 (Levy's theorem). *Let $\mu(t)=(\mu^1(t),\dots,\mu^m(t))$, $\mu(0)=0$, $\mu^k(t)\in \mathcal{LM}^c$, and $\langle\mu^k,\mu^j\rangle_t=\delta_{kj}t$. Then $\mu(t)$ is an m-dimensional Wiener process.*

Proof. Applying Itô's formula to the function $\eta(t)=e^{i(u,\mu(t))}$, we obtain

$$\eta(t)=\eta(s)+\int_s^t\eta(\theta)\left[i\sum_{k=1}^m u^k\,d\mu^k(\theta)-\tfrac{1}{2}|u|^2\,d\theta\right].$$

Let τ_n be a sequence of random times which reduces $\mu(t)$, $\mu_n(t)=\mu(t\wedge\tau_n)$ and $\mathfrak{F}_t^n=\mathfrak{F}_{t\wedge\tau_n}$. Then

$$\eta_n(t)=\eta_n(s)+\xi_n(t)-\xi_n(s)-\frac{|u|^2}{2}\int_{s\wedge\tau_n}^{t\wedge\tau_n}\eta(\theta)\,d\theta,$$

where

$$\xi_n(t)=i\int_0^{t\wedge\tau_n}\eta(\theta)(u,d\mu)=i\int_0^t\eta(\theta)(u,d\mu_n)$$

is a square integrable martingale (with respect to the current $\{\mathfrak{F}_t^n, t\geq 0\}$). Whence

$$J_n(t)\underset{\text{Def}}{=}\mathsf{E}\{\eta_n(t)|\mathfrak{F}_s^n\}=\eta_n(s)-\frac{|u|^2}{2}\mathsf{E}\{\textstyle\int_{s\wedge\tau_n}^{t\wedge\tau_n}\eta(\theta)\,d\theta|\mathfrak{F}_s^n\}. \tag{8}$$

Set $J(t)=\mathsf{E}\{\eta(t)|\mathfrak{F}_s\}$. We show that $J_n(t)\to J(t)$ in L_1. We have

$$\begin{aligned}\mathsf{E}|J_n(t)-J(t)|&\leq\mathsf{E}|\mathsf{E}\{\eta_n(t)-\eta(t)|\mathfrak{F}_s^n\}|+\mathsf{E}|\mathsf{E}\{\eta|\mathfrak{F}_s^n\}-\mathsf{E}\{\eta|\mathfrak{F}_s\}|\\&\leq\mathsf{E}|\eta_n(t)-\eta(t)|+\mathsf{E}|\mathsf{E}\{\eta|\mathfrak{F}_s^n\}-\mathsf{E}\{\eta|\mathfrak{F}_s\}|.\end{aligned}$$

Since $|\eta(t)|\leq 1$, $|\eta_n(t)|\leq 1$, and $\eta_n(t)\to\eta(t)$ with probability 1, we have $\mathsf{E}|\eta_n(t)-\eta(t)|\to 0$ as $n\to\infty$. We now verify that $\mathfrak{F}_s=\sigma\{\mathfrak{F}_s^n, n=1,2,\dots\}$. Choose an arbitrary set $A\in\mathfrak{F}_s$. Clearly $A=\bigcup_{n=1}^\infty(A\cap\{\tau_n>s\})$. Moreover $(A\wedge\{\tau_n>s\})\wedge\{\tau_n\wedge s\leq t\}\in\mathfrak{F}_t$ for any $t\geq 0$ so that $A\wedge\{\tau_n>s\}\in\mathfrak{F}_{\tau_n\wedge s}=\mathfrak{F}_s^n$. This implies that $A\in\sigma\{\mathfrak{F}_s^n, n=1,2,\dots\}$ and hence $\mathfrak{F}_s=\sigma\{\mathfrak{F}_s^n, n=1,2,\dots\}$. In view of a well-known theorem (Volume I, Chapter II, Section 2, Theorem 4) $\mathsf{E}\{\eta|\mathfrak{F}_s^n\}\to\mathsf{E}\{\eta|\mathfrak{F}_s\}$ with probability 1. Consequently $\mathsf{E}|\mathsf{E}\{\eta|\mathfrak{F}_s^n\}-\mathsf{E}\{\eta|\mathfrak{F}_s\}|\to 0$. Thus $\mathsf{E}|J_n(t)-J(t)|\to 0$ as $n\to\infty$.

Analogously, we verify that

$$\mathsf{E}\{\textstyle\int_{s\wedge\tau_n}^{t\wedge\tau_n}\eta(\theta)\,d\theta|\mathfrak{F}_s^n\}\to\mathsf{E}\{\int_s^t\eta(\theta)\,d\theta|\mathfrak{F}_s\}=\int_s^t J(\theta)\,d\theta$$

in L_1. Approaching the limit in (8) as $n\to\infty$, we obtain the equation

$$J(t)=\eta(s)-\frac{|u|^2}{2}\int_s^t J(\theta)\,d\theta. \tag{9}$$

This equation is equivalent to the differential equation $J'(t) = -(|u|^2/2)J(t)$, $J(s) = \eta(s)$, which implies that

$$J(t) = \eta(s) \exp\left\{-\frac{|u|^2}{2}(t-s)\right\}.$$

Therefore

$$\mathsf{E}\{\exp\{i(u, \mu(t) - \mu(s))\} \mid \mathfrak{F}_s\} = \exp\left\{-\frac{|u|^2}{2}(t-s)\right\}. \tag{10}$$

The relation obtained shows that the difference $\mu(t) - \mu(s)$ does not depend on σ-algebra $\mathfrak{F}_s$ and is normally distributed with mean 0 and covariance matrix $\delta_{kj}(t-s)$. □

Assume now that $\mu(t)$ is a one-dimensional process.

Theorem 4. *Let* $\mu(t) \in l\mathcal{M}^c$, $\mu(0) = 0$, *and* $\alpha(t) = \langle \mu, \mu \rangle_t \to \infty$ *as* $t \to \infty$ *with probability* 1. *Set*

$$\tau_t = \inf\{s : \alpha(s) \geq t\}, \qquad \eta(t) = \mu(\tau_t). \tag{11}$$

Then the process $\{\eta(t), \mathfrak{F}_{\tau_t}, t \geq 0\}$ *is a Wiener process.*

Proof. First observe that a random time τ_t reduces a local martingale $\mu(t)$.

Indeed, let $\{\sigma_n, n = 1, 2, \ldots\}$ be a sequence of random times which completely reduces $\mu(t)$. The process $\mu(\sigma_n \wedge \tau_t \wedge s)$, $s \geq 0$, is a square integrable martingale. As $n \to \infty$

$$\mathsf{E}|\mu(\sigma_n \wedge \tau_t) - \mu(\sigma_{n+m} \wedge \tau_t)|^2 = \mathsf{E}[\alpha(\sigma_{n+m} \wedge \tau_t) - \alpha(\sigma_n \wedge \tau_t)] \to 0,$$

since $\alpha(\sigma_n \wedge \tau_t) \to \alpha(\tau_t)$ with probability 1 and $\alpha(s \wedge \tau_t) \leq t$. Hence $\mu(\sigma_n \wedge \tau_t \wedge s)$ converges in $\mathcal{M}_2$ to a limit.

On the other hand, $\mu(\sigma_n \wedge \tau_t \wedge s) \to \mu(\tau_t \wedge s)$ with probability 1 so that $\mu(\tau_t \wedge s) \in \mathcal{M}_2^c$. Now let s and t be arbitrary and $s < t < N$. Then (cf. Corollary 3 to Theorem 6 in Section 1)

$$\begin{aligned}\mathsf{E}\{[\mu(\tau_t) - \mu(\tau_s)]^2 \mid \mathfrak{F}_{\tau_s}\} &= \mathsf{E}\{[\mu(\tau_N \wedge \tau_t) - \mu(\tau_N \wedge \tau_s)]^2 \mid \mathfrak{F}_{\tau_s}\} \\ &= \mathsf{E}\{\alpha(\tau_t) - \alpha(\tau_s) \mid \mathfrak{F}_{\tau_s}\} = t - s.\end{aligned}$$

In view of Levy's theorem (Theorem 3) the process $\mu(\tau_t)$ is a Wiener process. □

Bounds on the moments of continuous martingales. Let $\mu(t) \in l\mathcal{M}^c$, $\mu(0) = 0$. Assume that the characteristic $\alpha(t)$ of a local martingale $\mu(t)$ is absolutely continuous with respect to the Lebesgue measure and

(12) $$\langle\mu,\mu\rangle_t=\int_0^t\varphi(s)\,ds,\qquad \varphi(t)\geq 0.$$

Lemma 2. *If the function $\varphi(t)$, $t\in[0,T]$, is bounded with probability* 1 *and*

$$|\varphi(t)|\leq\sigma^2,\qquad t\in[0,T],$$

where σ^2 is a nonrandom constant, then $\mu(t)$ possesses moments of all orders.

Proof. Let $\tau=T$ if $\sup\{|\mu(t)|,t\leq T\}\leq n$ and $\tau=\inf\{t\colon|\mu(t)|\geq n,t\in[0,T]\}$ otherwise. Let $\mu_n(t)=\mu(t\wedge\tau)$. Since $|\mu_n(t)|\leq n$, it follows that $\mu_n(t)$ possesses moments of all orders. Moreover, $\mu_n(t)$ is a martingale. Applying Itô's formula to $\mu_n(t)$ and to function $f(x)=e^{ax}$, $a>0$, we obtain

$$e^{a\mu_n(t)}=1+\frac{a^2}{2}\int_0^t e^{a\mu_n(s)}\varphi(s)\,ds+a\int_0^t e^{a\mu_n(s)}\,d\mu_n(s).$$

Since the function $e^{a\mu_n(s)}$ is bounded with probability 1, the last summand in the right-hand side of the equality obtained is a martingale. Consequently,

$$\mathsf{E}\,e^{a\mu_n(t)}=1+\frac{a^2}{2}\int_0^t\mathsf{E}\,e^{a\mu_n(s)}\varphi(s)\,ds\leq 1+\frac{a^2\sigma^2}{2}\int_0^t\mathsf{E}\,e^{a\mu_n(s)}\,ds.$$

Setting $z_n=1+(a^2\sigma^2/2)\int_0^t\mathsf{E}\,e^{a\mu_n(s)}\,ds$ we deduce from the last relation that

$$\frac{z_n'(t)}{z_n(t)}\leq\frac{a^2\sigma^2}{2},\qquad z_n(0)=1,$$

which implies that $z_n(t)\leq e^{(a^2\sigma^2/2)t}$. Approaching the limit as $n\to\infty$ and utilizing Fatou's lemma we obtain the inequality

(13) $$\mathsf{E}\,e^{a\mu(t)}\leq e^{(a^2\sigma^2/2)t},$$

which implies the assertion of the lemma. □

Theorem 5. *If $\mu(t)\in\mathcal{LM}^c$, $\mu(0)=0$, condition* (12) *is satisfied, and for some $p>1$*

$$\int_0^T\mathsf{E}\varphi^p(s)\,ds<\infty,$$

then $\mathsf{E}|\mu(t)|^{2p}<\infty$ for $t\in[0,T]$ and

(14) $$\mathsf{E}|\mu(t)|^{2p}\leq p(2p-1)^p t^{p-1}\int_0^t\mathsf{E}\varphi^p(s)\,ds,\qquad t\leq T.$$

Proof. First let us assume that $\varphi(t)\leq\sigma^2$, where σ^2 is a constant. Then $\mu(t)$ possesses moments of all orders.

We now apply Itô's formula to $\mu(t)$ and to function $f(x)=|x|^{2p}$. We obtain

$$|\mu(t)|^{2p}=p(2p-1)\int_0^t|\mu(s)|^{2p-2}\varphi(s)\,ds+2p\int_0^t|\mu(s)|^{2p-2}\mu(s)\,d\mu(s)$$

and

$$\begin{aligned}\mathsf{E}|\mu(t)|^{2p}&=p(2p-1)\int_0^t\mathsf{E}|\mu(s)|^{2p-2}\varphi(s)\,ds\\&\leq p(2p-1)\Big(\int_0^t\mathsf{E}|\mu(s)|^{2p}\,ds\Big)^{(p-1)/p}\Big(\int_0^t\mathsf{E}\varphi^p(s)\,ds\Big)^{1/p}.\end{aligned}\tag{15}$$

Setting $z(t)=\int_0^t\mathsf{E}|\mu(s)|^{2p}ds$ we can rewrite the inequality obtained in the form

$$\frac{dz}{z^{(p-1)/p}}\leq p(2p-1)\Big(\int_0^t\mathsf{E}\varphi^p(s)\,ds\Big)^{1/p},$$

hence

$$\begin{aligned}z(t)&\leq(2p-1)^p\Big[\int_0^t\Big(\int_0^u\mathsf{E}\varphi^p(s)\,ds\Big)^{1/p}du\Big]^p\\&\leq[t(2p-1)]^p\int_0^t\mathsf{E}\varphi^p(s)\,ds.\end{aligned}$$

The last relation in conjunction with (15) yields inequality (14).

We now proceed to the general case. Set

$$\varphi_n(t)=(\varphi(t)\wedge n),\qquad \mu_n(t)=\int_0^t\sqrt{\frac{\varphi_n(s)}{\varphi(s)}}\,d\mu(s);$$

moreover, if $\varphi(s)=0$ we define $\varphi_n(s)/\varphi(s)=0$. Then $\mu_n(t)-\mu(t)$ is a locally square integrable martingale with characteristic

$$\int_0^t\Big(\sqrt{\frac{\varphi_n(s)}{\varphi(s)}}-1\Big)^2\varphi(s)\,ds=\int_0^t(\sqrt{\varphi_n(s)}-\sqrt{\varphi(s)})^2\,ds,$$

which with probability 1 tends to 0 as $n\to\infty$. Therefore a subsequence n_k exists such that $\mu_{n_k}(t)\to\mu(t)$ with probability 1. Applying Fatou's lemma to the inequality

$$\mathsf{E}|\mu_{n_k}(t)|^{2p}\leq p(2p-1)^pt^{p-1}\int_0^t\mathsf{E}\varphi^p(s)\,ds$$

we obtain (14) in the general case. □

Representation of martingales by means of stochastic integrals over a Wiener measure. If $\{w(t),\mathfrak{F}_t,t\geq 0\}$ is a Wiener process and $\varphi(t)$ is a process adopted to $\{\mathfrak{F}_t,t\geq 0\}$ such that

$$\alpha(t)=\int_0^t\varphi^2(s)\,ds<\infty$$

with probability 1 for all $t \geq 0$, then the stochastic integral

$$\mu(t) - \int_0^t \varphi(s)\, dw(s) \tag{16}$$

exists and is a continuous local martingale with characteristic $\alpha(t)$. We shall now establish how a local martingale can admit representation (16).

Theorem 6. *Let $\mu(t) \in \mathcal{LM}^c$ and let the characteristic $\alpha(t)$ of the process $\mu(t)$ be absolutely continuous with respect to the Lebesgue measure. Then a Wiener process $\{w(t), \mathfrak{F}_t^*, t \geq 0\}$ exists, where $\mathfrak{F}_t \subset \mathfrak{F}_t^*$ by means of which process $\mu(t)$ is representable in terms of formula* (16). *If $\varphi(t) > 0$ for every $t \geq 0$, then it may be assumed that $\mathfrak{F}_t = \mathfrak{F}_t^*$ for every $t \geq 0$.*

Proof. Suppose for a moment that $\varphi(t) > 0$, $t \geq 0$. Set

$$\zeta(t) = \int_0^t \frac{d\mu(s)}{\varphi(s)}.$$

Since $\int_0^t [\varphi^2(s)]^{-1}\, d\alpha(s) = t < \infty$, the integral on the right-hand side of the last equality exists and the characteristic of the process $\zeta(t)$, $\zeta(t) \in \mathcal{LM}^c$ is equal to t. In view of Levy's theorem $\zeta(t)$ is a Wiener process. Moreover, $\zeta(t)$ is adopted to the current of σ-algebras $\{\mathfrak{F}_t, t \geq 0\}$ and

$$\mu(t) = \int_0^t \varphi(s)\, d\zeta(s).$$

Thus the theorem is proved under the additional assumption that $\varphi(t) > 0$ for every $t > 0$.

Consider now the general case. Define a Wiener process $w^*(t)$, $t > 0$, independent of the current of σ-algebras $\{\mathfrak{F}_t, t > 0\}$ by extending—if necessary—the basic probability space $\{\Omega, \mathfrak{S}, \mathsf{P}\}$. Set $\eta_\varepsilon(t) = \mu(t) + \varepsilon w^*(t)$ and let $\mathfrak{F}_t^*$ be the minimal σ-algebra containing $\mathfrak{F}_t$ and $\sigma\{w^*(s), s \leq t\}$. It is easy to verify that the processes $\eta_\varepsilon(t)$, $\mu(t)$, and $w^*(t)$ are $\mathfrak{F}_t^*$-martingales. Therefore (cf. Theorem 17 in Section 1) the characteristic of the process $\eta_\varepsilon(t)$ is equal to

$$\langle \eta_\varepsilon, \eta_\varepsilon \rangle_t = \int_0^t (\varphi^2(s) + \varepsilon^2)\, ds.$$

It follows from the above that the process

$$\zeta_\varepsilon(t) = \int_0^t \frac{1}{\sqrt{\varphi^2(s) + \varepsilon^2}}\, d\eta_\varepsilon(s)$$

is a Wiener process.

We now show that as $\varepsilon \to 0$, $\zeta_\varepsilon(t)$ converges in the mean square to a limit. Indeed the difference

$$\zeta_\varepsilon(t)-\zeta_{\varepsilon'}(t) = \int_0^t \left(\frac{1}{\sqrt{\varphi^2(s)+\varepsilon^2}}-\frac{1}{\sqrt{\varphi^2(s)+\varepsilon'^2}}\right) d\mu + \int_0^t \left(\frac{\varepsilon}{\sqrt{\varphi^2(s)+\varepsilon^2}}-\frac{\varepsilon'}{\sqrt{\varphi^2(s)+\varepsilon'^2}}\right) dw^*$$

is a local martingale with characteristic

$$\int_0^t \left[\frac{(\varepsilon'^2-\varepsilon^2)^2\varphi^2(s)}{(\varphi^2(s)+\varepsilon^2)(\varphi^2+\varepsilon'^2)(\sqrt{\varphi^2(s)+\varepsilon^2}+\sqrt{\varphi^2(s)+\varepsilon'^2})^2} + \left(\frac{\varepsilon}{\sqrt{\varphi^2+\varepsilon^2}}-\frac{\varepsilon'}{\sqrt{\varphi^2+\varepsilon'^2}}\right)^2\right] ds. \tag{17}$$

The expression under the integral sign does not exceed 2 and approaches 0 as $\varepsilon, \varepsilon' \to 0$. Thus the characteristic of the process $\zeta_\varepsilon(t)-\zeta_{\varepsilon'}(t)$ converges to 0 with probability 1; therefore the limit $\lim \zeta_\varepsilon(t)=\zeta(t)$ exists for each t. Clearly a modification for the process $\zeta(t)$ exists which is a standard Wiener process and we shall retain the same notation for this modification. On the other hand,

$$\begin{aligned}\eta_\varepsilon(t) &= \mu(t)+\varepsilon w^*(t)=\int_0^t \sqrt{\varphi^2(s)+\varepsilon^2}\, d\zeta_\varepsilon(s)\\ &= \int_0^t \sqrt{\varphi^2(s)+\varepsilon^2}\, d\zeta(s)+\int_0^t \sqrt{\varphi^2(s)+\varepsilon^2}\, d(\zeta_\varepsilon(s)-\zeta(s)).\end{aligned}$$

Let $I_1(\varepsilon)$ and $I_2(\epsilon)$ denote the stochastic integrals appearing in the extreme right-hand side of the equality above. It follows from inequality (18) in Section 2 that

$$\text{P-lim}\, I_1(\varepsilon)=\int_0^t \varphi(s)\, d\zeta(s).$$

Next, taking (17) into account it is easy to verify that the characteristic of the local martingale $I^2(\varepsilon)$ is absolutely continuous and that its derivative possesses an integrable majorant and approaches 0 with probability 1 as $\varepsilon \to 0$. Therefore $I_2 \to 0$ as $\varepsilon \to 0$. Thus we obtain

$$\mu(t)=\int_0^t \varphi(s)\, d\zeta(s). \quad \square$$

An analogous result is also valid in the multidimensional case.

Before presenting this result we shall clarify the meaning of the integral

$$I(t)=\int_0^t \psi(t)\, dw(t),$$

where $w(t)=\{w'(t),\ldots, w^m(t)\}$ is an m-dimensional Wiener process and $\psi(t)$ is a matrix process, $\psi(t)=\{\psi_{kl}(t)\}$, $k=1,\dots,d$, $l=1,\dots,m$. We shall assume that the following conditions are satisfied:

a) a current of σ-algebras $\{\mathfrak{F}_t, t\geq 0\}$ with respect to which $w(t)$ is a Wiener process is given;

b) processes ψ_{kl} are adopted to the current $\{\mathfrak{F}_t, t\geq 0\}$ $k=1,\dots,d$, $l=1,\dots,m$, and

$$\int_0^t |\psi_{kl}(s)|^2\, ds<\infty \qquad \forall t>0.$$

Then the integral $I(t)$ is interpreted as a vector-valued process $\{I_1(t),\dots,I_d(t)\}$ whose components are defined by the equalities

$$I_k(t)=\sum_{l=1}^{m}\int_0^t \psi_{kl}(s)\, dw_l(s).$$

It follows from our assumptions that each of the integrals in the right-hand side of the last equality exists and is a continuous function.

Theorem 7. *Let $\mu^k(t)\in \mathcal{M}^c\{\mathfrak{F}_t, t\geq 0\}$ and the characteristics $\alpha^k(t)$ of the processes $\mu^k(t)$ be absolutely continuous with respect to the Lebesgue measure, $k=1,2,\dots,m$. Then an m-dimensional Wiener process $w(t)=\{w^1(t),\dots,w^m(t)\}$ exists and a matrix process $\psi(t)$, $t\geq 0$, adopted to $\{\mathfrak{F}^*_t, t\geq 0\}$, $\mathfrak{F}^*_t\supset\mathfrak{F}_t$, such that*

$$\mu(t)=\int_0^t \psi(s)\, dw(s).$$

Proof. Set $\alpha^{kj}(t)=\langle\mu^k,\mu^j\rangle_t$ and let

$$\alpha^k(t)=\alpha^{kk}(t)=\int_0^t \varphi^{kk}(s)\, ds,$$

where $\varphi^{kk}(s)\geq 0$. It follows from the inequality $|\Delta\alpha^{kj}|^2\leq\Delta\alpha^k\,\Delta\alpha^j$ that the functions $\alpha^{kj}(t)$ are with probability 1 absolutely continuous with respect to the Lebesgue measure. Therefore functions $\varphi^{kj}(t)$ exist such that

$$\alpha^{kj}(t)=\int_0^t \varphi^{kj}(s)\, ds.$$

Let z_k, $k=1,\dots,m$, be arbitrary real numbers. Then the process

$$\sum_{k,j=1}^{m}\alpha^{kj}(t)z_kz_j=\int_0^t\left(\sum_{k,j=1}^{m}\varphi^{kj}(s)z_kz_j\right)ds$$

is a characteristic of the martingale $\sum_{k=1}^m \mu^k(t)z_k$ and is therefore monotonically nondecreasing. Hence

$$\sum_{k,j=1}^{m}\varphi^{kj}(s)z_kz_j\geq 0$$

for any z_k for almost all s, i.e., the matrix $\Phi(s)=\{\varphi^{kj}(s)\}$ is nonnegative-definite for almost all s.

First let us assume that the matrix $\Phi(s)$ is uniformly nondegenerate, i.e.,

$$\sum_{k,j=1}^{m} \varphi^{kj}(s)z_k z_j \geqslant \varepsilon \sum_{j=1}^{m} z_j^2, \qquad \varepsilon>0 \quad \forall s>0.$$

It is known that a positively definite matrix $\Phi(s)$ may be represented in the form $U^*(s)D(s)U(s)$ where $U(s)$ is an orthogonal matrix, $U^*(s)$ is the conjugate of $U(s)$ and $D(s)$ is a diagonal matrix with diagonal entries $\lambda_j(s)$, where $\lambda_j(s)$ are the eigenvalues of the matrix $\Phi(s)$, $\lambda_j(s)\geqslant\varepsilon$.

Set $\Phi^{-1/2}(s)=U^*(s)D^{-1/2}(s)U(s)$, where $D^{-1/2}$ is the diagonal matrix with entries $\delta_{kj}\lambda_j^{-1/2}(s)$. The elements $\gamma_{kj}(s)$ of the matrix $\Phi^{-1/2}(s)$ are bounded (uniformly with respect to s and ω), i.e.,

$$|\gamma_{kj}(s)|=\left|\sum_{r=1}^{m} u_{rk}(s)\lambda_r^{-1/2}(s)u_{kr}(s)\right|\leqslant\frac{m}{\sqrt{\varepsilon}}.$$

Moreover, the matrix $\Phi^{-1/2}(s)$ is symmetric and

$$\Phi^{-1/2}(s)\Phi(s)\Phi^{-1/2}(s)=\mathbf{I},$$

where $\mathbf{I}$ is the identity matrix.

Consider the process

$$\zeta(t)=\int_0^t \Phi^{-1/2}(s)\,d\mu(s).$$

As it follows from the discussion above, the stochastic integrals which define this process do exist (it is easy to verify that $\gamma_{kj}(s)$ are Borel functions of the entries $\varphi_{kj}(s)$ of the matrix $\Phi(s)$). Moreover,

$$\begin{aligned}\langle\zeta^k,\zeta^j\rangle_t&=\left\langle\sum_i\int_0^t\gamma_{ki}(s)\,d\mu^i(s),\sum_r\int_0^t\gamma_{jr}(s)\,d\mu^r(s)\right\rangle_t\\&=\sum_{i,r}\int_0^t\gamma_{ki}(s)\,d\langle\mu^i,\mu^r\rangle_s\gamma_{jr}(s)\\&=\sum_{i,r}\int_0^t\gamma_{ki}(s)\varphi^{ir}(s)\gamma_{rj}(s)\,ds=t\delta_{kj}.\end{aligned}$$

Also, it follows from Levy's theorem (Theorem 3) that $\zeta(t)$ is an m-dimensional Wiener process. On the other hand, in view of Theorem 3 in Section 2

$$\int_0^t\Phi^{1/2}(s)\,d\zeta(s)=\int_0^t\Phi^{1/2}(s)\Phi^{-1/2}(s)\,d\mu(s)=\mu(t).$$

Thus the theorem is proved under the additional assumption on $\Phi(s)$ stipulated above.

The proof in the general case is analogous to the proof of Theorem 6. Let $w^*(t)$ be an m-dimensional Wiener process independent of $\{\mathfrak{F}_t, t\geq 0\}$ and let $\mathfrak{F}_t^*$ be a σ-algebra generated by $\mathfrak{F}_t$ and a family of random variables $\{w^*(s), s\leq t\}$. Set $\eta_\varepsilon(t)=\mu(t)+\varepsilon w^*(t)$. Clearly $\eta_\varepsilon^k(t)\in l\mathcal{M}^c\{\mathfrak{F}_t^k, t\geq 0\}$ and

$$\langle \eta_\varepsilon^k, \eta_\varepsilon^j\rangle_t=\langle \mu^k, \mu^j\rangle_t+t\varepsilon^2\delta_{kj}=\int_0^t \varphi_\varepsilon^{kj}(s)\,ds.$$

The matrix $\Phi_\varepsilon(s)=\{\varphi_\varepsilon^{kj}(s)\}$ is now uniformly nondegenerate, i.e.,

$$\sum_{k,j=1}^m \varphi_\varepsilon^{kj}(s)z_kz_j\geq \varepsilon^2\sum_{k=1}^m z_k^2,$$

and in view of the proof given above

$$\eta_\varepsilon(t)=\int_0^t \Phi_\varepsilon^{1/2}(s)\,d\zeta_\varepsilon(s),$$

where $\zeta_\varepsilon(t)$ is an m-dimensional Wiener process. Moreover,

$$\zeta_\varepsilon(t)=\int_0^t \Phi_\varepsilon^{-1/2}(s)\,d\eta_\varepsilon(s)=\int_0^t \Phi_\varepsilon^{-1/2}(s)\,d\mu(s)+\varepsilon\int_0^t \Phi_\varepsilon^{-1/2}(s)\,dw^*(s).$$

We now show that the process $\zeta_\varepsilon(t)$ converges in the mean square for each t to a limit $\zeta(t)$. Clearly, the process $\zeta(t)$ is also a Wiener process. It is sufficient to show that $\zeta_\varepsilon(t)$ satisfies the Cauchy condition as $\varepsilon\to 0$. Set

$$\zeta_\varepsilon(t)-\zeta_{\varepsilon'}(t)=I_1(t)+I_2(t),$$

where

$$I_1(t)=\int_0^t (\Phi_\varepsilon^{-1/2}(s)-\Phi_{\varepsilon'}^{-1/2}(s))\,d\mu(s),$$
$$I_2(t)=\int_0^t (\varepsilon\Phi_\varepsilon^{-1/2}(s)-\varepsilon'\Phi_{\varepsilon'}^{-1/2}(s))\,dw^*(s).$$

As above, let $\Phi(s)=U^*(s)D(s)U(s)$, where $U(s)$ is an orthogonal matrix, $D(s)$ is a diagonal matrix with entries $\delta_{kj}\lambda_j(s)$, $\lambda_j(s)\geq 0$.

Then $\Phi_\varepsilon^{-1/2}(s)=U^*(s)D_1^\varepsilon(s)U(s)$ and $D_1^\varepsilon(s)$ is a diagonal matrix with entries $\delta_{kj}(\varepsilon^2+\lambda_j(s))^{-1/2}$. The matrix characteristic of the local martingale $I_1(t)$ is of the form:

$$\langle I_1, I_1\rangle_t=\int_0^t (\Phi_\varepsilon^{-1/2}(s)-\Phi_{\varepsilon'}^{-1/2}(s))\Phi(s)(\Phi_\varepsilon^{-1/2}(s)-\Phi_{\varepsilon'}^{-1/2}(s))\,ds.$$

Since $\Phi_\varepsilon^{-1/2}(s)\Phi^{1/2}(s)=U^*(s)D_1^\varepsilon(s)D^{1/2}(s)U(s)$, it follows that

$$\begin{aligned}\langle I_1, I_1\rangle_t&=\int_0^t U^*(s)(D_1^\varepsilon(s)D^{1/2}(s)-D_1^{\varepsilon'}(s)D^{1/2}(s))^2U(s)\,ds\\&=\int_0^t U^*(s)D_2(s)U(s)\,ds,\end{aligned}$$

where $D_2(s)$ is a diagonal matrix with the entries

$$\delta_{kj}\left(\frac{\lambda_j^{1/2}(s)}{\sqrt{\varepsilon^2+\lambda_j(s)}}-\frac{\lambda_j^{1/2}(s)}{\sqrt{\varepsilon'^2+\lambda_j(s)}}\right)^2$$

$$=\delta_{kj}\frac{\lambda_j(s)(\varepsilon^2-\varepsilon'^2)^2}{(\varepsilon^2+\lambda_j(s))(\varepsilon'^2+\lambda_j(s))(\sqrt{\varepsilon^2+\lambda_j(s)}+\sqrt{\varepsilon'^2+\lambda_j(s)})^2}.$$

It follows directly from the expression presented above that the derivative of a matrix characteristic of a local martingale $I_1(t)$ is uniformly bounded and tends to 0 as ε and $\varepsilon'\to 0$ so that $\langle I_1, I_1\rangle_t\to 0$ with probability 1 for all $t>0$.

Next, the matrix characteristic of the local martingale $I_2(t)$ is of the form

$$\langle I_2, I_2\rangle_t=\int_0^t [\varepsilon\Phi_\varepsilon^{-1/2}(s)-\varepsilon'\Phi_{\varepsilon'}^{-1/2}(s)]^2\,ds,$$

where $\varepsilon\Phi_\varepsilon^{-1/2}(s)=U^*(s)D_3(s)U(s)$ and the entries of the matrix $D_3(s)$ are equal to $\delta_{kj}\varepsilon(\varepsilon^2+\lambda_j(s))^{-1/2}$ and are therefore uniformly bounded and tend to 0 as $\varepsilon\to 0$.

Thus we have shown that $\mathsf{E}|\zeta_\varepsilon(t)-\zeta_{\varepsilon'}(t)|^2\to 0$ as ε and $\varepsilon'\to 0$ for each $t>0$ and the limit $\lim \zeta_\varepsilon(t)=\zeta(t)$ exists.

In what follows $\zeta(t)$ will denote a continuous modification of the corresponding process.

We need only show that

$$\text{P-lim}\,\eta_\varepsilon(t)=\int_0^t \Phi^{1/2}(s)\,d\zeta(s).$$

We have

$$\eta_\varepsilon(t)=\int_0^t \Phi_\varepsilon^{1/2}(s)\,d\zeta(s)+\int_0^t \Phi_\varepsilon^{1/2}\,d(\zeta_\varepsilon(s)-\zeta(s))=I_3(t)+I_4(t).$$

Clearly, $I_3(t)\to\int_0^t \Phi^{1/2}(s)\,d\zeta(s)$ in probability as $\varepsilon\to 0$. On the other hand, it is easy to verify from the expression for the characteristic of $\zeta_\varepsilon(t)-\zeta_{\varepsilon'}(t)$ obtained above (equation (17)) that the matrix characteristic of the local martingale $I_4(t)$ is of the form

$$\int_0^t U^*(s)(\varepsilon^2\mathbf{I}+D_4(s))U(s)\,ds,$$

where $\mathbf{I}$ is the unit matrix and $D_4(s)$ is a diagonal matrix with the entries

$$\delta_{kj}\frac{\varepsilon^4\chi_j(s)}{(\sqrt{\lambda_j(s)}+\sqrt{\varepsilon^2+\lambda_j(s)})^2},$$

and $\chi_j(s)$ is the indicator of the set $\{s\colon \lambda_j(s)=0\}$. Thus

$$\text{P-}\lim_{\varepsilon\to 0} I_4(t)=0\qquad \forall t>0.\quad\square$$

Remark. In the case when the function $\varphi(s)>0$ or, correspondingly, the matrix $\Phi(s)$ is uniformly nondegenerate, the Wiener process constructed in Theorems 6 and 7 is adopted to the current $\{\mathfrak{F}_t, t\geq 0\}$. In particular, if $\mathfrak{F}_t^\mu=\sigma\{\mu(s), s\leq t\}$, $\mathfrak{F}_t^\zeta=\sigma\{\zeta(s), s\leq t\}$ then under the conditions stated above

$$\mathfrak{F}_t^\mu=\mathfrak{F}_t^\zeta. \tag{18}$$

If, however, these conditions are not fulfilled, Theorems 6 and 7 assure only the following result: one can construct a new probability space $\{\Omega^*, \mathfrak{S}^*, \mathsf{P}^*\}$, a current of σ-algebras $\{\mathfrak{F}_t^*, t\geq 0\}$, and find processes $\mu'(t), \zeta(t), \psi(t), t\geq 0$, defined on $\{\Omega^*, \mathfrak{S}^*, \mathsf{P}^*\}$ adopted to $\{\mathfrak{F}_t^*, t\geq 0\}$, where $\zeta(t)$ is a Wiener process and $\mu'(t)$ is a local martingale which is stochastically equivalent (in the wide sense) to the process $\mu(t)$ so that

$$\mu'(t)=\int_0^t \psi(s)\,d\zeta(s).$$

Corollary. *If a vector local martingale $\mu(t)\in \mathcal{LM}^c$ possesses the matrix characteristic with the entries*

$$\langle\mu^k, \mu^j\rangle_t=\int_0^t \sigma_{kj}(\xi(s))\,ds,$$

where $\sigma(x)=\{\sigma_{kj}(x)\}$ is a nonrandom nonnegative-definite Borel matrix function and $\xi(t)$ is a random process adopted to $\{\mathfrak{F}_t, t\geq 0\}$, then the process $\mu(t)$ admits representation

$$\mu^k(t)=\int_0^t \sum_{j=1}^m b_{kj}(\xi(s))\,dw^j(s),$$

where $w(t)=\{w^1(t),\ldots, w^m(t)\}$ is a Wiener process and $b(x)=\{b_{kj}(x)\}$ is a non-negative-definite symmetric matrix, $b^2(x)=\sigma(x)$.

Decomposition of a locally square integrable martingale into continuous and discontinuous components. Let $\xi(t)=\{\xi^1(t),\ldots,\xi^m(t)\}, t\geq 0$, be an m-dimensional local square integrable martingale adopted to a current of σ-algebras $\{\mathfrak{F}_t, t\geq 0\}$; moreover, let σ-algebra $\mathfrak{F}_0$ contain subsets of Ω of probability 0. In this subsection we shall construct a decomposition of the process $\xi(t)$ of the form

$$\xi(t)=\xi_c(t)+\xi_d(t)$$

where $\xi_c(t)\in \mathcal{LM}_2^c$ and the product $\eta(t)\xi_d^j(t)$ $(j=1,\ldots,m)$ is a local square integrable martingale for an arbitrary continuous local martingale $\eta(t)$. As before, we shall confine our discussion to the case of square integrable (local square integrable) martingales with continuous characteristics. First we shall consider processes $\xi(t)\in \mathcal{M}_2^r$ and then the results obtained are extended to the case of processes $\xi(t)\in \mathcal{LM}_2^r$.

Thus, let $\xi(t)\in \mathcal{M}_2^r(\mathfrak{F}_t, t\geq 0)$. Denote by $\mathfrak{B}_0^m$ the class of Borel sets in $\mathcal{R}^m$ whose closure does not contain point 0 and let $\nu(t, A)$ be the number of jumps of the function $\xi(s)$ on the interval $(0, t]$ whose values fall into the set A, $A\in\mathfrak{B}_0^m$. Since the sample functions of the process $\xi(t)$ belong with probability 1 to $\mathcal{D}^m[0,\infty)$, the process $\nu(t, A)$ is defined with probability 1 for all $t\geq 0$, $A\in\mathfrak{B}_0^m$. We extend its definition over the whole Ω, by setting $\nu(t, A)\equiv 0$ if $\xi(t,\omega)\notin\mathcal{D}[0,\infty)$. Clearly the process $\nu(t, A)$, $A\in\mathfrak{B}_0^m$, is adopted to the current of σ-algebras $\{\mathfrak{F}_t, t\geq 0\}$, and its sample functions are nonnegative, monotonically nondecreasing, continuous from the right, and take on integral values.

Consider a sequence of random times $\sigma=\tau_0\leq\tau_1\leq\cdots\leq\tau_n=\tau\leq T$ and set $\lambda=\max_{1\leq k\leq n}|\tau_k-\tau_{k-1}|$. Since

$$\sum_{\substack{|\delta\xi(s)|>\varepsilon\\ \sigma<s\leq\tau}}|\delta\xi(s)|^2\leq\lim_{\lambda\to 0}\sum_{k=1}^{n}|\xi(\tau_k)-\xi(\tau_{k-1})|^2 \pmod{\mathsf{P}},$$

where

$$\delta\xi(s)=\xi(s)-\xi(s-),$$

it follows that

$$\mathsf{E}\Big\{\sum_{\substack{|\delta\xi(s)|>\varepsilon\\ \sigma<s\leq\tau}}|\delta\xi(s)|^2\,|\,\mathfrak{F}_\sigma\Big\}\leq\mathsf{E}\{|\xi(\tau)-\xi(\sigma)|^2\,|\,\mathfrak{F}_\sigma\}, \tag{19}$$

and if $A\subset\{|x|:|x|\geq\varepsilon\}$

$$\begin{aligned}\mathsf{E}\{\nu(\tau,A)-\nu(\sigma,A)\,|\,\mathfrak{F}_\sigma\}&\leq\frac{1}{\varepsilon^2}\mathsf{E}\{|\xi(\tau)-\xi(\sigma)|^2\,|\,\mathfrak{F}_\sigma\}\\&=\frac{1}{\varepsilon^2}\mathsf{E}\{\alpha(\tau)-\alpha(\sigma)\,|\,\mathfrak{F}_\sigma\}<\infty\end{aligned}$$

with probability 1 where

$$\alpha(t)=\sum_{k=1}^{m}\langle\mu^k,\mu^k\rangle_t.$$

In particular, the process $\nu(t, A)$ is integrable for $A\in\mathfrak{B}_0^m$ and since the function $\alpha(t)$ is continuous the process $\nu(t, A)$ is regular, i.e., for any monotonically nondecreasing sequence of random times τ_n, $n=1,2,\ldots$ ($\tau_n\leq T$, $\lim\tau_n=\tau$),

$$\lim\mathsf{E}\nu(\tau_n, A)=\mathsf{E}\nu(\tau, A).$$

One can construct a measure from function $\nu(t, A)$ following the standard procedure. For this purpose set $\nu(\Delta\times A)=\nu(\Delta, A)=\nu(t+\Delta t, A)-\nu(t, A)$. The function $\nu(\Delta \times A)$ is additive on the semiring of the sets $\mathfrak{T}_0 \times \mathfrak{B}_0^m$ where $\mathfrak{T}_0$ is the semiring of intervals of the form $\Delta = (t, t+\Delta t]$. This function can be extended up to a measure on the σ-algebra of Borel sets of the space $[0, \infty) \times \mathscr{R}^m$.

The integral

$$\int_0^t \int_{\mathscr{R}^m} f(s, u)\nu(ds, du)$$

is well defined for an arbitrary Borel nonnegative function $f(s, u)$, and, moreover,

$$(20) \qquad \int_0^t \int_{\mathscr{R}^m} f(s, u)\nu(ds, du) = \sum_{\substack{s \leq t \\ \delta\xi(s) \neq 0}} f(s, \delta\xi(s)),$$

and the sum in the right-hand side of equation (20) contains at most a countable number of summands.

Setting $f(s, u) = \chi_\Delta(t)|u|^2$ we obtain

$$\int_{\mathscr{R}^m} |u|^2 \nu(\Delta, du) = \sum_{s \in (t, t+\Delta t]} |\delta\xi(s)|^2,$$

whence, in view of inequality (19), it follows that

$$\mathsf{E}\{\int_{\mathscr{R}^m} |u|^2 \nu(\Delta, du) | \mathfrak{F}_t\} \leq \mathsf{E}\{|\xi(t+\Delta t) - \xi(t)|^2 | \mathfrak{F}_t\},$$
$$\mathsf{E} \int_{\mathscr{R}^m} |u|^2 \nu(\Delta, du) \leq \mathsf{E}|\xi(t+\Delta t) - \xi(t)|^2 < \infty.$$

With no connection to the process $\xi(t)$ consider an arbitrary random measure $\nu(t, A)$ possessing the following properties:

1) function $\nu(t, A)$ is defined on $[0, \infty) \times \mathfrak{B}_0^m$ and takes on nonnegative integer values and for any $\varepsilon > 0$, $T > 0$, $\mathsf{E}\nu(T, \mathscr{R}^m \backslash S_\varepsilon) < \infty$, where S_ε is the sphere in $\mathscr{R}^m$ of radius ε with the center at point 0;

2) for a fixed t the variable $\nu(t, A)$ is $\mathfrak{F}_t$-measurable and for a fixed A this variable—as a function of the argument t—is monotonically nondecreasing and continuous from the right;

3) for an arbitrary monotonically nondecreasing sequence of random times τ_n ($\lim \tau_n = \tau \leq T$)

$$\lim \mathsf{E}\nu(\tau_n, A) = \mathsf{E}\nu(\tau, A).$$

In what follows, a function $\nu(t, A)$ possessing properties 1)–3) will be called an *integer-valued random measure.* The same term is preserved for a measure $\nu(\cdot)$ on the σ-algebra of Borel sets of the space $[0, \infty) \times \mathscr{R}^m$ defined by the equalities $\nu(\Delta \times A) = \nu(t+\Delta t, A) - \nu(t, A)$, where $\Delta = (t, t+\Delta t]$. The function $\nu(t, A)$, $t \geq 0$, is a regular $\mathfrak{F}_t$-submartingale for a fixed A. In view of Meyer's theorem (Theorem 9 in Section 1) $\nu(t, A)$ possesses a unique representation of the form

$$\nu(t, A) = \mu(t, A) + \pi(t, A),$$

where $\pi(t, A)$ is a continuous monotonically nondecreasing integrable process and $\mu(t, A)$ is a martingale.

Note that $\mu(t, A) \in l\mathcal{M}_2$. Indeed, set

$$\tau_n = \inf\{t: (\nu(t, A) \geq n) \cup (\pi(t, A) \geq n) \cup (t = T)\},$$
$$\nu_n(t, A) = \nu(t \cap \tau_n, A), \qquad \pi_n(t, A) = \pi(t \cap \tau_n, A),$$
$$\mu_n(t, A) = \mu(t \wedge \tau_n, A).$$

Then $\nu_n(t, A) \leq n$, $\pi_n(t, A) \leq n$, and $|\mu_n(t, A)| \leq n$. Thus $\mu(t, A) \in l\mathcal{M}_2$.

We now show that the characteristic of process $\mu(t, A)$ coincides with $\pi(t, A)$. First we verify that $\mu^2(t, A)$ is a regular submartingale.

Indeed, let τ'_n be a monotonically nondecreasing sequence of random times, $\lim \tau'_n = \tau' \leq T$,

$$\mu_*(t) = \mu_m(t, A), \qquad \nu_*(t) = \nu_m(t, A), \qquad \pi_*(t) = \pi_m(t, A).$$

Then

$$E(\mu_*^2(\tau') - \mu_*^2(\tau'_n)) \leq 2n\mathsf{E}[(\nu_*(\tau') - \nu_*(\tau'_n)) + \mathsf{E}(\pi_*(\tau') - \pi_*(\tau'_n))].$$

Since the function $\pi(t)$ is continuous and uniformly integrable $\mathsf{E}(\pi_*(\tau') - \pi_*(\tau'_n)) \to 0$ as $n \to \infty$ and, as it was shown above, $\mathsf{E}(\nu_*(\tau') - \nu_*(\tau'_n)) \to 0$. Thus $\mathsf{E}\mu_*^2(\tau'_n) \to \mathsf{E}\mu_*^2(\tau')$, i.e. $\mu_*^2(t)$ is a regular submartingale.

Theorem 13 in Section 1 is utilized again to conclude that the characteristic of the martingale $\mu_*(t)$ is continuous. Denote this characteristic by $\alpha_*(t)$. To prove the equality $\alpha_*(t) = \pi_*(t)$ we utilize Theorem 21 in Section 1, which asserts that

$$\alpha_*(t) = \lim_{|\delta| \to 0} \sum_{k=0}^{N-1} \mathsf{E}\{\Delta\mu_*^2(t_k) | \mathfrak{F}_{t_k}\}$$

in the sense of convergence in L_1, where $0 = t_0 < t_1 < \cdots < t_N = t$, $\Delta\mu_*^2(t_k) = \mu_*^2(t_{k+1}) - \mu_*^2(t_k)$, $|\delta| = \max(t_{k+1} - t_k)$. We have

$$\mathsf{E}\left|\sum_{k=0}^{N-1} \mathsf{E}\{\Delta\mu_*^2(t_k) | \mathfrak{F}_{t_k}\} - \pi_*(t)\right|$$
$$\leq \mathsf{E}\left|\sum_{k=0}^{N-1} \mathsf{E}\{\Delta\mu_*^2(t_k) - \Delta\nu_*(t_k) | \mathfrak{F}_{t_k}\}\right| + \mathsf{E}\left|\sum_{k=0}^{N-1} \mathsf{E}\{\Delta\nu_*(t_k) | \mathfrak{F}_{t_k}\} - \pi_*(t)\right|.$$

In view of the definition of $\pi_*(t)$ and Theorem 21 in Section 1, the second summand in the right-hand side of the obtained inequality tends to 0. The first summand can be bounded in the following manner:

$$\mathsf{E}\left|\sum_{k=0}^{N-1} \mathsf{E}\{\Delta\mu_*^2(t_k)-\Delta\nu_*(t_k)|\mathfrak{F}_{t_k}\}\right|$$

$$\leqslant \sum_{k=0}^{N-1} \mathsf{E}|[\Delta\mu_*(t_k)]^2-\Delta\nu_*(t_k)|$$

$$\leqslant \mathsf{E}\sum_{k=0}^{N-1}|(\Delta\nu_*(t_k))^2-\Delta\nu_*(t_k)-2\,\Delta\nu_*(t_k)\Delta\pi_*(t_k)+(\Delta\pi_*(t_k))^2|$$

$$\leqslant \mathsf{E}\left[\sum_{k=0}^{N-1}((\Delta\nu_*(t_k))^2-\Delta\nu_*(t_k))+2\max_k \Delta\pi_*(t_k)(\nu_*(t)+\pi_*(t))\right].$$

The expression appearing in square brackets is uniformly bounded (in N) (it does not exceed $\nu_*^2(t)+2\pi_*(t)(\nu_*(t)+\pi_*(t))\leqslant 5n^2$) and tends to 0 with probability 1.
Therefore

$$\mathsf{E}\left|\sum_{k=0}^{N-1} \mathsf{E}\{\Delta\mu_*^2(t_k)-\Delta\nu_*(t_k)|\mathfrak{F}_{t_k}\}\right|\to 0 \quad \text{as } |\delta|\to 0.$$

Thus $\alpha_*(t)=\pi_*(t)$. This implies that the characteristic of $\mu(t,A)$ is equal to $\pi(t,A)$.

Let $A_i\in\mathfrak{B}_0$, $i=1,2$, $A_1\cap A_2=\varnothing$. Since $\nu(t,A_1\cup A_2)=\nu(t,A_1)+\nu(t,A_2)$, in view of the uniqueness of the decomposition (21) $\pi(t,A_1\cup A_2)=\pi(t,A_1)+\pi(t,A_2)$. This implies that the characteristic of the sum of locally square integrable martingales $\mu(t,A_1)+\mu(t,A_2)$ is equal to $\pi(t,A_1)+\pi(t,A_2)$, which is possible if and only if the product $\mu(t,A_1)\mu(t,A_2)$ is a local martingale.

Definition. Processes $\mu_1(t)$ and $\mu_2(t)$, $\mu_i(t)\in l\mathcal{M}_2$ $(\mu_1(0)=\mu_2(0)=0)$ are called *orthogonal* if $\mu_1(t)\mu_2(t)$ is a local martingale.

This definition implies that if $\mu_1(t)$ and $\mu_2(t)$ are orthogonal and τ is a random time reducing $\mu_1(t)$ and $\mu_2(t)$ and $\mu_i^*(t)=\mu_i(t\wedge\tau)$, then $(\mathfrak{F}_t^*=\mathfrak{F}_{t\wedge\tau})$

$$\mathsf{E}\{\Delta\mu_1^*\,\Delta\mu_2^*\,|\mathfrak{F}_t^*\}=\mathsf{E}\{\Delta(\mu_1^*\mu_2^*)|\mathfrak{F}_t^*\}=0.$$

In particular,

$$\mathsf{E}\mu_1^*(t)\mu_2^*(t)=0 \qquad \forall t>0.$$

It follows from the discussion above that if $A_1\cap A_2=\varnothing$, $A_i\in\mathfrak{B}_0^m$, then $\mu(t,A_1)$ and $\mu(t,A_2)$ are orthogonal local square integrable martingales. We shall now reconsider the function $\pi(t,A)$. It was mentioned above that $\pi(t,A)$ is additive, as a function of A. Moreover, if B_n, $n=1,2,$ is a monotonically nondecreasing sequence of Borel sets in $\mathscr{R}^m$, $B_0=\bigcup_{n=1}^\infty B_n$ and $B_0\in\mathfrak{B}_0^m$ then it follows from the equality $\nu(t, B_0) = \lim \nu(t, B_n)$ that

$$\mathsf{E}\pi(t, B_n) = \mathsf{E}\nu(t, B_n) \to \mathsf{E}\nu(t, B_0) = \mathsf{E}\pi(t, B_0).$$

In that case, however, $\pi(t, B_n) \to \pi(t, B_0)$ in L_1 as well as with probability 1.

Note that if we set $\pi(t, \{0\}) = 0$, where $\{0\}$ is the singleton containing 0, then a modification of the random function $\pi(t, A)$ exists with realizations defined on $[0, \infty) \times \mathfrak{B}^m$ (possibly taking on the value $+\infty$) which are measures on $\mathfrak{B}^m$ for any $t \in [0, \infty)$ with probability 1 and are monotonically nondecreasing continuous functions of argument t for an arbitrary fixed $A \in \mathfrak{B}^m$.

The proof of this assertion can be obtained analogously to the proof of Theorem 3 in Section 1 of Volume I, Chapter I, which deals with the existence of regular conditional distributions of a random element.

In what follows $\pi(t, A)$ will denote just this type of modification of the random function (unless stated otherwise).

Definition. A family of martingales (local square integrable martingales) $\mu(t, A)$, $A \in \mathfrak{B}_0$, $t \geq 0$, $\mu(0, A) = 0$, adopted to the current $\{\mathfrak{F}_t, t \geq 0\}$ and satisfying the conditions

1) $\mu(t, A_1) + \mu(t, A_2) = \mu(t, A_1 \cup A_2)$, for $A_1 \cap A_2 = \varnothing$,
2) $\mu(t, A_1)\mu(t, A_2) \in \mathcal{LM}$, for $A_1 \cap A_2 = \varnothing$
3) $\langle \mu(\cdot, A), \mu(\cdot, A) \rangle_t = \pi(t, A)$,

where $\pi(t, A)$ is a random function which is, for fixed t with probability 1, a measure on $\mathfrak{B}^m$ and which is a continuous monotonically nondecreasing function of t for a fixed A, is called an *orthogonal martingale measure* (*orthogonal local martingale measure*).

The function $\pi(t, A)$ is called a *characteristic of the martingale measure* $\mu(t, A)$. The word "orthogonal" is sometimes omitted since martingale measures other than orthogonal are not discussed in what follows.

The reasoning above proves the following

Theorem 8. *An arbitrary integer-valued random measure* $\nu(t, A)$, $(t, A) \in [0, \infty) \times \mathfrak{B}_0^m$, *satisfying conditions* 1)–3) *can be represented in the form*

$$\nu(t, A) = \mu(t, A) + \pi(t, A) \tag{21}$$

where $\mu(t, A)$ *is an orthogonal local martingale measure with characteristic* $\pi(t, A)$.

The function $\pi(t, A)$ plays a dual role in the decomposition of measure $\nu(t, A)$ given by (21). On the one hand, the difference $\nu(t, A) - \pi(t, A)$ is a martingale (local martingale). In view of this property the measures $\pi(t, A)$ and $\nu(t, A)$ will be called *associated*. On the other hand, the function $\pi(t, A)$ is a characteristic of $\mu(t, A)$. The fact that the function $\pi(t, A)$ associated with $\nu(t, A)$ turns out to be at the same time a characteristic of the martingale $\mu(t, A) = \nu(t, A) - \pi(t, A)$ is a very important generalization of the following elementary fact: the mathematical expectation of a Poisson distribution is equal to its variance.

Since $\pi(\Delta, A)$ is a measure with probability 1, the integral

$$\int_0^\infty \int_{\mathscr{R}^m} f(t, u)\pi(dt, du)$$

is well defined for any nonnegative Borel function $f(t, u)$, $(t, u) \in [0, \infty) \times \mathscr{R}^m$.

We shall now find the relation between this integral and the integral over the measure $\nu(\Delta, A)$.

Recall that $\mathfrak{L}_0$ or $\mathfrak{L}_0(\mathfrak{T}_0 \times \mathfrak{B}_0^m)$ denotes the class of random functions that are simple relative to the semiring of sets $\mathfrak{T}_0 \times \mathfrak{B}_0^m$ and are bounded with probability 1 and are adopted to the current of σ-algebras $\{\mathfrak{F}_t, t \geq 0\}$, i.e., functions of the form

$$\varphi(t, u) = \sum_{k=1}^n \gamma_k \chi_{\Delta_k \times A_k}(t, u),$$

where $\Delta_k = (t_{k-1}, t_k]$, $0 \leq t_0 < t_1 < \cdots < t_n$, $A_k \in \mathfrak{B}_0$, and γ_k is an $\mathfrak{F}_{t_{k-1}}$-measurable bounded random variable with probability 1 ($|\gamma_k| \leq c$, $k = 1, \ldots, n$, where c is an absolute constant). It follows from formula (21) that if $\varphi \in \mathfrak{L}_0$ then

$$\mathsf{E} \int_0^\infty \int_{\mathscr{R}^m} \varphi(t, u)\nu(dt, du) = \mathsf{E} \int_0^\infty \int_{\mathscr{R}^m} \varphi(t, u)\pi(dt, du). \tag{22}$$

The last relation is carried over to arbitrary nonnegative functions $\varphi(t, u)$ jointly measurable in the variables (t, u, ω), adopted to the current of σ-algebras $\{\mathfrak{F}_t, t \geq 0\}$ and continuous from the left for a fixed (u, ω) for all $t > 0$. To verify this consider first functions $\varphi(t, u)$ satisfying yet additional conditions, namely, bounded with probability 1 and vanishing for either $t \geq T$ or for $u \in S_\varepsilon$. Then it is easy to see that equality (22) is carried over functions $\varphi_\varepsilon(t, u)$ of the form $\varphi_\varepsilon(t, u) = \varphi(t_{k-1}, u)$ for $t \in (t_{k-1}, t_k]$. Approaching the limit as $\max_k (t_k - t_{k-1}) \to 0$, we obtain

$$\mathsf{E} \int_0^\infty \int_{\mathscr{R}^m} \varphi(t-, u)\nu(dt, du) = \mathsf{E} \int_0^\infty \int_{\mathscr{R}^m} \varphi(t-, u)\pi(dt, du).$$

This implies the validity of equality (22) for the class of functions under consideration. A standard limiting transition over monotonically nondecreasing sequences of functions yields a proof of relation (22) also for arbitrary nonnegative functions satisfying the conditions stipulated above.

In particular, if an integer-valued random measure $\nu(t, A)$ satisfies condition

$$\mathsf{E} \int_{\mathscr{R}^m} |u|^2 \nu(t, du) < \infty$$

then for the associated measure $\pi(t, A)$

$$\mathsf{E} \int_{\mathscr{R}^m} |u|^2 \pi(t, du) < \infty \tag{23}$$

also.

Theorem 9. *Let* $\xi(t)\in\mathcal{M}_2^r$, $t\in[0,\infty)$. *Then*

$$\xi(t)=\xi_c(t)+\int_{\mathcal{R}^m} u\mu(t,du) \qquad \forall t>0, \tag{24}$$

where $\xi_c(t)\in\mathcal{M}_2^c$ *and* $\mu(t,A)$ *is an orthogonal martingale measure with characteristic* $\pi(t,A)$ *and, moreover,*

$$\mu(t,A)+\pi(t,A)=\nu(t,A),$$

where $\nu(t,A)$ *is the number of jumps of process* $\xi(s)$, $0\leqslant s\leqslant t$, *with values falling into the set* A $(A\in\mathfrak{B}_0)$, *and*

$$\mathsf{E}\int_{\mathcal{R}^m}|u|^2\pi(t,du)<\infty, \qquad \forall t>0.$$

Each component of the process

$$\xi_d(t)=\int_{\mathcal{R}^m} u\mu(t,du)$$

is orthogonal to an arbitrary continuous martingale adopted to a current of σ-*algebras* $\{\mathfrak{F}_t, t\in[0,T]\}$.

Proof. Let $\mu(t,A)$ and $\pi(t,A)$ be defined as indicated above. Since $\pi(t,A)$ is a characteristic of a martingale measure $\mu(t,A)$ and it follows from (23) that $u^k\in H_2^\pi$ $(u=(u^1,u^2,\ldots,u^m), k=1,\ldots,m)$, the integral

$$\xi_d(t)=\int_{\mathcal{R}^m} u\mu(t,du)$$

is well defined and is, viewed as a function of t, a square integrable martingale, and, moreover,

$$\alpha_d(t)\underset{\text{Def}}{=}\sum_{k=1}^m\langle\xi_d^{(k)},\xi_d^{(k)}\rangle_t=\int_{\mathcal{R}^m}|u|^2\pi(t,du).$$

Let S_ε be a sphere in $\mathcal{R}^m$ centered at 0 of radius $\varepsilon>0$, $\bar{S}_\varepsilon=\mathcal{R}^m\backslash S_\varepsilon$. Set

$$\xi_d^\varepsilon(t)=\int_{\bar{S}_\varepsilon} u\mu(t,du).$$

Here $\xi_d(t)$ and $\xi_d^\varepsilon(t)$ denote processes with sample functions belonging to $\mathcal{D}^m[0,T]$. Since

$$\begin{aligned}\mathsf{E}\sup_{0\leqslant t\leqslant T}|\xi_d(t)-\xi_d^\varepsilon(t)|^2&\leqslant 4\mathsf{E}|\xi_d(T)-\xi_d^\varepsilon(T)|^2\\&=4\mathsf{E}\int_{S_\varepsilon}|u|^2\pi(T,du)\to 0 \quad \text{as } \varepsilon\to 0,\end{aligned}$$

one can choose a sequence of values ε_n such that $\xi_d^{(n)}(t)=\xi_d^{\varepsilon_n}(t)$ converges to $\xi_d(t)$ uniformly in t with probability 1. On the other hand, since $\pi(t,A)$, viewed as a function of t, is continuous and the integral $\int_{\mathscr{R}^m} |u|^2 \pi(T, du)$ is finite (with probability 1) it follows that the integral

$$\int_{\bar{S}_\varepsilon} u\pi(t, du)$$

is with probability 1 a continuous function of the argument t. Since

$$\xi_d^\varepsilon(t) = \int_{\bar{S}_\varepsilon} u\nu(t, du) - \int_{\bar{S}_\varepsilon} u\pi(t, du),$$

the jumps of the functions $\xi_d^\varepsilon(t)$ and $\int_{\bar{S}_\varepsilon} u\nu(t, du)$ coincide for all $t \in [0, T]$ (mod P). Consequently, the difference $\xi(t) - \xi_d^\varepsilon(t)$ has no jumps with values in $\bar{S}_\varepsilon$.

Set $\xi_c(t) = \xi(t) - \xi_d(t)$. We have

$$\begin{aligned}\sup_t |\xi_c(t) - \xi_c(t-)| &\leq \sup_t \{|(\xi(t) - \xi_d^{(n)}(t)) - (\xi(t-) - \xi_d^{(n)}(t-))| \\ &\quad + |\xi_d(t) - \xi_d^{(n)}(t)| + |\xi_d(t-) - \xi_d^{(n)}(t-)|\} \\ &\leq \varepsilon_n + 2 \sup_t |\xi_d(t) - \xi_d^{(n)}(t)| \to 0 \quad (\text{mod P}).\end{aligned}$$

Thus $\xi_c(t) = \xi(t-)$ for all $t \in [0, T]$ with probability 1. We have thus proved the continuity of the process $\xi_c(t)$.

We now show that each component of the process $\xi_d(T)$ is orthogonal to an arbitrary continuous martingale (relative to a current of σ-algebras $\mathfrak{F}_t$). For this purpose we first verify that an arbitrary martingale $\eta(t)$ belonging to $\mathcal{M}_2^c$ is orthogonal to $\mu(t, A)$ $(A \in \mathfrak{B}_0^m)$. To compute the joint characteristic of the processes $\eta(t)$ and $\mu(t, A)$ we apply Theorem 21, Corollary 2, in Section 1.

We have

$$\langle \eta, \mu(\cdot, A)\rangle_t = \lim_{|\delta| \to 0} \sum_{k=0}^{n-1} \mathsf{E}\{\Delta\eta(t_k)\, \Delta\mu(t_k, A) | \mathfrak{F}_{t_k}\},$$

$$\delta = \max_{1 \leq k \leq n} |t_k - t_{k-1}|,$$

in the sense of convergence in L_1.

On the other hand,

$$\mathsf{E}\left|\sum_{k=0}^{n-1} \mathsf{E}\{\Delta\eta(t_k)\, \Delta\mu(t_k, A) | \mathfrak{F}_{t_k}\}\right| \leq \mathsf{E} \max_k |\Delta\eta(t_k)| (\nu(T, A) + \pi(t, A)).$$

Since $\max_k |\Delta\eta(t_k)| \to 0$ as $|\delta| \to 0$ and

$$\max_k \Delta\eta(t_k)(\nu(t, A) + \pi(t, A)) \leq 2 \max_{0 \leq t \leq T} \eta(t)(\nu(t, A) + \pi(t, A)),$$

and, moreover, the right-hand side of the last inequality is an integrable function, it follows that

$$\mathsf{E}\left|\sum_{k=0}^{n-1} \mathsf{E}\{\Delta\eta(t_k)\, \Delta\mu(t_k, A) | \mathfrak{F}_{t_k}\}\right| \to 0 \quad \text{as } |\delta| \to 0$$

and $\langle\eta, \mu(\cdot, A)\rangle_t \equiv 0$. Now set

$$\zeta_n(t) = \int_{\mathcal{R}^m} g_n(u)\mu(t, du),$$

where $g_n(u) = \sum c_k \chi_{A_k}(u)$, $A_k \in \mathfrak{B}_0^m$. It follows from the above that $\langle\eta, \zeta_n\rangle_t = 0$. Utilizing inequality $|\langle\eta, \zeta\rangle_t|^2 \leq \langle\eta, \eta\rangle_t\langle\zeta, \zeta\rangle_t$ (cf. (51) in Section 1) and passing to the limit, we easily obtain that $\langle\eta, \zeta\rangle_t \equiv 0$ for any martingale $\zeta(t)$ of the form

$$\zeta(t) = \int_{\mathcal{R}^m} g(u)\mu(t, du),$$

where $g(u)$ is a nonrandom function satisfying the condition

$$\mathsf{E} \int_{\mathcal{R}^m} |g(u)|^2 \pi(T, du) = \int_{\mathcal{R}^m} |g(u)|^2 m(T, du) < \infty,$$

and $m(t, A) = \mathsf{E}\pi(t, A)$ is a measure on $\mathfrak{B}^m$. Setting $\varphi(u) = u^k$ we obtain

$$\langle\eta, \zeta_{\mathrm{d}}^k\rangle_t \equiv 0, \qquad k = 1, 2, \ldots, m. \quad \square$$

Remark. Decomposition of the form

$$\xi(t) = \xi_{\mathrm{c}}(t) + \zeta(t)$$

where $\xi_{\mathrm{c}}(t) \in \mathcal{M}_2^{\mathrm{c}}$ and $\zeta(t)$ is a martingale with components which are orthogonal to every continuous martingale is unique.

Proof. To prove this assertion, it is sufficient to consider the one-dimensional case. If another decomposition $\xi(t) = \xi'_{\mathrm{c}}(t) + \zeta'(t)$ of the same type exists, then $\xi_{\mathrm{c}}(t) - \xi'_{\mathrm{c}}(t) = \zeta'(t) - \zeta(t)$. Since $\zeta'(t)$ and $\zeta(t)$ are orthogonal to the process $\xi_{\mathrm{c}}(t)$ as well as to $\xi'_{\mathrm{c}}(t)$, we have

$$\langle\xi_{\mathrm{c}} - \xi'_{\mathrm{c}}, \zeta' - \zeta\rangle_t = \langle\xi_{\mathrm{c}} - \xi'_{\mathrm{c}}, \xi_{\mathrm{c}} - \xi'_{\mathrm{c}}\rangle_t \equiv 0,$$

which implies that $\xi_{\mathrm{c}}(t) = \xi'_{\mathrm{c}}(t)$ (mod P) for every $t \in [0, T]$. Since both $\xi_{\mathrm{c}}(t)$ and $\xi'_{\mathrm{c}}(t)$ are continuous it follows that the equality $\xi_{\mathrm{c}}(t) = \xi'_{\mathrm{c}}(t)$ holds for all $t \in [0, T]$ with probability 1. $\square$

Corollary. *Let $\xi(t) \in l\mathcal{M}_2$, $t \in [0, T]$. Then a local martingale $\xi_{\mathrm{c}}(t) \in l\mathcal{M}$ and an orthogonal local martingale measure $\mu(t, A)$ on $\mathfrak{B}_0$ with characteristic $\pi(t, A)$ exist such that*

$$\xi(t) = \xi_{\mathrm{c}}(t) + \xi_{\mathrm{d}}(t), \qquad \xi_{\mathrm{d}}(t) = \int_{\mathcal{R}^m} u\mu(t, du),$$

$$\mu(t, A) + \pi(t, A) = \nu(t, A),$$

where $\nu(t, A)$ is defined as in Theorem 9. *Moreover, for an arbitrary $\eta(t) \in l\mathcal{M}_2^{\mathrm{c}}$*

$$\langle\eta, \xi_{\mathrm{d}}^k\rangle_t \equiv 0, \qquad k = 1, \ldots, m.$$

Proof. Let τ be an arbitrary random time reducing $\xi(t)$. In view of the preceding theorem

$$\xi(t\wedge\tau)=\xi_c^{(\tau)}(t)+\xi_d^{(\tau)}(t), \qquad \xi_d^{(\tau)}(t)=\int_{\mathcal{R}^m} u\mu_\tau(t, du);$$

moreover, $\mu_\tau(t, A)+\pi_\tau(t, A)=\nu(t\wedge\tau, A)$ and $\pi_\tau(t, A)$ is an increasing process associated with submartingale $\nu(t\wedge\tau, A)$. Therefore, $\pi_\tau(t, A)=\pi(t\wedge\tau, A)$ and $\mu_\tau(t, A)=\mu(t\wedge\tau, A)$. Taking these remarks into account the proof of the assertions stated in the corollary becomes obvious. □

Stochastic differentials of functions of discontinuous martingales. Let $\nu(t, A)$, $t\in[0, T]$, be an integer-valued random measure (we shall assume that it satisfies the conditions stipulated in the preceding subsection), $\mu(t, A)$ be a martingale measure associated with $\nu(t, A)$, $\pi(t, A)$ be its characteristic, and let

$$\mathsf{E}\int_{\mathcal{R}^m} u^2\pi(T, du)<\infty.$$

Denote by δ a subdivision of the interval $(0, T]$ into intervals $\Delta_k=(s_{k-1}, s_k]$, $k=1,\ldots,n$. Clearly for any $A\in\mathfrak{B}_0$ we have with probability 1

$$\lim_{|\delta|\to 0}\sum_{k=1}^{n}\nu^2(\Delta_k, A)=\nu(T, A).$$

The continuity of function $\pi(t, A)$ and its monotonicity with respect to t implies that with probability 1

$$\lim_{|\delta|\to 0}\sum_{k=1}^{n}\pi^2(\Delta_k, A)=0, \qquad \lim_{|\delta|\to 0}\sum_{k=1}^{n}\nu(\Delta_k, A)\pi(\Delta_k, A)=0.$$

If, however, $A_1\cap A_2=\emptyset$, $A_i\in\mathfrak{B}_0$, then

$$\lim_{|\delta|\to 0}\sum_{k=0}^{n}\nu(\Delta_k, A_1)\nu(\Delta_k, A_2)=0 \ (\mathrm{mod}\ \mathsf{P}).$$

The preceding equalities yield that

$$\sum_{k=0}^{n}\mu^2(\Delta_k, A)\to\nu(T, A)\ (\mathrm{mod}\ \mathsf{P}), \tag{25}$$

$$\sum_{k=0}^{n}\mu(\Delta_k, A_1)\mu(\Delta_k, A_2)\to 0\ (\mathrm{mod}\ \mathsf{P}). \tag{26}$$

Let $\gamma(t, u)\in\mathfrak{L}_0(\mathfrak{T}_0\times\mathfrak{B}_0)$ and

$$\zeta(t)=\zeta(t,\gamma)=\int_0^t\int_{\mathcal{R}^m}\gamma(s, u)\mu(ds, du).$$

Utilizing the preceding relations one can easily obtain the square variation $[\zeta, \zeta]$ of the process $\zeta(t)$:

$$[\zeta, \zeta]_T \underset{\text{Def}}{=} \mathsf{P}\text{-}\lim_{|\delta|\to 0} \sum_{r=1}^{n} (\zeta(t_r)-\zeta(t_{r-1}))^2. \tag{27}$$

Since the difference $\zeta(t_r)-\zeta(t_{r-1})\to 0$ in probability as $|\delta|\to 0$ one can assume in order to evaluate the limit (27) that points s_k are contained in subdivision δ. We apply relation (25) to the interval $(s_{k-1}, s_k]$ (instead of the interval $(0, T]$) and set $A = B_k$. We thus obtain

$$\mathsf{P}\text{-}\lim \sum_{s_{k-1}<t_k\leq s_k} (\zeta(t_r)-\zeta(t_{r-1}))^2 = \gamma_k^2 \nu(\Delta_k, B_k).$$

Summing up over k we have

$$[\zeta, \zeta]_T = \int_0^T \int_{\mathscr{R}^m} \gamma^2(s, u)\nu(ds, du). \tag{28}$$

Set

$$\zeta_i(t) = \zeta(t, \gamma_i), \qquad i = 1, 2, \qquad \gamma_i \in \mathfrak{L}_0(\mathfrak{T}_0, \mathfrak{B}_0^m),$$

and

$$[\zeta_1, \zeta_2]_T \underset{\text{Def}}{=} \mathsf{P}\text{-}\lim_{|\delta|\to 0} \sum_{r=1}^{n} (\zeta_1(t_r)-\zeta_1(t_{r-1}))(\zeta_2(t_r)-\zeta_2(t_{r-1})).$$

Formula (28) implies that

$$[\zeta_1, \zeta_2]_T = \int_0^T \int_{\mathscr{R}^m} \gamma_1(s, u)\gamma_2(s, u)\nu(ds, du). \tag{29}$$

We now establish for processes $\zeta_i(t)$ the formula of integration by parts.

First, however, we shall remark about the integrals which occur below.

Let $\{\eta(t), \mathfrak{F}_t, t\in[0, T]\}$ be a random process with sample functions belonging to $\mathscr{D}[0, T]$, and let $\zeta(t)$ be the process introduced above. The integral

$$\int_0^t \eta(t)\zeta(dt),$$

viewed as a stochastic integral over a local square integrable martingale, exists. Indeed, if for some subdivision δ of the interval $[0, T]$ we set $\eta^{(\delta)}(t) = \eta(t_{k-1})$ for $t\in(t_{k-1}, t_k)$ provided $|\eta(t_{k-1})| < 1/|\delta|$ and $\eta^{(\delta)}(t) = 0$ otherwise, then $\eta^{(\delta)}(t)\to \eta(t-)$ with probability 1 and, moreover, $\eta(t)-\eta(t-)\neq 0$ for at most a countable number of points and therefore $\eta(t)-\eta(t-) = 0$ for almost all (t, a) with probability 1 with respect to measure $\pi(\cdot, \cdot)$.

Since

$$\sup_{0\leq t\leq T} |\eta^{(\delta)}(t)| \leq \sup_{0\leq t\leq T} |\eta(t)| \leq \gamma < \infty,$$

it follows that

$$\int_0^T (\eta(t)-\eta^{(\delta)}(t))^2 \int \gamma^2(t, u)\pi(dt, du) \to 0 \pmod{\mathsf{P}}$$

and $\eta(t) \in H_2(\mathfrak{F}_t, \zeta(\cdot))$ (cf. Section 2). This proves the existence of the integral under consideration as well as the equality

$$\int_0^T \eta(t)\, d\zeta(t) = \mathsf{P}\text{-}\lim_{|\delta|\to 0} \int_0^T \eta^{(\delta)}(t)\, d\zeta(t) = \mathsf{P}\text{-}\lim_{|\delta|\to 0} \sum_{k=1}^{n} \eta^{(\delta)}(t_{k-1})\, \Delta\zeta(t_k),$$

where $\Delta\zeta(t_k) = \zeta(t_k) - \zeta(t_{k-1})$.

On the other hand, the function $\eta^{(\delta)}(t)\gamma(t, u)$ is a simple function in $\mathfrak{L}_0(\mathfrak{T}_0 \times \mathfrak{B}_0^m)$ and, as it was proved above,

$$\int_0^T \int_{\mathcal{R}^m} (\eta^{(\delta)}(t)\gamma(t, u) - \eta(t)\gamma(t, u))^2 \pi(dt, du) \to 0.$$

Therefore $\eta(t)\gamma(t, u) \subset \Pi_2^\pi$ and

(30) $$\int_0^T \eta(t)\zeta(dt) = \int_0^T \int_{\mathcal{R}^m} \eta(t)\gamma(t, u)\mu(dt, du),$$

where the integral on the right is an integral with respect to a local martingale measure.

In view of the discussion above we have for the process $\eta(t)$

$$\int_0^T \int_{\mathcal{R}^m} \eta(t)\gamma(t, u)\mu(dt, du) = \int_0^T \int_{\mathcal{R}^m} \eta(t-)\gamma(t, u)\mu(dt, du),$$

$$\int_0^T \int_{\mathcal{R}^m} \eta(t)\gamma(t, u)\pi(dt, du) = \int_0^T \int_{\mathcal{R}^m} \eta(t-)\gamma(t, u)\pi(dt, du).$$

However, an analogous equality does not hold for the integrals with respect to measure $\nu(dt, du)$ and the relationship between these three types of integrals is given by

(31) $$\int_0^T \int_{\mathcal{R}^m} \eta(t)\gamma(t, u)\mu(dt, du) + \int_0^T \int_{\mathcal{R}^m} \eta(t)\gamma(t, u)\pi(dt, du) = \int_0^T \int_{\mathcal{R}^m} \eta(t-)\gamma(t, u)\nu(dt, du).$$

We now return to processes $\zeta_i(t)$ introduced above. It follows from the previous arguments that

$$\int_0^T \zeta_1(t)\zeta_2(dt) = \mathsf{P}\text{-}\lim \sum_{k=1}^{n} \zeta_1(t_{k-1})\, \Delta\zeta_2(t_k).$$

Interchanging the positions of indices 1 and 2 in this equality and adding up the equalities obtained, we get

$$\begin{aligned}\int_0^T \zeta_1(t)\zeta_2(dt) + \int_0^t \zeta_2(t)\zeta_1(dt) &= \mathsf{P}\text{-}\lim\Big(\sum_{k=1}^{n} \zeta_1(t_k)\zeta_2(t_k) - \zeta_1(t_{k-1})\zeta_2(t_{k-1}) \\ &\quad - \Delta\zeta_1(t_k)\, \Delta\zeta_2(t_k)\Big) \\ &= \zeta_1(T)\zeta_2(T) - [\zeta_1, \zeta_2]_T \pmod{\mathsf{P}}.\end{aligned}$$

The last equality is clearly retained if we replace T by any fixed t, $t \in [0, T]$. Since the functions appearing in both sides of the equality are continuous from the right, this equality will be valid with probability 1 for all $t \in [0, T]$. Thus

$$
\begin{aligned}
\zeta_1(t)\zeta_2(t) = {} & \int_0^t \zeta_1(s)\zeta_2(ds) + \int_0^t \zeta_2(s)\zeta_1(ds) \\
& + \int_0^t \int_{\mathcal{R}^m} \gamma_1(s, u)\gamma_2(s, u)\nu(ds, du)
\end{aligned} \tag{32}
$$

for all $t \in [0, T]$ with probability 1. This can be written also in the differential form

$$
d(\zeta_1(t)\zeta_2(t)) = \zeta_1(t)\, d\zeta_2(t) + \zeta_2(t)\, d\zeta_1(t) + \int_{\mathcal{R}^m} \gamma_1(t, u)\gamma_2(t, u)\nu(dt, du).
$$

On the other hand, if $\alpha(t)$ is with probability 1 a continuous function of bounded variation on $[0, T]$ adopted to the current of σ-algebras $\{\mathfrak{F}_t, t \in [0, T]\}$, we then have for any $t \in [0, T]$

$$
\alpha(t)\zeta(t) = \int_0^t \zeta(s)\, d\alpha(s) + \int_0^t \alpha(s)\, d\zeta(s)
$$

or

$$
d(\alpha(t)\zeta(t)) = \zeta(t)\, d\alpha(t) + \alpha(t)\, d\zeta(t). \tag{33}
$$

Indeed, let $\Delta\alpha(t_k) = \alpha(t_k) - \alpha(t_{k-1})$, $\Delta\zeta(t_k) = \zeta(t_k) - \zeta(t_{k-1})$. Then

$$
\sum_{k=1}^n \zeta(t_{k-1})\, \Delta\alpha(t_k) + \alpha(t_{k-1})\, \Delta\zeta(t_k) = \alpha(T)\zeta(T) - \sum_{k=1}^n \Delta\alpha(t_k)\, \Delta\zeta(t_k). \tag{34}
$$

Since

$$
\begin{aligned}
\left| \sum_{k=1}^n \Delta\alpha(t_k)\, \Delta\zeta(t_k) \right| \le \max_k |\Delta\alpha(t_k)| \Big(& \int_0^T \int_{\mathcal{R}^m} |\gamma(s, u)|\, \nu(ds, du) \\
& + \int_0^T \int_{\mathcal{R}^m} |\gamma(s, u)|\, \pi(ds, du) \Big) \to 0 \quad \text{as} |\delta| \to 0,
\end{aligned}
$$

equality (34) implies relation (33) as $|\delta| \to 0$.

Formula (32) can be rewritten as follows:

$$
\begin{aligned}
\zeta_1(t)\zeta_2(t) = {} & \int_0^t \zeta_1(s)\zeta_2(ds) + \int_0^t \zeta_2(s)\zeta_1(ds) \\
& + \int_0^t \int_{\mathcal{R}^m} \gamma_1(s, u)\gamma_2(s, u)\mu(ds, du) \\
& + \int_0^t \int_{\mathcal{R}^m} \gamma_1(s, u)\gamma_2(s, u)\pi(ds, du).
\end{aligned} \tag{34*}
$$

In this form it can be easily extended to arbitrary fields $\gamma_i(t, u) \in H_2^\pi(\mathfrak{F}_t, \mu)$, with sample functions which are bounded for each ω with probability 1 on the interval $t \in [0, T]$. Indeed, let $\gamma_1(t, u) \in H_2^\pi$ and $\gamma_2(t, u) \in \mathfrak{L}_0(\mathfrak{T} \times \mathfrak{B}_0^m)$. Then one can find a sequence $\gamma_1^n(t, u) \in \mathfrak{L}_0(\mathfrak{T} \times \mathfrak{B}_0^m)$ such that

$$
\int_0^T \int_{\mathcal{R}^m} |\gamma_1(s, u) - \gamma_1^{(n)}(s, u)|^2 \pi(ds, du) \to 0
$$

as $n\to\infty$ with probability 1. In this case, however, we have

$$\int_0^t\int_{\mathscr{R}^m}\gamma_1^{(n)}\gamma_2\pi(ds,du)\to\int_0^t\int_{\mathscr{R}^m}\gamma_1\gamma_2\pi(ds,du),$$
$$\int_0^t\int_{\mathscr{R}^m}\gamma_1^{(n)}\gamma_2\mu(ds,du)\to\int_0^t\int_{\mathscr{R}^m}\gamma_1\gamma_2\mu(ds,du)$$

in the sense of convergence in probability. The first relation follows from the Cauchy–Schwarz inequality while the second relation holds because the characteristic of a local martingale—the difference between the left-hand and right-hand sides of this relation—is equal to

$$\int_0^t\int_{\mathscr{R}^m}(\gamma_1^{(n)}-\gamma_1)^2\gamma_2^2\pi(ds,du),$$

which $\to 0$ with probability 1. Analogously, using formulas (27) and (29) in Section 2 one can easily verify that

$$\int_0^t\zeta_1^{(n)}(s)\zeta_2(ds)\to\int_0^t\zeta_1(s)\zeta_2(ds),\qquad \int_0^t\zeta_2(s)\zeta_1^{(n)}(ds)\to\int_0^t\zeta_2(s)\zeta_1(ds)$$

in probability. Similarly, utilizing the boundedness of the sample function $\gamma_1(t,u)$ we can replace the functions $\gamma_2(t,u)\in\mathfrak{L}_0(\mathfrak{T}\times\mathfrak{B}_0^m)$ in formula (34*) by arbitrary functions belonging to H_2^{π}. It is also easy to verify that equalities (30) and (31) can be extended to be valid for the functions $\gamma(t,u)$ under consideration.

Formula (34*) is also applicable for processes $\eta_i(t)$ of the form

$$\eta_i(t)=\int_0^t\xi_i(s)\,d\zeta_i$$

where $\zeta_i(t)$ is defined as above, and $\xi_i(s)$ are processes continuous from the right. As it was stated above, these integrals are well defined and coincide with integrals in which $\xi_i(t)$ is replaced by $\xi_i(t-)$.

Set

$$\zeta_i^*(t)=\eta_i(t)+\alpha_i(t),$$
$$\alpha_i(t)=\int_0^t\int_{\mathscr{R}^m}\xi_i(s)\gamma_i(s,u)\pi(ds,du).$$

It follows from formula (32) that

$$d(\zeta_1^*(t)\zeta_2^*(t))=\zeta_1^*(t)\,d\zeta_2^*(t)+\zeta_2^*(t)\,d\zeta_1^*(t)$$
$$+\int_0^t\int_{\mathscr{R}^m}\prod_{i=1}^2\xi_i(s)\gamma_i(s,u)\mu(ds,du) \tag{35}$$
$$+\int_0^t\int_{\mathscr{R}^m}\prod_{i=1}^2\xi_i(s)\gamma_i(s,u)\pi(ds,du).$$

In view of (31)

$$\zeta_i^*(t)=\int_0^t\int_{\mathscr{R}^m}\xi_i(s-)\gamma(s,u)\nu(ds,du)$$

and we thus obtain that

$$
(36)\qquad d(\zeta_1^*(t)\zeta_2^*(t))=\int_{\mathcal{R}^m}\Big[\zeta_1^*(s-)\xi_2(s-)\gamma_2(s,u)+\zeta_2^*(s-)\xi_1(s-)\gamma_1(s,u)
+\prod_{i=1}^{2}\xi_i(s-)\gamma_i(s,u)\Big]\nu(ds,du).
$$

In particular, if we set

$$
(36^*)\quad \xi_i(t)=1,\qquad \zeta_1(t)=\zeta_2(t)=\zeta(t,\gamma),\qquad \psi(t)=\int_0^t\int_{\mathcal{R}^m}\gamma(s,u)\nu(ds,du),
$$

and assume for simplicity that $\gamma(t,u)\in\mathfrak{L}_0(\mathfrak{T}\times\mathfrak{B}_0^m)$, then

$$
\begin{aligned}
d\psi^2(t)&=2\psi(t-)\,d\psi(t)+\int_{\mathcal{R}^m}\gamma^2(t,u)\nu(dt,du)\\
&=\int_{\mathcal{R}^m}(2\psi(t-)+\gamma(t,u))\gamma(t,u)\nu(dt,du)\\
&=\int_{\mathcal{R}^m}[(\psi(t-)+\gamma(t,u))^2-\psi^2(t-)]\nu(dt,du).
\end{aligned}
$$

It is easy to verify using, for instance, an induction argument and formula (35) that for any integer n,

$$
(37)\qquad d\psi^n(t)=\int_{\mathcal{R}^m}([\psi(t-)+\gamma(t,u)]^n-\psi^n(t))\nu(dt,du).
$$

Indeed if we assume that formula (37) is valid for some n (it was shown above that it is valid for $n=2$) then it follows from (35) that

$$
\begin{aligned}
d\psi^{n+1}(t)&=\psi^n(t-)\,d\psi(t)+\psi(t-)\,d\psi^n(t)\\
&\quad+\int_{\mathcal{R}^m}([\psi(t-)+\gamma(t,u)]^n=\psi^n(t-))\gamma(t,u)\nu(dt,du)\\
&=\int_{\mathcal{R}^m}\{\psi^n(t-)\gamma(t,u)+[\psi(t-)\\
&\quad+\gamma(t,u)]^n-\psi^n(t-)+(\psi(t-)+\gamma(t,u))\}\nu(dt,du)\\
&=\int_{\mathcal{R}^m}([\psi(t-)+\gamma(t,u)]^{n+1}-\psi^{n+1}(t-))\nu(dt,du),
\end{aligned}
$$

which proves the validity of (37). Now (37) implies that for an arbitrary polynomial $P(x)$

$$
(38)\qquad dP(\psi(t))=\int_{\mathcal{R}^m}[P(\psi(t-)+\gamma(t,u))-P(\psi(t-))]\nu(dt,du).
$$

Now let $P(x_1,\dots,x_q)$ be a polynomial in q independent variables $\psi_j(t,u)\in\mathfrak{L}_0(\mathfrak{T}_0\times\mathfrak{B}_0^m)$, $j=1,\dots,q$, and $\psi_j(t)$ be a process defined by expression (36*) in which $\gamma(t,u)$ is replaced by $\gamma_j(t,u)$. We shall interpret sequences $\psi_j(t)$ and $\gamma_j(t,u)$, $j=1,\dots,q$, as vector-valued random functions with values in $\mathcal{R}^q$ and $P(x_1,\dots,x_q)=P(x)$, $x\in\mathcal{R}^q$, as a function in $\mathcal{R}^q$. If we set $\psi(t)=\{\psi_1(t),\dots,\psi_q(t)\}$, $\gamma(t,u)=\{\gamma_1(t,u),\dots,\gamma_q(t,u)\}$, formula (38) still remains valid. To prove this it is sufficient to consider the polynomial $P(x_1,\dots,x_q)$ of the form $P(x_1, \ldots, x_q) = \prod_{j=1}^q P_j(x_j)$. Now apply induction on the number of independent variables.

Let formula (38) be valid for the polynomials $Q(x_1, \ldots, x_j) = \prod_{k=1}^j P_k(x_k)$. Then in view of formula (35)

$$\begin{aligned} d(QP_{j+1}(\psi_{j+1})) &= P_{j+1}(\psi_{j+1})\, dQ + Q\, dP_{j+1}(\psi_{j+1}) \\ &\quad + \int_{\mathcal{R}^m} \left[\prod_{k=1}^{j} P_k(\psi_k(t-) + \gamma_k(t, u)) - \prod_{k=1}^{j} P_k(\psi_k(t-)) \right] \\ &\quad \times [P_{j+1}(\psi_{j+1}(t-) + \gamma_{j+1}(t, u)) - P_{j+1}(\psi_{j+1}(t-))] \nu(dt, du) \\ &= \int_{\mathcal{R}^m} \left[\prod_{k=1}^{j+1} P_k(\psi_k(t-) + \gamma_k(t, u)) - \prod_{k=1}^{j+1} P_k(\psi_k(t-)) \right] \nu(dt, du), \end{aligned}$$

which proves (38) in the case under consideration.

We note another generalization of this formula. Assume that $f(t, x_1, \ldots, x_q)$ is a polynomial in variables $x_1, \ldots, x_q$ with coefficients which are random functions of time t adopted to the current of σ-algebras $\{\mathfrak{F}_t, t \in [0, T]\}$ which are continuous and of bounded variation with probability 1. The it follows from formula (38) that

$$df(t, \psi(t)) = d_t f(t, \psi(t)) + \int_{\mathcal{R}^q} [f(t, \psi(t-) + \gamma(t, u)) - f(t, \psi(t-))] \nu(dt, du),$$

where the notation $d_t f(t, x)$ means that coefficients $\alpha(t)$ of the polynomial $P(t, x)$ ought to be replaced by $d\alpha$.

We now substitute the measure ν and the integrals $\psi_k(t)$ in formula (38) by the measure μ and the integrals $\zeta_k(t) = \psi_k(t) - \beta_k(t)$, where $\beta_k(t) = \int_0^t \int_{\mathcal{R}^m} \gamma(s, u) \pi(ds, du)$. Then this formula becomes

$$\begin{aligned} df(t, \zeta(t)) &= d_t f(t, \zeta(t)) + \int_{\mathcal{R}^m} L_d(f, \zeta) \pi(dt, du) \\ &\quad + \int_{\mathcal{R}^m} [f(t, \zeta(t) + \gamma(t, u)) - f(t, \zeta(t))] \mu(dt, du), \end{aligned} \tag{39}$$

where

$$L_d(f, \zeta) = L_d(f) = f(t, \zeta(t) + \gamma(t, u)) - f(t, \zeta(t)) - (\nabla f(t, \zeta(t)), \gamma(t, u)). \tag{40}$$

Here in the left-hand side of equalities (39) and (40) $\zeta(t-)$ is replaced by $\zeta(t)$ (without affecting the validity of the formulas) since the integrals under consideration remain well defined subject to this replacement and the values of both versions coincide with probability 1.

Two limit transitions are now carried out in the relationship obtained. First we replace polynomials in x by arbitrary differentiable functions $f(t, x)$ and secondly we substitute functions $\gamma(t, u) \in \mathfrak{L}_0$ by arbitrary functions $\gamma(t, u)$ belonging to H_2^π.

The first limit transition. Let $\gamma(t, u) \in \mathfrak{L}_0$. We write relation (39) in the integral form and assume that $P(t, x) = P_n(t, x) \to f(t, x)$. In the relation obtained, one can replace $P(t, x)$ by $f(t, x)$ provided, for instance, the following conditions are satisfied.

a) Polynomials $P_n(t, x)$ and the function $f(t, x)$ are differentiable with respect to t and $f(t, x)$ and $f'_t(t, x)$ are continuous with probability 1 on $[0, T]\times\mathscr{R}^q$ and $\partial/\partial t\, P_n(t, x)\to f'_t(t, x)$ for all values of x.

b) Function $f(t, x)$ is continuously differentiable with respect to x and $\nabla P_n(t, x)\to\nabla f(t, x)$ for all x with probability 1.

Clearly, the sequence of polynomials $P_n(t, x)$ satisfying conditions a) and b) exist provided $f(t, x)$ satisfies the requirements stipulated above, i.e., provided $f(t, x)$ is differentiable with respect to both t and x and its derivatives $f'_t(t, x)$, $\nabla f(t, x)$ are continuous with probability 1 on $[0, T]\times\mathscr{R}^q$.

The second limit transition. Let the function $f(t, x)$ be differentiable with respect to t and possess bounded and continuous partial derivatives with respect to x_k $(k=1, \dots, q)$ of the first and second orders $((t, x)\in[0, T]\times\mathscr{R}^m)$ and $\gamma(t, u)\in H_2^\pi$. Consider the sequence $\gamma_n(t, u)\in\mathfrak{L}_0$ $(n=1, \dots)$ and assume that

$$\int_0^T\int_{\mathscr{R}^m}|\gamma(t, u)-\gamma_n(t, u)|^2\pi(dt, du)\to 0.$$

The last relation implies that (cf. Section 2, (29))

$$\mathsf{P}\{\sup_{0\le t\le T}|\zeta(t)-\zeta_n(t)|>\varepsilon\}\to 0\quad\text{as } n\to\infty,\ \ \forall\varepsilon>0,$$

where $\zeta_n(t)=\zeta(t, \gamma_n)$. Therefore it may be assumed that $\zeta_n(t)$ approaches $\zeta(t)$ uniformly in $t\in[0, T]$ with probability 1.

Let S_ε be a sphere of radius ε centered at 0. Then

$$\int_0^T\int_{S_\varepsilon}L_d(f, \zeta_n)\pi(dt, du)\le C\int_0^T\int_{S_\varepsilon}|\gamma_n|^2\pi(dt, du),$$

and this quantity tends to 0 as $\varepsilon\to 0$ uniformly in n with probability 1. Moreover (cf. Section 2, (29)),

$$\begin{aligned}&\mathsf{P}\{|\textstyle\int_0^T\int_{S_\varepsilon}[f(t, \zeta_n+\gamma_n)-f(t, \zeta_n)]\mu(dt, du)|>\delta\}\\&\quad\le\frac{N}{\delta^2}+\mathsf{P}\{\textstyle\int_0^T\int_{S_\varepsilon}[f(t, \zeta_n+\gamma_n)-f(t, \zeta_n)]^2\pi(dt, du)>N\}\\&\quad\le\frac{N}{\delta^2}+\mathsf{P}\{\textstyle\int_0^T\int_{S_\varepsilon}C^2|\gamma_n|^2\pi(dt, du)>N\},\end{aligned}$$

which also tends to zero as $\varepsilon\to 0$ uniformly in n for any $\delta>0$.

Now it is easy to verify that

$$\int_0^T\int_{\mathscr{R}^m}L_d(f, \zeta_n)\pi(dt, du)\to\int_0^T\int_{\mathscr{R}^m}L_d(f, \zeta)\pi(dt, du)\tag{41}$$

and

$$\int_0^T\int_{\mathscr{R}^m}[f(t, \zeta_n+\gamma_n)-f(t, \zeta_n)]\mu(dt, du)\to\int_0^T\int_{\mathscr{R}^m}[f(t, \zeta+\gamma)-f(t, \zeta)]\mu(dt, du).\tag{42}$$

Indeed in view of the remarks above, in proving relations (41) and (42), one can replace the range of integration $\mathscr{R}^m$ by $\mathscr{R}^m \backslash S_\varepsilon$. In that case we have the following bounds:

$$\begin{aligned}
&\left|\int_0^T \int_{\mathscr{R}^m \backslash S_\varepsilon} (L_d(f, \zeta) - L_d(f, \zeta_n))\pi(dt, du)\right| \\
&\leq C \int_0^T \int_{\mathscr{R}^m \backslash S_\varepsilon} (|\zeta(t) - \zeta_n(t)| + |\gamma(t, u) - \gamma_n(t, u)| \\
&\qquad + |\gamma|\, |\nabla f(t, \zeta) - \nabla f(t, \zeta_n)|)\pi(dt, du) \\
&\leq C \sup_{0 \leq t \leq T} |\zeta(t) - \zeta_n(t)|\pi(T, \mathscr{R}^m \backslash S_\varepsilon) \\
&\qquad + C\left[\int_0^T \int_{\mathscr{R}^m} |\gamma(t, u) - \gamma_n(t, u)|^2 \pi(dt, du) \cdot \pi(T, \mathscr{R}^m \backslash S_\varepsilon)\right]^{1/2} \\
&\qquad + C \sup_{0 \leq t \leq T} |\nabla f(t, \zeta(t)) - \nabla f(t, \zeta_n(t))| \\
&\qquad \times \left[\int_0^T \int_{\mathscr{R}^m} |\gamma|^2 \pi(dt, du) \cdot \pi(T, \mathscr{R}^m \backslash S_\varepsilon)\right]^{1/2},
\end{aligned}$$

and this upper bound tends to 0 as $n \to \infty$ with probability 1.

This proves relation (41). An analogous method can be applied to verify relation (42).

In order to show the validity of (39) for the suitably differentiable functions $f(t, x)$ and $\gamma(t, u) \in H_2^\pi$ under consideration it is sufficient to show that

$$\int_0^T \int_{\mathscr{R}^m \backslash S_\varepsilon} [f(t, \zeta_n + \gamma_n) - f(t, \zeta_n) - f(t, \zeta + \gamma) + f(t, \zeta)]^2 \pi(dt, du) \to 0.$$

This follows from the arguments analogous to those presented above. Thus, formula (39) is valid for functions $f(t, x)$ possessing bounded and continuous partial derivatives of the first and second orders and $\gamma(t, u) \in H_2^\pi$.

Now one can, once again, use arguments analogous to those presented at the first stage and show that formula (39) is preserved for any function $f(t, x)$ which is continuously differentiable with respect to t and twice continuously differentiable with respect to x such that

$$\int_0^T \int_{\mathscr{R}^m} L_d(f, \zeta)\pi(dt, du) < \infty, \qquad \int_0^T \int_{\mathscr{R}^m} |f(t, \zeta + \gamma) - f(t, \zeta)|^2 \pi(dt, du) < \infty$$

with probability 1. We shall denote the class of these functions by E_ζ.

Theorem 10. *If $f(t, x) \in E_\zeta$, $\gamma(t, u) \in H_2^\pi$ then the function $f(t, \zeta(t))$ possesses the stochastic differential* (39).

A generalized Itô formula. Let $\xi(t)$ be a q-dimensional vector-valued process, $\xi(t) = (\xi^1(t), \xi^2(t), \ldots, \xi^q(t))$, with components of the form

$$\xi^k(t) = \alpha^k(t) + \beta^k(t) + \zeta^k(t), \tag{43}$$

where

$$\alpha^k(t)\in \mathcal{V}^c, \qquad \beta^k(t)\in l\mathcal{M}_2^c,$$

$$\zeta^k(t)=\int_0^t\int_{\mathcal{R}^m}\gamma(s,u)\mu(ds,du), \qquad \gamma\in H_2^\pi \quad k=1,\ldots,q,$$

and $\mu(\cdot,\cdot)$ be a local martingale measure associated with the integer-valued measure $\nu(t,A)$ with characteristic $\pi(t,A)$. As before, we shall assume that a current of σ-algebras $\{\mathfrak{F}_t, t\in[0,T)\}$ is fixed and that all the processes, martingales, and measures to be considered below are adopted to $\{\mathfrak{F}_t\}$.

Let $f(x)=f(x^1,\ldots,x^q)$ be a twice continuously differentiable function. Consider the process

$$\eta(t)=f(\xi(t)).$$

We show that the process $\eta(t)$ also admits decomposition of the form (43) and we obtain expressions for the corresponding components of this decomposition.

First we shall assume that $f(x)$ is thrice continuously differentiable and vanishes outside some compact set, and in addition that the function $\gamma(t,u)$ satisfies the condition

$$\int_0^T\int_{\mathcal{R}^m}|\gamma(s,u)|\pi(ds,du)<\infty$$

with probability 1. Then

$$\zeta^k(t)=\int_0^t\int_{\mathcal{R}^m}\gamma^k(s,u)\nu(ds,du)-\int_0^t\int_{\mathcal{R}^m}\gamma^k(s,u)\pi(ds,du),$$

where the first integral on the right-hand side of the equality is finite with probability 1 and the functions $\zeta^k(t)$ are of bounded variation with probability 1:

$$V_T^0(\zeta^k)=\sup_\delta\sum_{r=1}^n|\Delta\zeta^k(t_r)|\leq\int_0^T\int_{\mathcal{R}^m}|\gamma(s,u)|(\nu(ds,du)+\pi(ds,du)).$$

We now introduce a subdivision δ of the interval $[0,T]$ into intervals $\Delta_k=(t_{k-1},t_k]$, $k=1,2,\ldots,n$.

Set

$$f(\xi(T))-f(\xi(0))=S_1+S_2+S_3,$$

where

$$S_1=\sum_{k=1}^n f[\xi_c(t_k)+\zeta(t_{k-1})]-f(\xi(t_{k-1})),$$

$$S_2=\sum_{k=1}^n f[(\xi_c(t_{k-1})+\zeta(t_k)]-f(\xi(t_{k-1})),$$

$$S_3=\sum_{k=1}^n f(\xi(t_k))-f[\xi_c(t_{k-1})+\zeta(t_k)]-f[\xi_c(t_k)+\xi(t_{k-1})]+f[\xi(t_{k-1})],$$

and $\xi_c(t)=\alpha(t)+\beta(t)$. We show that P-lim $S_3=0$ as $|\delta|\to 0$.

The sum S_3 can be represented in the form

$$S_3 = \sum_{k=1}^{n} (\nabla^2 f[\tilde{\xi}_k + \tilde{\zeta}_k]\, \Delta\xi_k, \Delta\zeta_k),$$

where

$$\tilde{\xi}_k = \xi_c(t_{k-1}) + \theta_1\, \Delta\xi_{ck}, \qquad \tilde{\zeta}_k = \zeta(t_{k-1}) + \theta_2\, \Delta\zeta_k,$$

$$\Delta\xi_{ck} = \xi_c(t_k) - \xi_c(t_{k-1}), \qquad \Delta\zeta_k = \zeta(t_k) - \zeta(t_{k-1}),$$

$$0 \leq \theta_i \leq 1, \qquad i = 1, 2.$$

Hence

$$|S_3| \leq C \max_k |\Delta\xi_{ck}| \cdot |V_T^0(\zeta)|,$$

where C is a constant. The last inequality shows that $|S_3| \to 0$ as $|\delta| \to 0$ with probability 1.

Consider the sum S_1. It follows from (3) that

$$S_1 = \int_0^T \sum_{i=1}^{q} \nabla^i f(\zeta^{(\delta)}(t) + \xi_c(s))\, d\xi_c^i(s)$$

$$+ \frac{1}{2} \int_0^T \sum_{i,j=1}^{q} \nabla^i \nabla^j f(\xi_c(s) + \zeta^{(\delta)}(s))\, d\langle\beta^i, \beta^j\rangle_s,$$

where $\zeta^{(\delta)}(t) = \zeta(t_{k-1})$ for $t \in (t_{k-1}, t_k]$. Moreover, $\zeta^{(\delta)}(t) \to \zeta(t-)$ as $|\delta| \to 0$ with probability 1 and $\zeta(t) = \zeta(t-)$ everywhere except on a countable set of points. Since the derivatives $\nabla^i f(x)$ and $\nabla^i \nabla^j f(x)$ are bounded and the measures $d\alpha^k$ and $d\langle\beta^i, \beta^j\rangle_t$ are nonatomic, we have with probability 1

$$\lim S_1 = \int_0^T \sum_{i=1}^{q} \nabla^i f(\xi(s))\, d\xi_c^i(s) + \frac{1}{2} \int_0^T \sum_{i,j=1}^{q} \nabla^i \nabla^j f(\xi(s))\, d\langle\beta^i, \beta^j\rangle_s.$$

The sum S_2 is analyzed analogously. We have

$$S_2 = \int_0^T \int_{\mathscr{R}^m} [f(\xi_c^{(\delta)}(s) + \zeta(s) + \gamma(s, u)) - f(\xi_c^{(\delta)}(s) + \zeta(s))$$

$$- (\nabla f(\xi_c^{(\delta)}(s) + \zeta(s)), \gamma(s, u))]\pi(ds, du)$$

$$+ \int_0^T \int_{\mathscr{R}^m} [f(\xi_c^{(\delta)}(s) + \zeta(s) + \gamma(s, u)) - f(\xi_c^{(\delta)}(s) + \zeta(s))]\mu(ds, du),$$

where $\xi_c^{(\delta)}(t) = \xi_c(t_{k-1})$ for $t \in (t_{k-1}, t_k]$. Since the expressions under the integral sign possess with probability 1 majorants which are integrable in (t, u) (for a fixed ω) of the form $\varkappa|\gamma(t, u)|^2$ and $\varkappa|\gamma(t, u)|$, respectively, the passage to the limit as $|\delta| \to 0$ is justified in this case. We thus obtain

$$\lim S_2 = \int_0^T \int_{\mathscr{R}^m} L_d(f, \xi)\pi(ds, du) + \int_0^T \int_{\mathscr{R}^m} [f(\xi(t) + \gamma(s, u)) - f(\xi(t))]\mu(ds, du).$$

Evidently in the relationships obtained one can replace T by any $t\in[0, T]$. Thus for all $t\in[0, T]$

$$\begin{aligned} f(\xi(t)) &= f(\xi(0))+\int_0^t (\nabla f(\xi), d\alpha) \\ &+\frac{1}{2}\sum_{i,j=1}^q \nabla^i\nabla^j f(\xi)\, d\langle\beta^i, \beta^j\rangle_s+\int_0^t\int_{\mathcal{R}^m} L_{\mathrm{d}} f(\xi)\pi(ds, du) \\ &+\int_0^t (\nabla f(\xi), d\beta)+\int_0^t\int_{\mathcal{R}^m} [f(\xi+\gamma)-f(\xi)]\mu(ds, du). \end{aligned} \tag{44}$$

It is easy to observe that two limiting transitions can be carried out in the formula obtained.

Firstly, we set $\gamma=\gamma_n$, $\xi=\xi_n$ and let $\gamma_n\to\gamma$ in H_2^π. As was verified above, the sequence γ_n can be selected in a manner such that $\xi_n(t)\to\xi(t)$ with probability 1 uniformly in $t\in[0, T]$. Taking into account inequalities $|L_{\mathrm{d}} f(\xi_n)|\leq\varkappa|\gamma_n|^2$, $|f(\xi_n+\gamma_n)-f(\xi_n)|\leq\varkappa|\gamma_n|$, where $\varkappa$ is independent of s, u, and n, we see that one can approach the limit in formula (44) as $n\to\infty$. Thus, this formula remains valid for arbitrary $\gamma\in H_2^\pi$.

Secondly, the argument presented for an analogous case in the preceding subsection yields that the assumption that the function $f(x)$ vanishes outside a compact set can be relaxed and replaced by these requirements: $f\in E_\xi$, where E_ξ is the class of twice differentiable functions $f(x)$ such that

$$f(\xi(t)+\gamma(t, u))-f(\xi(t))-(\nabla f(\xi(t)), \gamma(t, u))$$

and

$$|f(\xi(t)+\gamma(t, u))-f(\xi(t))|^2$$

with probability 1 are integrable with respect to measure $\pi(dt, du)$.

Thus the following theorem has been proved.

Theorem 11. *Let $\alpha\in\mathcal{V}^c$, $\beta\in l\mathcal{M}^c$, and μ be a local martingale measure, $\gamma\in H_2^\pi$, the function $f(x)$ be twice continuously differentiable, and $f\in E_\xi$. Then the process possesses the stochastic differential*

$$d\eta=d\eta_{\mathrm{c}}+d\eta_{\mathrm{d}}$$

where

$$\begin{aligned} d\eta_{\mathrm{c}} &= (\nabla f(\xi), d\alpha)+\frac{1}{2}\sum_{i,j=1}^q \nabla^i\nabla^j f(\xi)\, d\langle\beta^i, \beta^j\rangle_t+(\nabla f(\xi), d\beta), \\ d\eta_{\mathrm{d}} &= \int_{\mathcal{R}^m} L_{\mathrm{d}}(f, \gamma)\pi(dt, du)+\int_{\mathcal{R}^m} [f(\xi+\gamma)-f(\xi)]\mu(dt, du). \end{aligned} \tag{45}$$

This formula is called *the generalized Itô formula.*

Corollary 1. *If $f(t, x)$, $x \in \mathcal{R}^q$, $t \in [0, T]$, is a function twice continuously differentiable in x and continuously differentiable in t and if, moreover, $f(t, x) \in E_\xi$, then*

$$df(t, \xi(t)) = d\eta_c + d\eta_d,$$

$$d\eta_c = f'_t(t, \xi(t))\, dt + (\nabla f(\xi(t)), d\xi_c) + \frac{1}{2} \sum_{i,j=1}^{q} \nabla^i \nabla^j f(t, \xi(t))\, d\langle \beta^i, \beta^j \rangle_t,$$

$$d\eta_d = \int_{\mathcal{R}^m} L_d(f, \gamma) \pi(ds, du) + \int_{\mathcal{R}^m} [f(t, \xi(t) + \gamma(t, u)) - f(t, \xi(t))] \mu(dt, du).$$

A generalization of formula (45) to the case of functions $f(t, x)$ dependent on time is justified in the same manner as in the case of continuous processes $\xi(t)$.

Corollary 2 (The rule for differentiation of a product).
Let

$$\xi^i(t) = \alpha^i(t) + \beta^i(t) + \int_0^t \int_{\mathcal{R}^m} \gamma^i(s, u) \mu(ds, du), \qquad i = 1, 2.$$

We apply formula (45) *to the functions of two variables $f(x, y) = xy$. After some simple manipulation we obtain*

$$d(\xi_1(t)\xi_2(t)) = \xi^2\, d\xi^1 + \xi^1\, d\xi^2 + d\langle \beta^1, \beta^2 \rangle_t + \int_{\mathcal{R}^m} \gamma^1(t, u) \gamma^2(t, u) \nu(dt, du). \tag{46}$$

Some corollaries from the generalized Itô formula: A generalization of Levy's theorem

Theorem 12. *Let $\xi(t)$ be a regular local square integrable martingale and let $\xi(t) = \xi_c(t) + \xi_d(t)$ be a decomposition of the process $\xi(t)$ into continuous and discontinuous parts. Assume that functions $\langle \xi_c^k, \xi_c^j \rangle_t$ and $\pi(\cdot, \cdot)$ are nonrandom. Then $\xi(t)$ is a process with independent increments.*

Proof. First we shall assume that $\xi(t) \in \mathcal{M}_2^r$. Set $f(x) = e^{i(z,x)}$, $z \in \mathcal{R}^m$, $x \in \mathcal{R}^m$, and apply the generalized Itô formula to the function $f(\xi(t))$. We obtain

$$f(\xi(t)) - f(\xi(t_0)) = \int_{t_0}^t e^{i(z,\xi(s))} \Big[-\frac{1}{2} \sum_{k,j=1}^{m} z^k z^j\, d\langle \xi_c^k, \xi_c^j \rangle_s + \int_{\mathcal{R}^m} [e^{i(z,u)} - 1 - i(z, u)] \pi(dt, du) + i(z, d\xi_c) + (e^{i(z,u)} - 1) \mu(ds, du) \Big],$$

where the last two summands possess finite second-order moments.

Setting $J(t) = \mathsf{E}\{e^{i(z,\xi(t))} \mid \mathfrak{F}_{t_0}\}$ we obtain from the preceding equality that

$$J(t) = \eta(t_0) + \int_{t_0}^t J(s)\, d\mathfrak{G}(s), \tag{47}$$

where $\mathfrak{G}(s)$ is a nonrandom function of bounded variation given by

$$\mathfrak{G}(s)=-\frac{1}{2}\sum_{k,j=1}^{m} z^k z^j \langle \xi_c^k, \xi_c^j\rangle_s + \int_{\mathcal{R}^m} (e^{i(z,u)}-1-i(z,u))\pi(s,du)$$

and $\eta(t_0)=e^{i(z,\xi(s_0))}$. Equation (47) has a unique solution which can be easily obtained by means of the method of iterations:

$$J(t)=\eta(t_0)\,e^{\mathfrak{G}(t)},$$

or

$$J(t)=e^{i(z,\xi(t_0))}\exp\left\{-\frac{1}{2}\sum_{k,j=1}^{m}\sigma^{kj}(t)z^k z^j+\int_{\mathcal{R}^m}(e^{i(z,u)}-1-i(z,u))\pi(t,du)\right\},$$

where $\sigma^{kj}(t)=\langle \xi_c^k, \xi_c^j\rangle_t$.

This equality shows in particular that the distribution of the vector $\xi(t)-\xi(t_0)$ does not depend on the σ-algebra $\mathfrak{F}_{t_0}$, i.e., the process $\xi(t)$ is a process with independent increments. Moreover, this equality gives us a general representation of a characteristic function of a regular process with independent increments and finite second-order moments. □

A generalization of the result obtained to the case of processes belonging to $l\mathcal{M}_2^r$ can be obtained by introducing a stopping of a local martingale $\xi(t)$ analogously to what was accomplished in the course of the proof of Levy's theorem (see Theorem 3 in this section).

Consider an arbitrary process $\xi(t)$, $t\in[0,T]$, with values in $\mathcal{R}^1$ and with trajectories which are with probability 1 constant everywhere, except possibly at a finite number of points at which they possess jumps of the unit size.

Let $\xi(0)=0$ and $\mathfrak{F}_t=\sigma\{\xi(s), s\in[0,t]\}$. Then $\{\xi(t),\mathfrak{F}_t, t\in[0,T]\}$ is a local submartingale.

We select the sequence $\tau_N=\inf\{t\colon \xi(t)\geq N\}$ (with the stipulation that $\inf\varnothing= T$) as the sequence of random times τ_N which completely reduces $\xi(t)$. We also assume that the process $\xi(t)$ is regular, i.e., for an arbitrary monotonically nondecreasing sequence of random times σ_n and $\sigma=\lim\sigma_n$

$$\lim \mathsf{E}\xi(\sigma_n\wedge\tau_N)=\mathsf{E}\xi(\sigma\wedge\tau_N),\qquad N=1,2,\ldots.$$

Such a process can be considered a random integer-valued measure $\nu(t,du)$ concentrated at the point $u=1$ ($\nu(t,A)\equiv 0$ if $1\notin A$). In view of the above there exists a monotonically nondecreasing continuous function $\alpha(t)$ such that $\xi(t)= \alpha(t)+\mu(t)$, where $\mu(t)\in l\mathcal{M}_2^r$ and $\langle\mu,\mu\rangle_t=\alpha(t)$. In particular, if $\xi(t)$ is a stochastically continuous Poisson process, then $\alpha(t)=\mathsf{E}\xi(t)$ is a nonrandom function and $\xi(t)$ is a regular submartingale. On the other hand, in view of Theorem 12 this condition is also sufficient for the process $\xi(t)$ to be a Poisson process. Thus we have the following corollary from Theorem 12.

Corollary. *In order that a process $\xi(t)$ with constant sample functions at all points (except possibly a finite number of points where these functions have jumps of unit size) be a stochastically continuous Poisson process, it is necessary and sufficient that it be regular and that the continuous process associated with $\xi(t)$ be nonrandom.*

The last assertion can be strengthened:

Theorem 13. *Let $\nu(t, A)$ be an integer-valued measure which satisfies the conditions stipulated above and let the measure $\pi(t, A)$ associated with it be nonrandom. Then $\nu(t, A)$ is a Poisson measure, i.e.,*

a) *for a fixed A, $\nu(t, A)$ is a Poisson process,*
b) *for any n, $A_1, \ldots, A_n$ ($A_k \in \mathfrak{B}_0^m$) the processes $\nu(t, A_1), \ldots, \nu(t, A_n)$ are mutually independent.*

Proof. For the proof, consider $\zeta(t) = \sum_{k=1}^n \lambda_k \nu(t, A_k)$, where λ_k are arbitrary constants. Applying Itô's formula to the function $\exp\{iz\zeta(t)\}$ we obtain (in the same manner as in the proof of Theorem 12) for the function $J(t) = \mathsf{E}\{\exp[iz\zeta(t)] \mid \mathfrak{F}_s\}$ the equation

$$J(t) = e^{iz\zeta(s)} + \int_s^t J(s)\Pi(ds), \qquad t \leq s,$$

where $\Pi(t) = \sum_{k=1}^n (e^{iz\lambda_k} - 1 - iz\lambda_k)\pi(t, A_k)$. A solution of this equation can be easily obtained. It is of the form

$$J(t) = \exp\left\{\sum_{k=1}^n (e^{iz\lambda_k} - 1 - iz\lambda_k)(\pi(t, A_k) - \pi(s, A_k)) + iz\zeta(s)\right\}.$$

It follows from the last formula that the variables $\nu(t, A_k) - \nu(s, A_k)$, $k = 1, \ldots, n$, are mutually independent and do not depend on σ-algebra $\mathfrak{F}_s$, i.e., they are processes with independent increments. Moreover, the variable $\nu(t, A)$ has the distribution which is identical to the distribution of the variable $\zeta - \pi(t, A)$, where ζ is a Poisson variable with the mean $\pi(t, A)$. □

Bounds on the moments of integrals with respect to a martingale measure. Let

$$\zeta(t) = \int_0^t \int_{\mathscr{R}} \gamma(s, u)\mu(ds, du), \tag{48}$$

where $\mu(\cdot, \cdot)$ is a martingale measure.

Consider the problem of the existence and bounds on the even order moments of the variable $\zeta(t)$. Assume that the characteristic $\pi(t, A)$ is absolutely continuous in t with probability 1 and let

$$\begin{aligned} \pi(t, A) &= \int_0^t \Pi(s, A)\, ds, \\ v_k(t) &= \int_{\mathscr{R}^m} |\gamma(t, u)|^k \Pi(t, du) < \infty, \qquad k = 2, \ldots, 2m, \end{aligned} \tag{49}$$

for all $t \in [0, T]$ with probability 1. Observe that if the condition of finiteness of the variables $v_k(t)$ is fulfilled for $k=2$ and $k=2m$ then these conditions are satisfied for all $k=2,3,\ldots,2m$. Indeed, if for $2<k<2m$ we set $|\gamma|^k=|\gamma|^\alpha|\gamma|^\beta$ where $\alpha=(2m-k)/(m-1)$, $\beta=(k-2)m/(m-1)$ and apply Hölder's inequality to the corresponding integral, we shall obtain

$$v_k(t)=\left(\int_{\mathcal{R}^m}\gamma^2(t,u)\Pi(t,du)\right)^{1/p}\left(\int_{\mathcal{R}^m}\gamma^{2m}(t,u)\Pi(t,du)\right)^{1/q},$$

where

$$p=\frac{2}{\alpha}=\frac{2m-2}{2m-k},\qquad q=\frac{p}{p-1}=\frac{2m-2}{k-2}.$$

Thus

$$v_k(t)\leq(v_2(t))^{(2m-k)/(2m-2)}(v_{2m}(t))^{(k-2)/(2m-2)}. \tag{50}$$

We now return to the question of the bounds on the moments of the stochastic integral (48).

Using formula (45) we have

$$\begin{aligned}\zeta^m(t)=&\int_0^t\int_{\mathcal{R}}[(\zeta(s)+\gamma(s,u))^m-\zeta^m(s)]\mu(ds,du)\\&+\int_0^t\int_{\mathcal{R}}[(\zeta(s)+\gamma(s,u))^m-\zeta^m(s)-m\gamma(s,u)\zeta^{m-1}(s)]\Pi(s,du)\,ds.\end{aligned}$$

Let

$$\tau=\tau_N=\inf\{t:|\xi(t)|\geq N\},$$

where $\inf\varnothing=T$. Clearly, in view of conditions (49) the random time τ reduces a local square integrable martingale $\zeta(t)$, i.e., $\zeta(t\wedge\tau)$ is a square integrable martingale.

Next,

$$\mathsf{E}\zeta^{2m}(t\wedge\tau)\leq 2(I_1+I_2),$$

where

$$\begin{aligned}I_1&=\mathsf{E}\left(\int_0^t\int_{\mathcal{R}}\chi_\tau(s)[(\zeta(s)+\gamma(s,u))^m-\zeta^m(s)]\mu(ds,du)\right)^2\\&=\mathsf{E}\int_0^t\int_{\mathcal{R}}\chi_\tau(s)[(\zeta(s)+\gamma(s,u))^m-\zeta^m(s)]^2\Pi(s,du)\,ds\\&\leq m\mathsf{E}\int_0^t\int_{\mathcal{R}}\chi_\tau(s)\sum_{k=1}^m(C_m^k)^2\zeta(s)^{2m-2k}\gamma^{2k}(s,u)\Pi(s,du)\,ds,\end{aligned}$$

$$I_2=\mathsf{E}\left(\int_0^t\int_{\mathcal{R}}\chi_\tau(s)\left(\sum_{k=2}^m C_m^k\zeta^{m-k}(s)\gamma^k(s,u)\right)\Pi(s,du)\,ds\right)^2,$$

and

$$\chi_\tau(s)=1 \quad \text{for } s\leqslant\tau, \qquad \chi_\tau(s)=0 \quad \text{for } s<\tau.$$

Set

$$v_k(s)=\int_{\mathscr{R}}|\gamma^k(s,u)|\,\Pi(s,du).$$

Then

$$\begin{aligned} I_1 &\leqslant m\mathsf{E}\int_0^t \chi_\tau(s)\sum_{k=1}^m (C_m^k)^2\left(\frac{(m-k)\zeta^{2m}(s)}{m}+\frac{kv_{2k}^{m/k}(s)}{m}\right)ds \\ &\leqslant K_m\int_0^t b_m(s)\,ds+A_m, \end{aligned}$$

where

$$b_m(t)=\mathsf{E}\zeta^{2m}(t\wedge\tau),$$

$$K_m=\sum_{k=1}^m (m-k)(C_m^k)^2, \qquad A_m=\mathsf{E}\int_0^T\sum_{k=1}^m (C_m^k)^2 kv_{2k}^{m/k}(s)\,ds.$$

I_2 can be bound analogously. We obtain

$$I_2\leqslant 2K_m\int_0^t b_m(s)\,ds+B_m, \qquad B_m=2\mathsf{E}\int_0^T\sum_{k=2}^m (C_m^k)^2 kv_k^{2m/k}(s)\,ds.$$

Thus

$$b_m(t)\leqslant 3K_m\int_0^t b_m(s)\,ds+(A_m+B_m)$$

and

$$b_m(t)\leqslant (A_m+B_m)\,e^{3K_m t}.$$

Utilizing Fatou's lemma we arrive at the following result:

Theorem 14. *If*

$$\mathsf{E}\int_0^T\left[\left(\int_{\mathscr{R}^q}|\gamma(s,u)|^2\Pi(s,du)\right)^m+\int_{\mathscr{R}^q}|\gamma(s,u)|^{2m}\Pi(s,du)\right]ds<\infty,$$

then the stochastic integral (48) *possesses finite moments up to the* $(2m)$th *order inclusively and*

$$\mathsf{E}|\zeta(t)|^{2m}\leqslant (A_m+B_m)\,e^{3K_m T}.$$

To prove this theorem it is sufficient to observe that (50) and Hölder's inequality imply that

$$\int_0^T v_k(t)^{2m/k}\,dt \le \int_0^T v_2(t)^{(2m-k)/(2m-2)\cdot(2m/k)} v_{2m}(t)^{(k-2)/(2m-2)\cdot(2m/k)}\,dt$$
$$\le \left(\int_0^T v_2^m(t)\,dt\right)^{(2/k)\cdot(2m-k)/(2m-2)} \left(\int_0^T v_{2m}(t)\,dt\right)^{(2m/k)\cdot(k-2)/(2m-2)} < \infty.$$

Solution of a simple stochastic differential equation. Consider the equation in stochastic differentials of the form

$$d\eta = \eta\,d\xi, \tag{51}$$

where $\xi(t)$ is a process of the form $\xi(t) = \alpha(t) + \beta(t) + \int_{\mathscr{R}} u\mu(t, du)$, $\alpha(t) \in \mathscr{V}^c$, $\beta(t)$ is a continuous local martingale, and μ is a local martingale measure associated with jumps of the process $\xi(t)$. We shall assume for simplicity that $\nu(t, (-\infty, -1]) = 0$. This assumption applied to equation (51) means that its solution—if it exists—cannot change its sign with a jump.

We shall seek the solution of equation (51) of the form

$$\eta(t) = \eta_0 \exp\{\gamma(t) + \int_0^t b\,d\beta + \int_0^t \int_{\mathscr{R}} \varphi(s, u)\mu(ds, du)\} = \eta_0 \exp\{\zeta(t)\}.$$

It follows from the generalized Itô formula that equation (51) is equivalent to the following:

$$d\gamma + \tfrac{1}{2}b^2\,d\langle\beta, \beta\rangle_t + b\,d\beta + \int_{\mathscr{R}} (e^{\varphi} - 1 - \varphi)\pi(dt, du) + \int_{\mathscr{R}} (e^{\varphi} - 1)\mu(dt, du)$$
$$= d\alpha + d\beta + \int_{\mathscr{R}} u\mu(dt, du),$$

whence we obtain

$$b = 1, \qquad e^{\varphi} - 1 = u,$$
$$d\gamma = d\alpha - \tfrac{1}{2}d\langle\beta, \beta\rangle_t - \int_{\mathscr{R}} (e^{\varphi} - 1 - \varphi)\pi(dt, du).$$

Thus

$$\eta(t) = \eta_0 \exp\{\xi(t) - \tfrac{1}{2}\langle\beta, \beta\rangle_t - \int_{\mathscr{R}} (u - \ln(1+u))\pi(t, du)$$
$$- \int_{\mathscr{R}} [u - \ln(1+u)]\mu(t, du)\}$$

or

$$\eta(t) = \eta_0 \exp\{\xi(t) - \tfrac{1}{2}\langle\beta, \beta\rangle_t - \int_{\mathscr{R}} [u - \ln(1+u)]\nu(t, du)\}. \tag{52}$$

The last equation can also be written in the form

$$\eta(t) = \eta_0 \exp\{\xi(t) - \tfrac{1}{2}\langle\beta, \beta\rangle_t\} \prod_{s \le t} (1 + \delta\xi(s))\,e^{-\delta\xi(s)}, \tag{53}$$

where $\delta\xi(s)$ is the jump of the function $\xi(t)$ at point $t = s$, $\delta\xi(s) = \xi(s) - \xi(s-)$.

The expressions obtained show that if the spectrum of the jumps of process $\xi(t)$ contains the region located to the left of point -1, then there are no solutions of equation (51) of the form $\eta = e^{-\zeta(t)}$. However, a simple examination of formula (53) shows that it remains valid also in the general case.

It follows from the general theory of stochastic differential equations developed below that the solution obtained of equation (51) is unique.

It follows from formula (53) that:

a) the solution of equation

$$\eta(t) = 1 + \int_0^t \eta(s)\beta(ds),$$

where $\beta(t) \in \mathcal{LM}^c$ is of the form

$$\eta(t) = \exp\{\beta(t) - \tfrac{1}{2}\langle\beta, \beta\rangle_t\}$$

and $\eta(t) \in \mathcal{LM}^c$;

b) the solution of the equation

$$\eta(t) = 1 + \int_0^t \eta(s)\, d\zeta(s),$$

where $\zeta(t) = \int_{\mathcal{R}} u\mu(t, du)$ and μ is a martingale measure associated with the measure of jumps of a certain process, can be represented in the form

$$\eta(t) = \exp\{\zeta(t)\} \prod_{s \leq t} (1 + \delta\zeta(s))\, e^{-\delta\zeta(s)}$$

and $\eta(t) \in \mathcal{LM}_2^r$.

Example—a multiplicative decomposition of a positive supermartingale. Let $\xi(t)$, $t \geq 0$, be a nonnegative square integrable regular nonvanishing supermartingale. Then $\xi(t-) > 0$ for all t and therefore $\inf \xi(t) > 0$ on each interval $[0, T]$. Consider Doob's decomposition of the process $\xi(t)$, $\xi(t) = \beta(t) - \alpha(t)$, where $\beta(t)$ is a local martingale and $\alpha(t)$ is an associated natural increasing process (see Section 1, Theorem 11). In the case under consideration the process $\alpha(t)$ is continuous and, as is easy to verify, $\beta(t)$ is a locally square integrable martingale.

Set

$$\zeta_1(t) = \int_0^t \frac{d\beta(s)}{\xi(s)}, \qquad \zeta_2(t) = \int_0^t \frac{d\alpha(s)}{\xi(s)}, \qquad \zeta(t) = \zeta_1(t) - \zeta_2(t).$$

Here $\zeta_1(t) \in \mathcal{LM}_2$, $\zeta_2(t)$ is a continuous increasing process, and $d\xi = \xi\, d\zeta$. Since submartingale $\xi(t)$ is positive it follows that the jumps of the process $\zeta_1(t)$ (and hence those of $\zeta(t)$) are greater than -1. Consequently,

$$\xi(t) = \xi_0 \exp\{\zeta(t) - \tfrac{1}{2}\langle\zeta_c, \zeta_c\rangle_t + \int_{\mathcal{R}} [\ln(1+u) - u]\nu_\zeta(t, du)\},$$

where ζ_c is the continuous component in the decomposition of $\zeta_1(t)$ into continuous and discrete parts and ν_ζ is the measure of jumps of process $\zeta(t)$.

The expression obtained can also be written in the form

$$\xi(t)=\eta_0(t)\eta_c(t)\eta_d(t), \tag{54}$$

where $\eta_0(t)$ is a continuous nonincreasing process,

$$\eta_0(t)=\xi(0)\exp\left\{-\int_0^t \frac{d\alpha(s)}{\xi(s)}\right\}, \qquad \eta_c(t)\in l\mathcal{M}_2^c,$$

$$\eta_c(t)=\exp\left\{\int_0^t \frac{d\beta_c(s)}{\xi(s)}-\frac{1}{2}\int_0^t \frac{d\langle\beta_c,\beta_c\rangle}{\xi^2(s)}\right\};$$

here $\beta_c(t)$ denotes the continuous component in the decomposition of the local martingale $\beta(t)$ into continuous and discrete parts,

$$\eta_d(t)=\exp\left\{\int_0^t\int_{\mathcal{R}} \frac{u\mu(ds,du)}{\xi(s)}\right\}\prod_{s\leq t}\left(1+\frac{\delta\xi(s)}{\xi(s-)}\right)e^{-(\delta\xi(s))/(\xi(s-))},$$

and also $\eta_d(t)\in l\mathcal{M}_2^r$. Here one must take into account that $\delta\zeta_1(t)=\delta\beta_1(t)/\xi(t-)=\delta\xi(t)/\xi(t-)$.

Theorem 15. *A positive square integrable regular supermartingale admits multiplicative decomposition* (54), *where $\eta_0(t)$ is a continuous increasing process, $\eta_c(t)$ is a positive continuous local martingale, and $\eta_d\in l\mathcal{M}_2^r$.*

Chapter II

Stochastic Differential Equations

§1. General Problems of the Theory of Stochastic Differential Equations

In the present section we introduce the notion of a stochastic differential equation and prove some general theorems concerning the existence and uniqueness of solutions of these equations. For this purpose it is necessary to generalize the notion of a stochastic integral introduced above. Generally speaking, our approach to stochastic differential equations is based on the following considerations.

Assume that we are dealing with a motion of a system S in the phase space $\mathscr{R}^m$ and let $\xi(t)$ denote the location of this system in $\mathscr{R}^m$ at time t ($\xi(t) = (\xi^1(t), \ldots, \xi^m(t))$). Assume also that a displacement of system S located at time t at point x during the time interval $(t, t+\Delta t)$ can be represented in the form

$$\xi(t+\Delta t)-\xi(t)=A(x, t+\Delta t)-A(x, t)+\delta. \tag{1}$$

Here $A(x, t)$ is, in general, a random function; the difference $A(x, t+\Delta t)-A(x, t)$ characterizes the action of an "external field of forces" at point x on S during the time period $(t, t+\Delta t)$ and δ is a quantity which is of a higher order of smallness in a certain sense than the difference $A(x, t+\Delta t)-A(x, t)$. If $A(x, t)$ as a function of t is absolutely continuous, then relation (1) can be replaced by the ordinary differential equation

$$\frac{d\xi}{dt}=A_t'(\xi(t), t). \tag{2}$$

Equation (2) defines the motion of S in $\mathscr{R}^m$ for $t>t_0$ under the initial condition $\xi(t_0)=\xi_0$ while $A_t'(x, t)$ determines the "velocity field" in the phase space at time t.

It is obvious that equation (2) cannot describe motions such as Brownian, i.e., motions which do not possess a finite velocity in the phase space or motions which possess discontinuities in the phase space. To obtain an equation which will describe a motion of systems of this kind it is expedient to replace relation (1) with an equation of the integral type. For this purpose, we visualize that the time interval $[t_0, t]$ is subdivided into subintervals by the subdividing points

$t_1, t_2, \ldots, t_n = t$. It then follows from (1) that

$$\xi(t)-\xi(t_0)=\sum_{i=0}^{n-1} A(\xi(t_i), t_{i+1})-A(\xi(t_i), t_i)+\sum_{i=1}^{n-1} \delta_i.$$

Since δ_i are of a small order, it is natural to assume that $\sum_{i=0}^{n-1} \delta_i \to 0$ as $n \to \infty$. In this case the last equality formally becomes

$$\xi(t)-\xi(t_0)=\int_{t_0}^{t} A(\xi(s), ds), \tag{3}$$

and the expression

$$\int_0^t A(\xi(s), ds)$$

can be called a stochastic integral in the random field $A(x, t)$ along the random curve $\xi(s)$, $s \in [t_0, t]$; the integral should be interpreted as the limit, in a certain sense, to be defined more precisely, of sums of the form

$$\sum_{i=0}^{n-1} [A(\xi(t_i), t_{i+1})-A(\xi(t_i), t_i].$$

Relation (3) is called a stochastic differential equation and is written in the form

$$d\xi = A(\xi(t), dt), \qquad \xi(t_0)=\xi_0, \qquad t \geqslant t_0.$$

Under sufficiently general assumptions, for example, if $A(x, t)$ is a quasi-martingale for each $x \in \mathcal{R}^m$, one can assume that

$$A(x, t)=\alpha(x, t)+\beta(x, t), \tag{4}$$

where $\beta(x, t)$ as a function of t is a local martingale and process $\alpha(x, t)$ is representable as the difference of two monotonically nondecreasing natural processes. In this connection it makes sense to suppose that the right-hand side of equation (3) can be represented according to formula (4) and introduce further restrictions on functions $\alpha(x, t)$ and $\beta(x, t)$ in various ways. For example, we may assume that the function $\alpha(x, t)$ appearing in expression (4) is an absolute continuous function of t while $\beta(x, t)$—as a function of t—is a local square integrable martingale. (Some more general assumptions concerning $\beta(x, t)$ are considered below.)

In what follows, equation (3) will be written in the form

$$\xi(t)=\xi_0+\int_{t_0}^{t} \alpha(\xi(s), s)\, ds+\int_{t_0}^{t} \beta(\xi(s), ds) \tag{5}$$

or

$$d\xi=\alpha(\xi(t), t)\, dt+\beta(\xi(t), dt), \qquad \xi(t_0)=\xi_0.$$

In the case when $\beta(x, t) \equiv 0$, equation (6) is called *an ordinary differential equation* (with a random right-hand side).

Often fields $\beta(x, t) = \{\beta^1(x, t), \ldots, \beta^m(x, t)\}$ of the form

$$\beta^k(x, t) = \int_0^t \sum_{j=1}^r \gamma_j^k(x, s)\, d\mu^j(s), \qquad k = 1, \ldots, m, \tag{6}$$

are considered. Here $\mu^j(s)$ are local mutually orthogonal square integrable martingales, $j = 1, \ldots, r$, and $\gamma_j^k(x, s)$ are random functions satisfying conditions which assure the existence of corresponding integrals. In this case, the second integral in equation (5) may be defined as the vector-valued integral with components

$$\int_{t_0}^t \beta^k(\xi(s), ds) = \int_{t_0}^t \sum_{j=1}^r \gamma_j^k(\xi(s), s)\, d\mu^j(s), \qquad k = 1, \ldots, m,$$

and the theory of stochastic integrals described in Section 2 of Chapter I can be utilized. However, if we confine ourselves to functions $\beta(x, t)$ of type (6) a substantial amount of generality is lost. This can be seen from the fact that the joint characteristic of processes $\beta^k(x, t)$ and $\beta^k(y, t)$ defined by formula (6) is of the form

$$\langle \beta^k(x, \cdot), \beta^k(y, \cdot) \rangle_t = \int_0^t \sum_{i=1}^m \gamma_i^k(x, s) \gamma_i^k(y, s)\, d\langle \mu^i, \mu^i \rangle_s,$$

while in the general case it is given by a function $\Gamma^k(x, y, t)$ which for a fixed t is an arbitrary nonnegative-defined kernel of arguments x and y:

$$\sum_{i,j=1}^N \Gamma^k(x_i, y_j, t) z_i z_j \geq 0 \quad \text{for all } z_j \in \mathcal{R}^1, \quad j = 1, \ldots, n, \quad n = 1, 2, \ldots.$$

For example (for simplicity we consider here the one-dimensional case) let functions $\gamma_j(x, t) = c_j(x, t)$, $j = 1, \ldots, m$, be nonrandom and $\mu_j(t) = w_j(t)$ be independent Wiener processes. In this case the correlation function $R(x, y, t)$ of the field

$$\beta(x, t) = \int_0^t \sum_{j=1}^m c_j(x, s)\, dw_j(s)$$

equals

$$R(x, y, t) = \mathsf{E}\beta(x, t)\beta(y, t) = \int_0^t \sum_{j=1}^m c_j(x, s) c_j(y, s)\, ds.$$

On the other hand, if we set $\beta(x, t) = w(x, t)$, where $w(x, t)$ is an arbitrary Gaussian field with independent increments in t, its correlation function

$R_w(x, y, t) = \mathsf{E}w(x, t)w(y, t)$ is then an arbitrary nonnegative-defined kernel (for a fixed t).

Thus, the restrictions, when considering stochastic integrals along a process $\xi(t)$ imposed by fields of the type (6) lead to a substantial narrowing of the class of problems under consideration. Therefore it is expedient to introduce a direct definition and investigate properties of the stochastic integral

$$\int_0^T \beta(\xi(s), ds)$$

by interpreting it in the simplest cases as the limit in probability of the sums

$$\sigma = \sigma(\xi) = \sum_{k=1}^{n} \beta(\xi(s_{k-1}), s_k) - \beta(\xi(s_{k-1}), s_{k-1}).$$

The sums σ are called *integral sums.*

It is appropriate to observe that remarks about the insufficient generality of random fields as given by relation (6) are not fully justified in the case when the stochastic processes are represented by equation (5), with $\alpha(x, t) = a(x, t)$ being a nonrandom function and $\beta(x, t)$ being a function with independent increments in t. Indeed, the increment $\Delta\xi(t)$ of a solution of equation (5) at each time t depends on $\xi(t)$ and on the value of the field $\beta(x, t)$ at the point $x = \xi(t)$ and is independent of the nature of the relationship between $\beta(x, t)$ and $\beta(y, t)$ at the point $y \neq \xi(t)$ (provided the probabilistic characteristics of the field $\beta(x, t)$ as a function of x are sufficiently smooth). Therefore one may expect that solutions of equations (5) will be stochastically equivalent for any two fields $\beta(x, t) = \beta_1(x, t)$ and $\beta(x, t) = \beta_2(x, t)$ under the condition that the joint distributions of the sequence of vectors

$$\{\beta_i(x, t_1), \beta_i(x, t_2), \ldots, \beta_i(x, t_N)\} \qquad \forall x \in \mathcal{R}^m, \quad \forall N = 1, 2, \ldots,$$

coincide for $i = 1$ and $i = 2$ and that the fields $\beta_i(x, t)$ possess independent increments in t.

For example, let $w(x, t)$ be an arbitrary Gaussian field possessing independent increments in t, $B(x, t) = \mathsf{E}w^k(x, t)w^j(x, t) = \{B_{jk}(x, t)\}$ and the functions $B_{jk}(x, t)$ be differentiable with respect to t, $b_{jk}(x, t) = (d/dt)B_{kj}(x, t)$. Denote by $\sigma(x, t)$ a symmetric matrix such that $\sigma^2(x, t) = b(x, t)$ and introduce independent Wiener processes $w_j(t), j = 1, \ldots, n$. Set

$$\beta_1(x, t) = \int_0^t \sigma(x, s)\, dw(s), \qquad w(t) = (w_1(t), \ldots, w_n(t)).$$

Then

$$\mathsf{E}\beta_1(x, t)\beta_1(x, t) = \int_0^t \sigma(x, s)\sigma(x, s)\, ds = \int_0^t b(x, s)\, ds,$$

and one can expect that the solutions of the differential equations

$$d\xi = a(\xi, t)\, dt + w(\xi, dt),$$
$$d\xi = a(\xi, t)\, dt + \sigma(\xi, t)\, dw(t)$$

will be stochastically equivalent, although the fields $w(x, t)$ and $\beta_1(x, t)$ in general are not.

Analogous observations can be made also in the case when $\beta(x, t)$ is an arbitrary field with independent increments in t with finite moments of the second order. Assume that the increment $\beta(x, t+\Delta t)-\beta(x, t)$ possesses the characteristic function

$$\mathsf{E}\exp\{i(z, \beta(x, t+\Delta t)-\beta(x, t))\}$$
$$=\exp\left\{-\frac{1}{2}\int_t^{t+\Delta t}(b(x, s)z, z)\,ds\right.$$
$$\left.+\int_t^{t+\Delta t}ds\int_{\mathscr{R}^m}[e^{i(z,c(x,s,u))}-1-i(z, c(x, s, u))]\Pi(s, du)\right\}$$

(if $\beta(x, t)$ possess finite moments of the second order and is absolutely continuous in t one can reduce an arbitrary characteristic function of a process with independent increments to this form). In this case, one can expect for sufficiently smooth functions $a(x, t)$, $b(x, t)$ and $c(x, t, u)$ that solutions of the stochastic equations

$$d\xi = a(\xi, t)\,dt + \beta(\xi, dt)$$

and

$$d\xi = a(\xi, t)\,dt+\sigma(\xi, t)\,dt+\int_{\mathscr{R}^m}c(\xi, t, u)\tilde{\nu}(dt, du)$$

will be stochastically equivalent. Here $\sigma(x, t)$ is a symmetric matrix, $\sigma^2(x, t)= b(x, t)$, $\tilde{\nu}(t, A)$ is a centered Poisson measure with $\operatorname{Var}\tilde{\nu}(t, A)=\int_0^t\Pi(s, A)\,ds$.

It is also clear that if the increments in t of the field $\beta(x, t)$ are dependent then the remarks above concerning the possibility of replacing the field $\beta(x, t)$ in equation (5) by a simpler field without restricting the class of obtained solutions are no longer valid.

The preceding outline of a definition of a stochastic differential equation is expedient to extend in yet another direction. At present, "feedback" systems play an important part in a number of scientific–engineering problems. For such systems the "exterior field of forces" acting on the system at a given time depends not only on the instantaneous location of the system in the phase space but also on its phase trajectory in "the past":

$$\xi(t+\Delta t)-\xi(t)=\alpha(\xi|_{t_0}^t, t+\Delta t)-\alpha(\xi|_{t_0}^t, t)+\delta_t, \tag{7}$$

where $\alpha(\varphi|_{t_0}^t, s)$, $s\geq t>t_0$ is a family of random functionals with values in $\mathscr{R}^m$ defined on a certain space of functions $\varphi(u)$, $u\in[t_0, t]$, with values in $\mathscr{R}^m$.

The notation $\alpha(\varphi|_{t_0}^t, s)$ will be inconvenient for our further discussions mainly due to the absence of a fixed region in which the arguments of the functional $\alpha(\cdot, s)$ vary. In order to avoid this difficulty one can proceed as follows.

Introduce the space $\mathscr{D}_T^m$ ($\mathscr{D}^m[a, b]$) of functions $\varphi(s)$ defined on $(-\infty, T]$ (on $[a, b]$), continuous from the right with values in $\mathscr{R}^m$ possessing—at each point of the domain of definition—right-hand and left-hand limits (and in the case of the space $\mathscr{D}_T^m$ also possessing the limit as $s \to -\infty$). Let $\mathscr{D}^m = \mathscr{D}_0^m$. Denote by θ_t $(t \le T)$ the mapping of $\mathscr{D}_t^m$ into $\mathscr{D}^m$ defined by the relation:

$$(\theta_t \varphi)(s) = \varphi(t+s), \qquad s \le 0.$$

Next let $\alpha(\varphi, t) = \alpha(\varphi, t, \omega)$ be a random function defined on $\mathscr{D}^m \times [0, T] \times \Omega$. Relation (7) can be rewritten as follows,

$$\xi(t+\Delta t) - \xi(t) = \alpha(\theta_t \xi, t+\Delta t) - \alpha(\theta_t \xi, t) + \delta_t,$$

and equation (5) can be represented by equation

$$\xi(t) = \xi_0 + \int_{t_0}^t \alpha(\theta_s \xi, s)\, ds + \int_{t_0}^t \beta(\theta_s \xi, ds), \qquad t > t_0. \tag{8}$$

Here it becomes necessary to define the process $\xi(t)$ over the whole "past," i.e., up to the time t_0. In this connection one should adjoin to equation (8) relation

$$\xi(t) = \varphi(t), \qquad t \le t_0, \tag{9}$$

which will be called from now on the *initial condition* for the stochastic differential equation (8).

The stochastic line integral. Let $\{\mathfrak{F}_t, t \in [0, T]\}$ be a current of σ-algebras on a fixed probability space $\{\Omega, \mathfrak{S}, \mathsf{P}\}$, $(\mathfrak{F}_t \subset \mathfrak{S})$, and $\beta(\varphi, t)$ be a random function adopted to $\{\mathfrak{F}_t\}$ with values in $\mathscr{R}^m$.

Two variants of theorems are considered below. One variant will refer to random processes with continuous sample functions (with probability 1), the other to processes with sample functions without discontinuities of the second kind (mod P). In this connection two sets of assumptions are introduced.

Let $\mathscr{C}_T^m$ $(\mathscr{C}^m, \mathscr{C}^m[a, b])$ be the subspace of the space $\mathscr{D}_T^m$ $(\mathscr{D}^m, \mathscr{D}^m[a, b])$ consisting of continuous functions. The space $\mathscr{C}^m$ is endowed with the uniform norm

$$\|\varphi\| = \sup_{s \le 0} |\varphi(s)|.$$

The space $\mathscr{D}^m$ will be assumed to be a metric space with the metric $\rho_{\mathscr{D}^m}$ of the space of functions without discontinuities of the second kind (Volume I, Chapter VI, Section 5). In order to simplify the study of discontinuous processes a simpler metric will be utilized in $\mathscr{D}^m$. (Further assumptions on the equations under consideration will be formulated in terms of this metric.) This metric is generated by the seminorm $\|\varphi\|_*$, defined by relation

$$\|\varphi\|_* = \{\int_{-\infty}^0 |\varphi(s)|^2 |K(ds)|\}^{1/2}, \tag{10}$$

where $K(\cdot)$ is a finite measure defined on Borel sets on the half-line $(-\infty, 0]$ and $K(-\infty, 0] = K < \infty$.

If, for example, *stochastic differential equations with a lag*, i.e., equations of the form

$$d\xi(t)=\alpha(\xi(t-h_1),\ldots,\xi(t-h_r),t)\,dt$$
$$+\beta(\xi(t-h_1),\xi(t-h_2),\ldots,\xi(t-h_r),dt),$$

are discussed, then the functions depending on values of $\varphi(s)$ at a finite number of points, i.e., functions of the form $\beta(\varphi(0),\varphi(-h_1),\ldots,\varphi(-h_r),t)$ should be chosen as functions $\beta(\varphi,t)$. In this case, it is natural to identify functions $\varphi(s)$ taking on the same values at points $0,-h_1,\ldots,-h_r$ and metrize $\mathscr{D}^m$ by means of the metric

$$\|\varphi-\psi\|_*=[(\varphi(0)-\psi(0))^2+(\varphi(-h_1)-\psi(-h_1))^2+\cdots+(\varphi(-h_r)-\psi(-h_r))^2]^{1/2},$$

i.e., using the seminorm (10) which corresponds to measure K concentrated at points $0=h_0,-h_1,\ldots,-h_r$ and taking on values $K(\{-h_k\})=1$ at these points.

Returning to functions $\beta(\varphi,t)$ we shall first assume that they satisfy one of the following two sets of conditions:

$\beta1$) a) Function $\beta(\varphi,s)=\beta(\varphi,s,\omega)$ is defined on $\mathscr{D}^m\times[0,T]\times\Omega$ and for each $t\leq T$ its contraction on the segment $s\in[0,t]$ is $\mathfrak{B}_{\mathscr{D}^m}\times\mathfrak{T}_t\times\mathfrak{F}_t$-measurable;

b) $\beta(\varphi,t)$ is a square integrable $\mathfrak{F}_t$-martingale for a fixed φ, with sample functions belonging to $\mathscr{D}^m[0,T]$ with probability 1 and the characteristics of the components of this $\mathfrak{F}_t$-martingale are continuous with probability 1.

Here $\mathfrak{B}_{\mathscr{D}^m}$ is the minimal σ-algebra of the subsets of $\mathscr{D}^m$ containing cylindrical sets in $\mathscr{D}^m$ and $\mathfrak{T}_t$ is the σ-algebra of Borel sets on the interval $[0,t]$.

$\beta2$) Function $\beta(\varphi,s)$ satisfies conditions which are obtained from $\beta1$) if $\mathscr{D}^m$, $\mathfrak{B}_{\mathscr{D}^m}$, and $\mathscr{D}^m[0,T]$ are replaced by $\mathscr{C}^m$, $\mathfrak{B}_{\mathscr{C}^m}$, and $\mathscr{C}^m[0,T]$, respectively.

A random function $\beta(\varphi,t)$ satisfying conditions $\beta1$)($\beta2$)) is called a *martingale field* in $\mathscr{D}^m$ (in $\mathscr{C}^m$) or simply a field.

If $\beta(\varphi,t)$ is a martingale field in $\mathscr{D}^m$ then a random function $\Lambda(\varphi,t)$ exists which is, for a fixed φ, a natural integrable monotonically nondecreasing process such that $\Lambda(\varphi,0)=0$ and for any $\Delta=(t,t+\Delta t]$

$$\mathsf{E}\{|\beta(\varphi,\Delta)|^2\,|\,\mathfrak{F}_t\}=\mathsf{E}\{\Lambda(\varphi,\Delta)\,|\,\mathfrak{F}_t\},$$

where $\Lambda(\varphi,\Delta)=\Lambda(\varphi,t+\Delta t)-\Lambda(\varphi,t)$.

We say that the field $\beta(\varphi,t)$ is *linearly bounded in a seminorm* or in *a norm* if

$$\Lambda(\varphi,\Delta)\leq(1+\|\varphi\|_*^2)\Lambda_0(\Delta) \tag{11}$$

or, correspondingly,

$$\Lambda(\varphi,\Delta)\leq(1+\|\varphi\|^2)\Lambda_0(\Delta).$$

Here $\Lambda_0(t)$ is a continuous integrable monotonically nondecreasing process adopted to the current of σ-algebras $\{\mathfrak{F}_t, t\in[0, T]\}$. If $\Lambda(\varphi, t)$ is a continuous function of t for each φ, then the condition of linear boundedness in the seminorm is equivalent to the requirement:

There exists a process $\Lambda_0(t)$ satisfying the preceding conditions such that for any $\Delta\subset[0, T]$

$$\mathsf{E}\{|\beta(\varphi, \Delta)|^2|\mathfrak{F}_t\}\leq(1+\|\varphi\|_*^2)\mathsf{E}\{\Lambda_0(\Delta)|\mathfrak{F}_t\}. \tag{12}$$

It is trivial to verify that (11) implies (12). The converse follows easily from Theorem 21 in Section 1. An analogous remark is valid also for the fields which are linearly bounded in the norm.

Similar remarks hold concerning the martingale $\beta(\varphi, t)-\beta(\psi, t)$. If for an arbitrary $N>0$ a monotonically nondecreasing, continuous and integrable process $\Lambda_N(t)$, $t\in[0, T]$, exists; adopted to the current of σ-algebras $\{\mathfrak{F}_t, t\in[0, T]\}$, which is independent of φ and ψ and such that

$$\mathsf{E}\{|\beta(\varphi, \Delta)-\beta(\psi, \Delta)|^2|\mathfrak{F}_t\}\leq\|\varphi-\psi\|_*^2\mathsf{E}\{\Lambda_N(\Delta)|\mathfrak{F}_t\} \tag{13}$$

for all $\varphi, \psi\in\mathscr{D}^m$ satisfying the conditions $\|\varphi\|_*\leq N$ and $\|\psi\|_*\leq N$, we then say that $\beta(\varphi, t)$ satisfies *a local Lipschitz condition* (relative to the seminorm). If there exists a process $\Lambda(t)$ such that one can set $\Lambda_N(t)=\Lambda(t)$ for all $N>0$, we then say that $\beta(\varphi, t)$ satisfies a *uniform Lipschitz condition* (relative to the seminorm). Analogous terminology is used also in the case when the seminorm $\|\varphi-\psi\|_*$ is replaced by the norm $\|\varphi-\psi\|$ in equality (13).

We now present the definition of a stochastic line integral

$$\int_0^T \beta(\theta_t\xi, dt).$$

(This definition will be somewhat generalized below.)

The following assumptions are made concerning the random processes $\xi(t)$, $t\in(-\infty, T]$:

$\xi1$) The process $\xi(t)$, $t\in[0, T]$, is adopted to the current of σ-algebras $\{\mathfrak{F}_t, t\in[0, T]\}$, the variables $\xi(s)$ are $\mathfrak{F}_0$-measurable for $s<0$, and the sample functions of the process $\xi(t)$ belong to $\mathscr{D}_T^m$ with probability 1; or

$\xi2$) The process $\xi(t)$, $t\leq T$, satisfies condition $\xi1$) and the sample functions of this process belong with probability 1 to $\mathscr{C}_T^m$.

Let δ be a subdivision of the interval $[0, T]$ with the subdividing points

$$t_0=0<t_1<t_2<\cdots<t_n=T,$$

$$|\delta|=\max_{1\leq k\leq n}\Delta t_k, \qquad \Delta t_k=t_k-t_{k-1}.$$

In what follows Δ or Δ_k denote the semiinterval $(t, t+\Delta t]$ or $(t_k, t_{k+1}]$.

Theorem 1. *Let $\xi(t)$, $t \in (-\infty, T]$, satisfy the condition $\xi 1)$, and the random function $\beta(\varphi, t)$ satisfy $\beta 1)$ and also the local Lipschitz condition* (13). *Then the limit*

$$\int_0^T \beta(\theta_t \xi, dt) \underset{\text{Def}}{=} \mathsf{P}\text{-}\lim_{|\delta| \to 0} \sum_{k=1}^{n} \beta(\theta_{t_{k-1}} \xi, t_k) - \beta(\theta_{t_{k-1}} \xi, t_{k-1}) \tag{14}$$

exists.

Definition. The limit on the right-hand side of relation (14) (provided it exists) is called *the stochastic line integral* or *the stochastic integral in the field $\beta(\varphi, t)$ along the curve $\xi(t)$*.

Proof of Theorem 1. Let

$$\tau_N = (\inf\{t: |\xi(t)| > N\}) \vee 0$$

(we assume that $\inf\{\varnothing\} = T$), $\tau'_L = \inf\{t: \Lambda_N(t) > L\}$, and $\xi_N(t) = \xi(t)$ for $t < \tau_N$, $\xi_N(t) = 0$ if $t \geq \tau_N$, $\tilde{\beta}(\varphi, t) = \tilde{\beta}_{LN}(\varphi, t) = \beta(\varphi, t \wedge \tau'_L)$. Using the theorem on the stopping of a martingale, it is easy to verify that for $\|\varphi\| \leq N$, $\|\psi\| \leq N$,

$$\mathsf{E}\{|\tilde{\beta}(\varphi, \Delta) - \tilde{\beta}(\psi, \Delta)|^2 \,|\, \mathfrak{F}_t\} \leq \|\varphi - \psi\|_*^2 \mathsf{E}\{\tilde{\Lambda}_N(\Delta) \,|\, \mathfrak{F}_t\}, \tag{15}$$

where

$$\tilde{\Lambda}_N(\Delta) = \Lambda_N[(t + \Delta t) \wedge \tau_L] - \Lambda_N(t \wedge \tau_L).$$

Consider two subdivisions δ_1 and δ_2 of the interval $[0, T]$, where δ_2 is a refinement of δ_1 ($\delta_2 < \delta_1$). Denote the subdividing points of δ_1 by t_k ($k = 0, 1, \ldots, n$) and that of δ_2 by t_{kj} ($t_k = t_{k0} < t_{k1} < \cdots < t_{ks_k} = t_{k+1}$).

Set

$$\Delta t_k = t_k - t_{k-1}, \qquad \Delta t_{kj} = t_{kj} - t_{kj-1},$$

$$\Delta_k = (t_{k-1}, t_k], \qquad \Delta_{kj} = (t_{kj-1}, t_{kj}],$$

$$\sigma_1(\xi) = \sum_{k=1}^{n} \beta(\theta_{t_{k-1}} \xi, \Delta_k) = \sum_{k=1}^{n} \beta(\theta_{t_{k-1}} \xi, t_k) - \beta(\theta_{t_{k-1}} \xi, t_{k-1}),$$

let $\sigma_2(\xi)$ be the integral sum which is analogous to $\sigma_1(\xi)$ but is constructed using subdivision δ_2, and let $\tilde{\sigma}_i(\xi)$ be integral sums constructed from the field $\tilde{\beta}(\varphi, t)$ using subdivisions δ_i ($i = 1, 2$). Then

$$\mathsf{P}\{|\sigma_1(\xi) - \sigma_2(\xi)| > \varepsilon\} \leq \mathsf{P}\{\tau_N \vee \tau'_L < T\} + \mathsf{P}\{|\tilde{\sigma}_1(\xi_N) - \tilde{\sigma}_2(\xi_N)| > \varepsilon\}.$$

Since the sample functions of the processes $\xi(t)$ ($t \in (-\infty, T]$) and $\Lambda_N(t)$ ($t \in [0, T]$) are bounded, the probability $\mathsf{P}\{\tau_N \vee \tau'_L < T\}$ can be made as small as desired for N and $L = L(N)$ sufficiently large.

We now bound the second summand in the right-hand side of the last inequality. Observe that

$$\tilde{\sigma}_1(\xi_N)-\tilde{\sigma}_2(\xi_N)=\sum_{k=0}^{n-1}\sum_{r=1}^{s_k}\tilde{\beta}(\theta_{t_{k-1}}\xi_N,\Delta_{kr})-\tilde{\beta}(\theta_{t_{kr-1}}\xi_N,\Delta_{kr}).$$

Using (15) we obtain

$$\mathsf{E}|\tilde{\sigma}_1(\xi_N)-\tilde{\sigma}_2(\xi_N)|^2=\sum_{k=0}^{n-1}\sum_{r=1}^{s_k}\mathsf{E}|\tilde{\beta}(\theta_{t_{k-1}}\xi_N,\Delta_{kr})-\tilde{\beta}(\theta_{t_{kr-1}}\xi_N,\Delta_{kr})|^2$$

(16)

$$\leqslant\mathsf{E}\sum_{k=0}^{n-1}\sum_{r=1}^{s_k}\|\theta_{t_{k-1}}\xi_N-\theta_{t_{kr-1}}\xi_N\|_*^2\tilde{\Lambda}_N(\Delta_{kr}).$$

Here we utilized the fact that

$$\begin{aligned}&\mathsf{E}|\tilde{\beta}(\theta_{t_{k-1}}\xi_N,\Delta_{kr})-\tilde{\beta}(\theta_{t_{kr-1}}\xi_N,\Delta_{kr})|^2\\&=\mathsf{E}\mathsf{E}\{\cdots|\mathfrak{F}_{t_{kr-1}}\}\\&=\mathsf{E}[\mathsf{E}\{|\tilde{\beta}(\varphi,\Delta_{kr})-\tilde{\beta}(\psi,\Delta_{kr})|^2|\mathfrak{F}_{t_{kr-1}}\}]_{\varphi=\theta_{t_{k-1}}\xi_N,\psi=\theta_{t_{kr-1}}\xi_N}.\end{aligned}$$

Note that the sum appearing under the expectation sign in the right-hand side of inequality (16) is uniformly bounded. This sum is bounded above by $4N^2K\tilde{\Lambda}_N(T)\leqslant 4N^2KL$, where $K=K(-\infty,0]$. Furthermore,

$$\|\theta_{t_{k-1}}\xi_N-\theta_{t_{kr-1}}\xi_N\|_*^2=\int_{-\infty}^0|\xi_N(t_{k0}+s)-\xi_N(t_{kr-1}+s)|^2K(ds),$$

so that

$$\mathsf{E}|\tilde{\sigma}_1(\xi_N)-\tilde{\sigma}_2(\xi_N)|^2=\mathsf{E}\int_{-\infty}^0\Big(\sum_{k,r}|\xi_N(t_{k-1}+s)$$

(17)

$$-\xi_N(t_{kr-1}+s)|^2\Lambda_N(\tilde{\Delta}_{kr})\Big)K(ds).$$

Using a method common in integral calculus it is easy to verify that the sum appearing under the sign of the integral in equation (17) tends to 0 as $|\delta|\to 0$ for all $s\leqslant 0$.

Indeed let $\varepsilon_1>0$ be an arbitrary given number. The function $\xi_N(u)$ possesses only a finite number of jumps of the size at least $\sqrt{\varepsilon_1/L}$ on the interval $[s,s+T]$. Let this be the points $s_1,\ldots,s_m$. We enclose each one of these points by intervals i_r of length h_0, where $h_0m<\varepsilon_1$. Omitting the intervals i_r from the segment $[s,s+T]$ we obtain a closed set S. We now choose ε_2 such that $|\xi_N(s')-\xi_N(s'')|^2<2\varepsilon_1/L$ provided $|s'-s''|<\varepsilon_2$, s' and $s''\in S$. It is easy to verify that such an ε_2 exists. Indeed, assuming otherwise, we construct sequences of points s'_n, s''_n, $n=1,2,\ldots$, such that $|s'_n-s''_n|<1/n$, $\lim s'_n=\lim s''_n=s_0\in S$, $|\xi_N(s'_n)-\xi_N(s''_n)|>2\varepsilon/L$, which is possible only if $|\xi_N(s_0-)-\xi_N(s_0)|\geqslant 2\varepsilon_1/L$ since the one-sided limits of function $\xi_N(s)$ exist. But the last inequality contradicts the fact that $s_0\in S$. If $|\delta_1|<h_0/2$

then each one of the intervals $\Delta_k=[t_{k-1},t_k]$ is either located in the interior of S or it contains one of the end-points of i_r as an interior point or is located in the interior of i_r. Denote by I_1, I_2 and I_3 the corresponding sets of intervals Δ_k. Let $|\delta_1|<((h_0/2)\wedge\varepsilon_2)$. Then

$$\begin{aligned} z&=\sum_{k,r}|\xi_N(t_{k0}+s)-\xi_N(t_{kr-1}+s)|^2\tilde{\Lambda}_N(\Delta_{k_r})\\ &\leqslant\sum_{I_1}+\sum_{I_2}+\sum_{I_3}\\ &\leqslant\frac{2\varepsilon_1}{L}\sum_{I_1}\tilde{\Lambda}_N(\Delta_{kr})+2m\cdot 4N^2\cdot\max_{\Delta_k\in I_2}\tilde{\Lambda}_N(\Delta_k)+4N^2\sum_{\Delta_k\in I_3}\tilde{\Lambda}_N(\Delta_k)\\ &\leqslant 2\varepsilon_1+8mN^2\max_{\Delta_k\in I_2}\tilde{\Lambda}_N(\Delta_k)+4N^2\sum_{r=1}^m\tilde{\Lambda}_N(i_r). \end{aligned}$$

Since the function $\tilde{\Lambda}_N(t)$ is continuous, one can, by fixing an arbitrary ε_0, choose ε_1 and then choose h_0 and ε_2 such that $z\leqslant\varepsilon_0$ for all δ_1 satisfying $|\delta_1|<(h_0/2\wedge\varepsilon_2)$ (for a given ω). Thus $z\to 0$ as $|\delta_1|\to 0$ with probability 1.

Noting that it is permissible in inequality (17) to approach the limit under the sign of the integral, we obtain

$$\mathsf{E}|\tilde{\sigma}_1(\xi_N)-\tilde{\sigma}_2(\xi_N)|^2\to 0\quad\text{as }|\delta_1|\to 0.$$

Thus, $\mathsf{P}\{|\tilde{\sigma}_1(\xi_N)-\tilde{\sigma}_2(\xi_N)|>\varepsilon\}\to 0$ as $|\delta_1|\to 0$. This easily implies that for arbitrary subdivisions δ_1 and δ_2 of the interval $[0,T]$ (i.e., when δ_2 is not necessarily a subdivision of δ_1) we have

$$\mathsf{P}\{|\sigma_1(\xi)-\sigma_2(\xi)|>\varepsilon\}\to 0\quad\text{as }|\delta_1|,|\delta_2|\to 0.$$

The theorem is thus proved. □

The preceding theorem can be somewhat refined in the case when integration over continuous processes is considered.

Theorem 2. *Assume that the process $\xi(t)$, $t\in(-\infty,T]$, satisfies condition ξ2) and the field $\beta(\varphi,t)$ satisfies condition β2) and the local Lipschitz condition (with respect to a uniform norm), i.e., for $\|\varphi\|\leqslant N$, $\|\psi\|\leqslant N$, and $t\in[0,T]$*

$$\mathsf{E}\{|\beta(\varphi,\Delta)-\beta(\psi,\Delta)|^2|\mathfrak{F}_t\}\leqslant\|\varphi-\psi\|^2\mathsf{E}\{\Lambda_N(\Delta)|\mathfrak{F}_t\},\tag{18}$$

where $\Lambda_N(t)$ is a continuous monotonically nondecreasing integrable process adopted to $\{\mathfrak{F}_t, t\in[0,T]\}$. Then the stochastic line integral (14) *exists.*

Proof. Using the same argument as in the proof of Theorem 1 we obtain

$$\mathsf{E}|\tilde{\sigma}_1(\xi_N)-\tilde{\sigma}_2(\xi_N)|^2\leqslant\mathsf{E}\sum_{k,r}\|\theta_{t_{k-1}}\xi_N-\theta_{t_{kr-1}}\xi_N\|^2\tilde{\Lambda}_N(\Delta_{kr});$$

moreover, the sum in the right-hand side of the last inequality is uniformly bounded in ω. It follows from the assumptions imposed on the structure of function $\xi(t)$ that

$$\|\theta_{t_{k-1}}\xi_N - \theta_{t_{kr-1}}\xi_N\|^2 = \sup_{s\leq 0} |\xi(t_k+s) - \xi(t_{kr-1}+s)|^2 \to 0$$

with probability 1 as $|\delta_1| \to 0$ uniformly in k and r. Therefore

$$\mathsf{E}|\tilde{\sigma}_1(\xi_N) - \tilde{\sigma}_2(\xi_N)|^2 \to 0$$

as $|\delta_1| \to 0$. This, as in the case of Theorem 1, implies the required assertion. □

Remark. If the conditions of Theorems 1 and 2 are satisfied and $\sup_{-\infty<t<T}|\xi(t)| \leq C$, where C is an absolute constant, then the convergence in relation (14) is not only in probability but in the mean square as well, and the stochastic integral possesses finite moments of the second order.

We shall now derive several bounds on the integrals introduced above.

Lemma 1. *Let the field $\beta(\varphi, t)$ satisfy condition $\beta 1$) and the local Lipschitz condition, let $\xi_k(t)$, $k = 1, 2$, satisfy condition $\xi 1$), and let*

$$\|\theta_T \xi_k\| = \sup\{|\xi_k(t)|, t \leq T\} \leq N, \qquad k = 1, 2.$$

Then

$$\text{(19)} \quad \mathsf{E}\{|\textstyle\int_0^T \beta(\theta_t\xi_1, dt) - \int_0^T \beta(\theta_t\xi_2, dt)|^2 |\mathfrak{F}_0\} \leq \mathsf{E}\{\int_0^T \|\theta_t\xi_1 - \theta_t\xi_2\|_*^2 \Lambda_N(dt)|\mathfrak{F}_0\}.$$

Proof. As in the proof of Theorem 1 we obtain the inequality

$$\mathsf{E}\{|\sigma(\xi_1) - \sigma(\xi_2)|^2 | \mathfrak{F}_0\} \leq \mathsf{E}\left\{\int_{-\infty}^0 \left(\sum_{k=1}^n |\xi_1(t_k+s) - \xi_2(t_k+s)|^2 \Lambda_N(\Delta_k)\right) K(ds) | \mathfrak{F}_0\right\}.$$

The sum appearing under the integral sign is uniformly bounded (in ω) and tends to the limit

$$\int_0^T |\xi_1(t+s) - \xi_2(t+s)|^2 \Lambda_N(dt)$$

as $|\delta| \to 0$. This fact easily implies the validity of inequality (19). □

Remark. If $\beta(\varphi, t)$ and $\xi_k(t)$, $k = 1, 2$, satisfy the conditions of Theorem 2 and $\|\theta_T \xi_k(t)\| \leq N$, then

$$\text{(20)} \quad \mathsf{E}\{|\textstyle\int_0^T \beta(\theta_t\xi_1, dt) - \int_0^T \beta(\theta_t\xi_2, dt)| |\mathfrak{F}_0\} \leq \mathsf{E}\{\int_0^T \|\theta_t\xi_1 - \theta_t\xi_2\|^2 \Lambda_N(dt) | \mathfrak{F}_0\}.$$

The proof of this assertion is analogous to the preceding one.

The following lemma can be proven in an analogous manner:

Lemma 2. 1. *If the conditions of Theorem* 1 *are satisfied and if*

a) $\|\theta_T \xi\| \leqslant N$, *where* N *is an absolute constant,*

b) *there exists a continuous monotonically nondecreasing and integrable process* $\Lambda_0(t)$ *adopted to the current of* σ*-algebras* $\{\mathfrak{F}_t, t \in [0, T]\}$ *such that*

(21)
$$\mathsf{E}\{|\beta(\varphi, \Delta)|^2 | \mathfrak{F}_t\} \leqslant (1 + \|\varphi\|_*^2) \mathsf{E}\{\Lambda_0(\Delta t) | \mathfrak{F}_t\},$$

then

(22)
$$\mathsf{E}\{|\textstyle\int_0^T \beta(\theta_t \xi, dt)|^2 | \mathfrak{F}_0\} \leqslant \mathsf{E}\{\Lambda_0(T) + \int_0^T \|\theta_t \xi\|_*^2 \Lambda_0(dt) | \mathfrak{F}_0\}.$$

2. *If the conditions of Theorem* 2 *and condition* a) *are satisfied and if*

c) (23)
$$\mathsf{E}\{|\beta(\varphi, \Delta)|^2 | \mathfrak{F}_t\} \leqslant (1 + \|\varphi\|^2) \mathsf{M}\{\Lambda_0(\Delta) | \mathfrak{F}_t\},$$

then

(24)
$$\mathsf{E}\{(|\textstyle\int_0^T \beta(\theta_t \xi, dt)|)^2 | \mathfrak{F}_0\} \leqslant \mathsf{E}\{\Lambda_0(T) + \int_0^t \|\theta_t \xi\|^2 \Lambda_0(dt) | \mathfrak{F}_0\}.$$

Lemma 3. *Let* $\beta(\varphi, t)$ *and* $\xi(t)$ *satisfy the conditions of either Theorems* 1 *or* 2. *Assume also that* τ *is random time on* $[0, T]$, $\beta_\tau(\varphi, t) = \beta(\varphi, t \wedge \tau)$, $\xi_\tau(t) = \xi(t)$ *for* $t < \tau$, *and* $\xi_\tau(t) = \xi(\tau -)$ *for* $t \geqslant \tau$.

Then

(25)
$$\textstyle\int_0^t \beta(\theta_s \xi, ds) = \int_0^t \beta_\tau(\theta_s \xi_\tau, ds) = \int_0^t \beta_\tau(\theta_s \xi, ds)$$

with probability 1 *on the set* $t \leqslant \tau$.

First we note that if the conditions of either Theorems 1 or 2 are satisfied for the time interval $[0, T]$, then they are also fulfilled for the narrower interval $[0, t]$, $t < T$, and in accordance with either Theorems 1 or 2 one can uniquely (mod P) define the integral

$$\textstyle\int_0^t \beta(\theta_s \xi, ds).$$

Moreover, functions $\beta_\tau(\varphi, t)$ and $\xi_\tau(t)$ also satisfy the conditions of these theorems. Therefore the quantities appearing in relation (25) are well defined. Equalities (25) now follow from the fact that the sums in terms of which the integrals in equalities (25) are defined coincide on the set $t \leqslant \tau$.

Remark. We emphasize that (25) holds (mod P) also for $t = \tau$.

Lemma 4. *Let* $\beta(\varphi, t)$ *satisfy condition* $\beta 1)$ *and also the uniform Lipschitz condition and let* $\xi_k(t)$, $k = 1, 2$, *satisfy condition* $\xi 1)$, *and, moreover,*

$$\mathsf{E} \textstyle\int_0^T \|\theta_t \xi\|_*^2 \Lambda(dt) < \infty.$$

Then inequality (19) *is valid with* $\Lambda_N = \Lambda$.

Proof. Let $\tau_N = \inf\{t: \bigcup_{k=1}^2 |\xi_k(t)| \geq N\}$ and $\xi_k^N(t) = \xi_k(t)$ for $t < \tau_N$, $\xi_k^N(t) = \xi_k(\tau_N -)$ for $t \geq \tau_N$.

It follows from Lemma 3 that

$$\int_0^T \beta(\theta_s \xi_k^N, ds) = \int_0^T \beta(\theta_s \xi_k, ds)$$

for all N sufficiently large. In view of Fatou's lemma

$$\begin{aligned}\mathsf{E}\{|\int_0^T \beta(\theta_s \xi_1, ds) - \int_0^T \beta(\theta_s \xi_2, ds)|^2\} &\leq \underline{\lim}\, \mathsf{E}|\int_0^T \beta(\theta_s \xi_1^N, ds) - \int_0^T \beta(\theta_s \xi_2^N, ds)|^2 \\ &\leq \underline{\lim}\, \mathsf{E} \int_0^T \|\theta_t \xi_1^N - \theta_t \xi_2^N\|_*^2 \Lambda(dt).\end{aligned}$$

Taking into account that $\xi_k^N(t) \to \xi_k(t)$ uniformly in t with probability 1, we obtain

$$\text{(26)} \qquad \mathsf{E}\{|\int_0^T \beta(\theta_s \xi_1, ds) - \int_0^T \beta(\theta_s \xi_2, ds)|^2\} \leq \mathsf{E} \int_0^T \|\theta_t \xi_1 - \theta_t \xi_2\|_*^2 \Lambda(dt). \quad \square$$

Lemma 5. *If the field $\beta(\varphi, t)$ is linearly bounded and satisfies the local Lipschitz condition, the process $\xi(t)$ satisfies condition ξ1), and if, moreover,*

$$\mathsf{E} \int_0^T \|\theta_T \xi\|_*^2 \Lambda_0(dt) < \infty,$$

then inequality (22) *is satisfied even without the restriction $\|\theta_T \xi\| \leq N$.*

The proof is analogous to the proof of Lemma 4.

Remark. Inequalities analogous to (22) and (26) are valid also in the case when $\beta(\varphi, t)$ and $\xi_k(t)$, $k = 1, 2$, satisfy the conditions of Theorem 2 and if, in addition, the field $\beta(\varphi, t)$ is linearly bounded (or correspondingly satisfies the uniform Lipschitz condition) and

$$\mathsf{E} \int_0^T \|\theta_t \xi_k\|^2 \Lambda_0(dt) < \infty \qquad (\mathsf{E} \int_0^T \|\theta_t \xi_k\|^2 \Lambda(dt) < \infty).$$

Lemma 6. *Assume that the random field $\beta(\varphi, t)$ satisfies the conditions of Theorem 1 and processes $\xi_k(t)$, $k = 1, 2$, that of Lemma 1. Then for any $\varepsilon > 0$ and $N > 0$*

$$\text{(27)} \quad \mathsf{P}\{|\int_0^T \beta(\theta_t \xi_1, dt) - \int_0^T \beta(\theta_t \xi_2, dt)| > \varepsilon\} \leq \frac{N}{\varepsilon^2} + \mathsf{P}\{\int_0^T \|\theta_t \xi_1 - \theta_t \xi_2|_*^2 \Lambda_N(dt) > N\}.$$

Proof. Let δ be a subdivision of the interval $[0, T]$ with the subdividing points t_k, $k = 1, 2, \ldots, n$. The sequence of sums

$$\sum_{j=1}^k \beta(\theta_{t_{j-1}} \xi_1, \Delta_j) - \beta(\theta_{t_{j-1}} \xi_2, \Delta_j), \qquad k = 0, 1, \ldots, n,$$

forms a square integrable martingale and in view of Lemma 10 in Section 1,

$$\mathsf{P}\left\{\left|\sum_{j=1}^{n} \beta(\theta_{t_{j-1}}\xi_1, \Delta_j) - \sum_{j=1}^{n} \beta(\theta_{t_{j-1}}\xi_2, \Delta_j)\right| \geq \varepsilon\right\}$$

$$\leq \frac{N}{\varepsilon^2} + \mathsf{P}\left\{\sum_{j=1}^{n} \|\theta_{t_{j-1}}\xi_1 - \theta_{t_{j-1}}\xi_2\|_*^{(2)} \Lambda_N(\Delta_j) \geq N\right\}.$$

Approaching the limit as $|\delta| \to 0$ and using the results obtained in the proof of Theorem 1 we obtain inequality (27). □

Remark. If the conditions of the remark following Lemma 1 are satisfied and $\|\theta_T\xi_1\| \leq N$, $\|\theta_T\xi_2\| \leq N$, then

$$\mathsf{P}\{|\textstyle\int_0^T \beta(\theta_t\xi_1, dt) - \int_0^T \beta(\theta_t\xi_2, dt)| > \varepsilon\}$$

(28)

$$\leq \frac{N}{\varepsilon^2} + \mathsf{P}\{\textstyle\int_0^T \|\theta_t\xi_1 - \theta_t\xi_2\| \Lambda_N(dt) > N\}.$$

Inequality (27) allows us to extend the definition of a stochastic line integral to a wider class of processes $\xi(t)$ than those appearing in Theorem 1. However, since, in what follows, stochastic line integrals are utilized only in the theory of those stochastic differential equations, whose solutions result in processes with sample functions belonging to $\mathscr{D}_T^m$ or $\mathscr{C}_T^m$, it is sufficient to confine ourselves to the definition of the integral given above and to classes of processes $\xi(t)$ introduced above for which these integrals exist.

The stochastic line integral as a function of the upper limit of integration. Let $\beta(\varphi, t)$ and $\xi(t)$ satisfy the conditions of either Theorems 1 or 2. If $0 \leq a < b \leq T$, then the corresponding conditions are satisfied when the interval $[0, T]$ is replaced by $[a, b]$.

Thus one can define the stochastic integral

$$\int_a^b \beta(\theta_s\xi, ds).$$

Clearly this integral is an $\mathfrak{F}_b$-measurable random variable and for $0 \leq a < b < c \leq T$

(29) $$\int_a^b \beta(\theta_s\xi, ds) + \int_b^c \beta(\theta_s\xi, ds) = \int_a^c \beta(\theta_s\xi, ds) \quad (\text{mod } \mathsf{P}).$$

Set

$$\eta(t) = \int_0^t \beta(\theta_s\xi, ds).$$

The process $\eta(t)$ is adopted to the current of σ-algebras $\{\mathfrak{F}_t, t \in [0, T]\}$ and is uniquely determined with probability 1 for each t.

As before one can utilize the fact that the process $\eta(t)$ is not unique, and in all cases we interpret $\eta(t)$ as the separable modification of this process.

Lemma 7. *Let the conditions of Theorem* 1 *be satisfied. Then:*

a) *The process* $\eta(t)$, $t\in[0, T]$, *is a local square integrable martingale and possesses a modification with sample functions belonging to* $\mathscr{D}[0, T]$ *with probability* 1.

b) *If* $\sup_{\infty<t\leq T}|\xi(t)|\leq N$, $N>0$, *and condition* b) *of Lemma* 2 *is satisfied, then* $\eta(t)$ *is a square integrable martingale and*

$$\mathsf{E}\{|\int_{t_1}^{t_2}\beta(\theta_s\xi, ds)|^2\,|\,\mathfrak{F}_{t_1}\}\leq\mathsf{E}\{\int_{t_1}^{t_2}(1+\|\theta_s\xi\|_*^2)\Lambda_0(dt)\,|\,\mathfrak{F}_{t_1}\}. \tag{30}$$

c) *If condition* b) *of Lemma* 2 *is satisfied with* $\Lambda_0(\Delta)=C_0\,\Delta t$, *where* C_0 *is an absolute constant and*

$$\int_0^T\mathsf{E}\|\theta_s\xi\|_*^2\,ds<\infty,$$

then $\eta(t)$ *is a square integrable martingale; moreover,*

$$\mathsf{E}\{|\int_{t_1}^{t_2}\beta(\theta_s\xi, ds)|^2\,|\,\mathfrak{F}_{t_1}\}\leq C_0\int_{t_1}^{t_2}(1+\mathsf{E}\{\|\theta_s\xi\|_*^2\,|\,\mathfrak{F}_{t_1}\})\,ds. \tag{31}$$

d) *Set*

$$\int_0^\tau\beta(\theta_s\xi, ds)\underset{\text{Def}}{=}\eta(\tau),$$

$\beta_\tau(\varphi, t)=\beta(\varphi, t\wedge\tau)$, *where* τ *is a random time on* $\{\mathfrak{F}_t, t\in[0, T]\}$; *then*

$$\int_0^\tau\beta(\theta_s\xi, ds)=\int_0^T\beta_\tau(\theta_s\xi, ds)\quad(\text{mod}\,\mathsf{P}). \tag{32}$$

e) *For any* $\varepsilon>0$ *and* $N>0$

$$\mathsf{P}\{\sup_{0\leq t\leq T}|\int_0^t\beta(\theta_s\xi, ds)|>\varepsilon\}\leq\frac{N}{\varepsilon^2}+\mathsf{P}\{\int_0^T(1+\|\theta_s\xi\|_*^2)\Lambda_0(ds)>N\}. \tag{33}$$

Proof. First we shall prove assertion b). Inequality (30) follows directly from Lemma 2. Since in this case the sum $\sum_{a\leq t_k\leq b}\beta(\theta_{t_k}\xi, \Delta_k)$ is uniformly integrable, one can approach the limit in the equality

$$\mathsf{E}\Big\{\sum_{a\leq t_k\leq b}\beta(\theta_{t_k}\xi, \Delta_k)\,|\,\mathfrak{F}_a\Big\}=0$$

as $|\delta|\to 0$ and obtain

$$\mathsf{E}\{\int_a^b\beta(\theta_s\xi, ds)\,|\,\mathfrak{F}_a\}=0.$$

Therefore $\eta(t)$ is a square integrable martingale. Assertion c) is proved analogously.

To prove assertion d) we first assume that $\sup_t|\xi(t)|\leqslant N$. Set

$$\sigma(t)=\sum_{k=1}^{j}\beta(\theta_{t_{k-1}}\xi,\Delta_k)+\beta(\theta_{t_j}\xi,t)-\beta(\theta_{t_j}\xi,t_j),$$

if $t\in(t_j,t_{j+1}]$. Then by definition,

$$\text{P-lim}\,\sigma(\tau)=\int_0^T\beta_\tau(\theta_s\xi,ds).$$

On the other hand, $|\eta(\tau)-\sigma(\tau)|\leqslant\sup_{0\leqslant t\leqslant T}|\eta(t)-\sigma(t)|$, and since $\eta(t)-\sigma(t)$ is a square integrable separable martingale, in view of the remark following Theorem 2 we have

$$\mathsf{E}\sup|\eta(t)-\sigma(t)|^2\leqslant 4\mathsf{E}|\eta(T)-\sigma(T)|^2\to 0.$$

Thus

$$\eta(\tau)=\text{P-}\lim_{\lambda\to 0}\sigma(\tau)=\int_0^T\beta_\tau(\theta_s\xi,ds),$$

which proves equality (32) in this particular case.

To study the general case we shall introduce the random time τ_N, which is the time of the first exit of the process $\xi(t)$ outside the sphere of radius N (here $\tau_N=T$ provided $|\xi(t)|\leqslant N$ for all $t<T$) and set $\xi_N(t)=\xi(t)$ for $t<\tau$ and $\xi_N(t)=\xi(\tau_N-)$ for $t\geqslant\tau$ and $\beta_{\tau_N}(\varphi,t)=\beta(\varphi,t\wedge\tau_N)$.

Utilizing Lemma 3 and formula (32) we have

$$\eta(t\wedge\tau_N)=\int_0^{t\wedge\tau_N}\beta(\theta_s\xi,ds)=\int_0^{t\wedge\tau_N}\beta_{\tau_N}(\theta_s\xi_N,ds)=\int_0^t\beta_{\tau_N}(\theta_s\xi_N,ds),$$

so that in view of b) $\eta(t\wedge\tau)$ is a square integrable martingale. Since $\lim_{N\to\infty}\tau_N=T$, it follows that $\eta(t)$ is a local square integrable martingale. The existence of a modification with sample functions belonging to $\mathscr{D}[0,T]$ with probability 1 follows from the general properties of local martingales.

Next, since $\tau_N=T$ with probability 1 for N sufficiently large, $\eta(\tau)=\text{P-lim}\,\eta(\tau\wedge\tau_N)$ for any random time τ. Consequently,

$$\eta(\tau)=\text{P-lim}\,\eta(\tau\wedge\tau_N)=\text{P-lim}\int_0^{\tau\wedge\tau_N}\beta_{\tau\wedge\tau_N}(\theta_s\xi_{\tau\wedge\tau_N},ds)=\text{P-lim}\int_0^T\beta_{\tau\wedge\tau_N}(\theta_s\xi,ds).$$

On the other hand, in view of Lemma 3

$$\mathsf{P}\Big\{\int_0^T\beta_{\tau\wedge\tau_N}(\theta_s\xi,ds)-\int_0^T\beta_\tau(\theta_s\xi,ds)\neq 0\Big\}\leqslant\mathsf{P}\{\tau_N<\tau\},$$

and since $\mathsf{P}(\tau_N<\tau)\to 0$ as $N\to\infty$ we arrive at formula (32).

Finally inequality 33) follows from the fact that $\eta(t)=\int_0^t\beta(\theta_s\xi,ds)$ is a local square integrable martingale and the remark following Lemma 10 in Section 1. □

Remark. If the conditions of Theorem 2 are satisfied, then a separable modification of the process $\eta(t)$ is a continuous process and the inequalities (30), (31), and (33) are satisfied when $\|\theta_s\xi\|_*$ is replaced in these inequalities by $\|\theta_s\xi\|$.

We shall now compute the characteristic of the stochastic integral $\eta(t)$.

Lemma 8. *Assume that the conditions of Theorem* 1 *are satisfied and that*

$$\langle\beta(\varphi,\cdot),\beta(\psi,\cdot)\rangle_t=\int_0^t b(\varphi,\psi,s)\,ds$$

and

$$|b(\varphi,\varphi,t)-2b(\varphi,\psi,t)+b(\psi,\psi,t)|\leqslant\lambda_N(t)\|\varphi-\psi\|_*^2$$

for $\|\varphi\|\leqslant N$ and $\|\psi\|\leqslant N$, where $\lambda_N(t)$ is a nonnegative random process adopted to the current of σ-algebras $\{\mathfrak{F}_t, t\in[0,T]\}$ and integrable with probability 1 *on the interval* $[0,T]$. *Set*

$$\eta_i(t)=\int_0^t\beta(\theta_s\xi_i, ds),$$

where $\xi_i(t)$ $(i=1,2)$ also satisfy the conditions of Theorem 1. *Then*

$$\langle\eta_1(\cdot),\eta_2(\cdot)\rangle_t=\int_0^t b(\theta_s\xi_1,\theta_s\xi_2,s)\,ds. \tag{34}$$

Proof. We introduce "integral sums with a variable limit of summation"

$$\sigma_i(t)=\sum_{k=1}^{j}\beta(\theta_{t_{k-1}}\xi_i,\Delta_k)+\beta(\theta_{t_j}\xi,t)-\beta(\theta_{t_j}\xi,t_j)\quad\text{for } t\in(t_j,t_{j+1}),$$

where $\Delta_k=(t_{k-1},t_k]$. It is easy to verify that

$$\langle\sigma_1,\sigma_2\rangle_t=\sum_{k=1}^{j}\int_{t_{k-1}}^{t_k}b(\theta_{t_{k-1}}\xi_1,\theta_{t_{k-1}}\xi_2,s)\,ds+\int_{t_j}^{t}b(\theta_{t_j}\xi_1,\theta_{t_j}\xi_2,s)\,ds.$$

Now observe that

$$\begin{aligned}&\Delta\langle(\beta(\varphi,\cdot)-\beta(\psi,\cdot)),(\beta(\varphi,\cdot)-\beta(\psi,\cdot))\rangle_t\\&=\int_t^{t+\Delta t}[b(\varphi,\varphi,s)-2b(\varphi,\psi,s)-b(\psi,\psi,s)]\,ds\\&\leqslant\|\varphi-\psi\|_*\int_t^{t+\Delta t}\lambda_N(s)\,ds\end{aligned}$$

and

$$\Delta\langle\beta(\varphi,\cdot),\beta(\psi_1,\cdot)-\beta(\psi_2,\cdot)\rangle_t=\int_t^{t+\Delta t}[b(\varphi,\psi_1,s)-b(\varphi,\psi_2,s)]\,ds.$$

This implies (for almost all s)

$$\begin{aligned}&|b(\varphi,\psi_1,s)-b(\varphi,\psi_2,s)|\\&\leqslant[b(\varphi,\varphi,s)[b(\psi_1,\psi_1,s)-2b(\psi_1,\psi_2,s)+b(\psi_2,\psi_2,s)]^{1/2}\\&\leqslant[b(\varphi,\varphi,s)\lambda_N(s)]^{1/2}\|\psi_1-\psi_2\|_*,\end{aligned}$$

and

$$|b(\varphi_1, \psi_1, s) - b(\varphi_2, \psi_2, s)|$$
$$\leqslant \lambda_N(s)^{1/2}(\|\varphi_1 - \varphi_2\|_* b(\psi_1, \psi_1, s)^{1/2} + \|\psi_1 - \psi_2\|_* b(\varphi_2, \varphi_2, s)^{1/2})$$

for almost all s. Thus the function $b(\varphi, \psi, s)$ is (for almost all s) a continuous function of arguments φ and ψ (relative to the seminorm $\|\cdot\|_*$). It is easy to verify—using the same arguments as in the proof of Theorem 1—that

$$\langle \sigma_1, \sigma_2 \rangle_t \to \int_0^t b(\theta_s \xi_1, \theta_s \xi_2, s)\, ds$$

with probability 1 as $|\delta| \to 0$. Therefore the convergence of $\sigma_i(t)$ to $\eta_i(t)$ implies equality (34). □

Remark. If we assume that the conditions of Theorem 2 and those of Lemma 8 are satisfied and that $\|\varphi - \psi\|_*$ is replaced by $\|\varphi - \psi\|$ in these conditions, then equality (34) is still valid.

The existence and uniqueness theorems of solutions of stochastic differential equations. Let a current of σ-algebras $\{\mathfrak{F}_t, t \in [0, T]\}$ and random function $\alpha(\varphi, t)$ and $\beta(\varphi, t)$ with values in $\mathcal{R}^m$, adopted to $\{\mathfrak{F}_t\}$ with $\varphi \in \mathcal{D}^m$, $t \in [0, T]$, be given. A random time τ, $0 < \tau \leqslant T$, on $\{\mathfrak{F}_t\}$ and a random process $\xi(t)$ defined for $t \in [0, \tau)$, $\mathfrak{F}_t \times \mathfrak{T}_t$-progressively measurable and satisfying with probability 1 relations

$$\xi(t) = \varphi(t) \quad \text{for } t < 0,$$
$$\xi(t) = \xi(0) + \int_0^t \alpha(\theta_s \xi, s)\, ds + \int_0^t \beta(\theta_s \xi, ds), \qquad t \geqslant 0, \tag{35}$$

for each $t < \tau$, is called a *solution of the stochastic differential equation*

$$d\xi = \alpha(\theta_t \xi, t)\, dt + \beta(\theta_t \xi, dt), \qquad t \geqslant 0, \tag{36}$$

satisfying the "initial condition"

$$\xi(t) = \varphi(s), \qquad s \leqslant 0.$$

It is assumed here that integrals in the right-hand side of (35) are well defined, the first as a Lebesgue integral and the second as a stochastic integral.

The random variable τ is called the *lifetime* of the process $\xi(t)$ (i.e., of a solution of a stochastic differential equation).

Equation (36) is called *regular* on $[0, T]$ provided it possesses a unique solution of the whole time interval $[0, T]$ (i.e., if a unique solution of equation (36) with $\tau = T$ exists).

We now introduce some general assumptions on functions $\alpha(\varphi, t)$ and $\beta(\varphi, t)$ under which the right-hand side of equality (35) will be well defined for a sufficiently wide class of processes $\xi(t)$. Observe that there is no point to consider the most general classes of processes $\xi(t)$ since the right-hand side of equation (35) represents a process which possesses a continuous modification or a modification with sample functions in $\mathcal{D}_T^m$ and hence the process $\xi(t)$ must also possess these properties.

We shall first discuss the function $\alpha(\varphi, t)$. We introduce two sets of assumptions:

$\alpha 1)$ a) The function $\alpha(\varphi, s) = \alpha(\varphi, s, \omega)$ is defined on $\mathcal{D}^m \times [0, T] \times \Omega$ and is $\mathfrak{B}_{\mathcal{D}^m} \times \mathfrak{T}_t \times \mathfrak{F}_t$-measurable for $s \in [0, t]$.

$\alpha 1)$ b) $\alpha(\varphi, \cdot) \in \mathcal{D}^m[0, T]$ with probability 1 for a fixed ω.

$\alpha 1)$ c) For a fixed ω the family of functions $\{\alpha(t, \cdot), t \in [0, T]\}$ of argument φ is uniformly continuous on $\mathcal{D}^m$ relative to metric $\rho_{\mathcal{D}}$.

$\alpha 2)$ The function $\alpha(\varphi, s)$ satisfies assumptions $\alpha 1)$ provided $\mathcal{D}^m$, $\mathfrak{B}_{\mathcal{D}^m}$, and $\mathcal{D}[0, T]$ are replaced by $\mathscr{C}^m$, $\mathfrak{B}_{\mathscr{C}^m}$, and $\mathscr{C}^m[0, T]$, respectively.

We say that the function $\alpha(\varphi, t)$ is *linearly bounded* (relative to a uniform norm or a seminorm if in the succeeding inequalities the norm $\|\cdot\|$ can be replaced by the seminorm $\|\cdot\|_*$) provided a continuous monotonically nondecreasing process $\lambda_0(t)$ exists adopted to $\{\mathfrak{F}_t, t \in [0, T]\}$ such that $\lambda_0(T) < \infty$ with probability 1 and

$$|\textstyle\int_a^b \alpha(\varphi, t)\, dt| \leq (1 + \|\varphi\|) \int_a^b \lambda_0(t)\, dt. \tag{37}$$

We say that $\alpha(\varphi, t)$ satisfies the *local Lipschitz condition* (in a uniform norm or seminorm) if for any $N > 0$ there exists a monotonically nondecreasing process $\lambda_N(t)$ adopted to $\{\mathfrak{F}_t, t \in [0, T]\}$ such that

$$|\textstyle\int_a^b [\alpha(\varphi, t) - \alpha(\psi, t)]\, dt| \leq \|\varphi - \psi\| \int_a^b \lambda_N(t)\, dt \tag{38}$$

for all φ and ψ satisfying $\|\varphi\| \leq N$ and $\|\psi\| \leq N$. If we choose the process $\lambda(t)$ which is independent of N in place of $\lambda_N(t)$ then we shall refer to the process $\alpha(\varphi, t)$ as one satisfying the *uniform Lipschitz condition.*

The class of processes $\alpha(\varphi, t)$ which satisfy conditions $\alpha 2)$, (37) and (38) will be denoted by $S_\alpha^c(\lambda_0, \lambda_N)$. We denote by $S_\alpha(\lambda_0, \lambda_N)$ the class of processes which satisfy $\alpha 1)$ and the conditions obtained from (37) and (38) when the uniform norm $\|\cdot\|$ is replaced by $\|\cdot\|_*$ in the corresponding inequalities. In the case when we shall be dealing with random functions $\alpha(\varphi, t)$ satisfying only one of the inequalities (37) or (38), for instance (37), we shall write $\alpha(\varphi, t) \in S_0^c(\lambda_0, \cdot)$ and analogously in other cases.

Note that $\varphi_t = \theta_t \psi$ ($\psi \in \mathcal{D}^m$, $t \in [0, T]$), with values in $\mathcal{R}^m$ is a Borel function in argument t.

Indeed if B is a cylinder in $\mathcal{D}^m$ with the basis $B = \prod_{i=1}^n B_i$ in coordinates $(s_1, \ldots, s_n)$, $s_k \leq 0$, then

$$\{t\colon \varphi_t \in B\} = \bigcap_{i=1}^n \{t\colon \varphi(t + s_i) \in B_i\}.$$

Since the sets $\{z: \varphi(z)\in B_i\}=Z_i$ are Borel sets provided $B_i\in\mathscr{R}^m$ are such, the set $\{t: \varphi_t\in B\}=\bigcap_{i=1}^n\{Z_i-s_i\}$, where $Z-s$ denotes the set $\{z: z+s\in Z\}$, will also be a Borel set. Thus, if $g(\varphi, t)$ is a $\mathfrak{B}_{\mathscr{D}^m}\times\mathfrak{T}$-measurable function in the arguments (φ, t) where $\mathfrak{B}_{\mathscr{D}^m}$ is the minimal σ-algebra generated by the cylinders in $\mathscr{D}^m$ and $\mathfrak{T}$ is a σ-algebra of Borel sets in $[0, T]$, then $g(\theta_t\varphi, t)$ will be a Borel function in argument t.

Consequently, if the field $\alpha(\varphi, t)$ satisfies conditions $\alpha 1$) or $\alpha 2$), then the integral

$$\int_0^T\alpha(\theta_t\psi, t)\,dt$$

exists with probability 1.

Next, if $\alpha(\varphi, t)\in S_\alpha(\lambda_0,\cdot)$, then we have for $0\le a<b\le T$,

$$|\int_a^b\alpha(\theta_t\psi, t)\,dt|\le\int_a^b(1+\|\theta_t\psi\|_*)\lambda_0(t)\,dt\quad(\text{mod P}).\tag{39}$$

The proof follows easily from the fact that $\theta_t\psi$, $t\in[0, T]$ is a continuous function with values in $\mathscr{D}^m$ with respect to metric $\rho_{\mathscr{D}}$ in $\mathscr{D}^m$, and $\alpha(\varphi, t)$ is a continuous function of argument φ (uniformly in t).

Analogously, if $\alpha(\varphi, t)\in S_\alpha^c(\lambda_0,\cdot)$, then

$$|\int_a^b\alpha(\theta_t\psi, t)\,dt|\le\int_a^b(1+\|\theta_t\psi\|)\lambda_0(t)\,dt.\tag{40}$$

If, however, $\alpha(\varphi, t)\in S_\alpha(\cdot,\lambda_N)$ and $\|\psi_1\|\vee\|\psi_2\|\le N$, then

$$|\int_a^b\alpha(\theta_t\psi, t)\,dt-\int_a^b\alpha(\theta_t\psi_2, t)\,dt|\le\int_a^b\|\theta_t(\psi_1-\psi_2)\|_*\lambda_N(t)\,dt,\tag{41}$$

and an analogous inequality is valid for $\alpha(\varphi, t)\in S_\alpha^c(\cdot,\lambda_N)$.

As far as the integral $\int_0^t\beta(\theta_s\xi, ds)$ is concerned, conditions for its existence and its properties were discussed in the preceding subsections.

We introduce the notation for the classes of fields $\beta(\varphi, t)$ analogous to the notation for the classes of functions $\alpha(\varphi, t)$ introduced above. Namely, we write $\beta(\varphi, t)\in S_\beta(\lambda_0, \lambda_N)$ (or $\beta(\varphi, t)\in S_\beta^c(\lambda_0,\lambda_N)$) provided the field $\beta(\varphi, t)$ satisfies conditions $\beta 1$) ($\beta 2$)), is linearly bounded, satisfies the local Lipschitz condition relative to a seminorm (norm), and the dominating processes $\Lambda_0(t)$ and $\Lambda_N(t)$ are absolutely continuous with $\lambda_0(t)=\Lambda_0'(t)$ and $\lambda_N(t)=\Lambda_N'(t)$.

Set

$$A(\varphi, t)=\int_0^t\alpha(\varphi, s)\,ds+\beta(\varphi, t),$$

and write $A(\varphi, t)\in S(\lambda_0,\lambda_N)$ ($S^c(\lambda_0,\lambda_N)$) provided

$$\alpha(\varphi, t)\in S_\alpha(\lambda_0,\lambda_N)\quad\text{and}\quad\beta(\varphi, t)\in S_\beta(\lambda_0,\lambda_N)$$
$$(\alpha(\varphi, t)\in S_\alpha^c(\lambda_0,\lambda_N)\quad\text{and}\quad\beta(\varphi, t)\in S_\beta^c(\lambda_0,\lambda_N)).$$

Let the process $\xi(t)$, $t\in[0, T]$, be adopted to a current of σ-algebras $\{\mathfrak{F}_t, t\in[0, T]\}$ with sample functions belonging with probability 1 to $\mathscr{D}^m[0, T]$. We complete the definition of $\xi(t)$ for $t \leq 0$ by setting $\xi(t) = \varphi(t)$ for $t \leq 0$, where $\varphi(t)$ is a given function in $\mathscr{D}^m$. Assume that $\alpha(\varphi, t)$ satisfies condition $\alpha 1$ a) and that $\beta(\varphi, t) \in S_\beta(\cdot, \lambda_N)$. In what follows these conditions will always be assumed to be valid unless specified otherwise.

Define a new process $\eta(t)$, $t \in (-\infty, T]$, by setting

$$\eta(t) = \varphi(t), \qquad t \leq 0$$

$$\eta(t) = \varphi(0) + \int_0^t A(\theta_s\xi, ds), \qquad t \geq 0,$$

where

$$\int_0^t A(\theta_s\xi, ds) \underset{\text{Def}}{=} \int_0^t \alpha(\theta_s\xi, s)\, ds + \int_0^t \beta(\theta_s\xi, ds), \qquad t \in [0, T]. \tag{42}$$

The stochastic line integral in the right-hand side of (42) is interpreted here as a modification with sample functions belonging to $\mathscr{D}^m[0, T]$. We shall denote by I the correspondence $\xi \to \eta$, i.e., $\eta(t) = I(t, \xi)$.

Lemma 9. If $A(\varphi, t) \in S(C, \lambda_N)$ where C is an absolute constant and $\sup_{0 \leq t \leq T} \mathsf{E}|\xi(t)|^2 < \infty$, then

$$\mathsf{E}\{\sup_{0 \leq h \leq a} |I(t+h, \xi) - I(t, \xi)|^2 \mid \mathfrak{F}_t\} \leq C'[a(1 + \|\varphi\|^2) + \int_t^{t+a} z(s)\, ds], \tag{43}$$

where C' depends on C, K, and T only, and

$$z(s) = \sup_{0 \leq u \leq s} \mathsf{E}\{|\xi(u)|^2 \mid \mathfrak{F}_t\}.$$

Proof. Since

$$\sup_{0 \leq h \leq a} |I(t+h, \xi) - I(t, \xi)|^2$$
$$\leq 2 \sup_{0 \leq h \leq a} \{|\int_t^{t+h} \alpha(\theta_s\xi, s)\, ds + \int_t^{t+h} \beta(\theta_s\xi, ds)|^2\},$$

and taking into account the fact that for separable square integrable martingales $\eta(t)$

$$\mathsf{E}\Big\{\sup_{t \leq s \leq t+a} |\eta(s)|^2 \mid \mathfrak{F}_t\Big\} \leq 4\mathsf{E}\Big\{|\eta(t+a)|^2 \mid \mathfrak{F}_t\Big\},$$

and utilizing inequalities (22) and (40), we obtain

$$\mathsf{E}\Big\{\sup_{0 \leq h \leq a} |I(t+h, \xi) - I(t, \xi)|^2 \mid \mathfrak{F}_t\Big\}$$
$$\leq 2(2C^2a^2 + 4C^2a + 2C^2\mathsf{E}\{(\int_t^{t+a} \|\theta_s\xi\|_* \, ds)^2 \mid \mathfrak{F}_t\}$$
$$+ 4C^2 \int_t^{t+a} \mathsf{E}\{\|\theta_s\|_*^2 \mid \mathfrak{F}_t\}\, ds$$
$$\leq C'(a + \int_t^{t+a} \mathsf{E}\{\|\theta_s\xi\|_*^2 \mid \mathfrak{F}_t\}\, ds),$$

where $C'=4C^2(T+2)$. On the other hand,

$$\mathsf{E}\{\|\theta_s\xi\|_*^2\,|\,\mathfrak{F}_t\}=\int_{-\infty}^0 \mathsf{E}\{|\xi(s+u)|^2\,|\,\mathfrak{F}_t\}K(du)\leq K(\|\varphi\|^2+z(s)),$$

which together with the preceding inequality proves the lemma. □

Remark. If $A(\varphi,t)\in S^c(C,\lambda_N)$ then

$$\mathsf{E}\sup_{0\leq h\leq a}|I(t+h,\xi)-I(t,\xi)|^2\leq C'[(1+\|\varphi\|)^2a+\int_t^{t+a}Z(s)\,ds], \tag{44}$$

where $Z(s)=\mathsf{E}\sup_{0\leq t\leq s}|\xi(t)|^2$.

The proof of inequality (44) is analogous to the proof of the preceding lemma.

The following lemma is proved in the same manner as Lemma 8.

Lemma 10. *If* $\sup_{0\leq t\leq T}\mathsf{E}|\xi_k(t)|^2<\infty$, $k=1,2$, *and* $A(\varphi,t)\in S(\cdot\,,C)$, *then*

$$\mathsf{E}\sup_{0\leq t\leq a}|I(t,\xi_1)-I(t,\xi_2)|^2\leq C''\int_0^a v(t)\,dt,$$

where C'' *is a constant depending on* C *and* T *only and*

$$v(t)=\sup_{0\leq s\leq t}\mathsf{E}|\xi_1(s)-\xi_2(s)|^2.$$

If, however, $A(\varphi,t)\in S^c(\cdot\ C)$, *then*

$$\mathsf{E}\sup_{0\leq t\leq a}|I(t,\xi_1)-I(t,\xi_2)|^2\leq C'''\int_0^a V(t)\,dt,$$

where $V(t)=\sup_{0\leq s\leq t}\mathsf{E}|\xi_1(s)-\xi_2(s)|^2$.

To simplify the writing, we introduce the following notation.

Denote by $H^*(H^c)$ the space of random processes satisfying conditions (ξ1 (ξ2)), and by H_2^* (H_2^c) the subspace H^* (H^c) consisting of processes satisfying the additional condition

$$\|\xi(\cdot)\|_2=\Big\{\sup_{0\leq t\leq T}\mathsf{E}|\xi(t)|^2\Big\}^{1/2}<\infty\quad(\mathsf{E}\sup_{0\leq t\leq T}|\xi(t)|^2<\infty).$$

We also note the following elementary lemma which will be used repeatedly below.

Lemma 11. *If* $z(t)$ *is a function bounded on* $[0,T]$ *and*

$$z(t)\leq A+B\int_0^t z(s)\,ds,\qquad B>0,$$

then

$$z(t)\leq A\,e^{Bt}.$$

Proof. Indeed, clearly

$$\begin{aligned}z(t) &\leq A+B\int_0^t (A+B\int_0^{t_1} z(s)\,ds)\,dt_1\\ &\leq A+ABt+AB^2\frac{t^2}{2}+\cdots+AB^n\frac{t^n}{n!}\\ &\quad +B^{n+1}\int_0^t\int_0^{t_1}\cdots\int_0^{t_n} z(s)\,ds\,dt_n\cdots dt_1.\end{aligned}$$

Approaching the limit as $n\to\infty$ we obtain the required assertion. □

Theorem 3. *Let $A(\varphi, t)\in S(C, C)$. A stochastic differential equation* (36) *under an arbitrary initial condition $\varphi\in\mathscr{D}_0^m$ is regular in H_2^*, i.e., it possesses a unique solution in H_2^* defined for all $t\in[0, T]$. This solution has the properties*

(45) $$\mathsf{E}\{\sup_{0\leq s\leq T}|\xi(s)|^2|\mathfrak{F}_0\}\leq A(1+\|\varphi\|^2),$$

(46) $$\mathsf{E}\{\sup_{t\leq s\leq t+h}|\xi(s)-\xi(t)|^2|\mathfrak{F}_t\}\leq B(1+\|\varphi\|^2+\sup_{a\leq u\leq t}|\xi(u)|^2)h,$$

where A and B are constants which depend on C, T and K only.

Proof. Introduce the norm

$$\|\xi(\cdot)\|_2=\{\sup_{0\leq t\leq T}\mathsf{E}|\xi(t)|^2\}^{1/2}$$

in the space H_2^* and consider H_2^* as a subset of the space H_2 of random functions $\xi(t)$, $t\in[0, T]$ with values in $\mathscr{R}^m$ adopted to the current of σ-algebras $\{\mathfrak{F}_t, t\geq 0\}$ such that $\|\xi(\cdot)\|_2<\infty$, where the norm is defined by the preceding relation. The space H_2 is a complete space (as opposed to the space H_2^*). As it follows from inequality (43) operator I maps H_2^* into itself and Lemma 9 shows that some power of operator I is a contracting operator. Starting with an arbitrary process $\xi_0(t)\in H_2^*$ $(\xi_0(0)=\varphi(0))$ we construct the successive approximations

$$\xi_{n+1}(t)=I(t, \xi_n),\qquad t\in[0, T],$$

and set

$$v_n(t)=\mathsf{E}\sup_{s\leq t}|\xi_{n+1}(s)-\xi_n(s)|,\qquad n=1, 2, \ldots,$$

$$v_0(t)=\sup_{0\leq s\leq t}\mathsf{E}|\xi_1(s)-\xi_0(s)|^2,\qquad V_0=v_0(T).$$

It follows from Lemma 9 that

$$v_1(t)\leq C''V_0t, \ldots, v_n(t)\leq(C'')V_0\frac{t^n}{n!}.$$

If we define

$$\varepsilon_n = \left[V_0 \frac{(C''T)^M}{n!} \right]^{1/3}$$

then the Chebyshev inequality implies that $\mathsf{P}\{\sup_{0\le t\le T} |\xi_{n+1}(t) - \xi_n(t)| > \varepsilon_n\} \le \varepsilon_n$, and since the series $\sum_{n=1}^{\infty} \varepsilon_n$ is convergent so is the series

$$\sum_{n=1}^{\infty} \sup_{0\le t\le T} |\xi_{n+1}(t) - \xi_n(t)|$$

with probability 1. Thus $\lim \xi_n(t) = \xi(t)$ exists with probability 1 and, moreover, uniformly in $t \in [0, T]$. We complete the definition of $\xi(t)$ for $t < 0$ by setting $\xi(t) = \varphi(t)$. Under this definition $\xi(t)$ satisfies the conditions $\xi 1$) and $\mathsf{E} \sup_{0\le t\le T} |\xi(t)|^2 \le \infty$. Moreover, $\mathsf{E} \sup_{0\le t\le T} |\xi(t) - \xi_n(t)|^2 \to 0$. Indeed,

$$\begin{aligned}
\mathsf{E} \sup_{0\le t\le T} |\xi(t) - \xi_n(t)|^2 &\le \mathsf{E} \lim_{m\to\infty} \sup_t |\xi_{n+m}(t) - \xi_n(t)|^2 \\
&\le \lim_{m\to\infty} \mathsf{E}\Big(\sum_{k=n}^{n+m-2} \sup_t |\xi_{k+1} - \xi_k| \cdot \sqrt{k(k-1)} \cdot \frac{1}{\sqrt{k(k-1)}} \Big)^2 \\
&\le \sum_n^{\infty} k(k-1) v_k(T) \cdot \sum_{k=n}^{\infty} \frac{1}{k(k-1)} \\
&\le \frac{V_0 C''^2 T^2}{n-1} \cdot \sum_n^{\infty} \frac{T^{k-2}}{(k-2)!} \to 0 \quad \text{as } n \to \infty.
\end{aligned}$$

It is now easy to justify the limit transition in the relation

$$\xi_{n+1}(t) = \varphi(0) + \int_0^t \alpha(\theta_s \xi_n, s)\, ds + \int_0^t \beta(\theta_s \xi_n, ds).$$

Indeed, it follows from the uniform convergence of $\xi_n(t)$ to $\xi(t)$ that $\theta_s \xi_n(u)$ converges to $\theta_s \xi(s)$ uniformly in u. Therefore $\alpha(\theta_s \xi_n, s) \to \alpha(\theta_s \xi, s)$ for all $s \in [0, T]$ with probability 1. Hence with probability 1

$$\int_0^t \alpha(\theta_s \xi_n, s)\, ds \to \int_0^t \alpha(\theta_s \xi, s)\, ds.$$

Furthermore, in view of the above,

$$\mathsf{E} \left| \int_0^t \beta(\theta_s \xi_n, ds) - \int_0^t \beta(\theta_s \xi, ds) \right|^2 \le C^2 \int_0^t \mathsf{E} |\xi_n(s) - \xi(s)|^2\, ds \to 0.$$

Thus, $\xi(t)$ is a solution of the equation

$$\xi(t) = \varphi(0) + \int_0^t \alpha(\theta_s \xi, s)\, ds + \int_0^t \beta(\theta_s \xi, ds)$$

with probability 1 for each t. Since the functions in the right-hand and left-hand sides of the equality are continuous from the right, this equality is valid with probability 1 for all $t \in [0, T]$.

We now verify inequalities (45) and (46). Set $z(t) = \mathsf{E}\{\sup_{0 \le s \le t} |\xi(s)|^2 | \mathfrak{F}_0\}$. Taking Lemma 9 into account, we obtain

$$\begin{aligned} z(t) &\le 2|\varphi(0)|^2 + 2\mathsf{E} \sup |I(s, \xi)|^2 \\ &\le 2|\varphi(0)|^2 + 2C_1^*[t(1 + \|\varphi\| + K \int_0^t z(s)\, ds] \\ &\le C_2(\|\varphi\|^2 + t + \int_0^t z(s)\, ds), \end{aligned}$$

where C_2 is a new constant which depends on C, K, and T only. The last inequality implies that

$$\frac{z(t)+1}{\|\varphi\|^2 + t + \int_0^t z(s)\, ds + (1/C_2)} \le C_2$$

or

$$z(t) \le A(\|\varphi\|^2 + 1)\, e^{C_2 t}.$$

Analogously for

$$z_1(t) = \mathsf{E}\{\sup_{a \le s \le t} |\xi(s) - \xi(a)|^2 | \mathfrak{F}_a\}$$

we have

$$z_1(t) \le C_3(t(1 + \|\varphi\|^2) + \int_0^t z_1(s)\, ds);$$

this implies that

$$z_1(t) \le (1 + \|\varphi\|^2 + 2a \sup_{0 \le u \le a} |\xi(u)|^2)(e^{C_3(t-a)} - 1).$$

We shall now prove the uniqueness of the solution of equation (36) in H_2^*. If there were two solutions $\xi(t)$ and $\eta(t)$ then in view of Lemma 10

$$V(t) \le C'' \int_0^t V(s)\, ds,$$

where

$$V(t) = \mathsf{E} \sup_{0 \le s \le t} |\xi(s) - \eta(s)|^2,$$

$$V(t) \le C''T \sup_{0 \le s \le T} \mathsf{E}|\xi(s) - \eta(s)|^2 = C'''.$$

Integrating the inequality just obtained, we have

$$\begin{aligned} V(t) &\le C''^2 \int_0^t dt_1 \int_0^{t_1} V(t_2)\, dt_2 \le \cdots \le C''^n \int_0^t \int_0^{t_1} \cdots \int_0^{t_{n-1}} V(t_n)\, dt_n\, dt_{n-1} \ldots dt_1 \\ &\le C''' C''^n \frac{t^n}{n!}. \end{aligned}$$

Thus, $V(t) = 0$.

The theorem is proved. □

Remark 1. Let the function $\alpha(\varphi, t)$ and $\beta(\varphi, t)$ satisfy the conditions of Theorem 3. Consider the equation

$$\xi(t)=\varphi(t)+\int_0^t \alpha(\theta_s\xi, s)\, ds+\int_0^t \beta(\theta_s\xi, ds), \tag{47}$$

where the function $\varphi(t)$ possesses the following properties: its contraction to the half-line $(-\infty, 0]$ is a fixed function in $\mathscr{D}_0^m$, and the contraction to the interval $(0, T]$ belongs to H_2^*.

It is clear that the part of the proof of Theorem 3 related to the existence and uniqueness of the solution for equation (36) in H_2^* is carried over without alterations to equation (47). Thus the following result is valid:

Under the previous assumptions equation (47) *possesses a unique solution in* H_2^*.

Remark 2. If $A(\varphi, t)\in S^c(C, C)$ and $\varphi(t)\in H_2^c$, then equation (47) possesses a unique solution in H_2^c defined for all $t\in[0, T]$ which also satisfies inequalities (45) and (46).

The proof differs only slightly from the proof of Theorem 3; one must just proceed from an initial approximation $\xi_0(t)$ such that $\mathsf{E}\sup_{0\leq t\leq T}|\xi_0(t)|^2<\infty$.

In order to generalize the theorem on the existence and uniqueness of the solution for equation (36) the following result is required.

Theorem 4. *Let* $A_i(\varphi, t)\in S(\cdot, \lambda_N)$ *or* $A_i(\varphi, t)\in S^c(\cdot, \lambda_N)$, *and for* $\|\varphi\|\leq N$ *and* $\|\psi\|\leq N$ *with* $t<\tau$, *where* τ *is an* $\mathfrak{F}_t$*-random time, let the following relations be satisfied*:

$$\alpha_1(\varphi, t)=\alpha_2(\varphi, t)=\alpha(\varphi, t),$$
$$\beta_1(\varphi, t)=\beta_2(\varphi, t)=\beta(\varphi, t).$$

Then if $\xi_i(t)$, $i=1, 2$, *are solutions of equations*

$$d\xi_i(t)=\alpha_i(\theta_t\xi_i, t)\, dt+\beta_i(\theta_t\xi_i, dt), \qquad t>0,$$
$$\xi_i(s)=\varphi(s), \qquad s\leq 0,$$

such that $\sup_{0\leq t\leq T}|\xi_i(t)|<\infty$ *with probability* 1, *then for* $\sup_{s\leq\tau}|\xi_i(s)|\leq N$, $i=1, 2$, *we have with probability* 1 *for all* $t<\tau$

$$\xi_1(t)=\xi_2(t).$$

Proof. We shall prove only the case when $A(\varphi, t)\in S(\cdot, \lambda_N)$. The second case is proven analogously. Let

$$\sigma=\inf\{t\colon t\geq\tau, |\xi_1(t)|>N, |\xi_2(t)|\geq N, \lambda_N(t)\geq L\},$$

if the set in the braces is nonvoid, and $\sigma = T$ otherwise; also set

$$\alpha_\sigma(\varphi, t) = \alpha(\varphi, t) \quad \text{for } t < \sigma, \qquad \alpha_\sigma(\varphi, t) = 0 \quad \text{for } t \geqslant \sigma,$$
$$\beta_\sigma(\varphi, t) = \beta(\varphi, t \wedge \sigma).$$

Then for $\|\varphi\| \leqslant N$ and $\|\psi\| \leqslant N$

$$\begin{aligned}\left|\int_t^{t+\Delta t} \alpha_{i\sigma}(\varphi, s)\, ds - \int_t^{t+\Delta t} \alpha_{i\sigma}(\psi, s)\, ds\right| &= \left|\int_{t\wedge\sigma}^{(t+\Delta t)\wedge\sigma} [\alpha_i(\varphi, s) - \alpha_i(\psi, s)]\, ds\right| \\ &\leqslant \|\varphi - \psi\|_* \int_{t\wedge\sigma}^{(t+\Delta t)\wedge\sigma} \lambda_N(s)\, ds \\ &\leqslant L\|\varphi - \psi\|_* \Delta t,\end{aligned}$$

and analogously applying the theorem on the characteristic of a stopping martingale we obtain

$$\begin{aligned}\mathsf{E}\{|\beta_{i\sigma}(\varphi, \Delta) - \beta_{i\sigma}(\psi, \Delta)|^2 | \mathfrak{F}_t\} &\leqslant \mathsf{E}\left\{\int_{t\wedge\sigma}^{(t+\Delta t)\wedge\sigma} \lambda_N(s)\, ds \|\varphi - \psi\|_* | \mathfrak{F}_t\right\} \\ &\leqslant L\|\varphi - \psi\|_*^2 \Delta t.\end{aligned}$$

Set $\xi(t) = \xi_1(t) - \xi_2(t)$, $\chi(t) = 1$ for $t < \sigma$ and $\chi(t) = 0$ for $t \geqslant \sigma$. Then

$$\begin{aligned}\mathsf{E}\chi(t)|\xi(t)|^2 &\leqslant 2\mathsf{E}\chi(t)\left|\int_0^t [\alpha_1(\theta_s\xi_1, s) - \alpha_2(\theta_s\xi_2, s)]\, ds\right|^2 \\ &\quad + 2\mathsf{E}\chi(t)\left|\int_0^t [\beta_1(\theta_s\xi_1, ds) - \beta_2(\theta_s\xi_2, ds)]\right|^2 \\ &= 2\mathsf{E}\chi(t)\left|\int_0^{t\wedge\sigma} [\alpha_{1\sigma}(\theta_s\xi_1, s) - \alpha_{1\sigma}(\theta_s\xi_2, s)]\, ds\right|^2 \\ &\quad + 2\mathsf{E}\chi(t)\left|\int_0^{t\wedge\sigma} \beta_{1\sigma}(\theta_s\xi_1, ds) - \beta_{1\sigma}(\theta_s\xi_2, ds)\right|^2 \\ &\leqslant 2\mathsf{E}\chi(t)L^2T \int_0^{t\wedge\sigma} \|\theta_s\xi_1 - \theta_s\xi_2\|_*^2\, ds + 2\mathsf{E}L \int_0^{t\wedge\sigma} \|\theta_s\xi_1 - \theta_s\xi_2\|_*^2\, ds \\ &\leqslant L'\mathsf{E}\int_0^t \chi(s)\|\theta_s\xi\|_*^2\, ds.\end{aligned} \tag{48}$$

Set $z(t) = \sup\{\mathsf{E}\chi(s)|\xi(s)|^2, s \leqslant t\}$. Since

$$\begin{aligned}\mathsf{E}\chi(s)\|\theta_s\xi\|_*^2 &= \mathsf{E}\chi(s)\int_{-\infty}^0 |\xi(s+u)|^2 K(du) \\ &\leqslant \mathsf{E}\int_{-\infty}^0 \chi(s+u)|\xi(s+u)|^2 K(du) \leqslant Kz(s),\end{aligned}$$

it follows from inequality (48) that

$$z(t) \leqslant L'K\int_0^t z(s)\, ds, \qquad 0 \leqslant t \leqslant T.$$

This implies that $z(t) = 0$ for all $t \in [0, T]$ or that $\chi(t)|\xi_1(t) - \xi_2(t)| = 0$ for each t with probability 1. Noting that $\xi_1(t)$ and $\xi_2(t)$ are continuous from the right we deduce that $\chi(t)|\xi_1(t) - \xi_2(t)| = 0$ for all t with probability 1 or that $\xi_1(t) = \xi_2(t)$ for all $t < \sigma$ with probability 1. For L approaching ∞ and taking into account the boundedness of $\lambda_N(t)$ we obtain that $\xi_1(t) = \xi_2(t)$ for all

$$t < \inf\{t\colon t \geqslant \tau, |\xi_1(t)| \geqslant N, |\xi_2(t)| \geqslant N\}. \quad \square$$

Theorem 5. *Assume that $A(\varphi, t) \in S(\lambda_0, \lambda_N)$ (or that $A(\varphi, t) \in S^c(\lambda_0, \lambda_N)$. The stochastic differential equation (36) possesses in H^* (H^c) a unique solution which is defined for all $t \in [0, T]$.*

Proof. As in the proof of Theorem 4 we shall confine ourselves to the case when $A(\varphi, t) \in S(\lambda_0, \lambda_N)$. We first establish the existence of a solution for equation (36).

Let $p > 0$. Introduce the functions $\alpha'_p(\varphi, t)$ and $\beta'_p(\varphi, t)$ such that $\alpha'_p(\varphi, t) \in S_\alpha(\lambda_0, \lambda')$, $\beta'_p(\varphi, t) \in S_\beta(\lambda_0, \lambda')$ and $\alpha'_p(\varphi, t) = \alpha(\varphi, t)$, $\beta'_p(\varphi, t) = \beta(\varphi, t)$ for $\|\varphi\| \leqslant r(p)$, where $\lambda' = \lambda'(t)$ does not depend on N and $r(p)$ is a quantity to be defined more precisely below. Set $\tau_p = \inf\{t: \lambda_0(t) \geqslant p, \lambda'(t) \geqslant l_p\}$ if the set in braces is nonempty and set $\tau_p = T$ otherwise. Also let

$$\alpha_p(\varphi, t) = \alpha'_p(\varphi, t) \quad \text{for } t < \tau_p, \qquad \alpha_p(\varphi, t) = 0 \quad \text{for } t \geqslant \tau_p,$$

$$\beta_p(\varphi, t) = \beta'_p(\varphi, t \wedge \tau_p\} \quad \text{for } t < \tau_p, \qquad \beta_p(\varphi, t) = 0 \quad \text{for } t \geqslant \tau_p.$$

(The values of the constants l_p will also be defined more precisely below.) In this case

$$\alpha_p(\varphi, t) = \alpha(\varphi, t), \qquad \beta_p(\varphi, t) = \beta(\varphi, t)$$

for $\|\varphi\| \leqslant r(p)$ and $t < \tau_p$ and $A_p(\varphi, t) = \int_0^t \alpha_p(\varphi, s)\, ds + \beta_p(\varphi, t) \in S(p, l_p)$ (an analogous fact was established in the course of the proof of the preceding theorem). Moreover, $\alpha_p(\varphi, t)$ and $\beta_p(\varphi, t)$ are $\mathfrak{F}_{\tau_p \wedge t}$-measurable functions and $\beta_p(\varphi, t)$ is for a fixed φ a martingale relative to $\{\mathfrak{F}_{\tau_p \wedge t}\}$. It follows from Theorem 3 that the equations

$$\text{(49)} \qquad d\xi_p = A_p(\theta_t \xi_p, dt), \quad t \in [0, T], \qquad \xi_p(s) = \varphi(s), \quad s < 0,$$

possess on the interval $[0, T]$ a solution $\xi_p(t) \in H_2^*$. In view of Theorem 4 we have for $r(p') > r(p)$ on the set $\Omega_p = \{\omega: \tau_p = T, \|\theta_T \xi\| \leqslant r(p)\}$ equality $\xi_p(t) = \xi_{p'}(t)$ for all t. Moreover, $\mathsf{P}(\bar{\Omega}_p) \leqslant \mathsf{P}(\tau_p < T) + \mathsf{P}(\|\theta_T \xi\| \geqslant r(p))$. It then follows from Theorem 3 and the Chebyshev inequality that

$$\mathsf{P}(\|\theta_T \xi\| \geqslant r(p)) \leqslant \frac{K(p)}{r^2(p)},$$

where $K(m)$ is a function which depends on m, $\|\varphi\|$, and T only (and does not depend on l_p).

Let $K(p)/r^2(p) \to 0$ as $p \to \infty$. Then for p sufficiently large

$$\mathsf{P}(\|\theta_T \xi\| \geqslant r(p)) \leqslant \frac{\varepsilon}{2}.$$

We now find values p and l_p such that

$$\mathsf{P}(\tau_p < T) = \mathsf{P}(\{\lambda_0(T) \geqslant p\} \cup \{\lambda'(T) \geqslant l_p\}) < \frac{\varepsilon}{2}.$$

Then $\mathsf{P}(\bar{\Omega}_p) < \varepsilon$. This implies that processes $\xi_p(t)$ converge to a limit $\xi(t)$ with probability 1 and, moreover, $\xi(t) = \xi_p(t)$ with probability 1 for all $t \in [0, T]$ starting with some value of $p = p_0(\omega)$. In particular, the sample functions of the process $\xi(t)$ can be assumed to have left-hand limits and be continuous from the right with probability 1 for all $t \in [0, T]$. Since $\xi_p(t)$ is $\mathfrak{F}_{\tau_p \wedge t}$-measurable, it follows that $\xi(t)$ is measurable with respect to the σ-algebra $\mathfrak{F}_t$. Finally, if $\omega \in \Omega_p$, then $\alpha_p(\varphi, t) = \alpha(\varphi, t)$, $\beta_p(\varphi, t) = \beta(\varphi, t)$, and in view of Theorem 5 we have on the set Ω_p

$$\xi(t) = \varphi(0) + \int_0^t A(\theta_s \xi, s)\, ds$$

with probability 1.

Consequently the last equality is valid with probability 1 on the whole set Ω. The existence of the solution is thus verified. The uniqueness of the solution easily follows from Theorems 3 and 4. □

Assume now that $A(\varphi, t) \in S(\cdot, \lambda_N)$ (or $A(\varphi, t) \in S^c(\cdot, \lambda_N)$). Construct fields $\alpha_p(\varphi, t)$, $\beta_p(\varphi, t)$ which coincide with $\alpha(\varphi, t)$ and $\beta(\varphi, t)$ correspondingly for $\|\varphi\| \leq p$, which satisfy the uniform Lipschitz condition with the dominating function $\lambda_p(t)$, and which vanish for $\|\varphi\| \geq p+1$. Let $\xi_p(t)$ be a solution of the equations

$$d\xi_p = A_p(\theta_t \xi_p, dt),$$

$$\xi_p(t) = \varphi(t), \qquad t \leq 0,$$

and $\tau_p = \inf\{t : \|\theta_t \xi_p\| \geq p, t \in [0, T]\}$ or $\tau_p = T$ if the set of values of t in the braces is void. The sequence of random times τ_p is monotonically nondecreasing. Set $\tau_\infty = \lim \tau_p$. As before $\xi_p(t) = \xi_{p'}(t)$ for $p' > p$ and $t < \tau_p$. Therefore the limit $\lim \xi_p(t) = \xi(t)$ exists with probability 1 for all $t < \tau_\infty$ and coincides with probability 1 with a function $\xi_p(t)$. Thus $\xi(t)$ possesses the left-hand limit and is continuous from the right for all $t < \tau_\infty$ (with probability 1).

Similarly, as in the proof of the preceding theorem, one can verify that $\xi(t)$ satisfies equation (36) for $t < \tau_\infty$.

Theorem 6. *If $A(\varphi, t) \in S(\cdot, \lambda_N)$ (or $(A(\varphi, t) \in S^c(\cdot, \lambda_N))$ then a random time τ_∞ exists on $\{\mathfrak{F}_t, t \in [0, T]\}$ and a random process $\xi(t)$ adopted to $\{\mathfrak{F}_t, t \in [0, T]\}$ defined for $t < \tau_\infty$ such that $\xi(t)$ satisfies* (36) *for all $t < \tau_\infty$, and the sample functions of $\xi(t)$ possess with probability* 1 *left-hand limits and are continuous from the right for all $t < \tau_\infty$. Moreover, $\mathsf{P}(\tau_\infty > 0) = 1$.*

Theorem 7. *If under the conditions of Theorem* 5 *one can set $\lambda_0 = C$, then the solution of equation* (36) *belongs to H_2^* (H_2^c) and*

$$\mathsf{E} \sup_{0 \leq t \leq T} |\xi(t)|^2 \leq B(1 + \|\varphi\|^2), \tag{50}$$

$$\mathsf{E} \sup_{t \leq s \leq t+h} |\xi(s) - \xi(t)|^2 \leq B(1 + \|\varphi\|^2)h, \tag{51}$$

where B is a constant which depends on C, K, and T only.

Proof. Let τ_p be a random time introduced in the proof of Theorem 5 and let $\xi_p(t)$ be a solution of equation (49). Then $\xi_p(t) \in H_2^*$ and $\mathsf{E} \sup_{0 \le t \le T} |\xi_p(t)|^2 < \infty$. It follows from Lemma 8 that

$$(52) \quad \mathsf{E}\{\sup_{0 \le h \le a} |\xi_p(t+h) - \xi_p(t)|^2 | \mathfrak{F}_t\} \le C'[a(1+\|\varphi\|^2) + \int_t^{t+a} \sup \mathsf{E}\{|\xi(s)|^2 | \mathfrak{F}_t\} \, ds],$$

where C' depends on C, K, and T only.

Set $z(t) = \mathsf{E}\{\sup_{0 \le s \le t} |\xi_p(s)|^2\}$. It follows from the preceding inequality that

$$z(t) \le 2|\varphi(0)|^2 + 2C'[T(1+\|\varphi\|^2) + \int_0^t z(s) \, ds]$$

so that in view of Lemma 11

$$z(t) \le C''(1+\|\varphi\|^2),$$

where C''—like C'—depends on C, K, and T only. Analogously one obtains the inequality

$$(53) \quad \mathsf{E}\{\sup_{t \le s \le t+h} |\xi_p(s)|^2 | \mathfrak{F}_t\} \le C''(1+\|\theta_t \xi_p\|^2),$$

which together with (52) yields:

$$(54) \quad \mathsf{E}\{\sup_{0 \le h \le a} |\xi_p(t+h) - \xi_p(t)|^2 | \mathfrak{F}_t\} \le C'''(1+\|\varphi\|^2 + \sup_{0 \le s \le t} |\xi(s)|^2)a.$$

In particular,

$$(55) \quad \mathsf{E} \sup_{0 \le h \le a} |\xi_p(t+h) - \xi_p(t)|^2 \le C^{\mathrm{IV}}(1+\|\varphi\|^2)a.$$

Taking into account that $\xi_p(t)$ converges with probability 1 as $p \to \infty$ to a solution $\xi(p)$ of equation (36) and utilizing Fatou's lemma we obtain the required result from inequalities (53) and (55). □

Remark. Analogous results can be obtained for equation (47). In this case one should consider the auxiliary equations

$$\xi_p(t) = \varphi(t) + \int_0^t \alpha_p(\theta_s \xi, s) \, ds + \int_0^t \beta_p(\theta_s \xi, ds).$$

If it is assumed that

$$(56) \quad \mathsf{E} \sup_{0 \le s \le T} |\varphi(s)|^2 = v < \infty,$$

then, in the same manner as in Theorem 7, the following assertion is obtained:

if the assumptions of Theorem 7 and condition (56) *are fulfilled, then the following bound is valid for a solution of equation* (47):

$$\mathsf{E} \sup_{0\le t\le T} |\xi(t)|^2 \le B(1+\|\varphi\|^2+v),$$

where B depends on C, K, and T only.

Bounds on the moments of solutions of stochastic differential equations. Consider equation (36) satisfying the conditions of Theorem 5. Assume first that for a fixed φ, $\beta(\varphi, t) \in l\mathcal{M}_2^c$. Furthermore, let the characteristic $\langle\beta^k, \beta^k\rangle_t$ of the process $\beta^k(\varphi, t)$ be absolutely continuous with respect to the Lebesgue measure:

$$\langle\beta^k, \beta^k\rangle_t = \int_0^t \beta^{kk}(\varphi, s)\, ds, \qquad k=1, \ldots, m, \tag{57}$$

and

$$|\beta^{kk}(\varphi, s)| \le \lambda^2(s)(1+\|\varphi\|^2). \tag{58}$$

It follows from here that the function $\langle\beta^k, \beta^j\rangle_t$ is also absolutely continuous with respect to the Lebesgue measure. Hence

$$\langle\beta^k, \beta^j\rangle_t = \int_0^t \beta^{kj}(\varphi, s)\, ds,$$

and inequality (51) in Section 1 implies that

$$|\beta^{kj}(\varphi, s)| \le \lambda^2(s)(1+\|\varphi\|^2).$$

Here $\lambda(t)$ is a nonnegative random process bounded on $[0, T]$ and adopted to the current of σ-algebras $\{\mathfrak{F}_t, t \in [0, T]\}$. Also assume that

$$|\alpha(\varphi, s)| \le \lambda(s)(1+\|\varphi\|). \tag{59}$$

We now apply Itô's formula to the function $f(x) = |x|^r$ and to the process $\xi(t)$ satisfying the equation

$$\begin{aligned} d\xi(t) &= \alpha(\theta_t\xi, t)\, dt + \beta(\theta_t\xi, dt), \qquad && t\in(0, T], \\ \xi(t) &= \varphi(t), && t \le 0. \end{aligned} \tag{60}$$

Since

$$\begin{aligned} \nabla f(x) &= r|x|^{r-2}x, \\ \nabla^2 f(x) &= r(r-2)|x|^{r-1}(x\times x) + r|x|^{r-2}E, \end{aligned}$$

where E is the unit matrix and $x \times x$ is the matrix with elements $x_{jk} = x^j x^k$

$(j, k=1,2,\ldots,m)$, then

$$|\xi(t)|^r=|\varphi(0)|^r+\int_0^t L_c(\xi,s)\,ds+\xi_r(t),$$

where

$$L_c(\xi,s)=r|\xi(s)|^{r-2}\Big[(\xi(s)\alpha(\theta_s\xi,s))+\sum_{k=1}^m\beta^{kk}(\theta_s\xi,s)\Big]$$
$$+r(r-2)|\xi(s)|^{r-4}\sum_{j,k=1}^m\beta^{jk}(\theta_s\xi,s)\xi^j(s)\xi^k(s)$$

and

$$\xi_r(t)=r\int_0^t|\xi(s)|^{r-2}(\xi(s)\beta(\theta_s\xi,ds)).$$

Set

$$\rho(t)=\sup_{0\le s\le t}|\xi(s)|,\qquad \tau=\inf\{t:\rho(t)>N_1,\lambda(t)>N\},$$

where N_1 and N are constants. Here it is assumed that $\inf\varnothing=T$.

The process $\xi_r(t)$ is the local square integrable martingale with characteristic

$$\langle\xi_r,\xi_r\rangle_t=r^2\int_0^t|\xi(s)|^{2r-4}\sum_{j,k=1}^m\xi^j(s)\xi^k(s)\beta^{jk}(\theta_s\xi,s)\,ds,$$

and in view of Lemma 2 in Section 3 of Chapter I, $\xi_r(t\wedge\tau)$ possesses moments of all orders. Consequently,

$$\mathsf{E}\rho^{2r}(t\wedge\tau)\le3\Big[|\varphi(0)|^{2r}+\mathsf{E}\sup_{0\le u\le t}\Big(\int_0^{u\wedge\tau}L_c(\xi,s)\,ds\Big)^2+\mathsf{E}\sup_{0\le u\le t}|\xi_r(u\wedge\tau)|^2\Big].$$

We now bound the terms on the right-hand side of the inequality obtained. We have

$$\sup_{0\le u\le t}\Big(\int_0^{u\wedge\tau}L_c(\xi,s)\,ds\Big)^2\le T\int_0^{t\wedge\tau}c_1\lambda^2(s)[c_2+\|\varphi\|^{2r}+\rho^{2r}(s)]\,ds,$$

where c_1 and c_2 are constants which depend on r and m only. In the last step we have used the inequality $\|\theta_s\varphi\|\le\|\varphi\|+\rho(s)$, which is self-evident. Furthermore, it follows from the inequality for martingales (Section 1, (4)) that

$$\mathsf{E}\sup_{0\le u\le t}|\xi_r(u\wedge\tau)|^2\le4\mathsf{E}|\xi_r(t\wedge\tau)|^2$$
$$\le4mr^2\mathsf{E}\int_0^{t\wedge\tau}\lambda^2(s)|\xi(s)|^{2r-2}(1+\|\theta_s\xi\|^2)\,ds$$
$$\le c_3\mathsf{E}\int_0^{t\wedge\tau}\lambda^2(s)(c_4+\|\varphi\|^{2r}+\rho^{2r}(s))\,ds,$$

where the constants c_3 and c_4 depend on r and m only. Thus

$$\mathsf{E}\rho^{2r}(t\wedge\tau)\leq 3\Big[|\varphi(0)|^{2r}+c'N^2(1+\|\varphi\|^{2r})t+c''N^2\int_0^{t\wedge\tau}\mathsf{E}\rho^{2r}(s)\,ds\Big].$$

and the constants c' and c'' depend only on r, m, and T. The inequality obtained and Lemma 11 imply that

$$\mathsf{E}\rho^{2r}(t\wedge\tau)\leq 3[|\varphi(0)|^{2r}+c'N^2(1+\|\varphi\|^{2r})]\,e^{3c''N^2t}. \tag{61}$$

Here $\tau=\tau(N_1, N)$. Let $N_1\to\infty$. Since the sample functions of process $\xi(t)$ are bounded with probability 1, it is easy to verify that

$$\rho(t\wedge\tau)=\sup_{0\leq s\leq T\wedge\tau}|\xi(s)|\to\sup_{t\leq T\wedge\tau_N}|\xi(s)|,$$

where $\tau_N=\inf\{t\colon\lambda(t)>N\}$. Moreover, in view of Fatou's lemma and inequality (61)

$$\mathsf{E}\rho^{2r}(t\wedge\tau_N)\leq\mathsf{E}\lim_{N_1\to\infty}\rho^{2r}(t\wedge\tau)\leq C'(\|\varphi\|^{2r},N^2)\,e^{3c''N^2t},$$

where $C'(\|\varphi\|^{2r}, N^2)$ is a constant which depends linearly on $\|\varphi\|^{2r}$, N^2, and T.

We have thus proved the following theorem:

Theorem 8. *A process $\xi(t)$ satisfying equation* (60) *and conditions* (57)–(59) *for $t\leq\tau_N=\inf\{t\colon\lambda(t)>N\}$ possesses moments of all orders.*

Corollary. *If*

$$|\alpha(\varphi,t)|\leq C(1+\|\varphi\|),$$
$$\mathsf{E}\{|\beta(\varphi,\Delta)|^2\,|\,\mathfrak{F}_t\}\leq C(1+\|\varphi\|^2)\,\Delta t,$$

where C is an absolute constant, then the solution of equation (60) *possesses finite moments of all orders for all $t\in[0,T]$.*

Consider now the problem of the existence of moments of solutions for a stochastic equation of the form

$$d\xi=\alpha(\theta_t\xi,t)\,dt+\beta(\theta_t\xi,dt)+\zeta(\theta_t\xi,dt), \tag{62}$$

where $\alpha(\varphi,t)$ and $\beta(\varphi,t)$ satisfy conditions (57)–(59) and

$$\zeta(\varphi,t)=\int_0^t\int_{\mathscr{R}^q}\gamma(\varphi,s,u)\mu(ds,du),$$

where $\mu(\cdot,\cdot)$ is a local martingale measure associated with an integer-valued measure $\nu(t,A)$ on $\mathscr{R}^q$ with characteristic $\pi(t,A)$ which is absolutely continuous with respect to the Lebesgue measure.

Also assume that

(63) $$\int_{\mathcal{R}^q} \gamma^2(\varphi, t, u)\pi(t, du) \leq \lambda^2(t)(1+\|\varphi\|^2)$$

and for any $N>0$, for $\|\varphi_i\| \leq N$, $i=1, 2$,

$$\mathsf{E}\{\int_t^{t+\Delta t}\int |\gamma(\varphi_1, s, u)-\gamma(\varphi_2, s, u)|^2 \pi(ds, du)|\mathfrak{F}_t\} \leq \|\varphi_1-\varphi_2\|^2 \mathsf{E}\{\int_t^{t+\Delta t} \lambda_N^2(s)\, ds\,|\mathfrak{F}_t\},$$

where $\lambda(s)$ and $\lambda_N(s)$ satisfy

(65) $$\int_0^T \lambda^2(s)\, ds < \infty, \qquad \int_0^T \lambda_N^2(s)\, ds < \infty$$

with probability 1. Under these assumptions the conditions of Theorem 5 are easily seen to be fulfilled and equation (62) possesses a unique solution on the interval $t \in [0, T]$.

Let $\tau = \tau(N_1, N) = \inf\{t: |\xi(t)| > N_1, \lambda(t) > N\}$. Then

$$\alpha_\tau(\varphi, t) = \alpha(\varphi, t \wedge \tau), \qquad \beta_\tau(\varphi, t) = \beta(\varphi, t \wedge \tau),$$
$$\zeta_\tau(\varphi, t) = \zeta(\varphi, t \wedge \tau)$$

are linearly bounded by an absolute constant and the process $\xi_\tau(t)$ satisfying the equation

$$d\xi_\tau = \alpha_\tau(\theta_t\xi_\tau, t)\, dt + \beta_\tau(\theta_t\xi_\tau, dt) + \zeta_\tau(\theta_t\xi_\tau, dt), \qquad t>0,$$
$$\xi_\tau(t) = \varphi(t), \qquad t \leq 0,$$

possesses finite second-order moments and, moreover, $\xi_\tau(t) = \xi(t)$ for $t \leq \tau$. In this connection consider first equation (62). Assume that for functions α, β, and ζ the corresponding majorant equals $\lambda(t) = N$ (although in the preceding discussion it could have happened that $\lambda(\tau) > N$, this is of no consequence since $\lambda(t) < N$ for $t < \tau$ and in the succeeding inequalities the value of the function $\lambda(t)$ at a single point is irrelevant).

We now utilize the generalized Itô formula (Chapter I, Section 3, (45)) with $f(x) = |x|^r$. We obtain

(66) $$|\xi(t)|^r = |\varphi(0)|^r + \int_0^t [L_c(\xi, s) + L_d(\xi, s)]\, ds + \xi_r(t) + \eta_r(t),$$

where $L_c(\xi, t)$ and $\xi_r(t)$ are as defined above while

$$L_d(\xi, t) = \int_{\mathcal{R}^q} \{|\xi(t) + \gamma(\theta_t\xi, t, u)|^r - |\xi(t)|^r$$
$$- r(\gamma(\theta_t\xi, t, u), \xi(t))|\xi(t)|^{r-2}\}\pi(t, du),$$

$$\eta_r(t) = \int_0^t \int_{\mathcal{R}^q} (|\xi(s) + \gamma(\theta_s\xi, s, u)|^r - |\xi(s)|^r)\mu(ds, du).$$

In addition to the preceding conditions we shall also assume that

$$\int_{\mathcal{R}^q} |\gamma(\varphi, t, u)|^{2r} \pi(t, du) \leqslant \lambda^2(t)(1+\|\varphi\|)^{2r}. \tag{67}$$

Taking into account the results in Section 2 of Chapter 1 concerning the finiteness of the moments of stochastic integrals it is easy to verify that the terms on the right-hand side of equation (66) possess finite moments of the second order.

Proceeding as above we obtain the inequality

$$\mathsf{E}\rho^{2r}(t) \leqslant 4\left(|\varphi(0)|^{2r} + \tilde{C}N^2(1+\|\varphi\|^{2r})t + \tilde{\tilde{C}}N^2 \int_0^t \mathsf{E}\rho^{2r}(s)\, ds\right),$$

where

$$\rho(t) = \sup_{0\leqslant s\leqslant t} |\xi(s)|,$$

and the bound

$$\mathsf{E}\rho^{2r}(t) \leqslant \tilde{C}(\|\varphi\|^{2r})\, e^{\tilde{\tilde{C}}N^2 t}.$$

Thus the following theorem is proved.

Theorem 9. *If for a stochastic equation* (62) *conditions* (57)–(59), (67), *and* (69) *are satisfied, then the solution of this equation possesses for* $t<\tau$ *finite moments up to the* $(2r)$th *order inclusively. If we also assume that* $\lambda(t)=N$, *where* N *does not depend on chance, then,*

$$\mathsf{E} \sup_{0\leqslant t\leqslant T} |\xi(t)|^{2r} \leqslant C_1(1+\|\varphi\|^{2r}), \tag{68}$$

where C_1 *is a constant which depends on* N, T, *and the dimensionality of the space only.*

Corollary. *Under the conditions of Theorem 9 and with* $\lambda(t)=N$ *we have*

$$\mathsf{E} \sup_{0\leqslant s\leqslant t} |\xi(s)-\varphi(0)|^{2r} \leqslant C_2(1+\|\varphi\|^{2r})t, \tag{69}$$

where C_2 *is a constant.*

Indeed, the preceding arguments yield inequality

$$\mathsf{E}\rho_1^{2r}(t) \leqslant \tilde{C}N^2(1+\|\varphi\|^2)t + \tilde{\tilde{C}}N^2 \int_0^t \mathsf{E}\rho^{2r}(s)\, ds,$$

where $\rho_1(t) = \sup_{0\leqslant s\leqslant t} |\xi(s)-\varphi(0)|^2$. The bound (69) follows from this inequality and (66).

Continuous dependence on a parameter of solutions of stochastic equations. Consider the equation of the form

$$(70)\qquad \begin{aligned} \eta_u(t) &= \varphi_u(t)+\int_0^t A_u(\theta_s\eta_u, ds), \qquad t\geq 0,\\ \eta_u(s) &= \varphi(s), \qquad s<0, \end{aligned}$$

where u is a scalar parameter, $u\in[0, u_0]$, the field

$$A_u(\psi, t)=\int_0^t \alpha_u(\psi, s)\,ds+\beta_u(\psi, t)$$

and the function $\varphi_u(t)$ both depend on the parameter u and the initial condition ($\varphi(s)$ for $s<0$) does not depend on u.

Theorem 10. *Assume that $A_u(\varphi, t)\in S(C, C)$ and, moreover, let*

a) $\sup_{0\leq t\leq T} \mathsf{E}|\varphi_u(t)|^2\leq C,$

b) $\lim_{u\to 0}\sup_{0\leq t\leq T} \mathsf{E}|\varphi_u(t)-\varphi_0(t)|^2=0 \qquad \forall t\in[0, T],$

c) $\mathsf{E}\{|\Delta\beta_u(\psi, t)-\Delta\beta_0(\psi, t)|^2|\mathfrak{F}_t\}\leq \mathsf{E}\{\int_t^{t+\Delta t}\gamma_u(\psi, s)\,ds\,|\mathfrak{F}_t\}$

and, for arbitrary $N>0$ and $t\in[0, T]$, let

$$\lim_{u\to 0}\mathsf{P}\{\sup_{\|\psi\|\leq N}|\alpha_u(\psi, t)-\alpha_0(\psi, t)|+\gamma_u(\psi, t))>\varepsilon\}=0.$$

Then

$$\lim_{u\to 0}\sup_{0\leq t\leq T}\mathsf{E}|\eta_u(t)-\eta_0(t)|^2=0.$$

Proof. The variables $\xi_u(t)$ possess finite moments of the second order since equations (70) satisfy the conditions of Theorem 7. We represent the difference $\eta_u(t)-\eta_0(t)$ in the form

$$\eta_u(t)-\eta_0(t)=\sigma_u(t)+\int_0^t A_u(\theta_s\eta_u, ds)-\int_0^t A_u(\theta_s\eta_0, ds),$$

where

$$\sigma_u(t)=\varphi_u(t)-\varphi_0(t)+\int_0^t A_u(\theta_s\eta_0, ds)-\int_0^t A_0(\theta_s\eta_0, ds).$$

It is easy to verify that

$$\begin{aligned} \mathsf{E}|\eta_u(t)-\eta_0(t)|^2 &\leq 3\mathsf{E}|\sigma_u(t)|^2+3C^2(T+1)\mathsf{E}\int_0^t\|\theta_s(\eta_u-\eta_0)\|_*^2\,ds\\ &= 3\mathsf{E}|\sigma_u(t)|^2+C'\int_0^t\int_{-s}^0\mathsf{E}|\eta_u(s+u)+\eta_0(s+u)|^2K(du)\,ds. \end{aligned}$$

Set $v_u(t)=\sup_{0\leq s\leq t}\mathsf{E}|\eta_u(s)-\eta_0(s)|^2$. It follows from the last inequality that

$$v_u(t)\leq 3\sup_{0\leq s\leq t}\mathsf{E}|\sigma_u(t)|^2+C'K\int_0^t v_u(s)\,ds.$$

In view of Lemma 11 we have

$$v_u(t) \leqslant C'' \sup_{0 \leqslant s \leqslant t} \mathsf{E}|\sigma_u(s)|^2,$$

where C'' does not depend on u. Furthermore,

$$\begin{aligned}\sup_{0 \leqslant t \leqslant T} \mathsf{E}|\sigma_u(t)|^2 \leqslant 3\Big\{ &\sup_{0 \leqslant t \leqslant T} \mathsf{E}|\varphi_u(t) - \varphi_0(t)|^2 \\ &+ \mathsf{E}\big(\textstyle\int_0^T |\alpha_u(\theta_s\eta_0, s) - \alpha_0(\theta_s\eta_0, s)|\, ds\big)^2 \\ &+ \sup_{0 \leqslant t \leqslant T} \mathsf{E}|\textstyle\int_0^t \beta_u(\theta_s\eta_0, ds) - \int_0^t \beta_0(\theta_s\eta_0, ds)|^2\Big\} \\ = 3(&I_1 + I_2 + I_3).\end{aligned}$$

By the condition, the quantity $I_1 \to 0$. Next

$$I_2 \leqslant T\mathsf{E} \int_0^t |\alpha_u(\theta_s\eta_0, s) - \alpha_0(\theta_s\eta_0, s)|^2\, ds,$$

where the integrand is dominated by the quantity

$$4C^2\big(1 + \textstyle\int_{-\infty}^0 |\eta_0(s+u)|^2 K(du)\big),$$

which is independent of u and is integrable with respect to the measure $d\mathsf{P} \times ds$. On the other hand,

$$|\alpha_u(\theta_s\eta_0, s) - \alpha_0(\theta_s\eta_0, s)| \to 0$$

in probability for each s and hence in measure $d\mathsf{P} \times ds$. Therefore $I_2 \to 0$ as $u \to 0$. Finally,

$$I_3 \leqslant 4\mathsf{E}|\textstyle\int_0^T \beta_u(\theta_s\eta_0, ds) - \int_0^T \beta_0(\theta_s\eta_0, ds)|^2 = 4\mathsf{E} \int_0^T \gamma_u(\theta_s\eta_0, s)\, ds,$$

and as in the case of quantity I_2 it is easy to verify that $I_3 \to 0$ as $u \to 0$. □

Remark. We shall strengthen the assumptions of Theorem 10 and assume that in addition to the conditions stipulated in the theorem

$$\lim_{u \to 0} \mathsf{E} \sup_{0 \leqslant t \leqslant T} |\varphi_u(t) - \varphi_0(t)|^2 = 0.$$

In this case

$$\lim_{u \to 0} \mathsf{E} \sup_{0 \leqslant t \leqslant T} |\eta_u(t) - \eta_0(t)|^2 = 0. \tag{72}$$

The proof of this assertion is analogous to the proof of Theorem 10.

Theorem 11. *Consider the stochastic equation.*

$$d\xi_u = A_u(\theta_s\xi_u, dt), \qquad \xi_u(s)=\varphi(s), \quad s\leq 0 \quad (u\in[0, u_0]),$$

satisfying the conditions of Theorem 5 and for all $N>0$ let

$$\overline{\lim_{p\to\infty}}\,\sup_u\,[\mathsf{P}\{\sup_{0\leq t\leq T}\lambda_0^u(t)\geq p\}+\mathsf{P}\{\sup_{0\leq t\leq T}\lambda_N^u(t)\geq p\}]=0$$

and also let condition c) *of Theorem* 10 *be satisfied. Then, for any $\varepsilon>0$*

$$\mathsf{P}\{\sup_{0\leq t\leq T}|\xi_u(t)-\xi_0(t)|>\varepsilon\}\to 0 \quad \text{as } u\to 0.$$

Proof. Let

$$\tau_p=\inf\{t\colon \lambda_0^u(t)\geq p, \lambda_N^u(t)\geq p, |\xi_u(t)|\geq N\} \qquad (\inf\varnothing = T),$$
$$\alpha_u^p(\varphi, t)=\alpha_u(\varphi, t) \quad \text{for } t<\tau_p,$$
$$\alpha_u^p(\varphi, t)=0 \quad \text{for } \tau\geq\tau_p, \qquad \beta_u^p(\varphi, t)=\beta_u(\varphi, t\wedge\tau_p),$$
$$A_u^p(\varphi, t)=\int_0^t \alpha_u^p(\varphi, s)\,ds+\beta_u^p(\varphi, t).$$

Theorem 10 (or the corresponding remarks for this theorem) is applicable to equations

$$d\xi_u^p = A_u^p(\theta_t\xi_u^p, dt), \qquad \xi_u^p(s)=\varphi(s), \quad s\leq 0.$$

On the other hand, in view of Theorem 4 $\xi_u^p(t)=\xi_u(t)$ with probability 1 for all $t<\tau_p$. Therefore for any $\varepsilon>0$

$$\mathsf{P}\{\sup_{0\leq t\leq T}|\xi_u(t)-\xi_u^p(t)|>\varepsilon\}\leq \mathsf{P}\{\tau_p<T\}.$$

Furthermore,

$$\begin{aligned}\mathsf{P}\{\sup_{0\leq t\leq T}|\xi_u(t)-\xi_0(t)|>\varepsilon\}&\leq \mathsf{P}\left\{\sup_{0\leq t\leq T}|\xi_0(t)-\xi_0^p(t)|>\frac{\varepsilon}{3}\right\}\\ &\quad+\mathsf{P}\left\{\sup_{0\leq t\leq T}|\xi_0^p(t)-\xi_u^p(t)|>\frac{\varepsilon}{3}\right\}\\ &\quad+\mathsf{P}\left\{\sup_{0\leq t\leq T}|\xi_u^p(t)-\xi_u(t)|>\frac{\varepsilon}{3}\right\}\\ &\leq 2\mathsf{P}\{\tau_p<T\}+\mathsf{P}\left\{\sup_{0\leq t\leq T}|\xi_0^p(t)-\xi_u^p(t)|>\frac{\varepsilon}{3}\right\}.\end{aligned}$$

The uniform stochastic boundedness of the processes $\lambda_0^u(t)$ and $\lambda_N^u(t)$ and Theorem 10 imply that one can first choose a sufficiently large value of p and an N such that $\mathsf{P}\{\tau_p < T\} < \varepsilon/3$ and then choose a $\delta > 0$ such that $\mathsf{P}\{\sup_{0\le t\le T} |\xi_0^p(t) - \xi_u^p(t)| > \varepsilon/3\}$ for $u \in [0, \delta]$. Consequently, for $u < \delta$

$$\mathsf{P}\{\sup_{0\le t\le T} |\xi_u(t) - \xi_0(t)| > \varepsilon\} < \varepsilon. \quad \square$$

Finite-difference approximations of solutions of stochastic equations. Consider the equation

$$(73) \qquad \begin{aligned} d\xi &= A(\theta_t\xi, dt), & t \in [0, T], \\ \xi(t) &= \varphi(t), & t \le 0, \end{aligned}$$

where $A(\varphi, t) \in S(\lambda_0, \lambda_N)$. Introduce an arbitrary subdivision $\delta = (0, t_1, t_2, \dots, t_n = T)$ of the interval $[0, T]$ and define random processes $\xi_\delta(t)$, $\zeta_\delta(t)$, $t \in [0, T]$, using the following recursive relations:

$$\begin{aligned} &\xi_\delta(t) = \zeta_\delta(t) = \varphi(t) && \text{for } t \le 0, \\ &\zeta_\delta(t) = \xi_\delta(t_k) && \text{for } t \in [t_k, t_{k-1}),\ k = 0, \dots, n-1, \\ &\xi_\delta(t) = \xi_\delta(t_k) + \textstyle\int_{t_k}^t A(\theta_s\zeta_\delta, ds) && \text{for } t \in (t_k, t_{k+1}]. \end{aligned}$$

The process $\xi_\delta(t)$ can be expressed in terms of the process $\zeta_\delta(t)$ in the following manner:

$$\xi_\delta(t) = \varphi(0) + \textstyle\int_0^t A(\theta_s\zeta_\delta, ds) \qquad \forall t \in [0, T].$$

This process $\xi_\delta(t)$ is called a *finite-difference approximation* of a solution of equation (73). We shall now show that $\xi_\delta(t)$ converges to process $\xi(t)$ as $|\delta| \to 0$.

Assume first that $A(\varphi, t) \in S(C_0, C)$. It follows directly from the general properties of the stochastic line integrals and the recurrent relations defining $\xi_\delta(t)$ and $\zeta_\delta(t)$ that the moments of the second order of the variables $\xi_\delta(t)$ and $\zeta_\delta(t)$ are finite. Furthermore, $\mathsf{E}\sup_{0\le t\le T} |\xi_\delta(t)|^2 < \infty$. Set

$$z_\delta(t) = \mathsf{E}\{\sup_{0\le t'\le T} |\xi(t') - \xi_\delta(t')|^2 \mid \mathfrak{F}_0\}.$$

Clearly

$$z_\delta(t) \le 2\mathsf{E}\{\sup_{0\le t'\le t} \left|\textstyle\int_0^{t'} [\alpha(\theta_s\xi, s) - \alpha(\theta_s\zeta_\delta, s)]\, ds\right|^2$$

$$+ \sup_{0\le t'\le t} \left|\textstyle\int_0^{t'} \beta(\theta_s\xi, ds) - \beta(\theta_s\zeta_\delta, ds)\right|^2 \mid \mathfrak{F}_0\}.$$

We shall now bound the terms appearing in the right-hand side of the inequality obtained using the methods which were repeatedly utilized above. We obtain

$$z_\delta(t) \leq C^2(8+2T)\mathsf{E}\{\int_0^t \|\theta_s(\xi-\zeta_\delta)\|_*^2\, ds \mid \mathfrak{F}_0\};$$

this implies that

$$\begin{aligned} z_\delta(t) &\leq C'\mathsf{E}\{\int_0^t \int_{-\infty}^0 (|\xi(s+u)-\xi_\delta(s+u)|^2 \\ &\quad + |\xi_\delta(s+u)-\zeta_\delta(s+u)|^2)K(du)\, ds \mid \mathfrak{F}_0\} \\ &= C'[z'(t)+z''(t)], \end{aligned}$$

where $C' = C'(C, T)$. Also, inequality

$$z'(t) \leq K \int_0^t z_\delta(s)\, ds$$

can be easily verified. On the other hand,

$$z''(t) \leq K \int_0^t \sup_{0 \leq t' \leq s} \mathsf{E}\{|\xi_\delta(t')-\zeta_\delta(t')|^2 \mid \mathfrak{F}_0\}\, ds.$$

Let

$$w(t) = \mathsf{E}\{|\xi_\delta(t)-\zeta_\delta(t)|^2 \mid \mathfrak{F}_0\}.$$

If $t \in (t_k, t_{k-1}]$, then

$$\begin{aligned} w(t) &= \mathsf{E}\{\mathsf{E}\{|\xi_\delta(t)-\zeta_\delta(t)|^2 \mid \mathfrak{F}_{t_k}\} \mid \mathfrak{F}_0\} \\ &= \mathsf{E}\{|\int_{t_k}^t A(\theta_s\zeta_\delta, ds)|^2 \mid \mathfrak{F}_0\} \\ &\leq C_0\mathsf{E}\{\int_{t_k}^t (1+\|\theta_s\zeta_\delta\|^2)\, ds \mid \mathfrak{F}_0\}, \end{aligned}$$

which yields the inequality

$$w(t) \leq C_0 K \int_{t_k}^t (1+\sup_{-\infty < t' \leq s} \mathsf{E}\{|\xi_\delta(t')|^2 \mid \mathfrak{F}_0\})\, ds.$$

We can bound the quantity $\mathsf{E}\{|\xi_\delta(t)|^2 \mid \mathfrak{F}_0\}$ in the same manner as the bound on $\mathsf{E}\{|\xi(t)|^2 \mid \mathfrak{F}_0\}$ was derived (cf. Theorem 7; cf. also Lemma 10, which implies inequality (74)). We have

$$\mathsf{E}\{|\xi_\delta(t)|^2 \mid \mathfrak{F}_0\} \leq C_0'(1+\|\varphi\|^2), \tag{74}$$

where C_0' depends on C, K, and T only. We thus obtain the bound

$$z''(t) \leq C_0''(1+\|\varphi\|^2)|\delta|,$$

where $|\delta| = \max \Delta t_k$. Hence

$$z_\delta(t) \leq C'K \int_0^t z_\delta(s)\,ds + C_0''(1+\|\varphi\|^2)|\delta|,$$

which implies that

$$z_\delta(t) \leq C_0''(1+\|\varphi\|^2)\,e^{C''t}|\delta|.$$

The following theorem is thus proved:

Theorem 12. *If $A(\varphi, t) \in S(C_0, C)$, then*

$$\mathsf{E}\{\sup_{0\leq t\leq T} |\xi(t)-\xi_\delta(t)|^2 \,|\, \mathfrak{F}_0\} \leq C_1(1+\|\varphi\|^2)|\delta|, \tag{75}$$

where C_1 depends on C_0, C, K, and T only.

Theorem 13. *If $A(\varphi, t) \in S(\lambda_0, \lambda_N)$, then*

$$\mathsf{P}\{\sup_{0\leq t\leq T} |\xi(t)-\xi_\delta(t)| > \varepsilon \,|\, \mathfrak{F}_0\} \to 0 \quad \text{as } |\delta| \to 0 \ (\text{mod } \mathsf{P}),$$

and the convergence is uniform in the class of all functions $A(\varphi, t)$ with fixed functions $\lambda_0(t)$ and $\lambda_N(t)$.

Proof. Set

$$\tau = \inf\{t: \lambda_0(t) > N\} \qquad (\inf \varnothing = T),$$

$$A^\tau(\varphi, t) = A(\varphi, t\wedge\tau),$$

and let $\xi_\delta^\tau(t)$ be a finite-difference approximation of the solution of equation

$$d\xi^\tau = A^\tau(\theta_t\xi^\tau, dt), \qquad t\in[0, T],$$

$$\xi^\tau(t) = \varphi(t), \qquad t\leq 0.$$

Then $\xi_\delta^\tau(t) = \xi_\delta(t)$ for $t<\tau$. Inequality (74) implies

$$\mathsf{P}\{\sup_{0\leq t\leq T} |\xi_\delta(t)| N_1 | \mathfrak{F}_0\} \leq \mathsf{P}\{\tau < T | \mathfrak{F}_0\} + \frac{C(N)(1+\|\varphi\|^2)|\delta|}{N_1^2}.$$

Thus $\mathsf{P}\{\sup_{0\leq t\leq T} |\xi_\delta(t)| > N_1 | \mathfrak{F}_0\} \to 0$ as $N_1\to\infty$ uniformly in δ (with probability 1).

First choose N_1 such that (for a given ω)

$$\mathsf{P}\{\sup_{0\leq t\leq T} (|\xi_\delta(t)\vee|\xi(t)|) > N_1 | \mathfrak{F}_0\} < \frac{\varepsilon}{4},$$

for all δ, where ε is an arbitrary positive number. Introduce a new random time (retaining the same designation τ):

$$\tau = \inf\{t: (\lambda_0(t) > N) \wedge (|\xi_\delta(t)| > N_1) \wedge (|\xi(t)| > N_1) \wedge (\lambda_{N_1}(t) > N_2)\}.$$

Then $\xi(t) = \xi^\tau(t)$ and $\xi_\delta^\tau(t) = \xi_\delta(t)$ for $t < \tau$.

Consequently if $\|\varphi\| < N_1$, then

$$\mathsf{P}\{\sup_{0 \le t \le T} |\xi(t) - \xi_\delta(t)| > \varepsilon \,|\, \mathfrak{F}_0\} \le \mathsf{P}(\tau < T) + \frac{C(N_1, N_2)(1 + \|\varphi\|^2)|\delta|}{\varepsilon^2},$$

and, moreover, for N and N_2 sufficiently large

$$\mathsf{P}\{\tau < T \,|\, \mathfrak{F}_0\} \le \mathsf{P}\{\lambda_0(t) > N \,|\, \mathfrak{F}_0\} + \mathsf{P}\{\lambda_{N_1}(t) > N_2 \,|\, \mathfrak{F}_0\} + \frac{\varepsilon}{4} < \frac{3\varepsilon}{4}.$$

Thus for $|\delta| < \varepsilon_0$ we have with probability 1

$$\mathsf{P}\{\sup_{0 \le t \le T} |\xi(t) - \xi_\delta(t)| > \varepsilon \,|\, \mathfrak{F}_0\} < \varepsilon,$$

where the choice of ε_0 depends only on functions $\lambda_0(t)$, $\lambda_N(t)$, and ε. The theorem is proved. □

Remark. The proof of Theorem 13 yields a somewhat stronger assertion. Actually, we have shown that the relation

$$\mathsf{P}\{\sup_{0 \le t \le T} |\xi(t) - \xi_\delta(t)| > \varepsilon\} \to 0$$

is uniformly fulfilled in the class of H functions $A(\varphi, t)$ such that

$$\lim_{C \to \infty} \sup_{A \in H} \mathsf{P}\{\sup_{0 \le t \le T} \lambda_0(t) > C\} = 0,$$

$$\lim_{C \to \infty} \sup_{A \in H} \mathsf{P}\{\sup_{0 \le t \le T} \lambda_N(t) > C\} = 0 \qquad \forall N > 0.$$

§ 2. Stochastic Differential Equations without an After-Effect

Solutions of stochastic differential equations without an after-effect as a Markov process. An equation of the form (36) in Section 1 is called *a stochastic differential equation without an after-effect* provided $A(\varphi, t+h) - A(\varphi, t)$ does not depend on the σ-algebra $\mathfrak{F}_t$ and on the values of $\varphi(s)$ for $s < 0$. Thus one can set $A(\varphi, t) = A(x, t)$ where $x = \varphi(0)$ and the process $A(x, t)$ for a fixed x is a process with independent increments.

Assume that $A(x, t)$ possesses finite moments of the second order and let

$$A(x, t)=\alpha(x, t)+\beta(x, t)$$

where $\beta(x, t)$ is a square integrable martingale with independent increments and $\alpha(x, t)$ is a nonrandom vector-valued function. In our case the condition $\alpha \in S_\alpha(\lambda_0, \lambda_N)$ implies that the function $\alpha(x, t)$ is a Borel function in arguments (x, t) and is differentiable with respect to t for almost all t. It is natural to assume that the function $\alpha_t'(x, t) \underset{\text{Def}}{=} a(x, t)$ exists for all t and to replace conditions (37) and (38) appearing in Section 1 by the following:

$$(1) \qquad |a(x, t)| \leq K(1+|x|) \qquad \forall x \in \mathcal{R}^m,$$

$$(2) \qquad |a(x, t)-a(y, t)| \leq C_N|x-y| \qquad \forall (x, y), |x| \leq N, |y| \leq N,$$

where K and C_N are constants. Thus in the case under consideration there is no point in distinguishing between classes $S_\alpha(\lambda_0, \lambda_N)$ and $S_\alpha(K, C_N)$. An analogous situation exists also in the case of the condition $\beta(x, t) \in S_\beta(\lambda_0, \lambda_N)$. We shall replace this condition by the following:

a) The function $\beta(x, t)$ is with probability 1 a Borel function in arguments (x, t) on each interval $t \in [0, s]$, is $\mathfrak{F}_t$-measurable as a function ω, and the sample functions belong for a fixed x with probability 1 to $\mathcal{D}^m[0, T]$.

b) $\mathsf{E}|\beta(x, \Delta)|^2 \leq K(1+|x|^2)\,\Delta t$ for each $x \in \mathcal{R}^m$, where $\Delta=(t, t+\Delta t]$.

c) For an arbitrary N there exists a constant C_N such that

$$\mathsf{E}|\beta(x, \Delta)-\beta(y, \Delta)|^2 \leq C_N|x-y|^2\,\Delta t, \qquad \forall (x, y), |x| \leq N, |y| \leq N,$$

and, as above, the classes $S_\beta(\lambda_0, \lambda_N)$ and $S_\beta(K, C_N)$ coincide.

If $\beta(x, t) \in S_\beta^c(\lambda_0, \lambda_N)$, then $\beta(x, t)$ satisfies conditions a)–c) and, moreover, the sample functions of $\beta(x, t)$ are continuous functions for a fixed x. Thus $\beta(x, t)$ is, in the case under consideration, a Gaussian process with independent increments (for a fixed x).

We shall agree to write $A(x, t) \in \bar{S}(K, C_N)$ provided $A(x, t)$ is a process with independent increments,

$$A(x, t)=\int_0^t a(x, s)\,ds+\beta(x, t),$$

$a(x, t)$ satisfies conditions (1), (2), and $\beta(x, t)$ is a square integrable martingale (on $[0, T]$) with independent increments satisfying conditions a)–c) stipulated above. We shall use the notation $A(x, t) \in \bar{S}^c(K, C_N)$ if $\beta(x, t)$ is, in addition, a Gaussian process for a fixed x.

Set $\tilde{B}(x, t)=\mathsf{E}\beta(x, t)\beta^*(x, t)$. The function $\tilde{B}(x, t)$ is the matrix characteristic of the field $\beta(x, t)$. Since

$$\tilde{B}(x, \Delta)=\tilde{B}(x, t+\Delta t)-\tilde{B}(x, t)=\mathsf{E}\beta(x, \Delta)\beta^*(x, \Delta),$$

it is easy to verify that condition b) is equivalent to the requirement that function $\tilde{B}(x, t)$ be absolutely continuous in t,

$$(3) \qquad \tilde{B}(x, t) = \int_0^t \tilde{b}(x, s)\, ds,$$

and its derivative $\tilde{b}(x, t)$ satisfies inequality

$$|\tilde{b}(x, t)| \leqslant K(1 + |x|^2).$$

We introduce the joint characteristic $\tilde{B}(x, y, t)$ of the processes $\beta(x, t)$ and $\beta(y, t)$, i.e., $\tilde{B}(x, y, t) = \mathsf{E}\beta(x, t)\beta^*(y, t)$. It follows from equation (3) that

$$\tilde{B}(x, y, t) = \int_0^t \tilde{b}(x, y, s)\, ds;$$

moreover, $\tilde{b}(x, x, t) = \tilde{b}(x, t)$ and $\tilde{b}(x, y, t) = \tilde{b}(y, x, t)$. Condition c) is equivalent to the following:

$$(4) \quad |\tilde{b}(x, x, t) - 2\tilde{b}(x, y, t) + \tilde{b}(y, y, t)| \leqslant C_N|x - y|^2 \qquad \forall (x, y), |x| \leqslant N, |y| \leqslant N.$$

We now state a number of previously obtained results concerning stochastic differential equations suitably adopted to the case under consideration.

Consider the stochastic differential equation

$$(5) \qquad d\xi(t) = A(\xi(t), dt) = a(\xi(t), t)\, dt + \beta(\xi(t), dt), \qquad t \geqslant s, \xi(s) = x,$$

where $a(x, t)$ is a nonrandom function with values in $\mathcal{R}^m$, $(x, t) \in \mathcal{R}^m \times [0, T]$, $\beta(x, t)$ is a family of processes with independent increments taking on values in $\mathcal{R}^m$ and possessing finite moments of the second order.

Theorem 1. *Assume that* $A(x, t) \in \tilde{S}(\cdot, C_N)$ *and the matrix function* $\tilde{B}(x, y, t)$ *is differentiable with respect to t.*

Then:

a) *A random time* τ *and a process* $\xi(t)$ *defined for* $s \leqslant t < \tau$ *exist such that* $\mathsf{P}(\tau > s) = 1$, *the process* $\xi(t)$ *satisfies equation* (5) *for* $s \leqslant t < \tau$, *and the sample functions of this process possess left-hand limits and are continuous on the right for all* t, $s \leqslant t < \tau$. *If* $\xi'(t)$ *is another solution of* (5) *with sample trajectories possessing the same property and defined for* $t < \tau'$, *then*

$$\mathsf{P}\{\exists t \colon \xi(t) \neq \xi'(t), s \leqslant t \leqslant \tau \wedge \tau'\} = 0.$$

b) *If* $A(x, t) \in \tilde{S}(K, C_N)$, *equation* (5) *possesses a solution defined for* $t \in [s, T]$ *possessing finite moments of the second order with sample functions belonging to* $\mathcal{D}^m[s, T]$ (mod P).

c) *If* $A(x, t) \in \tilde{S}^c(K, C_N)$, *equation* (5) *possesses a solution on the interval* $[s, T]$ *with sample functions belonging to* $\mathscr{C}^m$ (mod P). *This solution admits moments of all orders.*

Consider equation (5) and assume that for each $s \in [0, T]$ it possesses a unique solution on the interval $[s, T]$ satisfying the initial condition $\xi(s) = x$ with sample functions belonging to $\mathcal{D}^m[s, T]$. We denote this solution by $\xi_{xs}(t)$.

Let $\mathfrak{F}_t^s$ denote a completion of a σ-algebra generated by the random vectors $\beta(x, u) - \beta(x, s)$, $x \in \mathcal{R}^m$, $u \in (s, t]$, and let $\mathfrak{F}_t = \mathfrak{F}_t^0$. Clearly σ-algebras $\mathfrak{F}_{t_2}^{t_1}$ and $\mathfrak{F}_{t_3}^{t_2}$ are independent for $t_1 < t_2 < t_3$ and the variables $\xi_{xs}(t)$ are $\mathfrak{F}_t^s$-measurable.

We shall now derive several bounds which will be utilized below. These bounds are valid not only for equations without an after-effect and we shall therefore prove them for a more general case.

Denote by $\tilde{S}(\lambda_0, \lambda_N)$ ($\tilde{S}^c(\lambda_0, \lambda_N)$) the subclass of the class $S(\lambda_0, \lambda_N)$ ($S^c(\lambda_0, \lambda_N)$) consisting of random functions of the form

$$A(x, t) = \int_0^t \alpha(x, s)\, ds + \beta(x, t),$$

where $(x, t) \in \mathcal{R}^m \times [0, T]$. Equation (5) will be considered also in the case when $A(x, t) \in \tilde{S}(\lambda_0, \lambda_N)$. Results of Section 1, in particular the theorems concerning the existence and uniqueness of the solutions, are fully applicable in this case as well.

Lemma 1. *Let $A(x, t) \in \tilde{S}(K, C)$. Then for $0 \leq s \leq t \leq T$*

$$\mathsf{E}\{|\xi_{xs}(t) - \xi_{ys}(t)|^2 \mid \mathfrak{F}_s^0\} \leq \bar{C}|x - y|^2,$$

where the constant $\bar{C}$ depends on C and T only.

Proof. Since

$$\xi_{xs}(t) - \xi_{ys}(t) = x - y + \int_s^t [\alpha(\xi_{xs}(u), u) - \alpha(\xi_{ys}(u), u)]\, du$$
$$+ \int_s^t \beta(\xi_{xs}(u), du) - \beta(\xi_{ys}(u), du),$$

in view of Lemma 10 in Section 1 the function

$$v(t) = \mathsf{E}\{|\xi_{xs}(t) - \xi_{ys}(t)|^2 \mid \mathfrak{F}_s^0\}$$

satisfies the inequality

$$v(t) \leq 3\left(|x - y|^2 + C^2(T + 1) \int_s^t v(u)\, du\right).$$

It follows from Lemma 11 in Section 1 that $v(t) \leq \bar{C}|x - y|^2$ where $\bar{C}$ is a constant which depends on C and T only. □

Lemma 2. *Assume that $A(x, t) \in \tilde{S}(K, C)$. Then*

$$\mathsf{E}|\xi_{x_1 s_1}(t) - \xi_{x_2 s_2}(t)|^2 \leq C'(|x_1 - x_2|^2 + (1 + |x_2|^2)(s_2 - s_1)), \tag{6}$$

where C' is a constant which depends on K, C, and T only.

Proof. We have

$$\mathsf{E}|\xi_{x_1s_1}(t)-\xi_{x_2s_2}(t)|^2 \leqslant 2\mathsf{E}|\xi_{x_1s_1}(t)-\xi_{x_2s_1}(t)|^2+2\mathsf{E}|\xi_{x_2s_1}(t)-\xi_{x_2s_2}(t)|^2.$$

Furthermore,

$$\begin{aligned}\mathsf{E}|\xi_{x_2s_1}(t)-\xi_{x_2s_2}(t)|^2 &= \mathsf{E}\{\mathsf{E}\{|\xi_{x_2s_2}(t)-\xi_{\xi_{x_2s_1}(s_2)s_2}(t)|^2\,|\,\mathfrak{F}^0_{s_2}\}\}\\ &= \mathsf{E}\{(\mathsf{E}|\xi_{x_2s_2}(t)-\xi_{ys_2}(t)|^2)_{y=\xi_{x_2s_1}(s_2)}\,|\,\mathfrak{F}^0_{s_2}\};\end{aligned}$$

the last expression, in view of Lemma 1, is bounded by the quantity

$$A\mathsf{E}|x_2-\xi_{x_2s_1}(s_2)|^2.$$

In turn, it follows from Theorem 3 in Section 1 that

$$\mathsf{E}|x_2-\xi_{x_2s_1}(s_2)|^2 \leqslant B(1+|x_2|^2)(s_2-s_1).$$

Utilizing Lemma 1 once again to bound the quantity $\mathsf{E}|\xi_{x_1s_1}(t)-\xi_{x_2s_1}(t)|^2$ we obtain inequality (6). □

Corollary. *If $f(x)$, $x\in\mathscr{R}^m$, is a bounded and continuous function and the conditions of the preceding lemma are satisfied, then the function*

$$v(t,x)=\mathsf{E}f(\xi_{xt}(T))$$

is bounded and is jointly continuous in the variables (x,t). Moreover, if $f(x)$ is continuous and $|f(x)|\leqslant C(1+|x|^\rho)$ and $\mathsf{E}|\xi_{xt}(T)|^p$ is uniformly bounded on an arbitrary compact set of values (x,t), with $p>\rho$, then the function $v(t,x)$ is also continuous in (x,t).

Indeed if the function $f(x)$ is continuous, then in view of the corollary to Lemma 2, $f(\xi_{xt}(T))$ is a function continuous in probability in (x,t). The stipulated assumptions assure the possibility of a limit transition under the sign of the mathematical expectation.

The following formula will be utilized below.

Let $f(x,\omega)$ be a bounded $\mathfrak{B}\times\mathfrak{Y}$-measurable function, $(x,\omega)\in\mathscr{X}\times\Omega$, $\mathfrak{B}$ be a σ-algebra of Borel sets in the metric space $\mathscr{X}$, $(\Omega,\mathfrak{Y},\mathsf{P})$ be a probability space. Assume that $\zeta=\zeta(\omega)$ is an $\mathfrak{F}$-measurable mapping of Ω into $\mathscr{X}$, where $\mathfrak{F}\subset\mathfrak{Y}$. Then

$$\mathsf{E}\{f(\zeta,\omega)\,|\,\mathfrak{F}\}=g(\zeta),\quad \text{where } g(x)=\mathsf{E}\{f(x,\omega)\,|\,\mathfrak{F}\}.$$

To prove this assertion we introduce the class K of functions $f(x,\omega)$ for which the stated formula is valid. Clearly this class is linear and monotone (i.e., given an arbitrary monotonically nondecreasing sequence of nonnegative functions converging to a finite limit belonging to K, the limit function also belongs to K). Furthermore, K contains functions of the form $f(x,\omega)=\sum_{k=1}^n h_k(x)l_k(\omega)$, where $h_k(x)$ are bounded $\mathfrak{B}$-measurable and $l_k(\omega)$ are bounded $\mathfrak{Y}$-measurable functions.

Indeed,

$$\mathsf{E}\left\{\sum_{k=1}^{n} h_k(\zeta)l_k(\omega)\,|\,\mathfrak{F}\right\}=\sum_{k=1}^{n} h_k(\zeta)\alpha_k(\omega),$$

where $\alpha_k(\omega)=\mathsf{E}\{l_k(\omega)\,|\,\mathfrak{F}\}$ and

$$\mathsf{E}\left\{\sum_{k=1}^{n} h_k(x)l_k(\omega)\,|\,\mathfrak{F}\right\}=\sum_{k=1}^{n} h_k(x)\alpha_k(\omega).$$

It follows from the above stated properties of class K that it contains arbitrary Borel functions measurable with respect to the minimal σ-algebra which contains all sets of the form $B\times\Lambda$ where $B\in\mathfrak{B}$, $\Lambda\in\mathfrak{Y}$, i.e. which are $\mathfrak{B}\times\mathfrak{Y}$-measurable.

We now return to the process $\xi_{xs}(t)$. Set

$$P(s, x, t, A)=\mathsf{P}(\xi_{xs}(t)\in A)$$

where A is an arbitrary Borel set in $\mathscr{R}^m$. The function $P(s, x, t, A)$ is a stochastic kernel. The equality

$$\mathsf{E}\,f(\xi_{xs}(t))=\int_{\mathscr{R}^m} f(y)P(s, x, t, dy)$$

is valid for an arbitrary nonnegative Borel function $f(x)$. (This follows from the general rule of change of variables in integral calculus.)

Theorem 2. *The family of stochastic kernels $P(s, x, t, A)$, $0\leq s<t\leq T$, is a Markov family.*

Proof. To prove the theorem it is required to verify that the kernels $P(s, x, t, A)$ satisfy the Chapman–Kolmogorov equation (Volume II, Chapter I, Section 1). Let $s<u<t$. Then

$$\begin{aligned}\mathsf{E}f(\xi_{xs}(t))&=\mathsf{E}(\mathsf{E}\{f(\xi_{xs}(t))\,|\,\mathfrak{F}_u^s\})\\&=\mathsf{E}\mathsf{E}\{f(\xi_{\xi_{xs}(u)u}(t))\,|\,\mathfrak{F}_u^s\}\\&=\mathsf{E}(\mathsf{E}f(\xi_{yu}(t)))|_{y=\xi_{xs}(u)}\end{aligned}$$

or

$$\int_{\mathscr{R}^m} f(z)P(s, x, t, dz)=\int_{\mathscr{R}^m} P(s, x, u, dy)\int_{\mathscr{R}^m} f(z)P(u, y, t, dz).$$

This implies that for an arbitrary Borel set A

$$P(s, x, t, A)=\int_{\mathscr{R}^m} P(u, y, t, A)P(s, x, u, dy),$$

i.e., the kernels $P(s, x, t, A)$ indeed satisfy the Chapman–Kolmogorov equation. □

The proof of the theorem shows that the kernels $P(s, x, t, A)$ are transition probabilities of a certain Markov process (Volume II, Chapter I, Section 3).

We say that the stochastic differential equation under consideration generates a Markov process with transition probabilities $\mathsf{P}(\xi_{xs}(t)\in A)$. In this chapter we shall often identify this Markov process with the family of random processes $\xi_{xs}(t)$.

We now proceed to evaluate the generating operator of a Markov process $\xi_{xs}(t)$ generated by a stochastic differential equation without an after-effect.

Set

$$\xi'_{xs}(t)=x+\int_s^t a(x,u)\,du+\beta(x,t).$$

Lemma 3. *If $A(x,t)\in\bar{S}(K,C)$, then*

$$|\mathsf{E}[\xi_{xs}(t)-\xi'_{xs}(t)]|\leq C''(1+|x|)(t-s)^{3/2}, \tag{7}$$

$$\mathsf{E}|\xi_{xs}(t)-\xi'_{xs}(t)|^2\leq C'(1+|x|^2)(t-s)^2. \tag{8}$$

Proof. Denote

$$v(t)=|\mathsf{E}[\xi_{xs}(t)-\xi'_{xs}(t)]|,\qquad z(t)=\mathsf{E}|\xi_{xs}(t)-\xi'_{xs}(t)|^2.$$

Then

$$v(t)=|\mathsf{E}\int_s^t[a(\xi_{xs}(u),u)-a(x,u)]\,du|\leq C\int_s^t \mathsf{E}|\xi_{xs}(u)-x|\,du.$$

Taking the corollary to Theorem 9 in Section 1 into account we obtain the inequality

$$v(t)\leq CC^{(t)\frac{1}{2}}\int_s^t\sqrt{u-s}\,du=C''(t-s)^{3/2}.$$

Furthermore,

$$z(t)\leq 2\big(\mathsf{E}|\int_s^t[a(\xi_{xs}(u),u)-a(x,u)]\,du|^2$$
$$+\mathsf{E}|\int_s^t[\beta(\xi_{xs}(u),du)-\beta(x,du)]|^2\big).$$

Using Lemma 9 in Section 1 we arrive at

$$z(t)\leq 2(TC^2+C')\int_s^t\mathsf{E}|\xi_{xs}(u)-x|^2\,du,$$

which, together with the bound (69) in Section 1, yields inequality (8). □

Let $f(x)$, $x\in\mathcal{R}^m$, be an arbitrary thrice continuously differentiable function with bounded partial derivatives of the first, second, and third orders. We show that relation

$$z(s,t)=\frac{1}{t-s}\mathsf{E}[f(\xi_{xs}(t))-f(\xi'_{xs}(t))]$$

tends to zero uniformly on each compact set of the form $0 \leq s \leq t \leq T$, $|x| \leq N$, $N > 0$. Indeed, using Taylor's formula it is easy to establish the inequality of the form

$$(t-s)z(s,t) \leq K_1(|\mathsf{E}[\xi_{xs}(t) - \xi'_{xs}(t)]| \\ + \mathsf{E}|\xi_{xs}(t) - \xi'_{xs}(t)|\,|\xi'_{xs}(t) - x| + \mathsf{E}|\xi_{xs}(t) - \xi'_{xs}(t)|^2),$$

where the constant K_1 depends on the values of K, C only and the upper bounds on the derivatives of the first and second orders of the function $f(x)$. Clearly $\mathsf{E}|\xi'_{xs}(t) - x|^2 \leq C'(1+|x|^2)(t-s)$. It follows from Lemma 3 that

$$(t-s)z(s,t) \leq K'(1+|x|^2)(t-s)^{3/2}. \tag{9}$$

We now utilize the generalized Itô formula. Set

$$\beta(x,t) = \beta_c(x,t) + \zeta(x,t),$$

where $\beta_c(x,t)$ is a continuous component of the random function $\beta(x,t)$ and $\zeta(x,t)$ is its discontinuous martingale part, and let $\nu(x,t,A)$ be an integral-value measure constructed from the jumps of the process $\beta(x,t)$, $\mu(x,t,A)$ be the martingale measure associated with it, and $\pi(x,t,A)$ be its characteristic. Then

$$\zeta(x,t) = \int_{\mathcal{R}^m} u\mu(x,t,du).$$

Denote by $B(x,t)$ the matrix characteristic of the process $\beta_c(x,t)$. It follows from the orthogonality of $\beta_c(x,t)$ and $\zeta(x,t)$ that

$$\tilde{B}(x,t) = B(x,t) + \int_{\mathcal{R}_m} uu^* \pi(x,t,du).$$

Clearly the measure $\pi(x,t,A)$ is nonrandom. The condition $\beta(x,t) \in S_\beta(K, C_N)$ implies that $B(x,t)$ and the matrix function $\int_{\mathcal{R}^m} uu^*\pi(x,t,du)$ are absolutely continuous with respect to the Lebesgue measure. Set

$$B(x,t) = \int_0^t b(x,s)\,ds, \qquad \pi(x,t,A) = \int_0^t \Pi(x,s,A)\,ds,$$

where $\Pi(x,t,A)$ is a nonrandom function which is a measure on $\mathfrak{B}^m$ for a fixed (x,t). Moreover

$$\int_{\mathcal{R}^m} |u|^2 \Pi(x,t,du) < \infty, \qquad \forall (x,t) \in \mathcal{R}^m \times [0,T].$$

It follows from the generalized Itô formula (Chapter 1, Section 3, equation (45)) that:

$$f(\xi'_{xs}(t)) = f(x) + \int_s^t (L_c f(\xi'_{xs}(\theta)) + L_d f(\xi'_{xs}(\theta)))\,d\theta + \int_s^t (\nabla f(\xi'_{xs}(\theta)), \beta_c(x,d\theta)) \\ + \int_s^t \int_{\mathcal{R}^m} [f(\xi'_{xs}(\theta) + u) - f(\xi'_{xs}(\theta))]\mu(x,d\theta,du);$$

(observe that all the conditions for the applicability of this formula are fulfilled). Here

$$L_c f(\xi'_{xs}(\theta)) = (\nabla f(\xi'_{xs}(\theta)), a(x, \theta)) + \frac{1}{2} \sum_{k,j=1}^{m} \nabla^k \nabla^j f(\xi'_{xs}(\theta)) b^{kj}(x, \theta),$$
$$L_d f(\xi'_{xs}(\theta)) = \int_{\mathcal{R}^m} [f(\xi'_{xs}(\theta)+u) - f(\xi'_{xs}(\theta)) - (\nabla f(\xi'_{xs}(\theta)), u)] \Pi(x, \theta, du),$$

where $b^{kj}(x, t)$ are the entries of the matrix $b(x, t)$. From the assumptions on function $f(x)$ and the preceding bounds it is easy to obtain that for $t' \downarrow t$ and $s \uparrow t$

$$\lim_{\substack{t' \downarrow t \\ s \uparrow t}} \frac{\mathsf{E} f(\xi'_{xs}(t')) - f(x)}{t' - s} = (L_c + L_d) f(x)$$

uniformly in $(x, t) \in S_N \times [0, T]$ for any $N > 0$.

Finally, since

$$\lim_{\substack{t' \downarrow t \\ s \uparrow t}} \frac{\mathsf{E} f(\xi_{xs}(t')) - f(x)}{t' - s} = \lim_{\substack{t' \downarrow t \\ s \uparrow t}} \frac{z(s, t')}{t' - s} + \lim_{\substack{t' \downarrow t \\ s \uparrow t}} \frac{\mathsf{E} f(\xi_{xs}(t')) - f(x)}{t' - s},$$

we have

$$\lim_{\substack{t' \downarrow t \\ s \uparrow t}} \frac{\mathsf{E}(\xi_{xs}(t')) - f(x)}{t' - s} = (L_c + L_d) f(x)$$

$$= (\nabla f(x), a(x, t)) + \frac{1}{2} \sum_{k,j=1}^{m} \nabla^k \nabla^j f(x) b^{kj}(x, t) \tag{10}$$
$$+ \int_{\mathcal{R}^m} [f(x+u) - f(x) - (\nabla f(x), u)] \Pi(x, t, du).$$

The assumptions under which formula (10) was established can be somewhat weakened. Firstly it is sufficient to require only that $A(x, t) \in S(K, C_N)$.

Indeed, construct functions $a_N(x, t)$ and $\beta_N(x, t)$ such that they are linearly bounded, satisfy the uniform Lipschitz condition, and coincide with $a(x, t)$ and $\beta(x, t)$ on the sphere $S_N(x)$ of radius N with the center at point x. Let $\xi_N(t)$ be a solution of equation

$$d\xi_N(t) = A_N(\xi_N(t), dt), \qquad A_N(x, t) = \int_0^t a_N(x, s)\, ds + \beta_N(x, t).$$

Denote by τ_N the time of the first exit of function $\xi(t)$ from the sphere $S_N(x)$. Then $\xi_{xs}(t) = \xi_{N_{xs}}(t)$ for $t < \tau_N$. For an arbitrary bounded function $f(x)$ we have

$$\frac{1}{t'-s} |\mathsf{E}[f(\xi_{xs}(t')) - f(\xi_{N_{xs}}(t))]| \leqslant \frac{\mathsf{P}(\sup_{s \leqslant t \leqslant t'} |\xi_{N_{xs}}(t) - x| > N)}{t' - s}$$
$$\leqslant \frac{\mathsf{E} \sup_{s \leqslant t \leqslant t'} |\xi_{N_{xs}}(t) - x|^2}{(t'-s) N^2} \leqslant \frac{C}{N^2}.$$

Therefore applying relation (10) to process $\xi_{N_{xs}}(t)$ and then approaching the limit as $N\to\infty$, one verifies that this relation is preserved also for the classes of equations under consideration.

Analogously one can generalize equality (10) to arbitrary twice continuously differentiable and bounded functions $f(x)$ possessing bounded partial derivatives of the second order.

To show this, one constructs a sequence of functions $f_N(x)$ bounded and thrice continuously differentiable, with bounded partial derivatives up to the third order inclusive and such that these functions and their partial derivatives of the first and second order differ on the sphere $S_N(x)$ by at most $1/N$ from $f(x)$ and the corresponding partial derivatives of the function $f(x)$. Then

$$\frac{1}{t'-s}|\mathsf{E}f(\xi_{xs}(t'))-f(x)-[\mathsf{E}f_N(\xi_{xs}(t'))-f(x)]|$$

$$\leq\frac{1}{t'-s}|\mathsf{E}(f-f_N)(\xi_{xs}(t'))-(f-f_N)(x)|$$

$$\leq\frac{1}{t'-s}\left[\frac{C'}{N}\mathsf{E}|\xi_{xs}(t')-x|+C'\mathsf{P}(\tau_N<t)\right],$$

where C' is a constant independent of N and τ_N denotes as before the first exit time from the sphere $S_N(x)$. Since $\mathsf{P}(\tau_N<t)=N^{-2}\mathsf{E}\sup_{s\leq t\leq t'}|\xi_{xs}(t')-x|^2$, the quantity under consideration does not exceed

$$\frac{1}{t'-s}\frac{C''}{N^2}\mathsf{E}\sup_{s\leq t\leq t'}|\xi_{xs}(t')-x|^2\leq\frac{C^*}{N^2}.$$

It is also easy to verify that $L_cf-L_cf_N\to 0$ and $L_df-L_df_N\to 0$. Note that the boundedness of the second-order partial derivatives of function f is utilized only in the proof of relation $L_df-L_df_N\to 0$.

Theorem 3. *Equality* (10) *is valid for an arbitrary bounded and twice continuously differentiable function $f(x)$ with bounded partial derivatives of the second order and for a solution $\xi_{xs}(t)$ of the equation*

$$d\xi_{xs}(t)=A(\xi_{xs}(t),dt),\qquad \xi_{xs}(s)=x,$$

where $A(x,t)\in S(K,C_N)$.

If $A(x,t)$ is a continuous process, then

$$\lim_{\substack{t'\downarrow t\\ s\uparrow t}}\frac{1}{t'-s}[\mathsf{E}f(\xi_{xs}(t'))-f(x)]=(\nabla f(x),a(x,t))+\frac{1}{2}\sum_{k,j=1}^{m}b^{kj}(x,t)\frac{\partial^2 f(x)}{\partial x^k\,\partial x^j}$$

for an arbitrary twice continuously differentiable function which increases as $|x|\to\infty$ not faster than a power of $|x|$.

Proof. Only the second assertion requires a proof. Now let the function $f_N(x)$ coincide with the function $f(x)$ on $S_N(x)$ and be bounded together with its partial derivatives of the second order in $\mathscr{R}^m$. We apply relation (10) to this function. Note that

$$\begin{aligned}\frac{1}{t'-s}&|\mathsf{E}[f(\xi_{xs}(t'))-f(x)-(f_N(\xi_{xs}(t'))-f_N(x))]|\\ &\leq \frac{1}{t'-s} C\mathsf{E}\chi(\tau_N<t)(1+|\xi_{xs}(t')|^p)\\ &\leq \frac{C}{(t'-s)N^2}\mathsf{E}|\xi_{xs}(t')-x|^2(1+|\xi_{xs}(t')|^p).\end{aligned}$$

Using Itô's formula it is easy to verify, in the same manner as was done when the bounds on the moments of a solution of a stochastic differential equation were estimated, that for any $p>0$

$$\mathsf{E}|\xi_{xs}(t)-x|^2(1+|\xi_{xs}(t)|^p)\leq C(t-s),$$

where C is a constant. Thus

$$\frac{1}{t'-s}|\mathsf{E}f(\xi_{xs}(t'))-f_N(\xi_{xs}(t'))|\to 0$$

as $N\to\infty$. It is now obvious how to complete the proof of the theorem. □

Remark 1. In the case of a general equation belonging to the class $S(K, C_N)$ relation (10) can also be generalized to growing functions. One need only require the existence of moments of the process $\xi_{xs}(t)$ of a sufficiently high order.

Remark 2. Let function $f(t, x)$ and its partial derivatives with respect to x of the first and second order be uniformly bounded and continuous jointly in the variables (t, x). Then

$$\lim_{\substack{t'\downarrow t\\ s\uparrow t}}\frac{1}{t'-s}[\mathsf{E}f(t', \xi_{xs}(t'))-f(t', x)]=(L_c+L_d)f(t, x).$$

If $A(x, t)\in\bar{S}^c(K, C_N)$ it is sufficient to require, instead of the boundedness of $f(t, x)$ and its partial derivatives of the first and second orders with respect to x, that these derivatives increase as $x\to\infty$ not faster than a power of $|x|$.

This assertion is actually contained in the proof of Theorem 3.

Differentiability with respect to initial data of solutions of stochastic equations. Consider the problem of differentiability with respect to x of a solution $\xi_{xs}(t)$ of equation

$$d\xi_{xs}(t) = A(\xi_{xs}(t), dt), \qquad t > s, \tag{11}$$
$$\xi_{xs}(s) = x,$$

where $A(x, t) \in S(\tilde{C}, \lambda_N)$.

In what follows we shall interpret derivatives of random functions with respect to x in two different senses: as the ordinary derivatives existing with probability 1 and as mean square derivatives.

As far as the martingale field $\beta(x, t)$ is concerned, its derivative with respect to x would be interpreted in the sense of the mean square convergence.

Let δ_k be a vector with components $(\delta_{1k}, \delta_{2k}, \ldots, \delta_{mk})$, where $\delta_{kj} = 0$ for $k \neq j$ and $\delta_{kk} = 1$. Then

$$\frac{\partial}{\partial x^k} \beta(x, t) \underset{\text{Def}}{=} \text{l.i.m.} \frac{\beta(x + h\delta_k, t) - \beta(x, t)}{h}.$$

Several remarks connected with the differentiability of a square integrable martingale field $\beta(x, t)$ are in order. It easily follows from Theorem 16 in Section 1 of Chapter I that if the limit

$$\underset{h \to 0}{\text{l.i.m.}} \frac{\beta(x + hy, t) - \beta(x, t)}{h}$$

exists for $t = T$, then this limit also exists for any $t \in [0, T]$ and is a square integrable martingale.

Denote by $B(x, y, t)$ the joint matrix characteristic of martingales $\beta(x, t)$ and $\beta(y, t)$ and assume that it is absolutely continuous with respect to the Lebesgue measure:

$$B(x, y, t) = \int_0^t b(x, y, s)\, ds.$$

A necessary and sufficient condition for the existence of the mean square derivative $(\partial/\partial x^k)\beta^j(x, t)$ (see Volume I, Chapter IV, Section 1) is the existence of the limit

$$\lim_{h_1 \to 0} \lim_{h_2 \to 0} \mathsf{E} \frac{\beta^j(x + h_1\delta_k, t) - \beta^j(x, t)}{h_1} \frac{\beta^j(x + h_2\delta_k, t) - \beta^j(x, t)}{h_2}$$
$$= \lim_{h_1 \to 0} \lim_{h_2 \to 0} \mathsf{E} \frac{1}{h_1 h_2} [B^{jj}(x + h_1\delta_k, x + h_2\delta_k, t) \tag{12}$$
$$- B^{jj}(x + h_1\delta_k, x, t) - B^{jj}(x, x + h_2\delta_k, t) + B^{jj}(x, x, t)].$$

We shall require somewhat more, namely, that with probability 1 there exists for each s a continuous (with respect to x), generalized mixed derivative.

$$\frac{\partial^2}{\partial x^k\,\partial y^k} b^{jj}(x, x, s)$$

$$= \lim_{h_1\to 0}\lim_{h_2\to 0} \frac{1}{h_1 h_2}[b^{jj}(x+h_1\delta_k, x+h_2\delta_k, s)$$

$$-b^{jj}(x+h_1\delta_k, x, s)-b^{jj}(x, x+h_2\delta_k, s)+b^{jj}(x, x, s)], \qquad j=1,\ldots,m,$$

and, moreover,

$$\frac{\partial^2}{\partial x^k\,\partial y^k} b^{jj}(x, x, s)\le C, \qquad k, j=1,\ldots,m \pmod{d\mathsf{P}\times ds},$$

where C is a nonrandom constant.

Since $b^{ij}(x, y, s)$ is a nonnegative definite kernel the existence of the derivatives $(\partial^2/\partial x^k\,\partial y^k)b^{jj}(x, y, s)$ follows from the existence of the derivative $(\partial^2/\partial x^k\,\partial y^k)b^{jj}(x, x, s)$. The inequality

$$\left|\frac{\partial^2}{\partial x^k\,\partial y^k} b^{jj}(x, y, s)\right|\le C \pmod{d\mathsf{P}\times ds},$$

as well as the uniform boundedness of the expression appearing under the expectation sign in the right-hand side of equality (12) also follow from the existence of this derivative. This implies that in this case the condition for the existence of a mean square derivative $(\partial/\partial x^k)\beta^j(x, t)$ is fulfilled.

It is easy to verify that the joint characteristic of martingales $(\partial/\partial x^k)\beta^j(x, t)$ and $(\partial/\partial y^r)\beta^i(y, t)$ satisfies the relation

$$\left\langle \frac{\partial}{\partial x^k}\beta^j(x,\cdot), \frac{\partial}{\partial y^r}\beta^i(y,\cdot)\right\rangle_t = \frac{\partial^2}{\partial x^k\,\partial y^r}\int_0^t b^{ji}(x, y, s)\, ds,$$

while the existence of the corresponding derivatives and their continuity in x and y (mod $d\mathsf{P}\times ds$) follows from the preceding assumptions. Moreover,

$$\left\langle \beta^j(x,\cdot), \frac{\partial}{\partial y^r}\beta^k(y,\cdot)\right\rangle_t = \int_0^t \frac{\partial}{\partial y^r} b^{jk}(x, y, s)\, ds,$$

and the characteristic of the martingale

$$\tilde\beta_h^j(t)=\frac{\beta^j(x+hy, t)-\beta^j(x, t)}{h}-\nabla\beta^j(x, t)\cdot y$$

satisfies the following relationship:

$$\langle\tilde\beta^j_h, \tilde\beta^j_h\rangle_t = \int_0^t \Big\{\frac{1}{h^2}[b^{jj}(x+hy, x+hy, t) - 2b^{jj}(x+hy, x, t) + b^{jj}(x, x, s)] - \frac{2}{h}[\nabla_y b^{jj}(x+hy, x, s)\cdot y - \nabla_y b^{jj}(x, x, s)\cdot y] + \sum_{k,r=1}^{m} \frac{\partial^2}{\partial x^k\, \partial y^r} b^{jj}(x, x, s) y^k y^r\Big\}\, ds.$$

Utilizing Taylor's formula and the notation introduced we can rewrite the preceding relationship in the form

$$\langle\tilde\beta^j_h, \tilde\beta^j_h\rangle_t = \int_0^t [\nabla^2 b^{jj}(x, x, s) - \nabla^2 b^{jj}(x+\tilde h y, x+\tilde h y, s)]\cdot y\cdot y\, ds, \tag{13}$$

where $\tilde h$ is a number between 0 and h.

Theorem 4.

a) *Let the function $\alpha(x, t)$ be continuously differentiable with respect to x for a fixed t with probability* 1 *and $|\nabla\alpha(x, t)| \le C$.*

b) *Let the joint matrix characteristic $B(x, y, t)$ of the field $\beta(x, t)$ be differentiable with respect to t,*

$$B(x, y, t) = \int_0^t b(x, y, s)\, ds,$$

and the function $b(x, y, t)$, for a fixed t, possess with probability 1 *continuous and uniformly bounded derivatives $(\partial^2/\partial x^k\, \partial y^k) b(x, y, t)$, i.e.,*

$$\left|\frac{\partial^2}{\partial x^k\, \partial y^k} b(x, y, t)\right| \le C, \qquad k = 1, \ldots, m.$$

c) *Let the field $A(x, t) = \int_0^t \alpha(x, s)\, ds + \beta(x, t) \in \tilde S(C, C)$.*

Then $\xi_{xs}(t)$ is differentiable in the mean square with respect to x^k $(k = 1, \ldots, m)$ and $(\partial/\partial x^k)\xi_{xs}(t) = \eta_k(t)$ satisfies the linear stochastic differential equation

$$\eta_k(t) = \delta_k + \int_s^t \nabla A(\xi_{xs}(v), dv)\cdot \eta_k(v). \tag{14}$$

Proof. For simplicity of notation we set $s = 0$ and $\xi_{xs}(t) = \xi_x(t)$ and let

$$\eta_{h_k}(t) = \frac{1}{h}[\xi_{x+h_k}(t) - \xi_x(t)],$$

where $h_k = h\delta_k$ is a vector with components $h\delta_{kj}$, $j = 1, 2, \ldots, m$. The process $\eta_{h_k}(t)$ satisfies the equation

$$\eta_{h_k}(t) = \delta_k + \int_0^t A_h(\eta_{h_k}(s), ds),$$

where

$$A_h(y, t) = \int_0^t \frac{\alpha(\xi_x(x)+hy, s) - \alpha(\xi_x(s), s)}{h}\, ds$$

$$+\int_0^t \frac{\beta(\xi_x(s)+h_y, ds) - \beta(\xi_x(s), ds)}{h}$$

$$= \int_0^t \alpha_h(y, s)\, ds + \beta_h(y, t).$$

Denote

$$A_0(y, t) = \int_0^t \nabla\alpha(\xi_x(s), s) \cdot y\, ds + \int_0^t \nabla\beta(\xi_x(s), ds) \cdot y$$

$$= \int_0^t \alpha_0(y, s)\, ds + \beta_0(y, t).$$

We shall verify that the conditions of Theorem 11 in Section 1 are satisfied for the fields $A_h(y, t)$ and $A_0(y, t)$. It follows from the assumptions of Theorem 4 that

$$|\alpha_h(y, t)| \leq C|y|, \qquad \mathsf{E}\{|\Delta\beta_h(y, t)|^2 | \mathfrak{F}_t\} \leq C^2|y|^2\, \Delta t.$$

Moreover, in view of Lagrange's formula

$$|\alpha_h(y, t) - \alpha_0(y, t)| = |\nabla\alpha(\xi_x(t)+\tilde{h}y, t) - \nabla\alpha(\xi_x(t), t) \cdot y|,$$

where $|\tilde{h}| \leq |h|$. Since the function $\nabla\alpha(y, t)$ is continuous in y with probability 1 for any $t \in [0, T]$ we have

$$\mathsf{P}\{\sup_{|y|\leq N} |\alpha_h(y, t) - \alpha_0(y, t)| > \varepsilon\} \to 0 \quad \text{as } h \to 0 \quad \forall \varepsilon > 0.$$

Furthermore

$$\mathsf{E}\{|\Delta\beta_h - \Delta\beta_0|^2 | \mathfrak{F}_t\} = \mathsf{E}\{\int_t^{t+\Delta t} \gamma_h(y, s)\, ds \,|\, \mathfrak{F}_t\},$$

where in view of formula (13)

$$\gamma_h(y, t) = \sum_{j=1}^m (\nabla^2 b^{ij}[\xi(t)+\tilde{h}y, \xi(t+\tilde{h}y), t] - \nabla^2 b^{ij}[\xi(t), \xi(t), t]) \cdot y \cdot y$$

and $|\tilde{h}| \leq |h|$. Since the functions $(\partial^2/\partial x^k\, \partial y^r) b(x, y, t)$ are continuous with probability 1 jointly in the variables x, y, t, and $\sup |\xi(t)| < \infty$ with probability 1 it is easy to verify that $\mathsf{P}\{\sup_{|y|\leq N} |\gamma_h(y, t)| > \varepsilon\} \to 0$ as $h \to 0$. Thus, the conditions of Theorem 11 in Section 1 are fulfilled. Taking into account the remarks following Theorem 10 in Section 1, we obtain

$$\mathsf{E} \sup_{0\leq t\leq T} |\eta_{hk}(t) - \eta_{0k}(t)|^2 \to 0 \quad \text{as } h \to 0,$$

where $\eta_{0k}(t)$ is a solution of equation (14). The theorem is proved. □

If we strengthen the assumptions about the field $A(x, t)$ we can obtain theorems on the existence of derivatives of the second order of the function $\xi_{xs}(t)$ according to the initial data.

A formal differentiation of equation (14) leads to the relation

$$(15)\quad \eta_{kr}(t)=\int_s^t \nabla^2 A(\xi_{xs}(v), dv)\cdot \eta_k(v)\cdot \eta_r(v)+\int_s^t \nabla A(\xi_{xs}(v), dv)\cdot \eta_{kr}(v),$$

where

$$\eta_{kr}(t)=\frac{\partial^2}{\partial x^k\,\partial x^r}\xi_{xs}(t).$$

In order that the derivative $\eta_{kr}(t)$ possess finite moments of the second order it is natural to require the existence of moments of the fourth order for the variables $\eta_k(t)$ and a uniform, in a certain sense, boundedness of the fields $\nabla^2 A(x, t)$ in x.

First we state the conditions for the existence of moments of the fourth order for a solution of equation (14).

We utilize the generalized Itô formula. For this purpose we decompose the field $\beta(x, t)$ into continuous and discontinuous components, i.e.,

$$\beta(x, t)=\beta_c(x, t)+\zeta(x, t),$$

and let

$$\langle\beta_c(x,\cdot\,), \beta_c(y,\cdot\,)\rangle_t=\int_s^t b_c(x, y, s)\,ds,$$
$$\langle\zeta(x,\cdot\,), \zeta(y,\cdot\,)\rangle_t=\int_s^t b_d(x, y, s)\,ds.$$

Assume that matrices $b_c(x, y, t)$ and $b_d(x, y, t)$ possess, with probability 1, continuous mixed derivatives $(\partial^2/\partial x^k\,\partial y^r)(\cdot\,)$. Then the fields $\beta_c(x, t)$ and $\zeta(x, t)$ are mean square differentiable with respect to x^k $(k=1, \ldots, m)$. Set

$$A^\nabla(y, t)=\int_s^t \nabla A(\xi_{xs}(\theta), d\theta)\cdot y,$$

or, in more detail,

$$\begin{aligned}A^\nabla(y, t)&=\int_s^t \nabla\alpha(\xi_{xs}(\theta), \theta)\cdot y\,d\theta+\int_s^t \nabla\beta_c(\xi_{xs}(\theta), d\theta)\cdot y+\int_s^t \nabla\zeta(\xi_{xs}(\theta), d\theta)\cdot y\\&=\int_s^t \alpha^\nabla(y, \theta)\,d\theta+\beta_c^\nabla(y, t)+\zeta^\nabla(y, t).\end{aligned}$$

The matrix characteristic of the process $\beta_c^\nabla(y, t)$ is equal to

$$\langle\beta_c^\nabla(y,\cdot\,), \beta_c^\nabla(y,\cdot\,)\rangle_t=\int_s^t \sum_{k,r=1}^m \frac{\partial^2}{\partial x^k\,\partial y^r} b_c(\xi_{xs}(\theta), \xi_{xs}(\theta), \theta) y^k y^r\,d\theta$$

(Lemma 8, Section 1) and an analogous expression is valid for the characteristic of the process $\zeta^\nabla(y, t)$.

It follows from the assumptions in Theorem 4 that

$$\left|\frac{\partial^2}{\partial x^k\,\partial y^r} b_c(x, y, t)\right| + \left|\frac{\partial^2}{\partial x^k\,\partial y^r} b_d(x, y, t)\right| \le C,$$

$$|\alpha^\nabla(y, t)| \le C(1+|y|),$$

and hence the field $A^\nabla(y, t) \in S(C, C)$.

Consider the equation

$$\eta(t) = z + \int_0^t A^\nabla(\eta(s), ds), \tag{16}$$

and for simplicity set $s = 0$. It follows from Theorem 9 in Section 1 that if $A^\nabla(y, t) \in S(C, C)$ and if, moreover,

$$\int_{\mathscr{R}^m} |u|^4 \pi^\nabla(y, T, du) \le C(1+|y|^4), \tag{17}$$

where $\pi^\nabla(x, t, A)$ is a measure associated with the measure of jumps $\nu_y^\nabla(t, A)$ of the process $A^\nabla(y, t)$, then a solution of equation (16) possesses finite moments of the fourth order.

We now return to equation (15). We shall assume that the conditions of Theorem 4 and those given by (17) are satisfied. For simplicity we again set $s = 0$, $\xi_{xs}(t) = \xi_{x0}(t) = \xi_x(t)$. It is also necessary to assume the existence of the field $\nabla^2 A(x, t)$ and of the process

$$\varphi(t) = \int_0^t \nabla^2 A(\xi_x(v), dv) \cdot \eta_k(v) \cdot \eta_r(v).$$

Here the integral in the right-hand side of the equality is understood to be a line integral along the random curve $\xi_x(t)$ in the field

$$A_{kr}^{(2)}(x, t) = \int_0^t \nabla^2 A(x, dv) \cdot \eta_k(v) \cdot \eta_r(v).$$

We represent the random function $(\partial^2/\partial x^k\,\partial x^r)A(x, t)$ in the form

$$\frac{\partial^2}{\partial x^k\,\partial x^r} A(x, t) = \int_0^t \frac{\partial^2}{\partial x^k\,\partial x^r}\alpha(x, s)\,ds + \frac{\partial^2}{\partial x^k\,\partial x^r}\beta(x, t),$$

and assume the following:

a) $\alpha(x, t)$ is, with probability 1, twice continuously differentiable with respect to x for each $t \in [0, T]$ and

$$\left|\frac{\partial^2}{\partial x^k\,\partial x^r}\alpha(x, t)\right| \le C, \qquad k, r = 1, \ldots, m,$$

where C is a nonrandom constant.

b) There exists, with probability 1, for each $t \in [0, T]$ the partial derivative

$$\frac{\partial^4}{\partial x^k\, \partial y^k\, \partial x^r\, \partial y^r} b^{ij}(x, y, t), \tag{18}$$

continuous in x and y and bounded for all x, y, and t by an absolute constant C.

The derivative (18) is interpreted here as the mixed derivative $\partial^2/\partial x^r\, \partial y^r$ of the derivative $(\partial^2/\partial x^k\, \partial y^k) b^{ij}(x, y, t)$ in the sense described above.

If condition b) is satisfied, then it follows from the above that the derivative in the mean square

$$\frac{\partial^2}{\partial x^k\, \partial x^r}\beta(x, t) = \frac{\partial}{\partial x^r}\left(\frac{\partial}{\partial x^k}\beta(x, t)\right)$$

exists. We shall now discuss the function

$$\begin{aligned} A^{(2)}(x, t) &= \textstyle\int_0^t \nabla^2 \alpha(x, v) \cdot \eta_k(v) \cdot \eta_r(v)\, dv + \int_0^t \nabla^2 \beta(x, dv) \cdot \eta_k(v) \cdot \eta_r(v) \\ &= \textstyle\int_0^t \alpha^{(2)}(x, v)\, dv + \beta^{(2)}(x, t). \end{aligned}$$

Clearly the first integral exists with probability 1 and possesses finite moments of the second order. The second integral is a square integrable martingale field. The joint characteristics of its components can be expressed as:

$$\begin{aligned} &\langle \beta^{(2)p}(x, \cdot), \beta^{(2)q}(y, \cdot) \rangle_t \\ &\quad = \int_0^t \sum_{i,j,i',j'=1}^{m} \frac{\partial^4 b^{pq}(x, y, v)}{\partial x^i\, \partial y^j\, \partial x^{i'}\, \partial y^{j'}} \eta_k^i(v) \eta_r^j(v) \eta_k^{i'}(v) \eta_r^{j'}(v)\, dv. \end{aligned}$$

In view of these remarks the function $\varphi(t)$ exists and possesses finite moments of the second order. Also it follows from the available bounds that $\mathsf{E} \sup_{0 \le t \le T} |\varphi(t)|^2 < \infty$ as well. In that case, however, equation (15) possesses a unique solution and $\mathsf{E} \sup_{0 \le t \le T} |\eta_{kr}(t)|^2 < \infty$.

We now proceed to the problem of differentiability of the solution of equation (14) with respect to x. Denote the solution of equation (14) by $\eta(t, x)$ and set

$$\eta_h(t) = \frac{1}{h}[\eta(t, x + h\delta_r) - \eta(t, x)].$$

Function $\eta_h(t)$ satisfies the equation

$$\eta_h(t) = \varphi_h(t) + \int_0^t \nabla A(\xi_x(v), dv) \cdot \eta_h(v),$$

where

$$\varphi_h(t) = \frac{1}{h}\int_0^t [\nabla A(\xi_{x+h\delta_r}(v), dv) - \nabla A(\xi_x(v), dv)]\eta(v, x + h\delta_r).$$

Note that

$$\varphi_h(t)-\varphi(t)=\varphi'_h(t)+\varphi''_h(t),$$

$$\varphi'_h(t)=\int_0^t\left[\frac{1}{h}(\nabla A(\xi_{x+h\delta_r}(v),dv)-\nabla A(\xi_x(v),dv))\right.$$
$$\left.-\nabla^2A(\xi_x(v),dv)\eta_r(v)\right]\cdot\eta(v,x+h\delta_r),$$

$$\varphi''_h(t)=\int_0^t\nabla^2A(\xi_x(v),dv)\cdot\eta_r(v)\cdot[\eta(v,x+h\delta_r)-\eta(v,x)].$$

Utilizing the boundedness and continuity in x and y of function (18) as well as the expression for the joint characteristics of the field it is easy to arrive at relation

$$\mathsf{E}\sup_{0\le t\le T}|\varphi'_h(t)|^2\to 0.$$

Moreover,

$$\mathsf{E}\sup|\eta(t,x+\Delta x)-\eta(t,x)|^4=O(|\Delta x|^4).$$

This implies that

$$\mathsf{E}\sup_{0\le t\le T}|\varphi_h(t)-\varphi(t)|^2\to 0,$$

and in view of the remark following Theorem 11 in Section 1 we have

$$\mathsf{E}\sup_{0\le t\le T}|\eta_h(t)-\eta(t)|^2\to 0\quad\text{as } h\to 0.$$

Thus the following theorem is valid.

Theorem 5. *If the conditions of Theorem 4 are satisfied and inequality* (17) *and the above stated conditions* a) *and* b) *are fulfilled, then the derivatives* $\eta_{kr}(t)=(\partial^2/\partial x^k\,\partial x^r)\xi_{xs}(t)$ *exist in the mean square sense, satisfy equation* (15), *and are continuous in the mean square in* (x,s).

In the stated theorem only the continuity in the mean square of the second derivative of $\eta_{kr}(t)$ in (x,s) remains unproved. This can, however, be easily established analogously to Lemma 2 using the bounds obtained in Theorem 9 in Section 1.

Kolmogorov's equation. Let $\xi_{xs}(t)$ be a solution of equation (5) without an after-effect. It turns out that the function

$$F(t,x)=\mathsf{E}f(\xi_{xt}(T)),\qquad (t,x)\in[0,T)\times\mathcal{R}^m,$$

satisfies an important integro-differential equation of the form which does not depend on function $f(x)$. The dependence on $f(x)$ manifests itself in the boundary condition which should be adjoined to this equation.

Lemma 4. *Let the conditions of Theorem 4 be satisfied, $A(x, t) \in \bar{S}(C, C)$, and let the function $f(x)$ be twice continuously differentiable and its partial derivatives of the second order be uniformly bounded. Then the derivatives $\partial F(t, x)/\partial x^k$ exist, are continuous in (x, t), and*

$$\frac{\partial F(t, x)}{\partial x^k} = \mathsf{E}\left(\nabla f(\xi_{xt}(T)), \frac{\partial}{\partial x^k}\xi_{xt}(T)\right). \tag{19}$$

Proof. Indeed, let $h_k = h\delta_k$:

$$\left|\frac{F(t, x+h_k) - F(t, x)}{h} - \mathsf{E}\left(\nabla f(\xi_{xt}(T)), \frac{\partial}{\partial x^k}\xi_{xt}(T)\right)\right|$$

$$\leqslant \left|\mathsf{E}\left(\nabla f(\xi_{xt}(T)), \frac{\Delta\xi_{xt}}{h} - \frac{\partial}{\partial x^k}\xi_{xt}(T)\right)\right| + \left|\mathsf{E}\nabla^2 f(\xi_{xt}(T) + \theta\Delta\xi_{xt})\cdot\frac{\Delta\xi_{xt}}{h}\cdot\Delta\xi_{xt}\right|$$

$$\leqslant C\left[\mathsf{E}(1 + |\xi_{xt}(T)|^2)\cdot\mathsf{E}\left|\frac{\Delta\xi_{xt}}{h} - \frac{\partial}{\partial x^k}\xi_{xt}(T)\right|^2\right]^{1/2} + C\left[\mathsf{E}|\Delta\xi_{xt}|^2\cdot\mathsf{E}\left|\frac{\Delta\xi_{xt}}{h}\right|^2\right]^{1/2},$$

where C is constant depending only on $\sup|\nabla^2 f(x)|$, $\Delta\xi_{xt} = \xi_{x+h_k t}(T) - \xi_{xt}(T)$. The inequality thus obtained yields the assertion of the lemma. □

Remark. The lemma is also valid for a twice continuously differentiable function $f(x)$ increasing as $|x| \to \infty$ not faster than a power of x provided we stipulate additionally the finiteness of moments of a suitable order for the variables $\xi_{xt}(T)$.

Lemma 5. *Let function $f(x)$ be twice continuously differentiable and its partial derivatives of the second order be uniformly bounded and let the function $\xi_{xt}(T)$ possess mean square partial derivatives with respect to x^k of the first and second orders which are continuous in the mean square in variables (x, t).*

Then the function $F(t, x)$ possesses partial derivatives with respect to x of the second order,

$$\begin{aligned}\frac{\partial^2 F(t, x)}{\partial x^k\,\partial x^j} &= \left(\mathsf{E}\nabla^2 f(\xi_{xt}(T))\frac{\partial}{\partial x^k}\xi_{xt}(T), \frac{\partial}{\partial x^j}\xi_{xt}(T)\right)\\ &\quad + \left(\mathsf{E}\nabla f(\xi_{xt}(T)), \frac{\partial^2}{\partial x^k\,\partial x^j}\xi_{xt}(T)\right),\end{aligned} \tag{20}$$

and these derivatives are continuous in (t, x).

Proof. Set

$$\frac{1}{h}\left(\frac{\partial F(t,x+h_k)}{\partial x^i}-\frac{\partial F(t,x)}{\partial x^i}\right)-\left(\mathsf{E}\nabla^2 f(\xi_{xt}(T))\frac{\partial}{\partial x^k}\xi_{xt}(T),\frac{\partial}{\partial x^i}\xi_{xt}(T)\right)$$

$$-\left(\mathsf{E}\nabla f(\xi_{xt}(T)),\frac{\partial^2}{\partial x^k\,\partial x^i}\xi_{xt}(T)\right)=z_1+z_2+z_3+z_4,$$

where

$$z_1=\left(\mathsf{E}\nabla f(\xi_{xt}(T)),\frac{1}{h}\left[\frac{\partial}{\partial x^i}\xi_{x+h_kt}(T)-\frac{\partial}{\partial x^i}\xi_{xt}(T)-\frac{\partial^2}{\partial x^i\,\partial x^k}\xi_{xt}(T)\right]\right),$$

$$z_2=\left(\mathsf{E}[\nabla^2 f(\tilde\xi)-\nabla^2 f(\xi_{xt}(T))]\frac{\Delta\xi}{h},\frac{\partial}{\partial x^i}\xi_{x+h_kt}(T)\right),$$

$$z_3=\left(\mathsf{E}\nabla^2 f(\xi_{xt}(T))\left[\frac{\Delta\xi}{h}-\frac{\partial}{\partial x^k}\xi_{xt}(T)\right],\frac{\partial}{\partial x^i}\xi_{x+h_kt}(T)\right),$$

$$z_4=\left(\mathsf{E}\nabla^2 f(\xi_{xt}(T))\frac{\partial}{\partial x^k}\xi_{xt}(T),\left[\frac{\partial}{\partial x^i}\xi_{x+h_kt}(T)-\frac{\partial}{\partial x^i}\xi_{xt}(T)\right]\right).$$

Here $\tilde\xi$ denotes a point situated on the interval joining the points $\xi_{xt}(T)$ and $\xi_{x+h_kt}(T)$.

Since the function $\nabla f(x)$ increases as $|x|\to\infty$ not faster than $|x|$ and the mean square derivative $(\partial^2/\partial x^i\,\partial x^k)\xi_{xt}(T)$ exists, the quantity $z_1\to 0$ as $h\to 0$. For z_2 we have the following bound:

$$|z_2|\leqslant\left[\mathsf{E}\left(\frac{\Delta\xi}{h}\right)^2\right]^{1/2}\left[\mathsf{E}|\nabla^2 f(\tilde\xi)-\nabla^2 f(\xi_{xt}(T))|^2\left|\frac{\partial}{\partial x^i}\xi_{xt}(T)\right|^2\right]^{1/2}$$

$$+\left[\mathsf{E}|\nabla^2 f(\tilde\xi)-\nabla^2 f(\xi_{xt}(T))|^2\left|\frac{\partial}{\partial x^i}\xi_{x+h_kt}(T)-\frac{\partial}{\partial x^i}\xi_{xt}(T)\right|^2\right]^{1/2}.$$

It is easily seen that this bound implies that $|z_2|\to 0$ as $h\to 0$. Analogously one can verify that $|z_3|\to 0$ and $|z_4|\to 0$ as $h\to 0$. We thus establish the existence of partial derivatives $(\partial^2/\partial x^k\,\partial x^i)F(t,x)$ and have shown the validity of formula (20).

Formula (20) shows that the continuity is mean square in (x,t) of the derivatives $(\partial/\partial x^k)\xi_{xt}(T)$ and $(\partial^2/\partial x^k\,\partial x^i)\xi_{xt}(T)$ implies the continuity in (t,x) of the derivatives $(\partial^2/\partial x^k\,\partial x^i)F(t,x)$. □

Let the conditions of the lemma be fulfilled and let $0\leqslant t'<t<t''<T$. Then

$$F(t',x)=\mathsf{E}f(\xi_{\xi_{xt'}(t'')}(t'')(T))$$
$$=\mathsf{E}\{[\mathsf{E}f(\xi_{yt''}(T))]_{y=\xi_{xt'}(t'')}\}=\mathsf{E}F(t'',\xi_{xt'}(t'')).$$

Consequently,

$$\frac{F(t',x)-F(t'',x)}{t''-t'}=\frac{1}{t''-t'}\mathsf{E}\{F(t'',\xi_{xt'}(t''))-F(t'',x)\}.$$

Since function $F(t,x)$ is twice continuously differentiable, Theorem 3 and Remark 2 are applicable to this theorem.

We thus obtain the following:

Theorem 6

a) *Let* $A(x,t)\in\bar{S}(C,C_N)$ *and the solution* $\xi_{xs}(t)$ *of the stochastic differential equation* (5) *possess mean square partial derivatives of the first and second order with respect to* (x,s) *and be continuous in the mean square in* (x,s).

b) *Let the function* $f(x)$ *be twice continuously differentiable and uniformly bounded and the partial derivatives of the first and second orders of* $f(x)$ *also be uniformly bounded.*

Then the function

$$F(t,x)=\mathsf{E}f(\xi_{xt}(T)) \tag{21}$$

is twice continuously differentiable with respect to x, differentiable with respect to t, and satisfies the equation

$$\begin{aligned}&\frac{\partial F(t,x)}{\partial t}+(\nabla F(t,x),a(x,t))+\frac{1}{2}\sum_{k,j=1}^{m}\nabla^k\nabla^j F(t,x)b^{kj}(x,t)\\&+\int_{\mathscr{R}^m}[F(t,x+u)-F(t,x)-(\nabla F(t,x),u)]\Pi(x,t,du)=0\end{aligned}\tag{22}$$

and the boundary condition

$$\lim_{t\to T}F(t,x)=f(x). \tag{23}$$

Corollary 1

a) *Let a nonrandom function* $a(x,t)$ *be continuous and twice continuously differentiable with respect to x and the partial derivatives with respect to x of the first and second orders be uniformly bounded.*

b) *Let a random function* $\beta(x,t)$ *for a fixed x be a process with independent increments with finite moments of the second order*

$$\mathsf{E}\beta(x,t)=0,\quad \mathsf{E}\beta(x,t)\beta^*(y,t)=\int_0^t b(x,y,s)\,ds,$$

and let $b(x,y,t)$ *possess uniformly bounded mixed partial derivatives of the second and fourth orders of the form*

$$\frac{\partial^2}{\partial x^k\,\partial y^r}b^{ii}(x,y,t),\qquad \frac{\partial^4}{\partial x^k\,\partial y^k\,\partial x^r\,\partial y^r}b^{ii}(x,y,t).$$

c) *Let the discontinuous component of process* $\nabla\beta(x,t)$ *satisfy condition* (17).

d) *Let the function $f(x)$ be twice continuously differentiable with respect to x and uniformly bounded; also, let its partial derivatives of the first and second order be uniformly bounded.*

Then the function $F(t, x) = \mathsf{E}f(\xi_{xt}(T))$, where $\xi_{xt}(s)$ is a Markov process defined by the stochastic equation

$$\begin{aligned} d\xi(s) &= a(\xi(s), s)\, ds + \beta(\xi(s), ds), \qquad s \geq t, \\ \xi(t) &= x, \end{aligned} \tag{24}$$

satisfies equation (22) *and the boundary condition* (23).

We now show how one can arrive at a very general equation of the form (22) starting from the simplest probabilistic objects such as standard Wiener processes and the Poisson measure.

Corollary 2. *Assume that $w_1(t), \ldots, w_q(t)$ are mutually independent Wiener processes and $\nu(A, \Delta)$ is the Poisson measure on $\mathcal{R}^q \times [0, T]$ independent of the Wiener processes $w_j(t)$, $j = 1, \ldots, q$,*

$$\mathsf{E}\nu(A, [0, T]) = \Pi(A)t, \qquad \tilde{\nu}(A, \Delta) = \nu(A, \nu) - \Pi(A)\,\Delta t.$$

Let $a(x, t)$, $\sigma_k^j(x, t)$, $g^j(x, t, u)$ be nonrandom functions, $j = 1, \ldots, m$, $k = 1, 2, \ldots, q$, $(x, t, u) \in \mathcal{R}^m \times [0, T] \times \mathcal{R}^q$, satisfying the following conditions:

a) *Functions $a^j(x, t)$, $\sigma_k^j(x, t)$, and $g^j(x, t, u)$, $j = 1, \ldots, m$, $k = 1, \ldots, q$, are continuous in x, t and twice continuously differentiable with respect to x.*

b) *Partial derivatives with respect to x of the first and second orders of functions $a^j(x, t)$ and $\sigma_k^j(x, t)$ are uniformly bounded.*

c)
$$\int (|\nabla g|^2 + |\nabla g|^4 + |\nabla^2 g|^2)\Pi(du) \leq C,$$

where C does not depend on (x, t).

Denote by $\xi_{xt}(s)$ the solution of the stochastic differential equation

$$\begin{aligned} d\xi(s) &= a(\xi(s), s)\, ds + \sum_{k=1}^{q} \sigma_k(\xi(s), s)\, dw^k(s) \\ &\quad + \int_{\mathcal{R}^q} g(\xi(s), s, u)\tilde{\nu}(du, ds), \\ \xi(t) &= x. \end{aligned}$$

Then the function $F(t, x) = \mathsf{E}f(\xi_{xt}(T))$, where $f(x)$ satisfies the conditions of Theorem 6, *is a solution of equation*

$$\begin{aligned} &\frac{\partial F(t, x)}{\partial t} + (a(t, x), \nabla F(t, x)) + \frac{1}{2} \sum_{k,j=1}^{m} b^{kj}(x, t)\nabla^k \nabla^j F(t, x) \\ &\quad + \int_{\mathcal{R}^q} [F(t, x + g(t, x, u)) - F(t, x) \\ &\quad - (g(t, x, u), \nabla F(t, x))]\Pi(du) = 0, \end{aligned} \tag{25}$$

where $b^{kj}(x, t) = \sum_{r=1}^{q} \sigma_r^k(x, t)\sigma_r^j(x, t)$.

To prove this assertion we observe that the field

$$\beta_c(x, t) = \sum_{k=1}^{q} \int_0^t \sigma_k(x, s)\, dw(s)$$

possesses the joint characteristic defined by relation

$$\mathsf{E}\{\beta_c^j(x, \Delta)\beta_c^k(y, \Delta) \mid \mathfrak{F}_t\} = \mathsf{E}\beta_c^j(x, \Delta)\beta_c^k(y, \Delta)$$

$$= \int_t^{t+\Delta t} \sum_{r=1}^{q} \sigma_r^j(x, s)\sigma_r^k(y, s)\, ds,$$

where $\mathfrak{F}_t$ is the completion of the σ-algebra generated by the random variables $w^k(s)$, $\nu(A, s)$, $s \le t$, $A \in \mathfrak{B}^m$, $k = 1, \ldots, q$.

Analogously for the field

$$\zeta(x, t) = \int_0^t \int_{\mathcal{R}^q} g(x, s, u)\tilde{\nu}(du, ds)$$

we have

$$\mathsf{E}\{\zeta^j(x, \Delta)\zeta^k(y, \Delta) \mid \mathfrak{F}_t\} = \int_t^{t+\Delta t} \int_{\mathcal{R}^q} g^j(x, s, u) g^k(y, s, u)\Pi(du)\, ds.$$

It follows easily from the stipulated assumptions that the field

$$A(x, t) = \int_0^t a(x, s)\, ds + \beta_c(x, t) + \zeta(x, t)$$

satisfies the conditions of Theorem 6. Moreover,

$$\Pi(x, t, B) = \Pi\{u\colon g(x, t, u) \in B\}.$$

If we replace the variable of integration $u \to g(t, x, u)$ in equation (22) for the function $F(t, x)$, then equation (22) becomes (25).

Formula (21) can be viewed as a probabilistic representation of the solution of Cauchy's problem for an integro-differential equation in partial derivatives (22). On the one hand, equation (22) can be used, for example, for defining the transition probabilities of a Markov process $\xi_{xt}(s)$ or for a study of the analytical properties of these probabilities. On the other hand, if it is required to obtain a numerical or an approximate solution of equation (22) (or that of (25)), then one can utilize expression (21) for a probabilistic modeling of this solution (the Monte Carlo method). Theorems proved above dealing with the convergence of finite-difference approximations of solutions of stochastic differential equations serve, in particular, as a basis and as a justification for a simple finite-difference approximation procedure for solving equation (22) (or (25)).

One can extend the class of integro-differential equations in partial derivatives associated with solutions of stochastic differential equations. For this purpose consider the problem of determining the distribution of the random vector

$$\int_t^T h(\xi_{xt}(s), s)\, ds,$$

where $h(x, t)$, $(x, t) \in \mathcal{R}^m \times [0, T]$ is a function with values in $\mathcal{R}^q$ continuous and twice continuously differentiable with respect to x with uniformly bounded partial derivatives of the first and second orders.

To solve this problem we proceed as follows. We adjoin relations

$$d\eta(s) = h(\xi_{xt}(s), s)\, ds, \qquad s \geq t,$$
$$\eta(t) = y$$

to equation (24) and interpret them as a single stochastic differential equation

$$\begin{aligned} &d\zeta_{zt}(s) = B(\zeta_{zt}(s), ds); \\ &\zeta_{zt}(s) = (\xi_{xt}(s); \eta(s)), \qquad \zeta_{zt}(s) = z, \quad z = (x, y). \end{aligned} \tag{26}$$

Set

$$\bar{F}(t, x, y) = \bar{F}(t, z) = \mathsf{E}\bar{f}(\zeta_{zt}(T)),$$
$$\bar{f}(z) = \bar{f}(x, y) = f(x) \exp\{i(\lambda, x) + i(\mu, y)\},$$

where $f(x)$ is a twice continuously differentiable function with uniformly bounded partial derivatives of the first and second orders, λ is an m-dimensional vector and μ is a q-dimensional vector. Theorem 6 is applicable to equation (26); hence $F(z, t)$ satisfies the equation

$$\begin{aligned} &\frac{\partial}{\partial t}\bar{F} + (a, \nabla_x \bar{F}) + (h, \nabla_y \bar{F}) + \frac{1}{2} \sum_{k,j=1}^{m} b^{kj} \nabla_x^k \nabla_x^j \bar{F} \\ &+ \int_{\mathcal{R}^m} [\bar{F}(t, x+u, y) - \bar{F}(t, x, y) - (u, \nabla_x \bar{F})] \Pi(x, t, du) = 0. \end{aligned} \tag{27}$$

Here ∇_x denotes the gradient with respect to x and ∇_y with respect to y. Since $\eta(T) = y + \int_t^T h(s, \xi_{xt}(s))\, ds$, it follows that $\nabla_y \bar{F} = i\mu\bar{F}$. Replacing $\nabla_y \bar{F}$ by $i\mu\bar{F}$ in formula (27) and setting $y = 0$ we obtain

$$\begin{aligned} F(t, x) &= \bar{F}(t, x, 0) \\ &= \mathsf{E} f(\xi_{xt}(T)) \exp\{i(\lambda, \xi_{xt}(T)) + i(\mu, \int_t^T h(s, \xi_{xt}(s))\, ds)\}. \end{aligned}$$

Function $F(t, x)$ satisfies the equation

$$\begin{aligned} &\frac{\partial}{\partial t} F + (a, \nabla F) + \frac{1}{2} \sum_{k,j=1}^{m} b^{kj} \nabla^k \nabla^j F + i(\mu, h) F \\ &+ \int_{\mathcal{R}^m} [F(t, x+u) - F(t, x) - (u, \nabla F(t, x))] \Pi(x, t, du) = 0 \end{aligned} \tag{28}$$

and the boundary condition

$$F(T, x) = f(x)\, e^{i(\lambda, x)}.$$

If we set $f(x) = 1$ then the joint characteristic function of the distribution of random vectors $(\xi_{xt}(T), \int_t^T h(\xi_{xt}(s), s)\, ds)$ will satisfy equation (28). If we set $F(T, x) = 1$ we obtain an equation for the characteristic function of the distribution of the additive functional under consideration on the solution $\xi_{xt}(s)$ for the stochastic differential equation (24). Equation (28) differs from equation (22) by the presence of the additional term

$$i(\mu, h(x, t))F(t, x).$$

Example. Distribution of an additive functional on a Wiener process. We present several remarks concerning the evaluation of the distribution of homogeneous additive functionals (of an integral type) on a homogeneous Wiener process.

In the case under consideration $\xi_{xt}(s) = x + w(s) - w(t)$, $s \geq t$, and

$$\eta(T) = \int_t^T h(x + w(s) - w(t))\, ds.$$

The function

$$F(t, x) = \mathsf{E}\, e^{i\mu\eta(T)} f(x + w(T) - w(t))$$

satisfies the equation

$$\frac{\partial F(t, x)}{\partial t} + \frac{1}{2} \frac{\partial^2 F(t, x)}{\partial x^2} + i\mu h(x) F(t, x) = 0, \qquad t < T,$$

and the boundary condition

$$F(T, x) = f(x).$$

Setting $v(T - t, x) = F(t, x)$ we obtain the following equation for the function $v(t, x)$

$$\frac{\partial v(t, x)}{\partial t} = \frac{1}{2} \frac{\partial^2 v(t, x)}{\partial x^2} + i\mu h(x) v(t, x), \tag{29}$$

with the initial condition $v(0, x) = f(x)$. Moreover, the function $v(t, x)$ can be represented in the form

$$v(t, x) = \mathsf{E} \exp\{i\mu \int_0^t h(w(s) + x)\, ds\} f(x + w(t)).$$

Since the process $w(s)$ is stochastically equivalent to the process $\sqrt{t}\, w(s/t)$ the last expression for $v(t, x)$ can be replaced by the following:

(30) $$v(t, x) = \mathsf{E} \exp\{i\mu t \int_0^1 h(\sqrt{t}\, w(s) + x)\, ds\} f(x + \sqrt{t}\, w(1)).$$

Formula (30) yields a solution of the Cauchy problem for a parabolic equation (29) using the "quadrature" method (in the present case a "quadrature" is interpreted as an integral of a functional defined on $\mathscr{C}[0, 1]$ and the integration is carried over the standard Wiener measure, i.e., the measure generated in $\mathscr{C}[0, 1]$ by a Wiener process).

In what follows we shall set $f(x) \equiv 1$. In other words, we shall be dealing with the evaluation of the characteristic function of the distribution of variable $\eta(T)$. Equation (29) can be solved utilizing the Laplace transform with respect to t. Set

$$z(p, x) = \int_0^\infty e^{-pt} v(t, x)\, dt,$$

where p is a nonnegative number. Multiplying equation (29) by e^{-pt} and integrating with respect to t from 0 to ∞ we obtain

(31) $$pz(p, x) - 1 = \frac{1}{2} \frac{\partial^2}{\partial x^2} z(p, x) + i\mu h(x) z(p, x).$$

We show that equation (31) is also valid in the case when $h(x)$ is a piecewise continuous bounded function. We choose a sequence of uniformly bounded functions $h_n(x)$ such that these functions converge for each x to $h(x)$ and each one of them is twice continuously differentiable and possesses bounded derivatives of the first and second orders. Let

$$z_n(p, x) = \int_0^\infty e^{-pt}\, \mathsf{E} \exp\{i\mu \int_0^t h_n(x + w(s))\, ds\}\, dt.$$

Then $|z_n(p, x)| \leq 1/p$ and $z_n(p, x) \to z(p, x)$ as $n \to \infty$. Functions $z_n(p, x)$ satisfy equation (31). It can be seen from this equation that the derivatives $(\partial^2/\partial x^2) z_n(p, x)$ are uniformly bounded and converge to the limit $2(pz(p, x) - 1 - i\mu h(x) z(p, x))$ as $n \to \infty$. This implies the following.

Theorem 7. *If the function $h(x)$ is bounded and piecewise continuous, then the function $z(p, x)$ is continuously differentiable, possesses at all points of continuity of function $h(x)$ a second derivative, and satisfies equation* (31).

We shall utilize Theorem 7 to evaluate the distribution of the variable

$$\eta(t) = \int^t \operatorname{sgn} w(s)\, ds.$$

In the case under consideration equation (31) becomes

$$z''(p, x) + 2(i\mu \operatorname{sgn} x - p) = -2.$$

Solving this equation separately in the regions $x>0$ and $x<0$ we obtain

$$z(p,x)=\frac{1}{p-i\mu}+C_1 e^{\sqrt{2p-2i\mu}\,x}+C_2 e^{-\sqrt{2p-2i\mu}\,x}, \qquad x>0,$$

$$z(p,x)=\frac{1}{p+i\mu}+C_3 e^{\sqrt{2p+2i\mu}\,x}+C_4 e^{-\sqrt{2p+2i\mu}\,x}, \qquad x<0.$$

Since $z(p,x)$ is bounded as $x\to\pm\infty$ it follows that $C_1=C_4=0$. Utilizing the continuity at $x=0$ of functions $z(p,x)$ and $z'_x(p,x)$ we obtain the equalities

$$\frac{1}{p-i\mu}+C_2=\frac{1}{p+i\mu}+C_3,$$

$$-C_2\sqrt{2p-2i\mu}=C_3\sqrt{2p+2i\mu},$$

which imply that

$$C_3=\frac{1}{p+i\mu}\left[-1+\sqrt{\frac{p+i\mu}{p-i\mu}}\right].$$

To determine the distribution of the variable $\eta(T)$ it is sufficient to have the values of $z(p,0)$. For $|\mu|<p$

$$z(p,0)=\frac{1}{\sqrt{p^2+\mu^2}}=\frac{1}{p}\left(1+\frac{\mu^2}{p^2}\right)^{-1/2}=\frac{1}{p}\sum_{n=0}^{\infty}(-1)^n\frac{(2n-1)!!}{(2n)!!}\left(\frac{\mu}{p}\right)^{2n}.$$

Since

$$\int_0^\infty t^n e^{-pt}\,dt=\frac{n!}{p^{n+1}},$$

$$\int_{-\pi/2}^{\pi/2}\sin^k\varphi\,d\varphi=\begin{cases}0 & \text{for an odd } k,\\ \dfrac{(2n-1)!!}{(2n)!!}\pi, & k=2n,\end{cases}$$

we have

$$z(p,0)=\int_0^\infty\left(\sum_{n=0}^{\infty}(-1)^n\frac{(2n-1)!!}{(2n)!!}\frac{\mu^{2n}t^{2n}}{(2n)!}\right)e^{-pt}\,dt$$

$$=\int_0^\infty e^{-pt}\left(\sum_{k=0}^{\infty}\frac{1}{\pi}\int_{-\pi/2}^{\pi/2}\sin^k\varphi\frac{(i\mu t)^k}{k!}\,d\varphi\right)dt$$

$$=\frac{1}{\pi}\int_0^\infty e^{-pt}\int_{-\pi/2}^{\pi/2}e^{i\mu t\sin\varphi}\,d\varphi\,dt.$$

Thus

$$\mathsf{E}\exp\left\{i\mu\frac{1}{t}\int_0^t \operatorname{sgn} w(s)\,ds\right\}=\frac{1}{\pi}\int_{-\pi/2}^{\pi/2} e^{i\mu\sin\varphi}\,d\varphi=\int_{-\infty}^{\infty} e^{i\mu x}f(x)\,dx,$$

where

$$f(x)=\begin{cases}0 & \text{for } |x|>1\\ \dfrac{1}{\pi}\dfrac{1}{\sqrt{1-x^2}} & \text{for } |x|<1.\end{cases}\tag{32}$$

We thus obtain the following result: the variable $(1/t)\int_0^t \operatorname{sgn} w(s)\,ds$ possesses density (32).

The random variable

$$\tau_t=\int_0^t \frac{1+\operatorname{sgn} w(s)}{2}\,ds$$

has an intuitive meaning. It equals the time spent by the process $w(s)$ on the positive semiaxis during the (time) interval $(0, t)$. Using the density (32) one can find the distribution of the variable τ_t. Indeed,

$$\mathsf{P}\{\tau_t<xt\}=\mathsf{P}\left\{\frac{1}{t}\int_0^t \operatorname{sgn} w(s)\,ds<2x-1\right\}$$

$$=\frac{1}{\pi}\left(\arcsin(2x-1)+\frac{\pi}{2}\right).$$

Usually the formula obtained is written somewhat differently. Observe that

$$\arcsin(2x-1)+\frac{\pi}{2}=\arccos(1-2x).$$

If we set $\frac{1}{2}\arccos(1-2x)=z$, then

$$1-2x=\cos 2z,\qquad x=\frac{1-\cos 2z}{2}=\sin^2 z,\quad z=\arcsin\sqrt{x}.$$

Consequently

$$\mathsf{P}(\tau_t<x)=\frac{2}{\pi}\arcsin\sqrt{\frac{x}{t}},\qquad 0\leqslant x\leqslant t.\tag{33}$$

The result obtained is known as the *arcsine law*.

§ 3. Limit Theorems for Sequences of Random Variables and Stochastic Differential Equations

Let a sequence of series of random vectors

(1) $$\xi_{n0}, \xi_{n1}, \ldots, \xi_{nm_n}, \qquad n = 1, 2, \ldots$$

with the values in $\mathscr{R}^m$ be given. Assume that the increments $\Delta\xi_{nk} = \xi_{nk+1} - \xi_{nk}$ are small random variables. A classical problem in probability theory is to describe the class of possible limit distributions of the variable ξ_{nm_n} as $n \to \infty$ under varying assumptions on the variables $\Delta\xi_{nk}$. In the case when $\Delta\xi_{nk}$, $k = 0, 1, \ldots, m_n - 1$, are independent we are dealing with the thoroughly investigated problem of summation of the independent variables.

In this subsection we shall consider the general problem of investigating the limit distribution of a sequence of series of random variables (1) from the aspect of the theory of random processes or, more precisely, from its relation to the theory of stochastic differential equations.

A sequence of a series of random vectors (1) will correspond to a sequence of random processes $\xi_n(t)$, to be called processes associated with or generated by the sequence of series (1). To define these processes we must also specify a sequence of real numbers

$$0 = t_{n0} < t_{n1} < \cdots < t_{nm_n - 1} < t_{nm_n} = T, \qquad n = 1, 2, \ldots,$$

and then we set

$$\xi_n(t) = \xi_{nk} \quad \text{if } t \in [t_{nk}, t_{nk+1}).$$

If for $n \to \infty$ $\max_{1 \le k \le m_n} \Delta t_{nk} \to 0$ and all the variables $\{\Delta\xi_{nk}, k = 0, \ldots, m_n - 1\}$ are close in a certain sense—which will be described below—to the variables $\{\Delta\xi(t_{nk}), k = 0, 1, \ldots, m_n - 1\}$, where $\xi(t)$, $t \in [0, T]$, is a random process and $\Delta\xi(t_{nk}) = \xi(t_{nk+1}) - \xi(t_{nk})$, then one might expect that the distribution of the variable ξ_{nm_n} converges to the distribution of the variable $\xi(T)$ and, moreover, for continuous functionals $f[x(\cdot)]$ defined on $\mathscr{D}^m[0, T]$ the distribution of variables $f[\xi_n(\cdot)]$ will be close to the distribution of the variable $f[\xi(\cdot)]$.

Thus we would like to incorporate the problem under investigation into the general framework of limit theorems for random processes discussed in Chapter VI of Volume I.

In accordance with the results obtained in Volume I, when studying limit theorems for random processes, two problems can be distinguished: a) the investigation of conditions for the weak convergence of marginal distributions of random processes and the characterization of limiting distributions; and b) the determination of criteria of weak compactness of a sequence of measures corresponding to random processes in an appropriate functional space. General criteria for weak compactness of measures in functional spaces were established in Chapter VI of Volume I. In this section based on results established above we present some sufficient conditions for weak compactness of measures which, for the problems under consideration, are more convenient for verification. Next, we consider the weak convergence of marginal distributions of processes, which are either constructed from the sequence of series (1) or which are solutions of stochastic equations, to the marginal distributions of solutions of stochastic differential equations. In conclusion we present examples of the application of general theorems to more particular models and specific problems.

On a weak compactness of measure in $\mathscr{D}$ associated with sequences of series of random variables. In this section we shall use the letter $\mathscr{D}$ to denote the space $\mathscr{D}^m[0, T]$ and the measure defined on the σ-algebra generated by cylindrical sets in $\mathscr{D}$ will be referred to as the measure in $\mathscr{D}$.

Let $\xi_n(t)$, $n=1,2,\ldots, t\in[0,T]$, be a sequence of random processes with values in $\mathscr{R}^m$ and with sample functions belonging to $\mathscr{D}$ with probability 1. The process $\xi_n(t)$ generates on $\mathscr{D}$ a measure q_n, to be called the *measure associated in $\mathscr{D}$ with the process $\xi_n(t)$*, defined on cylindrical sets of the space $\mathscr{D}$ by relations

$$q_n(C_{t_1t_2\cdots t_r}(A^r))=\mathsf{P}\{(\xi_n(t_1),\ldots,\xi_n(t_r))\in A^r\}.$$

Here A^r is a Borel set in the space $\mathscr{R}^m\times\cdots\times\mathscr{R}^m=\mathscr{R}^{mr}$ and $C_{t_1,\ldots,t_r}(A^r)=\{x(\cdot): x(\cdot)\in\mathscr{D}, (x(t_1),\ldots,x(t_r))\in A^r\}$ is a cylindrical set with basis A^r over coordinates $(t_1,t_2,\ldots,t_r)$. We shall now be concerned with the conditions under which the sequence of measures $q_n(\cdot)$ converges weakly to a limit. The importance of this problem was clarified in Volume I, Chapter VI. Recall, for instance, that if a sequence $q_n(\cdot)$ is weakly convergent to $q(\cdot)$, where $q(\cdot)$ is the measure associated in $\mathscr{D}$ with a certain process $\xi(t)$, then for any bounded functional $f[x(\cdot)]$ q-almost everywhere continuous (in the metric of the space $\mathscr{D}$) the distribution of the random variable $\zeta_n=f[\xi_n(\cdot)]$ converges weakly to the distribution of the variable $\zeta=f[\xi(\cdot)]$.

In what follows we shall confine ourselves to the derivation of conditions for the weak convergence of measures in $\mathscr{D}$. Results related to the weak convergence of measures in $\mathscr{C}=\mathscr{C}^m[0,T]$ can be obtained as particular cases. The weak convergence of a sequence of measures $q_n(\cdot)$ associated with random processes $\xi_n(\cdot)$ is equivalent to the weak compactness of measures and weak convergence of all the marginal distributions of processes $\xi_n(t)$. Therefore in this subsection conditions for weak compactness of a sequence of measures are studied.

Recall the basic limit theorem for processes without discontinuities of the second kind (Volume I, Chapter VI, Section 5, Theorem 2).

Let $\xi_n(t)$, $t\in[0,T]$, $n=0,1,2,\ldots$ be a sequence of processes with sample functions belonging to $\mathscr{D}$ and let the finite-dimensional distributions of $\xi_n(t)$ converge weakly to finite-dimensional distributions of the process $\xi_0(t)$. Then for weak convergence of measures $q_n(\cdot)$ in $\mathscr{D}$ associated with random processes $\xi_n(t)$ to measure $q_0(\cdot)$, it is necessary and sufficient that

$$\lim_{c\to 0}\overline{\lim_{n\to\infty}}\,\mathsf{P}\{\Delta_c(\xi_n(\cdot))>\varepsilon\}=0, \tag{2}$$

where

$$\Delta_c(x(\cdot)) = \sup_{t-c\leqslant t'\leqslant t\leqslant t''\leqslant t+c} \{|x(t')-x(t)| \wedge |x(t)-x(t'')|\}$$
$$+ \sup_{0\leqslant t\leqslant c} |x(t)-x(0)| + \sup_{T-c\leqslant t\leqslant T} |x(T)-x(t)|.$$

In view of Volume I, Chapter VI, Section 5, Theorem 3, condition (2) is satisfied provided for some $\beta > 0$ and for $0 \leqslant t_1 < t_2 < t_3 \leqslant T$,

$$\mathsf{E}|\xi_n(t_2)-\xi_n(t_1)|^\beta |\xi_n(t_3)-\xi_n(t_2)|^\beta \leqslant H(t_3-t_1)^{1+\alpha},$$

where $\alpha > 0$ and the constant H does not depend on n. We shall require the following refinement of this result.

Assume that

$$\lim_{N\to\infty} \overline{\lim_{n\to\infty}} \mathsf{P}\{\sup_{0\leqslant t\leqslant T} |\xi_n(t)| > N\} = 0. \tag{3}$$

Let

$$\tau_n = \inf\{t: \sup_{0\leqslant t\leqslant T} |\xi_n(t)| > N\} \quad (\inf \varnothing = T),$$

and set

$$\xi_n^N(t) = \xi_n(t) \quad \text{for } t < \tau_n \quad \text{and} \quad \xi_n^N(t) = \xi(\tau_n-) \quad \text{for } t \geqslant \tau_n.$$

Then

$$\mathsf{P}\{\Delta_c(\xi_n(\cdot)) > \varepsilon\} \leqslant \mathsf{P}\{\tau_n < T\} + \mathsf{P}\{\Delta_c(\xi_n^N(\cdot)) > \varepsilon\}.$$

Thus the following assertion is valid.

Theorem 1. *If*

a) *a sequence of random processes* $\xi_n(t)$ *with sample functions in* $\mathscr{D}$ *satisfies condition* (3),

b) *for some* $\beta > 0$ *and any* $N > 0$

$$\mathsf{E}|\xi_n^N(t_2)-\xi_n^N(t_1)|^\beta |\xi_n^N(t_3)-\xi_n^N(t_2)|^\beta \leqslant H_N(t_3-t_1)^{1+\alpha}, \tag{4}$$

c) *finite-dimensional distributions of processes* $\xi_n(t)$ *converge weakly to the corresponding distributions of the process* $\xi_0(t)$,

then a sequence of measures $q_n(\cdot)$ *in* $\mathscr{D}$ *associated with the random processes* $\xi_n(t)$, $n = 0, 1, \ldots$ *is weakly convergent to* $q_0(\cdot)$.

Remark. Conditions (2) and (3) are necessary and sufficient for the weak compactness of sequences of measures $q_n(\cdot)$ in $\mathscr{D}$ associated with processes $\xi_n(t)$.

The proof of this assertion is actually contained in theorems presented in Volume I, Chapter VI, Section 5.

We now proceed to a discussion of processes $\xi_n(t)$ constructed from a sequence of series (1). We correspond to them the current of σ-algebras $\{\mathfrak{F}_{nk}, k = 0, 1, \ldots, m_n\}$, $n = 1, 2, \ldots$, where $\mathfrak{F}_{nk}$ is the σ-algebra generated by the random vectors $\xi_{n0}, \xi_{n1}, \ldots, \xi_{nk}$. It is understood here that the variables ξ_{nk} appearing in a given single series are defined on the same probability space, while distinct series are, in general, defined on different probability spaces.

Assume that variables ξ_{nk} possess finite moments of the second order. Set

$$\mathsf{E}\{\Delta\xi_{nk} \mid \mathfrak{F}_{nk}\} = \alpha_{nk}\,\Delta t_{nk},$$

$$\mathsf{E}\{(\Delta\xi_{nk} - \alpha_{nk}\,\Delta t_{nk})(\Delta\xi_{nk} - \alpha_{nk}\,\Delta t_{nk})^* \mid \mathfrak{F}_{nk}\} = \beta_{nk}^2\,\Delta t_{nk}.$$

Here the quantities Δt_{nk} are chosen arbitrarily, subject only to the following restrictions: $\Delta t_{nk} \to 0$; $\sum_{k=1}^{m_n-1} \Delta t_{nk} = T$ (T is fixed and is not a random quantity, $\max_k \Delta t_{nk} \to 0$ as $n \to \infty$). As far as random vectors α_{nk} and matrices β_{nk}^2 are concerned, these are uniquely determined for a chosen sequence of Δt_{nk} by the preceding equalities. Evidently, the matrix β_{nk}^2 is symmetric and nonnegative-definite. Denote by β_{nk} "the nonnegative-definite square root" of matrix β_{nk}^2. This quantity is also a symmetric and nonnegative-definite matrix. In what follows we shall assume that matrices β_{nk}^2 are nonsingular (with probability 1) so that their inverse β_{nk}^{-1} exists.

We represent the variable $\Delta\xi_{nk}$ in the form

$$\Delta\xi_{nk} = \alpha_{nk}\,\Delta t_{nk} + \beta_{nk}\,\Delta\psi_{nk},$$

where

$$\Delta\psi_{nk} = \beta_{nk}^{-1}(\Delta\xi_{nk} - \alpha_{nk}\,\Delta t_{nk})$$

and

$$\psi_{n0} = 0, \qquad \psi_{nk} = \sum_{j=0}^{k-1} \Delta\psi_{nj} = \sum_{j=0}^{k-1} \beta_{nj}^{-1}(\Delta\xi_{nj} - \alpha_{nj}\,\Delta t_{nj}), \qquad k = 1, \ldots, m_n.$$

Set

$$\varphi_{n0} = 0, \qquad \varphi_{nk} = \sum_{j=0}^{k-1} \beta_{nj}\,\Delta\psi_{nj} = \sum_{j=0}^{k-1} (\Delta\xi_{nj} - \alpha_{nj}\,\Delta t_{nj}), \qquad k = 1, \ldots, m_n.$$

Sequences $\{\psi_{nk}, k = 0, 1, \ldots, m_n\}$, $\{\varphi_{nk}, k = 0, 1, \ldots, m_n\}$ are $\mathfrak{F}_{nk}$-martingales. Moreover,

$$\mathsf{E}\{\Delta\psi_{nk}\,\Delta\psi_{nk}^* \mid \mathfrak{F}_{nk}\} = \boldsymbol{I}\,\Delta t_{nk}, \qquad \mathsf{E}\{\Delta\varphi_{nk}\,\Delta\varphi_{nk}^* \mid \mathfrak{F}_{nk}\} = \beta_{nk}^2\,\Delta t_{nk}.$$

Since α_{nk} and β_{nk} are $\mathfrak{F}_{nk}=\sigma(\xi_{n0},\xi_{n1},\ldots,\xi_{nk})$-measurable, there exist nonrandom Borel functions $a_{nk}(x_0,x_1,\ldots,x_k)$, $b_{nk}(x_0,x_1,\ldots,x_k)$, $x_j\in\mathcal{R}^m$, $j=0,1,\ldots,k$, $k=1,\ldots,m_n$, such that

$$\alpha_{nk}=a_{nk}(\xi_{n0},\xi_{n1},\ldots,\xi_{nk}),\qquad \beta_{nk}=b_{nk}(\xi_{n0},\xi_{n1},\ldots,\xi_{nk}).$$

Here the functions $a_{nk}(x_0,\ldots,x_k)$ take on values in $\mathcal{R}^m$, while $b_{nk}(x_0,\ldots,x_k)$ are matrix-valued functions.

Lemma 1. *Assume that functions $a_{nk}(x_0,\ldots,x_k)$ and $b_{nk}(x_0,\ldots,x_k)$ satisfy the condition*

$$|a_{nk}(x_0,\ldots,x_k)|+|b_{nk}(x_0,\ldots,x_k)|\leq C(1+\sup_{0\leq j\leq k}|x_j|), \tag{5}$$

where C is a constant independent of n. Then there are constants C_1 and C_2 which also do not depend on n such that

$$\mathsf{E}\{\sup_{0\leq j\leq k}|\xi_{nj}|^2\,|\,\mathfrak{F}_{n0}\}\leq C_1(1+|\xi_{n0}|^2), \tag{6}$$

$$\mathsf{E}\{\sup_{s\leq j\leq r}|\xi_{nj}-\xi_{ns}|^2\,|\,\mathfrak{F}_{ns}\}\leq C_2(1+|\xi_{ns}|^2)(t_{nr}-t_{ns}). \tag{7}$$

Proof. Since

$$\xi_{nk+1}=\xi_{n0}+\sum_{j=0}^{k}\alpha_{nj}\,\Delta t_{nj}+\sum_{j=0}^{k}\beta_{nj}\,\Delta\psi_{nj},\dagger$$

it follows that

$$\begin{aligned}\sup_{0\leq j\leq k+1}&|\xi_{nj}|^2\\ &\leq 3\left[|\xi_{n0}|^2+\sup_{0\leq j\leq k}\left|\sum_{r=0}^{j}\alpha_{nr}\,\Delta t_{nr}\right|^2+\sup_{0\leq j\leq k}\left|\sum_{r=0}^{j}\beta_{nr}\,\Delta\psi_{nr}\right|^2\right]\\ &\leq 3\left[|\xi_{n0}|^2+t_{nk}\sum_{r=0}^{k}|\alpha_{nr}|^2\,\Delta t_{nr}+\sup_{j\leq k}\left|\sum_{r=0}^{j}\beta_{nr}\,\Delta\psi_{nr}\right|^2\right].\end{aligned}$$

Set $v_{nk}=\mathsf{E}\{\sup_{0\leq j\leq k}|\xi_{nj}|^2\,|\,\mathfrak{F}_{n0}\}$. The preceding inequality implies that

$$v_{nk+1}\leq 3\left[|\xi_{n0}|^2+2TC^2\sum_{r=0}^{k}(1+v_{nr})\,\Delta t_{nr}+\mathsf{E}\left\{\sup_{j\leq k}\left|\sum_{r=0}^{j}\beta_{nr}\,\Delta\psi_{nr}\right|^2\,\middle|\,\mathfrak{F}_{n0}\right\}\right].$$

† In this equation (and in a number of succeeding expressions throughout this chapter) notation $\xi_{nk+1}(v_{nk+1},z_{nz+1},t_{nk+1}$ etc. should be interpreted as $\xi_{n\,k+1}(v_{n\,k+1},z_{n\,z+1},t_{n\,k+1}$ etc. respectively).

Noting that the sums $\sum_{r=0}^{j} \beta_{nr} \Delta\psi_{nr}$ form a martingale and utilizing Doob's inequality we obtain

$$\mathsf{E}\left\{\sup_{j\leq k}\left|\sum_{r=0}^{j} \beta_{nr}\Delta\psi_{nr}\right|^2 \Big| \mathfrak{F}_{n0}\right\} \leq 4\mathsf{E}\left\{\left|\sum_{r=0}^{k} \beta_{nr}\Delta\psi_{nr}\right|^2 \Big| \mathfrak{F}_{n0}\right\}$$

$$= 4\mathsf{E}\left\{\sum_{r=0}^{k} |\beta_{nr}\Delta\psi_{nr}|^2 | \mathfrak{F}_{n0}\right\}$$

$$= 4\mathsf{E}\left\{\sum_{r=0}^{k} \operatorname{sp} \beta_{nr}^2 \Delta t_{nr} | \mathfrak{F}_{n0}\right\}.$$

Thus

$$v_{nk+1} \leq 3|\xi_{n0}|^2 + C'\left(t_{nk} + \sum_{r=0}^{k} v_{nr}\Delta t_{nr}\right),$$

where C' is a constant which depends on T only.

We now introduce a piecewise constant function $v_n(t)$ by setting $v_n(t) = v_{nk}$ for $t \in [t_{nk}, t_{nk+1})$. The last inequality implies that

$$v_n(t) \leq 3|\xi_{n0}|^2 + C' \int_0^t (1 + v_n(s))\, ds.$$

Utilizing Lemma 1 in Section 1 we obtain

$$v_n(t) \leq (3|\xi_{n0}|^2 + C't)\, e^{C't}.$$

From here relation (6) follows.

To prove inequality (7) we proceed analogously. Inequality

$$\xi_{nk+1} - \xi_{ns} = \sum_{j=s}^{k} \alpha_{nj}\Delta t_{nj} + \sum_{j=s}^{k} \beta_{nj}\Delta\psi_{nj}$$

implies

$$\sup_{s\leq j\leq k+1} |\xi_{nj} - \xi_{ns}|^2 \leq 2\left[\sup_{s\leq j\leq k}\left|\sum_{r=s}^{j} \alpha_{nr}\Delta t_{nr}\right|^2 + \sup_{s\leq j\leq k}\left|\sum_{r=s}^{j} \beta_{nr}\Delta\psi_{nr}\right|^2\right].$$

Set

$$z_{nr} = \mathsf{E}\{\sup_{s\leq j\leq r} |\xi_{nj} - \xi_{ns}|^2 | \mathfrak{F}_{ns}\}.$$

From the preceding relation one easily obtains

$$z_{nr+1} \leq 4TC^2 \sum_{j=s}^{r} \mathsf{E}\{1 + \sup_{s\leq k\leq j} |\xi_{nk}|^2 | \mathfrak{F}_{ns}\}\Delta t_{nj} + 8\mathsf{E}\left\{\sum_{j=s}^{r} |\beta_{nj}\Delta\psi_{nj}|^2 | \mathfrak{F}_{ns}\right\}.$$

This yields the following inequality for the expectations z_{nr},

$$z_{nr+1} \leq C'' \sum_{k=s}^{r} (1+v'_{nk})\,\Delta t_{nk};$$

here C'' is a constant which depends on C only and $v'_{nk} = \mathsf{E}\{\sup_{s\leq j\leq k} |\xi_{nj}|^2 \mid \mathfrak{F}_{ns}\}$. Variables v'_{nr} can be bounded using inequality (6), which implies that $v_{nk} \leq C_1(1+|\xi_{ns}|^2)$. The second assertion of the lemma thus follows from the bounds obtained. □

Theorem 2. *If a sequence of series* (1) *satisfies the condition*

$$
\begin{aligned}
&|a_{nk}(x_0, x_1, \ldots, x_k)| + |b_{nk}(x_0, x_1, \ldots, x_k)| \leq C(1+\sup_{0\leq j\leq k} |x_j|),\\
&n = 1, 2, \ldots, \qquad k = 0, 1, \ldots, m_n,
\end{aligned}
\tag{8}
$$

where C is a constant independent of n and $\sup_n \mathsf{E}|\xi_{n0}|^2 < \infty$, then the sequence of measures $q_n(\cdot)$ in $\mathscr{D}$ is weakly compact.

Proof. Theorem 2 is a corollary to the remark following Theorem 1 and Lemma 1. Indeed, Lemma 1 and Chebyshev's inequality imply that

$$\mathsf{P}\{\sup_{0\leq t\leq T} |\xi_n(t)| > N\} \leq \frac{C\mathsf{E}(1+|\xi_{n0}|^2)}{N^2};$$

thus condition (3) of Theorem 1 is fulfilled in our case. Next let $\mathfrak{F}_n(t) = \mathfrak{F}_{nk}$ for $t \in [t_{nk}, t_{nk+1})$. Then

$$\mathsf{E}\{|\xi_n^N(t_3) - \xi_n^N(t_2)|^2 | \mathfrak{F}_n(t_2)\} \leq \chi_N(t_2)\mathsf{E}\{\sup_{t_2\leq t\leq t_3} |\xi_n(t) - \xi_n(t_2)|^2 | \mathfrak{F}_n(t_2)\},$$

where $\chi_N(t)$ is the indicator of the event $\{\tau_n > t\}$. Utilizing Lemma 1 once again we obtain

$$\mathsf{E}\{\sup_{t_2\leq t\leq t_3} |\xi_n^N(t) - \xi_n^N(t_2)|^2 | \mathfrak{F}_n(t_2)\} \leq \chi_N(t_2) C_2(1+|\xi_n^N(t_2)|^2)(t_3 - t_2).$$

Finally, we have

$$
\begin{aligned}
&\mathsf{E}|\xi_n^N(t_3) - \xi_n^N(t_2)|^2 \, |\xi_n^N(t_2) - \xi_n^N(t_1)|^2\\
&\quad \leq \mathsf{E}(\mathsf{E}\{|\xi_n^N(t_3) - \xi_n^N(t_2)|^2 \mid \mathfrak{F}_n(t_2)\} |\xi_n^N(t_2) - \xi_n^N(t_1)|^2)\\
&\quad \leq C_2^2(1+N^2)\mathsf{E}(1+|\xi_{n0}|^2)(t_3 - t_1)^2.
\end{aligned}
$$

Thus, the conditions of Theorem 1 are fulfilled and therefore Theorem 2 is proved. □

We note yet the following application of Theorem 1 to a sequence of series (1) which are square integrable martingales.

Let $\mathsf{E}\{\Delta\xi_{nk}|\mathfrak{F}_{nk}\}=0$. Set

$$\mathsf{E}\{|\Delta\xi_{nk}|^2|\mathfrak{F}_{nk}\}=\gamma_{nk}\,\Delta t_{nk},\qquad k=0,\ldots,m_n-1,\tag{9}$$

and let $\rho_n=\inf\{r;\ \gamma_{nr}\geq N\}$ ($\inf\varnothing=m_n$). Then ρ_n is a random time on $\{\mathfrak{F}_{nr},\ r=0,\ldots,m_n\}$. Let $\xi_n^N(t)=\xi_n(t\wedge t_{n\rho_n})$. The process $\xi_n^N(t)$ is also a martingale. Moreover,

$$\mathsf{P}\{\Delta_c(\xi_n(\cdot))>\varepsilon\}\leq\mathsf{P}\{\rho_n<m_n\}+\mathsf{P}\{\Delta_c(\xi_n^N(\cdot))>\varepsilon\}.$$

Furthermore, if the variables t_2 and t_3 are of the form $t_2=t_{nj}$, $t_3=t_{nr}$, then

$$\mathsf{E}\{|\xi_n^N(t_3)-\xi_n^N(t_2)|^2|\mathfrak{F}_n(t_2)\}\leq\sum_{k=j\wedge\rho_n}^{n\wedge\rho_n}\gamma_{nk}\,\Delta t_{nk}\leq N(t_3-t_2).$$

Thus (for $t_1<t_2<t_3$)

$$\mathsf{E}|\xi_n^N(t_3)-\xi_n^N(t_2)|^2|\xi_n^N(t_2)-\xi_n^N(t_1)|^2\leq N^2(t_3-t_1)^2.$$

The additional assumption that t_1, t_2 and t_3 are of the form t_{ni}, t_{nj}, and t_{nr} respectively is inessential and we arrive at the following theorem.

Theorem 3. *If each series in the sequence of the series is a martingale and if*

$$\lim_{N\to\infty}\overline{\lim_{n\to\infty}}\,\mathsf{P}\{\sup_{0\leq r\leq m_n-1}\gamma_{nr}>N\}=0,$$

where the variables γ_{nr} are defined by relation (9), *then the sequences of measures in $\mathcal{D}$ associated with processes $\xi_n(\cdot)$ is weakly compact.*

Corollary. *A sequence of measures in $\mathcal{D}$ associated with processes $\psi_n(t)$ such that $\mathsf{E}\{\Delta\psi_{nk}|\mathfrak{F}_{nk}\}=0$ and $\mathsf{E}\{\Delta\psi_{nk}\cdot\Delta\psi_{nk}^*|\mathfrak{F}_{nk}\}=\boldsymbol{I}\,\Delta t_{nk}$ is weakly compact.*

Theorem 2 can easily be generalized to the case of sequences of series of random vectors without finite moments of the second order. For this purpose we introduce on the current of σ-algebras $\{\mathfrak{F}_{nk},\ k=1,\ldots,m_n\}$ a random time j_n by setting $j_n=\min\{k\colon|\xi_{nk}|>N\}$ (or $j_n=m_n+1$ if the set $\{k\colon|\xi_{nk}|>N\}$ is void). Consider now for each $N>0$ a sequence of the series $\{\xi_{nk}^N,\ k=0,\ldots,m_n\}$, $n=1,2,\ldots$, where $\xi_{nk}^N=\xi_{nk}$ for $k<j_n$ and $\xi_{nk}^N=\xi_{nj_{n-1}}^N$ for $k\geq j_n$. Observe that vectors ξ_{nk}^N possess moments of all orders. Let $a_{nk}^N(x_0,x_1,\ldots,x_k)$ and $b_{nk}^N(x_0,x_1,\ldots,x_k)$ be constructed from $\{\xi_{nk}^N,\ k=0,1,\ldots,m_n\}$ in the same manner as $a_{nk}(x_0,\ldots,x_k)$ and $b_{nk}(x_0,\ldots,x_k)$ were constructed from the sequence $\{\xi_{nk},\ k=0,1,\ldots,m_n\}$.

Theorem 4. *If*

$$\lim_{N\to\infty}\overline{\lim_{n\to\infty}}\,\mathsf{P}\{\max_{0\le k\le m_n}|\xi_{nk}|>N\}=0$$

and

$$|a^N_{nk}(x_0,x_1,\dots,x_k)|+|b^N_{nk}(x_0,\dots,x_k)|\le C^N(1+\max_{0\le j\le k}|x_j|),$$

where C^N is a constant dependent possibly on N but independent of n and k, then the sequence of measures $q_n(\cdot)$ associated with processes $\xi_n(t)$ is weakly compact in $\mathscr{D}$.

An assertion analogous to Theorem 2 holds also for measures associated with solutions of stochastic differential equations. Consider the family of equations

$$(10)\qquad \begin{aligned} d\xi_\alpha&=A_\alpha(\theta_t\xi_\alpha,dt), &&t\ge 0,\\ \xi_\alpha(t)&=\varphi(t), &&t\le 0,\end{aligned}$$

depending on the parameter α.

Theorem 5. *Let $A_\alpha(\varphi,t)\in S(\lambda^\alpha_0,\lambda^\alpha_N)$ (or $A_\alpha(\varphi,t)\in S^c(\lambda^\alpha_0,\lambda^\alpha_N)$ and*

$$\lim_{N\to 0}\sup_\alpha\mathsf{P}\{\sup_{0\le t\le T}|\lambda^\alpha_0(t)|>N\}=0.$$

Then the family of measures $q_\alpha(\cdot)$ in $\mathscr{D}[0,T]$ associated with solutions $\xi_\alpha(t)$ of equations (10) *is weakly compact.*

The proof is analogous to the proof of preceding theorems. It is based on Theorem 1 and utilizes Theorem 7 (and Theorem 4) in Section 2 instead of Lemma 1.

Conditions for convergence to a Wiener process. We now proceed to study conditions for convergence of a sequence of processes $\{\xi_n(t),t\in[0,T]\}$, $n=1,2,\dots$, constructed from a sequence of series of random vectors (1) to a Wiener process.

For an arbitrary $\varepsilon>0$ set

$$(11)\qquad \mathsf{P}\{|\Delta\xi_{nk}|\ge\varepsilon\mid\mathfrak{F}_{nk}\}=\rho'_{nk}\,\Delta t_{nk},$$

$$(12)\qquad \mathsf{E}\{\chi_{nk},\Delta\xi^*_{nk}\mid\mathfrak{F}_{nk}\}=\rho''_{nk}\,\Delta t_{nk},$$

$$(13)\qquad \mathsf{E}\{\chi_{nk}\,\Delta\xi_{nk}\,\Delta\xi^*_{nk}\mid\mathfrak{F}_{nk}\}=(\boldsymbol{I}+\rho'''_{nk})\,\Delta t_{nk}.$$

Here

$$\Delta t_{nk}=t_{nk+1}-t_{nk},\qquad 0=t_{n0}<t_{n1}<\dots<t_{nm_0}=T,$$

and the numbers T and t_{nk} are arbitrary subject only to the condition

$$\max_{0 \leq k \leq m_n - 1} \Delta t_{nk} \to 0 \quad \text{as } n \to \infty$$

and the fulfillment of the subsequent assumptions: ρ'_{nk} are scalar, ρ''_{nk} are vector, and ρ'''_{nk} are matrix random variables defined by the corresponding equations (11)–(13). Finally, $\chi_{nk} = \chi_\varepsilon(\Delta\xi_{nk}) = 1$ if $|\Delta\xi_{nk}| < \varepsilon$ and $\chi_{nk} = 0$ if $|\Delta\xi_{nk}| \geq \varepsilon$.

For an m-dimensional Wiener process $\{w(t), \mathfrak{F}_t, t \geq 0\}$ conditional probabilities and expectations (11)–(13) coincide with the unconditional ones and are of order

$$\mathsf{P}\{|\Delta w| \geq \varepsilon\} = O\left[\left(\frac{\varepsilon^2}{\Delta t}\right)^{(m-2)/2} e^{-\varepsilon^2/2\Delta t}\right],$$

$$|\mathsf{E}\{\chi_\varepsilon(\Delta w)\,\Delta w\}| = O\left(\left(\frac{\varepsilon^2}{\Delta t}\right)^{(m-1)/2} e^{-\varepsilon^2/2\Delta t}\right),$$

$$\mathsf{E}\{\chi_\varepsilon(\Delta w)|\Delta w|^2\} = I\Delta t + O\left(\left(\frac{\varepsilon^2}{\Delta t}\right)^{m/2} e^{-\varepsilon^2/2\Delta t}\right).$$

One would expect that if ρ'_{nk}, ρ''_{nk}, and ρ'''_{nk} for an arbitrary fixed ε are "sufficiently" small then the marginal distributions of the process $\xi_n(t) - \xi_n(0)$ should weakly converge to a Wiener process as $n \to \infty$.

First we shall establish conditions for convergence of the distribution of the variable $\xi_n(T) - \xi_n(0) = \xi_{nm_n} - \xi_{n0}$ to the distribution of $w(T)$. For this purpose, consider the difference between conditional characteristic functions

$$\begin{aligned}\sigma_n &= \mathsf{E}\{\exp\{i(\xi_n(T) - \xi_n(0), z)\} \mid \mathfrak{F}_{n0}\} - \mathsf{E}\{\exp\{i(w(T), z)\} \mid \mathfrak{F}_0\} \\ &= \mathsf{E}\left\{\exp\{i(\xi_n(T) - \xi_n(0), z)\} - \exp\left\{-\frac{|z|^2 T}{2}\right\} \,\middle|\, \mathfrak{F}_{n0}\right\}.\end{aligned}$$

Set

$$\chi^{n0} = 1, \qquad \chi^{nk} = \prod_{j=0}^{k-1} \chi_{nj}.$$

Represent σ_n in the form:

$$\sigma_n = \sum_{k=0}^{m_n - 1} \mathsf{E}\{\sigma_{nk} \mid \mathfrak{F}_{n0}\} + \mathsf{E}\{(1 - \chi^{nm_n}) \exp\{i(\xi_n(T) - \xi_n(0), z)\} \mid \mathfrak{F}_{n0}\},$$

where

$$\sigma_{nk}=\chi^{nk+1}\exp\{i(\xi_n(t_{nk+1})-\xi_n(0),z)\}\exp\left\{-\frac{|z|^2(T-t_{nk+1})}{2}\right\}$$

$$-\chi^{nk}\exp\{i(\xi_n(t_{nk})-\xi_n(0),z)\}\exp\left\{-\frac{|z|^2(T-t_{nk})}{2}\right\}$$

$$=\chi^{nk}\exp\{i(\xi_n(t_{nk})-\xi_n(0),z)\}\exp\left\{-\frac{|z|^2(T-t_{nk+1})}{2}\right\}\tilde{\sigma}_{nk},$$

$$\tilde{\sigma}_{nk}=\chi_{nk}\exp\{i(\Delta\xi_{nk},z)\}-\exp\left\{-\frac{|z|^2\,\Delta t_{nk}}{2}\right\}.$$

We now bound the quantity $\gamma_{nk}=\mathsf{E}\{\tilde{\sigma}_{nk}\,|\,\mathfrak{F}_{nk}\}$. Utilizing Taylor's formula we represent γ_{nk} as

$$\gamma_{nk}=\delta_{nk}^{(1)}+\delta_{nk}^{(2)}+\delta_{nk}^{(3)}+\delta_{nk}^{(4)}+\delta_{nk}^{(5)},$$

where

$$\delta_{nk}^{(1)}=\mathsf{E}\{\chi_{nk}\,|\,\mathfrak{F}_{nk}\}-1,$$

$$\delta_{nk}^{(2)}=i\mathsf{E}\{\chi_{nk}(\Delta\xi_{nk},z)\,|\,\mathfrak{F}_{nk}\},$$

$$\delta_{nk}^{(3)}=\left(\frac{|z|^2}{2}\Delta t_{nk}-\tfrac{1}{2}\mathsf{E}\{\chi_{nk}(\Delta\xi_{nk},z)^2\,|\,\mathfrak{F}_{nk}\}\right),$$

$$\delta_{nk}^{(4)}=\frac{|z|^2\,\Delta t_{nk}}{2}\left(\exp\left\{-\frac{|z|^2\,\Delta t_{nk}}{2}\right\}-1\right),$$

$$\delta_{nk}^{(5)}=\tfrac{1}{2}\mathsf{E}\{\chi_{nk}(\Delta\xi_{nk},z)^2(\exp\{i\theta(\Delta\xi_{nk},z)\}-1)\,|\,\mathfrak{F}_{nk}\}.$$

Clearly,

$$|\delta_{nk}^{(1)}|=\rho'_{nk}\,\Delta t_{nk},\qquad |\delta_{nk}^{(2)}|\leq|z|\,|\rho''_{nk}|\,\Delta t_{nk},$$

$$|\delta_{nk}^{(3)}|\leq\tfrac{1}{2}|z|^2|\rho''_{nk}|\,\Delta t_{nk},\qquad |\delta_{nk}^{(4)}|\leq|\Delta t_{nk}|^2|z|^2,$$

$$|\delta_{nk}^{(5)}|\leq\frac{|z|^3}{2}|\boldsymbol{I}+\rho'''_{nk}|\varepsilon\,\Delta t_{nk}.$$

Here $|A|$ denotes the operator norm of matrix A. Observe that

$$|\mathsf{E}\{(1-\chi^{nm_n})\exp\{i(\xi_n(T)-\xi_n(0),z)\}\,|\,\mathfrak{F}_{n0}\}$$

$$\leq\mathsf{P}\{\chi^{nm_n}=0\,|\,\mathfrak{F}_{n0}\}\leq\mathsf{E}\left\{\sum_{k=1}^{m_n}\mathsf{P}(|\Delta\xi_{nk}|\geq\varepsilon\,|\,\mathfrak{F}_{nk})\,|\,\mathfrak{F}_{n0}\right\}$$

$$=\mathsf{E}\left\{\sum_{k=1}^{m_n}\rho'_{nk}\,\Delta t_{nk}\,|\,\mathfrak{F}_{n0}\right\}.$$

Thus for an arbitrary $\varepsilon > 0$ and $\max_k \Delta t_{nk}$ sufficiently small,

$$\sigma_n \leq C(z)\left[\mathsf{E}\left\{\sum_{k=0}^{m_n}(\rho'_{nk}+|\rho''_{nk}|+|\rho'''_{nk}|)\,\Delta t_{nk}\,|\,\mathfrak{F}_{n0}\right\}+\varepsilon T\right]. \tag{14}$$

We arrive at the following result:

Theorem 6. *If a sequence of series* (1) *is such that*

$$\mathsf{E}\left\{\sum_{k=0}^{m_n}(\rho'_{nk}+|\rho''_{nk}|+|\rho'''_{nk}|\,\Delta t_{nk})\right\}\to 0 \tag{15}$$

and $\max_k \Delta t_{nk} \to 0$, *where* ρ'_{nk}, ρ''_{nk}, *and* ρ'''_{nk} *are defined by equations* (11)–(13), *then the conditional distribution of the variable* $\xi_{nm_n} - \xi_{n0}$ *converges to the Gaussian distribution with mean* $\mathbf{0}$ *and covariance matrix* TI.

Remark 1. Modify the conditions of Theorem 6 as follows: set $t_{nm_n} = T_n$ and

$$\mathsf{E}\{\chi_{nk}\,\Delta\xi_{nk}\,\Delta\xi^*_{nk}\,|\,\mathfrak{F}_{nk}\} = (B+\rho'''_{nk})\,\Delta t_{nk},$$

where B is a constant matrix and $T_n \to T$ as $n\to\infty$. If we retain all the other conditions, then the distribution of the difference $\xi_{nm_n} - \xi_{n0}$ converges to the Gaussian distribution with mean $\mathbf{0}$ and covariance matrix TB.

Remark 2. If the assumptions of Theorem 6 are fulfilled and the distribution of the variable ξ_{n0} is weakly convergent to a measure $F(\cdot)$ on $\mathfrak{B}^m$, then the distribution of the variable ξ_{nm_n} is weakly convergent to the distribution with the density

$$\int_{\mathcal{R}^m} \frac{1}{\sqrt{2\pi}\,T}\, e^{-|x-y|^2/2T} F(dy).$$

Theorem 6 can be easily generalized.

Theorem 7. *Let the conditions of Theorem* 6 *be satisfied and let* $t_{nk_j} \to t_j$ *as* $n\to\infty$, $j = 1, 2, \ldots, r$, $0 \leq t_1 < t_2 < \cdots < t_r < T$. *Then the joint distribution of the random vectors*

$$\xi_{nk_1}-\xi_{n0},\ \xi_{nk_2}-\xi_{nk_1},\ \ldots,\ \xi_{nk_r}-\xi_{nk_{r-1}}$$

converges weakly to the joint distribution of the sequence

$$w(t_1)-w(0),\ w(t_2)-w(t_1),\ \ldots,\ w(t_r)-w(t_{r-1}),$$

where $w(t)$ *is an m-dimensional Wiener process.*

Proof. To prove this assertion consider the difference

$$\sigma_n = \mathsf{E}\Big\{\exp\Big\{i\sum_{j=0}^{r-1}(\xi_n(t_{nk_{j+1}})-\xi_n(t_{nk}), z_j)\Big\}$$
$$-\exp\Big\{-\frac{1}{2}\sum_{j=0}^{r-1}|z_j|^2(t_{j+1}-t_j)\Big\}\Big|\mathfrak{F}_{n0}\Big\},$$

where z_j, $j=0,\ldots,r$ are arbitrary vectors in $\mathscr{R}^m$, and express it as

$$\sigma_n = \sum_{k=0}^{r-1}\mathsf{E}\Big\{\exp\Big\{i\sum_{j=0}^{k}(\xi_n(t_{nk_{j+1}})-\xi_n(t_{nk_j}), z_j)-\frac{1}{2}\sum_{j=k+1}^{r-1}|z_j|^2(t_{j+1}-t_j)\Big\}$$
$$-\exp\Big\{i\sum_{j=0}^{k-1}(\xi_n(t_{nk_{j+1}})-\xi_n(t_{nk_j}), z)-\frac{1}{2}\sum_{j=k}^{r-1}|z_j|^2(t_{j+1}-t_j)\Big\}\Big|\mathfrak{F}_{n0}\Big\}$$
$$=\sum_{k=0}^{r-1}\mathsf{E}\{\mathsf{E}(\sigma_{nk}\,|\,\mathfrak{F}_{nk})\,|\,\mathfrak{F}_{n0}\}.$$

The quantity $\mathsf{E}(\sigma_{nk}\,|\,\mathfrak{F}_{nk})$ appearing in the last equality can be bounded as above using inequality (14) (with obvious modifications). □

Corollary. *If the sequence* $\{\xi_{nk}, \mathfrak{F}_{nk}, k=1,\ldots,m_n\}$ *is a martingale possessing finite moments of the second order and if, moreover,*

$$\mathsf{E}\{\Delta\xi_{nk}\,\Delta\xi_{nk}^*\,|\,\mathfrak{F}_{nk}\} = \boldsymbol{I}\Delta t_{nk} \tag{16}$$

and

$$\mathsf{E}\sum_{k=1}^{m_n-1}(1-\chi_{nk})|\,\Delta\xi_{nk}|^2 \to 0 \quad \textit{as } n\to\infty, \tag{17}$$

then the sequence of measures $q(\cdot)$ *in* $\mathscr{D}$ *associated with the processes* $\xi_n(t)-\xi_n(0)$ *converges weakly to a Wiener measure.*

Condition (17) is the classical Lindeberg condition in the central limit theorem for a sum of independent random variables. It is easy to verify using Chebyshev's inequality that (17) implies (15).

On the other hand, in the case under consideration one can apply the corollary to Theorem 3 which implies that a family of measures associated with processes $\xi_n(t)$ constructed from the sequence of series $\{\xi_{nk}, k=1,\ldots,m_k\}$, $n=1,2,\ldots$, is weakly compact.

It is of interest to generalize the preceding theorem by allowing the times t_{nk} to be chosen in a random manner. In this case, one should firstly require that the choice of variables $\Delta t_{nk}=t_{nk+1}-t_{nk}$ could not anticipate "the future," i.e., that the variables Δt_{nk} be $\mathfrak{F}_{nk}$-measurable. Under certain additional assumptions to be presented immediately below, the calculations and bounds utilized in the proof of Theorem 6 will be only slightly altered.

In this manner we obtain the following assertion.

Theorem 8. *Assume that*

a) *times* Δt_{nk} *are* $\mathfrak{F}_{nk}$*-measurable random variables,* $k=1,2,\ldots,m_n-1$,

b) $\mathsf{E}|t_{nm_n}-T|\to 0$ *as* $n\to\infty$, *where* T *is not a random quantity,*

c) *conditions* (16) *and* (17) *are satisfied.*

Then the distribution of the vector $\xi_{nm_n}-\xi_{n0}$ *converges weakly to the distribution of* $w(T)$ *where* $w(t)$ *is an m-dimensional Wiener process. If, moreover,* $t_{nm_n}=T$, *i.e.,* t_{nm_n} *is not a random quantity, then the joint distribution of the variables*

$$\xi_n(t_{nk_1})-\xi_n(t_{n0}),\ldots,\xi_n(t_{nk_r})-\xi_n(t_{nk_{r-1}})$$

is weakly convergent to the distribution of

$$w(t_j)-w(0),\ldots,w(t_r)-w(t_{r-1})$$

as $t_{nk_j}\to t_j$, $j=1,2,\ldots,r$, *in probability (where* t_j *is not a random quantity). Moreover, the measures in* $\mathscr{D}$ *associated with processes* $\xi_n(t)$ *converge weakly to a Wiener measure.*

Proof. To prove this assertion we return to Theorem 6 and examine the modifications which are needed in its proof in order that it will be applicable for the case under consideration. Since now $t_{nm_n}\neq T$ in general, the expression for σ_N will contain an additional summand of the form

$$\mathsf{E}\{\chi^{nm_n}\exp\{i(\xi_n(t_{nm_n})-\xi_n(0),z)\}\Big(\exp\Big\{-\frac{|z|^2}{2}(T-t_{nm_n})\Big\}-1\Big)\Big|\mathfrak{F}_{n0}\Big\},$$

which tends to zero in probability.

A bound on the sum $\sum_{k=1}^{m}\delta_{nk}^{(4)}$ will also require a different approach. In the present case it can be bounded by means of the inequality

$$\mathsf{E}\Big\{\sum_{k=1}^{m_n}\delta_{nk}^{(4)}\Big|\mathfrak{F}_{n0}\Big\}\leq\frac{|z|^4}{4}\varepsilon\mathsf{E}\{t_{nm_n}|\mathfrak{F}_{n0}\}+|z|^2\mathsf{E}\Big\{\sum_{k=1}^{m_n}\Delta t_{nk}(1-\chi_\varepsilon(\Delta t_{nk}))\Big|\mathfrak{F}_{n0}\Big\}.$$

Since the function $|t|(1-\chi_\varepsilon(|t|))$ is convex downward and $\Delta t_{nk}=(1/m)\mathsf{E}\{|\Delta\xi_{nk}|^2|\mathfrak{F}_{nk}\}$, it follows that

$$\Delta t_{nk}(1-\chi_\varepsilon(\Delta t_{nk}))\leq\mathsf{E}\frac{|\Delta\xi_{nk}|^2}{m}\Big(1-\chi_\varepsilon\Big(\frac{|\Delta\xi_{nk}|^2}{m}\Big)\Big).$$

Therefore condition (17) implies that for any $\varepsilon>0$

$$\lim_{n\to\infty}\mathsf{E}\sum_{k=1}^{m_n}\Delta t_{nk}(1-\chi_\varepsilon(\Delta t_{nk}))=0.$$

Taking into account condition a) of the theorem it is not difficult to verify that all the other transformations and inequalities used in the proof of Theorem 6 are applicable in the case under consideration. □

Let $\{\xi_n, \mathfrak{F}_n, n=0, 1, \ldots\}$ be a martingale with finite second-order moments. Set

$$\gamma_n^2 = \mathsf{E}\{(\xi_n - \xi_{n-1})^2 \mid \mathfrak{F}_{n-1}\}$$

and assume that there exists an $\mathfrak{F}_0$-measurable function $\varphi(n)$ such that

$$\frac{1}{\varphi(n)} \sum_{k=1}^{n} \gamma_k^2 \to 1 \tag{18}$$

in the sense of convergence in L_1.

Set

$$\xi_{nk} = \frac{1}{\sqrt{\varphi(n)}} \xi_k, \qquad k = 0, 1, \ldots, n, \qquad \Delta t_{nk} = \frac{\gamma_{k+1}^2}{\varphi(n)}.$$

Then the variables $t_{nk} = \Delta t_{n0} + \cdots + \Delta t_{nk-1}$ are $\mathfrak{F}_{k-1}$-measurable and

$$\mathsf{E}\{\xi_{nk+1} - \xi_{nk} \mid \mathfrak{F}_k\} = 0,$$

$$\mathsf{E}\{(\xi_{nk+1} - \xi_k)^2 \mid \mathfrak{F}_k\} = \frac{\gamma_{k+1}^2}{\varphi(n)} = \Delta t_{nk}.$$

Theorem 8 is thus applicable. Hence we arrive at the following assertion.

Theorem 9. *If the martingale $\{\xi_n, \mathfrak{F}_n, n = 1, 2, \ldots\}$ satisfies condition* (18) *(in the sense of convergence in L_1) and if for any $\varepsilon > 0$*

$$\frac{1}{\varphi(n)} \sum_{k=1}^{n} \int_{\{(\xi_k - \xi_{k-1})^2 > \varepsilon^2 \varphi(n)\}} (\xi_k - \xi_{k-1})^2 \, d\mathsf{P} \to 0$$

with probability 1 *as $n \to \infty$, then the conditional distribution of the variable $(1/\sqrt{\varphi(n)})(\xi_n - \xi_0)$ is asymptotically normal* (0, 1).

Remark. The random process $\xi_n(t)$ constructed for the sequence $\xi_{n0}, \ldots, \xi_{nk}$ considered in Theorem 9 cuts off at the random time t_{nn}. Since $\varphi(n) \to \infty$ with probability 1 we can extend the construction of the process $\xi_n(t)$ by means of variables ξ_{nk} for $k > n$ in order that it be defined on a fixed time interval, e.g., [0, 1]. It then follows from Theorem 8 that measures in $\mathscr{D}[0, 1]$ associated with processes $\xi_n(t)$ converge weakly to the Wiener measure.

Conditions for convergence to an arbitrary process with independent increments. Recall first that if ζ_h is a family of random vectors, $\zeta_h \to 0$ as $h \to 0$ and if the limit of

$(1/\Delta(h))\mathsf{E}(e^{i(\zeta_h, z)}-1)$ exists then it is of the form (cf. Volume I, Chapter III)

$$\lim_{h\to 0}\frac{1}{\Delta(h)}\mathsf{E}(e^{i(\zeta_h, z)}-1)$$

$$=i(a, z)-\tfrac{1}{2}(bz, z)+\int_{\mathscr{R}^m}\left(e^{i(z,u)}-1-\frac{i(u, z)}{1+|z|^2}\right)\frac{1+|u|^2}{|u|^2}\Pi(du),$$

where $\Pi(\cdot)$ is a finite measure continuous at point 0. The parameter $\Delta(h)$ can be interpreted as a natural local time corresponding to the random vector ζ_h.

In this connection we shall assume that one can correspond a positive nonrandom quantity Δt_{nk} to each vector $\Delta\xi_{nk}=\xi_{nk+1}-\xi_{nk}$ such that

$$\frac{1}{\Delta t_{nk}}\mathsf{E}\{e^{i(\Delta\xi_{nk}, z)}-1\,|\,\mathfrak{F}_{nk}\}=L(t_{nk}, z)+\rho_{nk},$$

where

$$L(t, z)=i(a(t), z)-\tfrac{1}{2}(b(t)z, z)+\int_{\mathscr{R}^m}\left(e^{i(z,u)}-1-\frac{i(u, z)}{1+|z|^2}\right)\frac{1+|u|^2}{|u|^2}\Pi(t, du),$$

$t_{nk}=\Delta t_{n0}+\cdots+\Delta t_{nk}$, $a(t)$, $b(t)$, and $\Pi(t, A)$ are nonrandom quantities, $a(t)$ is a vector function, and $b(t)$ is a nonnegatively definite matrix; also, $\Pi(t, A)$ is a finite measure on $\mathfrak{B}^m$ with $\Pi(t, \{0\})=0$.

Furthermore, we assume that $t_{nm_n}=T$, $\max_k \Delta t_{nk}\to 0$ as $n\to\infty$, and the function $L(t, z)$ is Riemann integrable on the segment $[0, T]$.

Theorem 10. *If the preceding assumptions are fulfilled and if*

$$\mathsf{E}\left(\sum_{k=0}^{m_n-1}|\rho_{nk}|\,\Delta t_{nk}\right)\to 0\quad \text{as } n\to\infty, \tag{19}$$

then the distribution of the vector $\xi_{nm_n}-\xi_{n0}$ *is weakly convergent (as* $n\to\infty$*) to the distribution with the characteristic function*

$$J(z)=\exp\left\{\int_0^T L(t, z)\,dt\right\}.$$

Proof. The proof of this theorem is analogous to the proof of Theorem 6. We introduce the quantity

$$\sigma_n=\mathsf{E}\{\exp\{i(\xi_{nm_n}-\xi_{n0}, z)\}-\exp\{\textstyle\int_0^T L(t, z)\,dt\}\,|\,\mathfrak{F}_{n0}\}$$

and represent it in the form

$$\sigma_n=\mathsf{E}\left\{\sum_{k=1}^{m_n-1}\exp\{i(\xi_{nk}-\xi_{n0}, z)\}\exp\{\textstyle\int_{t_{nk+1}}^T L(t, z)\,dt\}\mathsf{E}\{\tilde\sigma_{nk}\,|\,\mathfrak{F}_{nk}\}\,\Big|\,\mathfrak{F}_{n0}\right\},$$

where

$$\tilde{\sigma}_{nk} = \exp\{i(\Delta\xi_{nk}, z)\} - \exp\{\int_{t_{nk}}^{t_{nk+1}} L(t, z)\, dt\}.$$

Observe that

$$|\mathsf{E}\{\tilde{\sigma}_{nk} \mid \mathfrak{F}_{nk}\}| \leq |\rho_{nk}|\, \Delta t_{nk} + |\int_{t_{nk}}^{t_{nk+1}} [L(t, z) - L(t_{nk}, z)]\, dt$$
$$+ \int_{t_{nk}}^{t_{nk+1}} L(t, z)\, dt\, (\exp\{\theta \int_{t_{nk}}^{t_{nk+1}} L(t, z)\, dt\} - 1)|.$$

Taking into account the fact that $\exp\{\int_a^T L(t, z)\, dt\}$ is a characteristic function of a distribution we obtain the following bound for σ_n:

$$\sigma_n \leq \mathsf{E} \sum_{k=0}^{m_n-1} |\rho_{nk}|\, \Delta t_{nk} + \sum_{k=0}^{m_n-1} (\delta_{nk}\, \Delta t_{nk} + C(z)\, \Delta t_{nk}^2);$$

here δ_{nk} is the oscillation of the function $L(t, z)$ on the interval $[t_{nk}, t_{nk+1}]$, $C(z)$ is a constant dependent on T and $\sup_t |L(t, z)|$ only. The inequality obtained proves the theorem. □

In the same manner as in the case of convergence to a Wiener process one can easily deduce the following result from the theorem just proved.

Theorem 11. *If the conditions of Theorem* 10 *are satisfied and if* $t_{nk_j} \to t_j$ $(j = 1, \ldots, r)$, *then the joint distribution of the differences*

$$\xi_{nk_1} - \xi_{n0},\ \xi_{nk_2} - \xi_{nk_1}, \ldots, \xi_{nk_r} - \xi_{nk_{r-1}}$$

converges weakly to the joint distribution of the vectors

$$\xi(t_1) - \xi(0),\ \xi(t_2) - \xi(t_1), \ldots, \xi(t_r) - \xi(t_{r-1}),$$

where $\xi(t)$ *is an m-dimensional process with independent increments such that the distribution of the random variable* $\xi(s+h) - \xi(s)$ *possesses the characteristic function*

$$J(s, s+h, z) = \exp\{\int_s^{s+h} L(t, z)\, dt\}.$$

Limit theorems for sequences of series of random vectors with finite moments of the second order. We now investigate the conditions for convergence of a sequence of series (1) to processes more general than the processes with independent increments. In accordance with the constructions above we shall assume that a sequence of nonrandom times $0 < t_{n0} < t_{n1} < \cdots < t_{nm_n} = T$ corresponds to the sequence $\xi_{n0}, \xi_{n1}, \ldots, \xi_{nm_n}$ and we shall introduce the following representation for the variables

$$\Delta\xi_{nk} = a_{nk}(\xi_{n0}, \xi_{n1}, \ldots, \xi_{nk})\, \Delta t_{nk} + b_{nk}(\xi_{n0}, \xi_{n1}, \ldots, \xi_{nk})\, \Delta\psi_{nk},$$

where $\{\psi_{nk}, k=0, 1, \ldots, m_n\}$ is a martingale and

$$\mathsf{E}\{\Delta\psi_{nk}\,\Delta\psi_{nk}^*\,|\,\mathfrak{F}_{nk}\}=\boldsymbol{I}\,\Delta t_{nk}.$$

Define on $[0, T]\times\mathscr{D}[0, T]$ functions

$$a_n(t, x(\cdot)), \qquad b_n(t, x(\cdot)), \qquad t\in[0, T], \quad x(\cdot)\in\mathscr{D}[0, T],$$

by setting

$$a_n(t, x(\cdot))=a_{nk}(x(0), x(t_{n1}), \ldots, x(t_{nk})) \quad \text{for } t\in[t_{nk}, t_{nk+1}),$$
$$k=0, 1, \ldots, m_n-1,$$
$$a_n(T, x(\cdot))=a_{nm_n-1}(x(0), x(t_{n1}), \ldots, x(t_{nm_n-1})).$$

The quantities $b_n(t, x(\cdot))$ are defined analogously. It follows from the definition that if $x(t)=y(t)$ for $t\in[0, s]$, then $a_n(t, x(\cdot))=a_n(t, y(\cdot))$ and $b_n(t, x(\cdot))=b_n(t, y(\cdot))$ for all $t\in[0, s]$. Our basic assumption is now as follows: the functions $a_n(t, x(\cdot))$ and $b_n(t, x(\cdot))$ converge in $[0, T]\times\mathscr{D}[0, T]$ as $n\to\infty$ to the functions $a(t, x(\cdot))$ and $b(t, x(\cdot))$, respectively. More precisely, we shall assume that the following condition is satisfied:

(20)
$$\lim_{n\to\infty}\sup_{\substack{t\in[0, T]\\ x(\cdot)\in\mathscr{D}[0, T]}}\{[1+\|x(\cdot)\|]^{-1}[|a_n(t, x(\cdot))-a(t, x(\cdot))| + |b_n(t, x(\cdot))-b(t, x(\cdot))|]\}=0,$$

where $\|x(\cdot)\|=\sup_{0\le t\le T}|x(t)|$.

In accordance with our general aim we would like now to establish that the process $\xi_n(t)$ constructed from the sequence of series of random vectors (1) converges to the process $\xi(t)$ which is a solution of the stochastic differential equation

(21)
$$d\xi(t)=a(t, \xi(\cdot))+b(t, \xi(\cdot))d\psi(t),$$

where $\psi(t)$ is the limit process for $\psi_n(t)$ constructed from the martingale ψ_{nk}, $k=1, \ldots, m_n$. To achieve this, several bounds are required.

Together with the system of random vectors (1) consider the sequence of series $\{\eta_{nk}, k=0, 1, \ldots, m_n\}$, $n=1, 2, \ldots$, defined by the recurrent sequence of relations

(22)
$$\eta_{n0}=\xi_{n0},$$
$$\Delta\eta_{nk}=\eta_{nk+1}-\eta_{nk}=a(t_{nk}, \eta_n(\cdot))\,\Delta t_{nk}+b(t_{nk}, \eta_n(\cdot))\,\Delta\psi_{nk},$$

where $\eta_n(t)=\eta_{nk}$ for $t\in[t_{nk}, t_{nk+1})$, $k=0, 1, \ldots, m_n$. Such a definition makes sense since to evaluate the values $a(t_{nk}, \eta_n(\cdot))$ and $b(t_{nk}, \eta_n(\cdot))$ only the values of $\eta_{n0}, \eta_{n1}, \dots, \eta_{nk}$ are needed.

Lemma 2. *Assume that conditions* (5) *and* (20) *are satisfied and, moreover, let*

(23) $$|a(t, x(\cdot))-a(t, y(\cdot))|+|b(t, x(\cdot))-b(t, y(\cdot))|\leq C\|x(\cdot)-y(\cdot)\|.$$

Then

$$\mathsf{E}\{\sup_{0\leq k\leq r} |\eta_{nk}-\xi_{nk}|^2|\mathfrak{F}_{n0}\}\leq \varepsilon_n(1+|\xi_{n0}|^2)t_{nr},$$

where ε_n is a nonrandom quantity which tends to 0 as $n\to\infty$.

Proof. We represent the difference $\eta_{nk+1}-\xi_{nk+1}$ in the form

$$\begin{aligned}\eta_{nk+1}-\xi_{nk+1} &= \sum_{j=0}^{k} [a(t_{nj}, \eta_n(\cdot))-a(t_{nj}, \xi_n(\cdot))]\,\Delta t_{nj}\\ &\quad+\sum_{j=0}^{k} [b(t_{nj}, \eta_n(\cdot))-b(t_{nj}, \xi_n(\cdot))]\,\Delta\psi_{nj}\\ &\quad+\sum_{j=0}^{k} [a(t_{nj}, \xi_n(\cdot))-a_n(t_{nj}, \xi_n(\cdot))]\,\Delta t_{nj}\\ &\quad+\sum_{j=0}^{k} [b(t_{nj}, \xi_n(\cdot))-b_n(t_{nj}, \xi_n(\cdot))]\,\Delta\psi_{nj}\\ &=\textstyle\sum_k'+\sum_k''+\sum_k'''+\sum_k^{\mathrm{IV}}.\end{aligned}$$

Set

$$v_{nk}=\mathsf{E}\{\sup_{0\leq j\leq k} |\eta_{nj}-\xi_{nj}|^2|\mathfrak{F}_{n0}\}.$$

We now bound the sums $\sum_k', \dots, \sum_k^{\mathrm{IV}}$ using methods analogous to those applied in the proof of Lemma 1. For instance, utilizing the fact that $\sum_k''$ is a martingale we obtain

$$\begin{aligned}\mathsf{E}\{\sup_{j\leq k} |\textstyle\sum_j''|^2|\mathfrak{F}_{no}\}&\leq 4\mathsf{E}\{|\textstyle\sum_k''|^2|\mathfrak{F}_{n0}\}\\ &\leq 4\sum_{j=0}^{k} \mathsf{E}\{|b(t_{nj}, \eta_n(\cdot))-b(t_{nj}, \xi_n(\cdot))|^2\,\Delta t_{nj}\,|\mathfrak{F}_{n0}\}.\end{aligned}$$

Applying inequality (23) we observe that the quantity to be bound does not exceed

$$4C\sum_{j=0}^{k} v_{nj}\,\Delta t_{nj}.$$

Using (20) one easily obtains the inequality

$$\mathsf{E}\{\sup_{j\le k} |\Sigma_j^{\mathrm{IV}}|^2 \,|\, \mathfrak{F}_{n0}\} \le \varepsilon_n \sum_{j=0}^{k} \mathsf{E}\{(1+\sup_{r\le j} |\xi_{nr}|^2)\,\Delta t_{nj} \,|\, \mathfrak{F}_{n0}\},$$

where $\varepsilon_n \to 0$ as $n \to \infty$. The quantities $\sup_{j\le k} |\Sigma'_j|^2$ and $\sup_{j<k} |\Sigma'''_j|^2$ are estimated analogously.

Utilizing Lemma 1 we obtain the relation

$$v_{nk+1} \le C' \sum_{j=0}^{k} v_{nj}\,\Delta t_{nj} + \varepsilon_n t_{nk+1}(1+|\xi_{n0}|^2),$$

where C' is a constant which depends on C and T only. This implies that

$$v_{nk+1} \le \varepsilon_n (1+|\xi_{n0}|^2)(e^{C't}-1).$$

The lemma is proved □

It follows from Lemma 2 that the marginal distributions of processes $\xi_n(t)$ and $\eta_n(t)$ can weakly converge only simultaneously and that the corresponding limits coincide. Now it would be more convenient to study the limiting behavior of processes $\eta_n(t)$.

Let η'_{nk} and η''_{nk}, $k=0, 1, \ldots, m_n$, be sequences constructed from formulas (22) under the distinct initial conditions $\eta'_{n0} = \xi'$ and $\eta''_{n0} = \xi''$. Analogously to Lemma 2 the following lemma can be proven:

Lemma 3. *If the conditions of Lemma 2 are satisfied, then*

$$\mathsf{E}\{\sup_{0\le j\le k} |\eta'_{nj} - \eta''_{nj}|^2 \,|\, \mathfrak{F}_{n0}\} \le e^{c' t_{nk}} |\xi' - \xi''|^2,$$

where c' is a constant.

Above we introduced finite-difference approximations for stochastic differential equations and showed that these converge to solutions of stochastic differential equations (Section 1, Theorems 12 and 13). We shall now verify analogous assertions for the processes $\eta_n(t)$. The role of finite-difference approximations for processes $\eta_n(t)$ is taken here by processes $\zeta_n(t)$ defined as follows. Choose some values $t_{nk_1}, t_{nk_2}, \ldots, t_{nk_r}$ where r is a fixed integer. For brevity of notation set $t_{nk_j} = s_j$, $j = 1, 2, \ldots, r$, $s=0$, $s_{r+1} = T$, and for $t \in (s_j, s_{j+1}]$, $j = 0, 1, \ldots, r-1$, let

$$\zeta_n(0) = \xi_{n0},$$

$$\zeta_n(t) = \zeta_n(s_j) + a(s_j, \zeta_n(\cdot))(t-s_j) + b(s_j, \zeta_n(\cdot))[\psi_n(t) - \psi_n(s_j)].$$

We now bound the quantity

$$v_n(t) = \mathsf{E}\{\sup_{0\le s = t_{nj} < t} |\eta_n(s) - \zeta_n(s)|^2\}.$$

We have

$$(24) \qquad v_n^{\bullet}(T) \leqslant 2\mathsf{E}\sup|\Sigma_t'|^2 + 2\mathsf{E}\sup|\Sigma_t''|^2,$$

where

$$\Sigma_t' = \sum_{k=0}^{j-1} [a(s_{nk}, \zeta_n(\cdot)) - a(t_{nk}, \eta_n(\cdot))]\, \Delta t_{nk} + \sigma_n'(t),$$

$$\Sigma_t'' = \sum_{k=0}^{j-1} [b(s_{nk}, \zeta_n(\cdot)) - b(t_{nk}, \eta_n(\cdot))]\, \Delta\psi(t_{nk}) + \sigma_n''(t),$$

and

$$\sigma_n'(t) = a(s_{nj}, \zeta_n(\cdot))(t - t_{nj}),$$
$$\sigma_n''(t) = b(s_{nj}, \zeta_n(\cdot))[\psi_n(t) - \psi_n(t_{nj})],$$

for $t \in [t_{nj}, t_{nj+1})$ and $s_{nk} = s_i$ if $t_{nk} \in [s_i, s_{i+1})$.

Observe that if $t_{nk} \in [s_i, s_{i+1})$, then

$$(25) \qquad \begin{aligned} &|a(s_{nk}, \zeta_n(\cdot)) - a(t_{nk}, \eta_n(\cdot))| \\ &\leqslant |a(s_i, \zeta_n(\cdot)) - a(s_i, \eta_n(\cdot))| + |a(s_i, \eta_n(\cdot)) - a(t_{nk}, \eta_n(\cdot))| \\ &\leqslant C \sup_{s \leqslant s_i} |\eta_n(s) - \zeta_n(s)| + \rho(t_{nk} - s_i)(\sup_{s \leqslant t_{nk}} |\eta_n(s)| + 1). \end{aligned}$$

Here we introduce the following condition: for $t > s$

$$(26) \qquad |a(s, x(\cdot)) - a(t, x(\cdot))| \leqslant \rho(t-s)(1 + \sup_{0 \leqslant t' \leqslant t} |x(t')|),$$

where $\rho(t)$, $t > 0$, is a nonnegative monotonically nondecreasing function and $\rho(0+) = 0$.

Assume that the same inequality is valid for the matrix-valued function $b(t, x(\cdot))$:

$$(27) \qquad |b(s, x(\cdot)) - b(t, x(\cdot))| \leqslant \rho(t-s)(1 + \sup_{0 \leqslant t' \leqslant t} |x(t')|).$$

Then an inequality analogous to (25) will hold also for the differences $|b(s_{nk}, \zeta_n(\cdot)) - b(t_{nk}, \eta_n(\cdot))|$.

Also set $w_n(t) = \mathsf{E}\sup_{0 \leqslant s \leqslant t} |\eta_n(s)|^2$. It is easy to verify that

$$\mathsf{E}\sup_t |\Sigma_t'|^2 \leqslant 2T \sum_k C^2 \mathsf{E} \sup_{s < s_i} |\eta_n(s) - \zeta_n(s)|^2\, \Delta t_{nk}$$
$$+ 2T \sum_k \rho(t_{nk} - s_i)(1 + w_n(t_{nk}))\, \Delta t_{nk}$$

or

$$\mathsf{E}\sup_t |\Sigma_t'|^2 \leq 2TC^2 \sum_{i=0}^{r} v_n(s_i)(s_{i+1}-s_i)+2T^2(1+w_n(T))\rho(|\delta|),$$

where $|\delta| = \max_{0\leq i<r}(s_{i+1}-s_i)$.

We now bound the second summand in the right-hand side of inequality (24). For this purpose observe that the sum Σ_t'' as a function of t is a square integrable martingale. Therefore

$$\mathsf{E}\sup_t |\Sigma_t''|^2 \leq 4\mathsf{E}|\Sigma_T''|^2 = 4\mathsf{E}\sum_k |b(s_{nk}, \zeta_n(\cdot))-b(t_{nk}, \eta_n(\cdot))|^2\,\Delta t_{nk}.$$

From this relation we obtain analogously to the above that

$$\mathsf{E}\sup_t |\Sigma_t''|^2 \leq 8C^2 \sum_{i=0}^{r} v_n(s_i)(s_{i+1}-s_i)+8T\rho(|\delta|)(1+w_n(T)).$$

Thus

(28) $$v_n(T)\leq 2C^2(T+4)\sum_{i=0}^{r} v_n(s_i)(s_{i+1}-s_i)+\rho(|\delta|)C'(w_n(T)+1).$$

The bound on $w_n(T)$ can be deduced from Lemma 1:

$$w_n(T)\leq C_1(1+\mathsf{E}|\xi_{n0}|^2).$$

Observe that the function $v_n(t)$ is monotonically nondecreasing.

Let $\bar{v}_n(t) = v_n(s_i)$ for $t\in[s_i, s_{i+1})$. One may replace the function $v_n(s)$ in inequality (28) by $\bar{v}_n(s)$ and T by any $t\in[0, T]$. This yields the following integral inequality,

$$\bar{v}_n(t)\leq C_1\int_0^t \bar{v}_n(s)\,ds + C_2\rho(|\delta|),$$

which implies that

$$\bar{v}_n(t)\leq C_2\rho(|\delta|)\,e^{C_1T}\leq C_3\rho(|\delta|).$$

Here C_3 is a constant of the form $C_3 = C'(1+\mathsf{E}|\xi_{n0}|^2)$ and C' depends on C and T only. We have thus proved the following lemma:

Lemma 4. *If the conditions of Lemma 2 and inequalities* (26) *and* (27) *are satisfied, then*

$$\mathsf{E}\{\sup_{0\leq t\leq T}|\eta_n(t)-\zeta_n(t)|^2\,|\,\mathfrak{F}_{n0}\}\leq C'\rho(|\delta|)(1+\mathsf{E}|\xi_{n0}|^2).$$

Up until now no assumptions have been imposed concerning the convergence of processes $\psi_n(t)$. Recall that processes $\psi_n(t)$, $t \in [0, T]$, are $\mathfrak{F}_{nt}$-martingales, where $\mathfrak{F}_{nt} = \mathfrak{F}_{nk}$ for $t \in [t_{nk}, t_{nk+1})$ and, moreover,

$$\mathsf{E}\{\Delta\psi_{nk}\,\Delta\psi_{nk}^* \mid \mathfrak{F}_{nk}\} = \boldsymbol{I}\,\Delta t_{nk}.$$

It is natural to suppose that the limiting process $\psi(t)$ is also a square integrable martingale with respect to a current of σ-algebras $\{\mathfrak{F}_t, t \in [0, T]\}$ and that

$$\mathsf{E}\{\Delta\psi\,\Delta\psi^* \mid \mathfrak{F}_s\} = \boldsymbol{I}(t-s), \tag{29}$$

where $\Delta\psi = \psi(t) - \psi(s)$, $s < t$. The conditions for convergence to martingales with independent increments were discussed above.

We thus assume that the following condition is fulfilled:

Ψ_1: Measures $Q_n(\cdot, \cdot)$ in the space $\mathcal{R}^m \times \mathcal{D}$ generated by random vectors ξ_{n0} and random processes $\psi_n(t)$, $t \in [0, T]$, converge weakly to the measure $Q(\cdot, \cdot)$ associated with the random vector ξ_0 and a square integrable martingale $\psi(t)$.

We now proceed to the proof of the weak convergence of marginal distributions of the process $\xi_n(t)$ to the marginal distributions of a solution of stochastic equation (21). Note that if the conditions of Lemma 2 are satisfied, equation (21) has then a unique solution with finite second-order moments (Theorem 3 in Section 1).

Let $\xi_\delta(t)$ denote a finite-difference approximation of the solution of equation (21) constructed from the subdivision

$$\delta = \{0 = s_0, s_1, \ldots, s_r, s_{r+1} = I\} \quad \text{of the interval } [0, T].$$

Here we shall modify slightly the definition of the process $\xi_\delta(t)$. Namely, for $t \in (s_i, s_{i+1}]$ we set

$$\xi_\delta(t) = \xi_\delta(s_i) + a(s_i, \xi_\delta(\cdot))(t - s_i) + b(s_i, \xi_\delta(\cdot))(\psi(t) - \psi(s_i)),$$

and $\xi_\delta(0) = \xi_0$. It is easy to verify that if inequalities (26) and (27) are fulfilled, then the assertion of Theorem 12 in Section 1 remains valid under this modification as well, so that

$$\mathsf{E}\{\sup_{0 \le t \le T} |\xi_\delta(t) - \xi(t)|^2\} \to 0.$$

Using induction it is easy to show that $\xi_\delta(t)$ is a continuous function in the argument ξ_0 for each t and is a continuous functional on $\psi(s)$, $0 \le s \le t$, $\xi_\delta(t) = g_{(t,\delta)}(\xi_0, \psi(\cdot))$. Also, $\zeta_n(t)$ is expressible in terms of ξ_{n0} and $\psi_n(s)$ in the same manner: $\zeta_n(t) = g_{(t,\delta)}(\xi_{n0}, \psi_n(\cdot))$.

Now choose an arbitrary sequence $\{t_k, k = 1, \ldots, p\}$, $t_k \in [0, T]$, and an arbitrary continuous bounded function $f(x_0, \ldots, x_p)$, $x_k \in \mathcal{R}^m$, and bound the difference

$$r_n = \mathsf{E}f(\xi_0, \xi(t_1), \ldots, \xi(t_p)) - \mathsf{E}f(\xi_{n0}, \eta_n(t_1), \ldots, \eta_n(t_p)).$$

We have

$$\begin{aligned}|r_n| &\leq \mathsf{E}|f(\xi_0, \xi(t_1), \ldots, \xi(t_p)) - f(\xi_0, \xi_\delta(t_1), \ldots, \xi_\delta(t_p))| \\ &+ |\mathsf{E}f(\xi_0, \xi_\delta(t_1), \ldots, \xi_\delta(t_p)) - \mathsf{E}f(\xi_{n0}, \zeta_n(t_1), \ldots, \zeta_n(t_p))| \\ &+ \mathsf{E}|f(\xi_{n0}, \zeta_n(t_1), \ldots, \zeta_n(t_p)) - f(\xi_{n0}, \eta_n(t_1), \ldots, \eta_n(t_p))| \\ &= r' + r_n'' + r_n'''.\end{aligned}$$

Given an arbitrary $\varepsilon > 0$ we first choose a δ such that $r' < \varepsilon/3$ and $r_n''' < \varepsilon/3$ for sufficiently large n. We need now only note that in view of the preceding remarks

$$f(\xi_0, \xi_\delta(t_1), \ldots, \xi_\delta(t_p)) = F_{(t,\delta)}(\xi_0, \psi(\cdot)), \qquad f(\xi_{n0}, \zeta_n(t_1), \ldots, \zeta_n(t_p)) = F_{(t,\delta)}(\xi_{n0}, \psi_n(\cdot)),$$

where $F_{(t,\delta)}(x, \psi(\cdot))$ is a bounded continuous function in x and ψ. Thus if the conditions Ψ_1 are satisfied $r_n'' \to 0$ as $n \to \infty$. Hence we have proved the following theorem:

Theorem 12. *Assume that a sequence of series of random vectors* (1) *satisfies conditions* (5), (20), *and* Ψ_1 *and the functionals* $a(t, x(\cdot)), b(t, x(\cdot))$ *satisfy conditions* (23), (26), *and* (27). *Then the marginal distributions of the processes* $\xi_n(t)$ *constructed from the sequence* (1) *converge weakly to the corresponding marginal distributions of a solution of the stochastic equation* (21).

An important particular case of the model just discussed is one in which the coefficients $a(t, x(\cdot))$ and $b(t, x(\cdot))$ in equation (21) are independent of the "past", i.e.,

$$a(t, x(\cdot)) = a(t, x(t)), \qquad b(t, x(\cdot)) = b(t, x(t)). \tag{30}$$

Equation (21) then becomes a stochastic differential equation without delay,

$$d\xi = a(t, \xi(t))\,dt + b(t, \xi(t))\,d\psi(t), \qquad \xi(0) = \xi_0, \tag{31}$$

and the functions $a(t, x)$ and $b(t, x)$, where $(t, x) \in [0, T] \times \mathscr{R}^m$, satisfy the conditions

$$|a(t, x)| + |b(t, x)| \leq C(1 + |x|), \tag{32}$$

$$|a(t, x) - a(t, y)| + |b(t, x) - b(t, y)| \leq C|x - y|, \tag{33}$$

$$|a(s, x) - a(t, x)| + |b(s, x) - b(t, x)| \leq \rho(t - s)(1 + |x|), \tag{34}$$

where $\rho(t)$ is as defined above.

If, moreover, the martingale $\psi(t)$ turns out to be a process with independent increments, then equation (31) becomes an equation without an after-effect and

$$A(x, t)=\int_0^t a(t, x)\,dt+\int_0^t b(t, x)\,d\psi(t)\in \bar{S}(C, C).$$

Theorem 12 can be somewhat strengthened in the case under consideration by noting that $\xi_\delta(t)$ is a continuous function in arguments ξ_0, $\Delta\psi(s_k)=\psi(s_{k+1})-\psi(s_k)$, $k=0, 1, \ldots, i-1$, and $\psi(t)-\psi(s_i)$ for $t\in(s_i, s_{i+1}]$. Thus the functionals $F_{(t,\delta)}(x, x(\cdot))$ introduced above become continuous functions in x and in a finite number of differences of the form $x(t_{k+1})-x(t_k)$. Hence we can weaken condition Ψ_1 and replace it by the following:

Ψ_2: For any r and t_k, $t_k\in[0, T]$, $k=1, \ldots, r$, the joint distribution of the random vectors

$$\xi_n, \psi_n(t_1), \ldots, \psi_n(t_r)$$

converges weakly to the joint distribution of random vectors

$$\xi_0, \psi(t_1), \ldots, \psi(t_r)$$

defined on a certain probability space, where $\psi(t)$ is a square integrable martingale satisfying condition (29).

Theorem 13. *Assume that a sequence of series* (1) *satisfies condition* (20) *and let the functionals $a(t, x(\cdot))$ and $b(t, x(\cdot))$ be functions of the form* (30) *satisfying conditions* (26), (27), (32), *and* (33), *and martingales $\psi_n(t)$ satisfy condition Ψ_2. Then the marginal distributions of processes $\xi_n(t)$ converge weakly to the corresponding marginal distributions of a solution $\xi(t)$ of the stochastic differential equation* (31).

If Lindeberg's condition is satisfied, i.e., for any $\varepsilon>0$

$$\sum_{k=1}^{m_n-1} \mathsf{E}\bar{\chi}_{nk}(\varepsilon)|\Delta\psi_{nk}|^2\to 0 \quad \text{as } n\to\infty,$$

where $\bar{\chi}_{nk}(\varepsilon)=1$ if $|\Delta\xi_{nk}|\geq\varepsilon$ and $\bar{\chi}_{nk}(\varepsilon)=0$ otherwise, then condition Ψ_2 is satisfied, $\psi(t)$ is a Wiener process, and the measures $q_n(\cdot)$ associated with the random processes $\xi_n(t)$ converge weakly in $\mathscr{D}$ to the measure associated with a solution of the stochastic equation (31).

Limit theorems for stochastic differential equations. Consider stochastic differential equations

$$d\xi_u=A_u(\xi_u, dt), \qquad \xi_u(0)=\xi_u^0, \qquad t\in[0, T], \tag{35}$$

dependent on a parameter $u \in [0, u_0]$. Here

$$A_u(x, t) = \int_0^t \alpha_u(x, s)\, ds + \beta_u(x, t),$$

and $A_u \in S(\lambda_0^u, \lambda_N^u)$. One limit theorem for these equations was considered above (Theorem 11 in Section 1). In this subsection we shall discuss conditions for weak convergence of measures $q_u(\cdot)$ generated by solutions of equations (35) in $\mathscr{D}$ ($\mathscr{D} = \mathscr{D}^m[0, T]$). We shall prove the following theorem.

Theorem 14. *Let the following conditions be satisfied*:

(36) a)
$$\lim_{N\to\infty} \sup_{u\in[0,u_0]} \mathsf{P}\{\sup_{0\le t\le T} |\lambda_0^u(t)| > N\} = 0;$$

b) *For any* $N_1 > 0$

(37)
$$\lim_{N\to\infty} \sup_{u\in[0,u_0]} \mathsf{P}\{\sup_{0\le t\le T} \lambda_{N_1}^u(t) > N\} = 0;$$

c) *Marginal distributions of random functions*

$$\int_0^t \alpha_u(x, s)\, ds, \qquad \beta_u(x, t),$$

converge weakly as $u \to 0$ *to the corresponding distributions of random functions*

$$\int_0^t \alpha_0(x, s)\, ds \qquad \beta_0(x, t).$$

d) *The distribution of vector* ξ_u^0 *as* $u \to 0$ *converges weakly to the distribution of the vector* ξ^0.

Then the measures $q_u(\cdot)$ *converge weakly to* $q(\cdot)$.

Note that it is not assumed in Theorem 14 that functions $A_u(t, x)$ are defined on the same probability space.

In the course of the proof of this theorem the following lemma on small perturbations of stochastic differential equations will be used (cf. Theorem 10, Section 1).

Lemma 5. *Let*

$$d\tilde{\xi}_u = A_u(\tilde{\xi}_u(t), dt) + A_{\delta u}(\tilde{\xi}_u(t), dt), \qquad \delta > 0,$$
$$\tilde{\xi}_u(0) = \xi_u^0, \qquad u \in [0, u_0],$$

and let the following conditions be satisfied:

a) $A_u(x, t) \in S(\lambda_0^u, \lambda_N^u)$, *where the functions* λ_0^u *and* λ_N^u *satisfy conditions* (36) *and* (37).

b) $A_{\delta u}(x, t) \in S(\lambda_0^u, \tilde{\lambda}_N^{\delta u})$, *where* $\lambda_0^u(t)$ *is as in condition* a) *and, moreover,*

$$|\alpha_{\delta u}(x, t)| \leq \gamma_{\delta u}(x, t),$$
$$\mathsf{E}\{|\Delta\beta_{\delta u}(x, t)|^2 \,|\, \mathfrak{F}_t\} \leq \mathsf{E}\{\textstyle\int_t^{t+\Delta t} \gamma_{\delta u}^2(x, s)\, ds \,|\, \mathfrak{F}_t\},$$
$$\lim_{\delta\to 0} \sup_{u \in (0, u_0]} \mathsf{P}\{\sup_{\substack{t\in[0,T]\\ |x|\leq N}} \gamma_{\delta u}^2(x, t) > \varepsilon\} = 0$$

for any $\varepsilon > 0$, $N > 0$.

c) *The distribution of the initial vector* ξ_u^0 *as* $u \to 0$ *converges weakly to a limit. Then*

$$\mathsf{P}\{\sup_{0\leq t\leq T} |\tilde{\xi}_u(t) - \xi_u(t)| > \varepsilon\} \to 0 \quad \text{as } \delta \to 0$$

uniformly in u.

Proof. Let ε' be an arbitrary given positive number,

$$\tau = \inf\{t: \lambda_0^u(t) \geq N, \quad \lambda_{N_1}^u(t) \geq N_2, \quad \sup\nolimits_{|x|\leq N_1} \gamma_{\delta u}(x, t) \geq \varepsilon'\},$$

if the set of values of t in the braces is nonvoid and $\tau = T$ otherwise. Here N, N_1, and N_2 are positive numbers to be specified below. For the time being we note only that inequality

$$\mathsf{P}\{\tau < T\} \leq \mathsf{P}\{\sup_{0\leq t\leq T} \lambda_0^u(t) \geq N, \sup_{0\leq t\leq T} \lambda_{N_1}^u(t) \geq N_2\} + \mathsf{P}\{\sup_{\substack{0\leq t\leq T\\ |x|\leq N_1}} \gamma_{\delta u}(x, t) > \varepsilon'\}$$

and the assumptions of the lemma imply that for any N_1 and $\varepsilon > 0$ there exists a δ_0 sufficiently small independent of u, N, and N_2, and N^0, N_2^0 sufficiently large, independent of u, ε', and δ, such that

$$\mathsf{P}\{\tau < T\} < \varepsilon$$

for $N \geq N^0$, $N_2 \geq N_2^0$ and $\delta < \delta_0$.

We now construct functions $\alpha'_u(x, t)$ and $\beta'_u(x, t)$ in such a manner that they coincide with $\alpha_u(x, t)$ and $\beta_u(x, t)$, respectively, for $|x| \leq N_1$ and belong to the class $S(\lambda_0^u, \lambda^{u'})$, where $\lambda^{u'} = \lambda^{u'}(t)$ does not depend on N and $\lambda^{u'}(t) \leq 1 + \lambda_{N_1}^u(t)$. Set $\bar{\alpha}_u(x, t) = \alpha'_u(x, t)$ for $t \leq \tau$, $\bar{\alpha}_u(x, t) = 0$ for $t > \tau$, $\bar{\beta}_u(x, t) = \beta'_u(x, t \wedge \tau)$, $\bar{\alpha}_{\delta u}(x, t) = \alpha_{\delta u}(x, t)$ *for* $t \leq \tau$, $\bar{\alpha}_{\delta u}(x, t) = 0$ *for* $t > \tau$, $\bar{\beta}_{\delta u}(x, t) = \beta_{\delta u}(x, t \wedge \tau)$. Moreover, $\bar{A}_u(x, t) \in S(N, N_1 + 1)$, $\bar{A}_{\delta u}(x, t) \in S(N, \tilde{\lambda}_N^{\delta u})$, where

$$\bar{A}_u(x, t) = \int_0^t \bar{\alpha}_u(x, s)\, ds + \bar{\beta}(x, t)$$

and $\bar{A}_{\delta u}(x, t)$ is defined analogously via $\bar{\alpha}_{\delta u}(x, t)$.

Consider the equations

$$d\eta(t)=\bar{A}_u(\eta(t), dt),$$
$$d\tilde{\eta}(t)=\bar{A}_u(\tilde{\eta}(t), dt)+\bar{A}_{\delta u}(\tilde{\eta}(t), dt),$$
$$\eta(0)=\tilde{\eta}(0)=\xi_u^0.$$

Solutions for these equations do exist and, furthermore, $\eta(t)=\xi_u(t)$ as long as $t<\tau$ and $|\eta(t)|<N_1$ analogously $\tilde{\eta}(t)=\tilde{\xi}_u(t)$ as long as $t<\tau$ and $\sup_{s<t}|\tilde{\eta}(s)|<N_1$.

We now bound the difference $\eta(t)-\tilde{\eta}(t)$.

Note that

$$\mathsf{P}\{\sup_{0\leq t\leq T}|\eta(t)-\tilde{\eta}(t)|>\varepsilon\}\leq \mathsf{P}\{\sup_{0\leq t\leq T}|\tilde{\eta}(t)|\geq N_1\}$$
$$+\mathsf{P}\{\sup_{0\leq t\leq T}|\eta(t\wedge\tau_1)-\tilde{\eta}(t\wedge\tau_1)|>\varepsilon\},$$

where $\tau_1=\inf\{t: |\tilde{\eta}(t)|\geq N_1\}$ ($\inf\varnothing=T$).

It follows from the previous results (in Theorem 7, Section 1) that

$$\mathsf{E}\{\sup_{0\leq t\leq T}|\tilde{\eta}(t)|^2\,|\,\mathfrak{F}_0\}\leq(1+|\xi_u^0|^2)C(N).$$

Consequently for any $C_0>0$

$$\mathsf{P}\{\sup_{0\leq t\leq T}|\tilde{\eta}(t)|\geq N_1\}\leq\frac{(1+C_0^2)C(N)}{N_1^2}+\mathsf{P}\{|\xi_u^0|>C_0\}.$$

Set

$$v_u(t)=\mathsf{E}\sup_{0\leq s\leq t}|\eta(s\wedge\tau_1)-\tilde{\eta}(s\wedge\tau_1)|^2.$$

Since

$$|\eta(t)-\tilde{\eta}(t)|^2\leq 3\Big(\Big|\int_0^t[\bar{\alpha}_u(\eta(s), s)-\bar{\alpha}_u(\tilde{\eta}(s), s)]\,ds\Big|^2$$
$$+\Big|\int_0^t\bar{\beta}_u(\eta(s), ds)-\bar{\beta}_u(\tilde{\eta}(s), ds)\Big|^2+\Big|\int_0^t\bar{A}_{\delta u}(\tilde{\eta}(s), ds)\Big|^2\Big),$$

applying the methods utilized above we obtain

$$v_u(t)\leq 12\mathsf{E}\sup_{0\leq s\leq t}\Big|\int_0^{s\wedge\tau_1}\bar{A}_{\delta u}(\tilde{\eta}(s), ds)\Big|^2+C(N_2)\int_0^t v_u(s)\,ds,$$

where $C(N_2)$ depends on T and N_2 only. It follows from the last integral inequality that

$$v_u(t)\leq C'(N_2)\mathsf{E}\sup_{0\leq s\leq t}\Big|\int_0^{s\wedge\tau_1}\bar{A}_{\delta u}(\tilde{\eta}(s), ds)\Big|^2.$$

Furthermore, we have

$$\mathsf{E}\sup_{0\leqslant s\leqslant t}\Big|\int_0^{s\wedge\tau_1}\bar{A}_{\delta u}(\tilde{\eta}(s),ds)\Big|^2$$

$$\leqslant 2\Big(t\mathsf{E}\int_0^{t\wedge\tau_1}|\bar{a}_{\delta u}(\tilde{\eta}(s),s)|^2\,ds+\mathsf{E}\sup_{0\leqslant s\leqslant t}\Big|\int_0^{s\wedge\tau_1}\tilde{\beta}_{\delta u}(\tilde{\eta}(s),ds)\Big|^2\Big)$$

$$\leqslant(2t+8)\mathsf{E}\int_0^{t\wedge\tau_1\wedge\tau}\gamma^2_{\delta u}(\tilde{\eta}(s),s)\,ds.$$

Taking the definition of random times τ and τ_1 into account we obtain

$$v_u(t)\leqslant\varepsilon' C''(N_2).$$

This implies that

$$\mathsf{P}\{\sup_{0\leqslant t\leqslant T}|\eta(t\wedge\tau_1)-\tilde{\eta}(t\wedge\tau_1)|>\varepsilon\}<\frac{\varepsilon' C''(N_2)}{\varepsilon^2}$$

and

$$\mathsf{P}\{\sup_{0\leqslant t\leqslant T}|\eta(t)-\tilde{\eta}(t)|>\varepsilon\}\leqslant\mathsf{P}\{|\xi_u^0|>C_0\}+\frac{(1+C_0^2)C(N)}{N_1^2}+\frac{\varepsilon' C''(N_2)}{\varepsilon^2}.$$

Furthermore,

$$\begin{aligned}\mathsf{P}\{\sup_{0\leqslant t\leqslant T}|\xi_u(t)-\tilde{\xi}_u(t)|>\varepsilon\}\leqslant{}&\mathsf{P}\{\sup_{0\leqslant t\leqslant T}|\xi_u(t)-\eta(t)|>0\}\\&+\mathsf{P}\{\sup_{0\leqslant t\leqslant T}|\eta(t)-\tilde{\eta}(t)|>\varepsilon\}\\&+\mathsf{P}\{\sup_{0\leqslant t\leqslant T}|\tilde{\eta}(t)-\tilde{\xi}_u(t)|>0\};\end{aligned}$$

moreover,

$$\mathsf{P}\{\sup_{0\leqslant t\leqslant T}|\xi_u(t)-\eta(t)|>0\}\leqslant\mathsf{P}(\tau<T)+\mathsf{P}\{\sup_{0\leqslant t\leqslant T}|\eta(t)|\geqslant N_1\},$$

with an analogous inequality valid for the probability $\mathsf{P}\{\sup_{0\leqslant t\leqslant T}|\tilde{\xi}_u(t)-\tilde{\eta}(t)|>0\}$. Hence

$$\begin{aligned}\mathsf{P}\{\sup_{0\leqslant t\leqslant T}|\xi_u(t)-\tilde{\xi}(t)|>\varepsilon\}\leqslant{}&2\mathsf{P}(\tau<T)+3\mathsf{P}\{|\xi_u^0|>C_0\}\\&+\frac{3(1+C_0^2)C(N)}{N_1^2}+\frac{\varepsilon' C''(N_2)}{\varepsilon^2}.\end{aligned}$$

Given an ε one can now choose constants C_0, N, N_1, N_2, and ε' independent of u

in the following manner. Choose C_0 first from the condition $3\mathsf{P}\{|\xi_u^0|>C_0\}<\varepsilon/4$ and then N such that

$$2\mathsf{P}\{\sup_{0\le t\le T}\lambda_0^u(t)>N\}<\frac{\varepsilon}{12};$$

next, determine N_1 so that the inequality

$$3(1+C_0^2)C(N)/N_1^2<\varepsilon/4$$

will be fulfilled. Furthermore, let N_2 be such that

$$2\mathsf{P}\{\sup_{0\le t\le T}\lambda_{N_1}^u(t)>N_2\}<\frac{\varepsilon}{12};$$

now determine ε' from the condition $\varepsilon' C''(N)/\varepsilon^2<\varepsilon/4$ and, finally, choose δ_0 in such a manner that for $\delta<\delta_0$

$$2\mathsf{P}\{\sup_{\substack{0\le t\le T\\ |x|\le N_1}}\gamma_{\delta u}(x,t)>\varepsilon'\}<\frac{\varepsilon}{12}.$$

We then obtain that $\mathsf{P}\{\sup_{0\le t\le T}|\xi_u(t)-\tilde{\xi}_u(t)|>\varepsilon\}<\varepsilon$ for all $\delta<\delta_0$ and for any $u\in[0,u_0]$. The lemma is thus proved. □

Proof of Theorem 14. In view of the fact that condition a) of the theorem assures weak compactness of measures in $\mathscr{D}$ which are associated with solutions of equations (35), to prove the theorem it is sufficient to verify the convergence of marginal distributions of processes $\xi_u(t)$ to the corresponding marginal distributions of the process $\xi_0(t)$. We shall first prove this for the fields $\alpha_u(x,t)$ and $\beta_u(x,t)$ of a particular form and then proceed to the general case.

Let

$$\alpha_u(x,t)=\sum_{k=1}^r a_k(x)\alpha_u^k(t),\qquad \beta_u(x,t)=\sum_{k=1}^r b_k(x)\beta_u^k(t),$$

where $a_k(x)$ and $b_k(x)$, $k=1,\dots,r$, are nonrandom scalar differentiable functions with uniformly bounded derivatives, let $\beta_u^k(t)$ be square integrable martingales, and let

$$|\alpha_u^k(t)|\le\lambda_0^u(t),\qquad k=1,\dots,r,$$
$$\mathsf{E}\{|\Delta\beta_u^k(t)|^2\,|\,\mathfrak{F}_t\}\le\mathsf{E}\{\textstyle\int_t^{t+\Delta t}(\lambda_0^u(s))^2\,ds\,|\,\mathfrak{F}_t\}.$$

Assume that the function $\lambda_0^u(t)$ satisfies condition (36), that marginal distributions of the compound process

$$\left(\textstyle\int_0^t\alpha_u^1(s)\,ds,\dots,\int_0^t\alpha_u^r(s)\,ds,\quad \beta_u^1(t),\dots,\beta_u^r(t)\right)$$

are weakly convergent as $u \to 0$ to the corresponding distributions of a process

$$\left(\int_0^t \alpha_u^1(s)\,ds, \ldots, \int_0^t \alpha_0^r(s)\,ds, \qquad \beta_0^1(t), \ldots, \beta_0^r(t)\right)$$

and that the random vector ξ_u^0 converges in distribution to the vector ξ_0.

We shall prove Theorem 14 under these additional assumptions.

Let t_k, $t_k \in [0, T]$, $k = 1, \ldots, s$, be a given sequence of numbers. Introduce finite-difference approximations $\xi_{\delta u}(t)$ of solutions of equations (35) and assume that the points t_k, $k = 1, \ldots, s$, are included in the subdivision δ. Since the conditions stipulated in the Remark following Theorem 13 in Section 1 are fulfilled in our case, given an arbitrary $\varepsilon > 0$ one can find δ_0 such that for $|\delta| < \delta_0$

$$\mathsf{P}\{\sup_{0 \le t \le T} |\xi_u(t) - \xi_{\delta u}(t)| > \varepsilon\} \le \varepsilon$$

for all $u \in [0, u_0]$. Let $f(x_1, \ldots, x_s)$ be an arbitrary continuous and bounded function ($x_k \in \mathscr{R}^m$) and let its partial derivatives of the first order be bounded also. We have

$$\begin{aligned}
&|\mathsf{E}f[\xi_u(t_1), \ldots, \xi_u(t_s)] - \mathsf{E}f[\xi_0(t_1), \ldots, \xi_0(t_s)]| \\
&\le |\mathsf{E}(f[\xi_u(t_1), \ldots, \xi_u(t_s)] - f[\xi_{\delta u}(t_1), \ldots, \xi_{\delta u}(t_s)])| \\
&\quad + |\mathsf{E}f[\xi_{\delta u}(t_1), \ldots, \xi_{\delta u}(t_s)] - \mathsf{E}f[\xi_{\delta 0}(t_1), \ldots, \xi_{\delta 0}(t_s)]| \\
&\quad + |\mathsf{E}(f[\xi_{\delta 0}(t_1), \ldots, \xi_{\delta 0}(t_s)] - f[\xi_0(t_1), \ldots, \xi_0(t_s)])| \\
&= I_1 + I_2 + I_3.
\end{aligned}$$

Moreover,

$$I_1 \le C[\varepsilon + \mathsf{P}\{\sup_{0 \le t \le T} |\xi_{\delta u}(t) - \xi_u(t)| > \varepsilon\}],$$

$$I_3 \le C[\varepsilon + \mathsf{P}\{\sup_{0 \le t \le T} |\xi_{\delta 0}(t) - \xi_0(t)| > \varepsilon\}],$$

where C is a constant. Thus we have for $|\delta| < \delta_0$, $I_1 + I_3 \le 4C\varepsilon$ independently of the values of u. Furthermore, it is easy to verify that $f[\xi_{\delta u}(t_1), \ldots, \xi_{\delta u}(t_s)]$ is a continuous and bounded function of the quantities

$$\int_{s_j}^{s_{j+1}} \alpha_u(s)\,ds, \qquad \beta^k(s_{j+1}) - \beta^k(s_j), \qquad j = 0, 1, \ldots, l, \qquad k = 1, \ldots, r,$$

where s_j are the points which form the subdivision δ. Therefore for a δ chosen as above, we have

$$I_2 = \mathsf{E}f[\xi_{\delta u}(t_1), \ldots, \xi_{\delta u}(t_s)] - \mathsf{E}f[\xi_{\delta 0}(t_1), \ldots, \xi_{\delta 0}(t_s)] \to 0$$

as $u \to 0$. Thus the weak convergence as $u \to 0$ of the marginal distributions of processes $\xi_u(t)$ is proved for the particular case under consideration.

We now turn to the proof of Theorem 14 in the general case.

We introduce fields $\tilde{A}_{\delta u}(x, t)$ approximating $A_u(x, t)$, $u \in [0, u_0]$. For this purpose we construct for each $\delta > 0$ in the sphere $\{x: |x| \leq 1/\delta\}$ a δ-net $x_1, x_2, \ldots, x_{n_\delta}$ and a system of functions $g_j(x)$, $j = 1, \ldots, n_\delta$, satisfying the following conditions: $g_j(x) \geq 0$ and $g_j(x) = 0$ for $|x - x_j| \geq \delta$, $\sum_{j=1}^{n_\delta} g_j(x) = 1$, $|x| \leq 1/\delta$, and the functions $g_j(x)$ are continuously differentiable.

Set

$$\tilde{\alpha}_{\delta u}(x, t) = \sum_{j=1}^{n_\delta} g_j(x)\alpha_u(x_j, t),$$

$$\hat{\beta}_{\delta u}(x, t) = \sum_{j=1}^{n_\delta} g_j(x)\beta_u(x_j, t),$$

$$\tilde{A}_{\delta u}(x, t) = \int_0^t \tilde{\alpha}_{\delta u}(x, s)\, ds + \tilde{\beta}_{\delta u}(x, t),$$

$$A_{\delta u}(x, t) = \tilde{A}_{\delta u}(x, t) - A_u(x, t) - \int_0^t \alpha_{\delta u}(x, s)\, ds + \beta_{\delta u}(x, t).$$

Introduce stochastic differential equations

$$(38) \qquad d\eta_u(t) = \tilde{A}_{\delta u}(\eta_u(t), dt), \qquad \eta_u(0) = \xi_u^0, \qquad t \in [0, T].$$

Note that if the conditions of Theorem 14 are satisfied, then equation (38) satisfies the conditions stipulated in the particular case above for any fixed δ.

As before, let $f(x_1, \ldots, x_s)$ denote an arbitrary continuous and continuously differentiable function which is bounded and possesses bounded partial derivatives. Set

$$\mathsf{E}f[\xi_u(t_1), \ldots, \xi_u(t_s)] - \mathsf{E}f[\xi_0(t_1), \ldots, \xi_0(t_s)] = J_1 + J_2 + J_3,$$

where

$$J_1 = \mathsf{E}(f[\xi_u(t_1), \ldots, \xi_u(t_s)] - f[\eta_u(t_1), \ldots, \eta_u(t_s)]),$$

$$J_2 = \mathsf{E}f[\eta_u(t_1), \ldots, \eta_u(t_s)] - \mathsf{E}f[\eta_0(t_1), \ldots, \eta_0(t_s)],$$

$$J_3 = \mathsf{E}(f[\eta_0(t_1), \ldots, \eta_0(t_s)] - f[\xi_0(t_1), \ldots, \xi_0(t_s)]).$$

It follows from the particular case of Theorem 14 discussed above that $J_2 \to 0$ for any fixed δ as $u \to 0$. Thus to prove the theorem in the general case it is sufficient to show that $J_1 + J_3 \to 0$ as $\delta \to 0$ uniformly in u. Now for any $\varepsilon > 0$

$$|J_1| + |J_3| \leq C(2\varepsilon + \mathsf{P}\{\sup_{0 \leq t \leq T} |\xi_u(t) - \eta_u(t)| > \varepsilon\} + \mathsf{P}\{\sup_{0 \leq t \leq T} |\xi_0(t) - \eta_0(t)| > \varepsilon\}),$$

where C is a constant which depends on the function $f(x_1, \ldots, x_s)$ only.

We show that Lemma 5 is applicable to equations (35) and (38). To do this we observe that for $|x|\leq N_0<1/\delta$

$$\begin{aligned}|\alpha_{\delta u}(x,t)|&=|\tilde{\alpha}_{\delta u}(x,t)-\alpha_u(x,t)|\\&\leq\sum_{j=1}^{n_\delta}g_j(x)|\alpha_u(x_j,t)-\alpha_u(x,t)|\\&\leq\sum_{j:|x_j-x|<\delta}g_j(x)|\alpha_u(x_j,t)-\alpha_u(x,t)|\leq\delta\lambda_{N_0}(t).\end{aligned}$$

Analogously,

$$\begin{aligned}\mathsf{E}\{|\Delta\beta_{\delta u}(x,t)|^2|\mathfrak{F}_t\}&\leq\sum_{j=1}^{n_\delta}g_j(x)\sum_{j=1}^{n_\delta}\mathsf{E}\{g_j(x)|\,\Delta\beta_u(x_j,t)-\Delta\beta_u(x,t)|^2|\mathfrak{F}_t\}\\&\leq\delta^2\mathsf{E}\{\textstyle\int_t^{t+\Delta t}\lambda_{N_0}^2(s)\,ds\,|\mathfrak{F}_t\}.\end{aligned}$$

Moreover, it is easy to verify that

$$|\tilde{\alpha}_{\delta u}(x,t)|\leq 2(1+|x|)\lambda_0(t),$$

$$\mathsf{E}\{|\Delta\tilde{\beta}_{\delta u}(x,t)|^2|\mathfrak{F}_t\}\leq 2(1+|x|^2)\mathsf{E}\{\textstyle\int_t^{t+\Delta t}\lambda_0^2(s)\,ds\,|\mathfrak{F}_t\}.$$

The bounds obtained do not depend on u and we observe that the conditions of Lemma 5 are indeed satisfied; moreover,

$$\sup_{\substack{t\in[0,T]\\|x|\leq N}}\gamma_{\delta u}^2(x,t)\leq\delta^2\sup_{t\in[0,T]}\lambda_N^2(t).$$

Hence there exists δ_0 independent of u such that $|J_1|+|J_3|<4C\varepsilon$ for $\delta<\delta_0$. □

Example. *Oscillations with a small nonlinearity.* Consider an oscillation equation with small nonlinear terms

$$\frac{d^2x}{dt^2}+\omega^2x=\varepsilon f_1\left(x,\frac{dx}{dt}\right)+\sqrt{\varepsilon}f_2\left(x,\frac{dx}{dt}\right)\dot{w}(t),\tag{39}$$

where $x=x(t)$, $f_1(x,y)$ and $f_2(x,y)$ are scalar functions, ε is a small parameter, and $w(t)$ is a Wiener process.

Equation (39) should be interpreted as a system of two stochastic differential equations of the form

$$\begin{aligned}d\dot{x}&=(-\omega^2x+\varepsilon f_1(x,\dot{x}))\,dt+\sqrt{\varepsilon}f_2(x,\dot{x})\,dw(t),\\dx&=\dot{x}\,dt.\end{aligned}\tag{40}$$

We shall apply the standard methodology for investigating nonlinear oscillations to system (40). Introduce the change of variables

$$x = a\cos\psi, \quad \dot{x} = -a\omega\sin\psi, \quad \psi = \omega t + \theta$$

or

$$a = \sqrt{x^2 + \frac{1}{\omega^2}\dot{x}^2}, \qquad \psi = -\arctan\frac{\dot{x}}{\omega x}.$$

Now we utilize Itô's formula to obtain an equation for quantities a and θ. The following relations will result:

$$da = \varepsilon\left(-\frac{\sin\psi}{\omega}\tilde{f}_1(a,\theta) + \frac{\cos^2\psi}{2\omega^2 a}\tilde{f}_2^2(a,\theta)\right)dt - \frac{\sqrt{\varepsilon}\sin\psi}{\omega}\tilde{f}_2(a,\theta)\,dw,$$

$$d\theta = \varepsilon\left(-\frac{\cos\psi}{\omega a}\tilde{f}_1(a,\theta) + \frac{\sin 2\psi}{2\omega^3 a^2}\tilde{f}_2^2(a,\theta)\right)dt - \frac{\sqrt{\varepsilon}\cos\psi}{2\omega a}\tilde{f}_2(a,\theta)\,dw.$$

These equations differ from the equations which would have been obtained if the process $w(t)$ were differentiable by the presence of the additional terms $\varepsilon(\cos^2\psi/2\omega a)\tilde{f}_2^2(a,\theta)$ in the first equations and $\varepsilon(\sin 2\psi/2\omega^3 a^2)\tilde{f}_2^2(a,\theta)$ in the second. Here

$$\tilde{f}_i(a,\theta) = f_i(a\cos\psi, -\omega a\sin\psi), \qquad i = 1, 2.$$

Introduce a new time by replacing $t \to t/\varepsilon$ and set

$$a_\varepsilon(t) = a\left(\frac{t}{\varepsilon}\right), \qquad \theta_\varepsilon(t) = \theta\left(\frac{t}{\varepsilon}\right), \qquad \xi_\varepsilon(t) = (a_\varepsilon(t), \theta_\varepsilon(t)).$$

Then

$$d\xi_\varepsilon(t) = A_\varepsilon(\xi_\varepsilon(t), dt),$$
$$A_\varepsilon^i(a,\theta,t) = \int_0^t \alpha_\varepsilon^i(a,\theta,s)\,ds + \beta_\varepsilon^i(a,\theta,t), \qquad i = 1, 2,$$

where

$$\alpha_\varepsilon^1(a,\theta,t) = -\frac{\sin\left(\frac{\omega}{\varepsilon}t+\theta\right)}{\omega}\tilde{f}_1\left(a, \frac{\omega}{\varepsilon}t+\theta\right) + \frac{\cos^2\left(\frac{\omega}{\varepsilon}t+\theta\right)}{2\omega^2 a}\tilde{f}_2^2\left(a, \frac{\omega}{\varepsilon}t+\theta\right),$$

$$\alpha_\varepsilon^2(a,\theta,t) = -\frac{\cos\left(\frac{\omega}{\varepsilon}t+\theta\right)}{\omega a}\tilde{f}_1\left(a, \frac{\omega}{\varepsilon}t+\theta\right) + \frac{\sin^2\left(\frac{\omega}{\varepsilon}t+\theta\right)}{2\omega^3 a^2}\tilde{f}_2^2\left(a, \frac{\omega}{\varepsilon}t+\theta\right),$$

$$\beta_\varepsilon^1(a,\theta,t) = -\frac{\sqrt{\varepsilon}}{\omega}\int_0^{t/\varepsilon}\tilde{f}_2(a, \omega s+\theta)\sin(\omega s+\theta)\,dw(s),$$

$$\beta_\varepsilon^2(a,\theta,t) = -\frac{\sqrt{\varepsilon}}{\omega a}\int_0^{t/\varepsilon}\tilde{f}_2(a, \omega s+\theta)\cos(\omega s+\theta)\,dw(s).$$

It is easy to verify that as $\varepsilon \to 0$ under quite general assumptions on functions $f_1(x, \dot{x})$ and $f_2(x, \dot{x})$,

$$\int_0^t \alpha_\varepsilon^1(a, \theta, s)\, ds \to t\bar{f}_1(a),$$
$$\int_0^t \alpha_\varepsilon^2(a, \theta, s)\, ds \to t\bar{f}_2(a),$$

where

$$\bar{f}_1(a) = \frac{1}{2\pi}\int_0^{2\pi}\Big[-f_1(a\cos\psi, -a\omega\sin\psi)\frac{\sin\psi}{\omega} + f_2^2(a\cos\psi, -a\omega\sin\psi)\frac{\cos^2\psi}{2\omega^2 a}\Big]\, d\psi,$$

$$\bar{f}_2(a) = \frac{1}{2\pi}\int_0^{2\pi}\Big[-f_1(a\cos\psi, -a\omega\sin\psi)\frac{\cos\psi}{\omega a} + f_2^2(a\cos\psi, -a\omega\sin\psi)\frac{\sin 2\psi}{2\omega^3 a^2}\Big]\, d\psi.$$

On the other hand $\beta_\varepsilon(a, \theta, t)$ is a Gaussian field with independent (time) increments and, moreover,

$$\mathsf{E}(\beta_\varepsilon^1(a, \theta, t))^2 = \frac{\varepsilon}{\omega}\int_0^{t/\varepsilon} \tilde{f}_2^2(a, \omega s + \theta)\sin^2(\omega s + \theta)\, ds,$$

$$\mathsf{E}\beta_\varepsilon^1(a, \theta, t)\beta_\varepsilon^2(a, \theta, t) = \frac{\varepsilon}{\omega a}\int_0^{t/\varepsilon} \tilde{f}_2^2(a, \omega s + \theta)\sin(\omega s + \theta)\cos(\omega s + \theta)\, ds,$$

$$\mathsf{E}(\beta_\varepsilon^2(a, \theta, t))^2 = \frac{\varepsilon}{\omega^2 a^2}\int_0^{t/\varepsilon} \tilde{f}_2^2(a, \omega s + \theta)\cos^2(\omega s + \theta)\, ds.$$

The correlation matrix of the field $\beta_\varepsilon(a, \theta, t)$ converges to the following limit as $\varepsilon \to 0$ $(h > 0)$:

$$\lim R_\varepsilon(t, t+h) = \lim R_\varepsilon(t, t) = t\begin{pmatrix} b_{11}(a) & b_{12}(a) \\ b_{12}(a) & b_{22}(a) \end{pmatrix},$$

where

$$b_{11}(a) = \frac{1}{2\pi\omega}\int_0^{2\pi} f_2^2(a\cos\psi, -\omega a\sin\psi)\sin^2\psi\, d\psi,$$

$$b_{12}(a) = \frac{1}{2\pi\omega a}\int_0^{2\pi} f_2^2(a\cos\psi, -\omega a\sin\psi)\sin\psi\cos\psi\, d\psi,$$

$$b_{22}(a) = \frac{1}{2\pi\omega^2 a^2}\int_0^{2\pi} f_2^2(a\cos\psi, -\omega a\sin\psi)\cos^2\psi\, d\psi.$$

The matrix

$$B(a)=\{b_{ik}(a)\}$$

is symmetric and nonnegative-definite. We now construct a nonnegative-definite and symmetric matrix

$$\sigma(a)=\{\sigma_{ik}\}, \qquad i,k=1,2,$$

such that

$$\sigma^2(a)=B(a).$$

Marginal distributions of the field $\beta_\varepsilon(a,t)$ converge weakly to the marginal distributions of the field

$$\beta_0(a,t)=\{\beta_0^1(a,t), \beta_0^2(a,t)\}$$

which can be defined by means of the relations:

$$\beta_0^1(a,t)=\sigma_{11}(a)w_1(t)+\sigma_{12}(a)w_2(t),$$
$$\beta_0^2(a,t)=\sigma_{12}(a)w_1(t)+\sigma_{22}(a)w_2(t).$$

Here $w_1(t)$ and $w_2(t)$ are two independent Wiener processes.

Thus if the remaining conditions of Theorem 14 dealing with the regularity properties of functions $f_1(x,\dot{x})$ and $f_2(x,\dot{x})$ are fulfilled, one can assert the following.

A solution of equation (39) *can be represented as*

$$x\left(\frac{t}{\varepsilon}\right)=a_\varepsilon(t)\cos\left(\frac{\omega t}{\varepsilon}+\theta(t)\right),$$

$$\dot{x}\left(\frac{t}{\varepsilon}\right)=-\omega a_\varepsilon(t)\sin\left(\frac{\omega t}{\varepsilon}+\theta_\varepsilon(t)\right),$$

and, moreover, the distributions of the random processes $(a_\varepsilon(t), \theta_\varepsilon(t))$ *are weakly convergent as* $\varepsilon\to 0$ *to the measure associated with the process* $(\bar{a}(t), \bar{\theta}(t))$ *which is a solution of the stochastic differential equation*

$$d\bar{a}=\bar{f}_1(\bar{a})\,dt+\sigma_{11}(\bar{a})\,dw_1+\sigma_{12}(\bar{a})\,dw_2,$$
$$d\theta=\bar{f}_2(\bar{a})\,dt+\sigma_{12}(\bar{a})\,dw_1+\sigma_{22}(\bar{a})\,dw_2.$$

Chapter III

Stochastic Differential Equations for Continuous Processes and Continuous Markov Processes in $\mathscr{R}^m$

§ 1. Itô Processes

When studying solutions of stochastic differential equations we had occasion to encounter processes possessing the Itô stochastic differential, i.e., representable by means of a stochastic integral over a Wiener process. Such processes are called Itô processes and we shall discuss their basic properties in this section.

Definition and some properties. We shall assume a fixed probability space $\{\Omega, \mathfrak{A}, \mathsf{P}\}$ and a current of σ-algebras $\{\mathfrak{F}_t, t \geq 0\}$ on this space. Let $w(t)$ be a Wiener process on this space with values in $\mathscr{R}^m$ adopted to the current $\mathfrak{F}_t$, i.e., $w(t)$ is an $\mathfrak{F}_t$-measurable variable and $w(s)-w(t)$ for $s>t$ are jointly independent of σ-algebra $\mathfrak{F}_t$. Denote by $\mathfrak{M}_1[0, T]$ the set of measurable functions $f(s, \omega)$ which are $\mathfrak{F}_s$-measurable as functions of ω for all $s \in [0, T]$ and such that

$$\mathsf{P}\{\int_0^T |f(s, \omega)|\, ds < \infty\} = 1.$$

The process $\eta(t)$, $t \in [0, T]$, with values in $\mathscr{R}^n$ is called the *Itô process with respect to* $(w(t), \mathfrak{F}_t)$ provided the following objects exist: a) an $\mathfrak{F}_0$-measurable variable η_0; b) a measurable function $a(s, \omega)$ with values in $\mathscr{R}^n$ which is also $\mathfrak{F}_s$-measurable as a function of ω for all s; c) a measurable function $B(s, w)$ whose values are linear operators from $\mathscr{R}^m$ into $\mathscr{R}^n$ (which is also $\mathfrak{F}_s$-measurable as a function of ω for all s) such that

$$\eta(t) = \eta_0 + \int_0^t a(s, \omega)\, ds + \int_0^t B(s, \omega)\, dw(s), \qquad 0 < t, \tag{1}$$

and, moreover,

$$|a(s, \omega)| + \mathrm{Sp}\, B(s, \omega) B^*(s, \omega) \in \mathfrak{M}_1[0, T].$$

The last condition is clearly necessary and sufficient for the existence of integrals in the left-hand side of (1). In what follows we shall consider mainly Itô processes in $\mathscr{R}^m$; denote by $\mathfrak{L}(\mathscr{R}^m)$ the set of linear operators in $\mathscr{R}^m$. We shall assume here that all the random functions are adopted to the current $\{\mathfrak{F}_t\}$.

One of the first questions which arises from the definition of an Itô process is whether it is possible to define from this process the functions $a(s, \omega)$ and $B(s, \omega)$ and how this can be accomplished. To solve this problem we need an important characteristic property of one-dimensional Wiener processes due to P. Levy.

Theorem 1. *If $\xi(t)$ is a continuous process in $\mathcal{R}^1$ and there exists a current of σ-algebras $\mathfrak{F}_t$ such that $(\xi(t), \mathfrak{F}_t)$ and $(\xi^2(t)-t, \mathfrak{F}_t)$ are martingales, then $\xi(t)$ is a Wiener process with respect to current $\{\mathfrak{F}_t\}$.*

The proof of this theorem is included in Chapter I, Section 3, Theorem 3, in the subsection *Some applications of the Itô formula.*

We now state a few corollaries to this theorem.

Corollary 1. *If $b(s, \omega) \in \mathcal{R}^m$ and $|b(s, \omega)| = 1$, then the process*

$$w_1(t) = \int_0^t (b(s, \omega), dw(s))$$

is a Wiener process.

Corollary 2. *If $\xi(t)$ is a continuous process with values in $\mathcal{R}^m$ and if for some current of σ-algebras $\mathfrak{F}_t$ and all $z \in \mathcal{R}^m$, $h > 0$,*

$$\mathsf{E}(\xi(t+h) - \xi(t) \mid \mathfrak{F}_t) = 0,$$
$$\mathsf{E}((\xi(t+h) - \xi(t), z)^2 \mid \mathfrak{F}_t) = h|z|^2,$$

then $\xi(t)$ is a Wiener process in $\mathcal{R}^m$.

This result follows from the fact that for each $z \in \mathcal{R}^m$ the process $(\xi(t), z)/|z|$ is a Wiener process in $\mathcal{R}^m$.

Corollary 3. *Let $B(s, \omega)$ be a unitary operator in $\mathcal{R}^m$ for all s and ω. Then*

$$\xi(t) = \int_0^t B(s, \omega)\, dw(s)$$

is a Wiener process in $\mathcal{R}^m$.

Indeed $(\xi(t), \mathfrak{F}_t)$ is a martingale and

$$\mathsf{E}((\xi(t+h) - \xi(t), z)^2 \mid \mathfrak{F}_t) = \mathsf{E}\left(\left(\int_t^{t+h} (B(s, \omega)z, dw(s))\right)^2 \Big| \mathfrak{F}_t\right)$$
$$= \mathsf{E}\left(\int_t^{t+h} |B(s, \omega)z|^2\, ds \,\Big|\, \mathfrak{F}_t\right) = h|z|^2.$$

Corollary 4. *Let $w_1(t)$ be a Wiener process in $\mathcal{R}^1$ with respect to the current $\{\mathfrak{F}_t\}$ and $\alpha(s, \omega)$ be a numerical function, and for all $t > 0$ let*

$$\int_0^t \alpha^2(s, \omega)\, ds < \infty, \qquad \int_0^\infty \alpha^2(s, \omega)\, ds = \infty$$

and τ_t be defined from the equality

$$t = \int_0^{\tau_t} \alpha^2(s, \omega)\, ds.$$

Then the process

$$\xi(t) = \int_0^{\tau_t} \alpha(s, \omega)\, dw(s)$$

is a Wiener process with respect to the current of σ-algebras $\hat{\mathfrak{F}}_t = \mathfrak{F}_{\tau_t}$, $t \geq 0$.

To verify this assertion we note that τ_t are Markov times; hence $(\xi(t), \hat{\mathfrak{F}}_t)$ is a martingale and, moreover,

$$\mathsf{E}[(\xi(t+h) - \xi(t))^2 \mid \hat{\mathfrak{F}}_t] = \mathsf{E}\left(\int_{\tau_t}^{\tau_{t+h}} \alpha^2(s, \omega)\, ds \mid \hat{\mathfrak{F}}_t\right) = h.$$

Theorem 2. *An Itô process $\eta(t)$ of the form* (1) *defines functions $a(s, \omega)$ and $B(s, \omega)$ uniquely almost everywhere in s and ω with respect to the product of the Lebesgue measure on the line and measure* **P**.

Proof. It is sufficient to show that if $\eta(t)$ as defined by (1) is zero then the functions $a(s, \omega)$ and $B(s, \omega)$ vanish almost everywhere.

For $t \in [0, T]$ let

$$0 = \int_0^t a(s, \omega)\, ds + \int_0^t B(s, \omega)\, dw(s). \tag{2}$$

It follows easily from (2) that

$$0 = \int_0^t (a(s, \omega), z)\, ds + \int_0^t (B(s, \omega)z, dw(s)). \tag{3}$$

Let $f(s, \omega)$ be an arbitrary bounded measurable function adopted to the current $\{\mathfrak{F}_t\}$. Equation (3) implies the equality

$$\int_0^t (f(s, \omega)B(s, \omega)z, dw(s)) = -\int_0^t f(s, \omega)(a(s, \omega), z)\, ds.$$

Set $f(s, \omega) = \operatorname{sgn}(a(s, \omega), z)$. Then

$$\int_0^t (f(s, \omega)B(s, \omega)z, dw(s)) = -\int_0^t |(a(s, \omega), z)|\, ds. \tag{4}$$

Now set

$$\zeta_N = \begin{cases} T & \text{if } \int_0^T |B(s, \omega)z|^2\, ds \leq N, \\ \inf\,[t\colon \int_0^t |B(s, \omega)|^2\, ds > N] & \text{if } \int_0^T |B(s, \omega)z|^2\, ds > N. \end{cases}$$

ζ_N is a Markov time with respect to the current $\{\mathfrak{F}_t\}$ and

$$\mathsf{E} \int_0^{\zeta_N} (f(s, \omega)B(s, \omega)z, dw(s)) = 0,$$

since

$$\mathsf{E}\left[\int_0^{\zeta_N} (f(s,\omega)B(s,\omega)z, dw(s))\right]^2 \leq N.$$

Substituting in (4) the quantity ζ_N in place of t and taking the mathematical expectation we obtain

$$\mathsf{E}\int_0^{\zeta_N} |(a(s,\omega), z)|\, ds = 0.$$

However, $\zeta_N \uparrow T$ as $N \to \infty$. Hence

$$\mathsf{E}\int_0^T |(a(s,\omega), z)|\, ds = 0.$$

The last relation yields

$$\int_0^T |a(s,\omega)|\, ds = 0$$

with probability 1. Thus the first summand in (3) equals 0. Hence we have

$$\int_0^t (B(s,\omega)z, dw(s)) = 0$$

for all t.

This implies that

$$\left[\int_0^{\zeta_N} (B(s,\omega)z, dw(s))\right]^2 = 0$$

with probability 1.

Applying the mathematical expectation and then approaching the limit as $N \to \infty$ we arrive at

$$\mathsf{E}\int_0^T |B(s,\omega)z|^2\, ds = 0.$$

This completes the proof of the theorem. □

We now obtain conditions under which an Itô process is a Wiener process. To do this we shall first prove the following two lemmas.

Lemma 1. *Let*

$$\xi(t) = \int_0^t \alpha(s,\omega)\, ds + \int_0^t \beta(s,\omega)\, dw_1(s),$$

where $w_1(t)$ is a Wiener process. If $\xi(t)$ is a martingale, $\mathsf{E}\xi^2(T) < \infty$, then $\alpha(s,\omega) = 0$ almost everywhere with respect to the product of the Lebesgue measure on the line and measure $\mathbf{P}$.

Proof. Set

$$\zeta_N = \begin{cases} T & \text{if } \int_0^T \beta^2(s,\omega)\,ds \leq N, \\ \inf[t: \int_0^t \beta^2(s,\omega)\,ds > N] & \text{if } \int_0^T \beta^2(s,\omega)\,ds > N, \end{cases}$$

$$\xi_N(t) = \xi(t \wedge \zeta_N).$$

Since $\xi_N(t)$ is a square integrable martingale, we have for a bounded measurable function $\psi(s, \omega)$

$$0 = \mathsf{E} \int_0^T \psi(s,\omega)\,d\xi_N(s) = \mathsf{E} \int_0^{\zeta_N} \psi(s,\omega)\alpha(s,\omega)\,ds + \mathsf{E} \int_0^{\zeta_N} \beta(s,\omega)\,dw_1(s).$$

Hence

$$0 = \mathsf{E} \int_0^{\zeta_N} \operatorname{sgn} \alpha(s,\omega) \cdot \alpha(s,\omega)\,ds = \mathsf{E} \int_0^{\zeta_N} |\alpha(s,\omega)|\,ds.$$

Approaching the limit as $N \to \infty$ we obtain the required assertion. □

Lemma 2. *If*

$$\xi(t) = \int_0^t \beta(s,\omega)\,dw_1(s) \quad \text{and} \quad \mathsf{E} \sup_{0 \leq t \leq T} \xi^2(t) < \infty,$$

then

$$\mathsf{E} \int_0^T \beta^2(s,\omega)\,ds < \infty.$$

Proof. We show that $\xi(t)$ is a martingale. Let

$$h_N = \sup\left[\delta < h: \int_t^{t+\delta} \beta^2(s,\omega)\,ds < N\right].$$

Then

$$\mathsf{E}\left(\int_t^{t+h_N} \beta(s,\omega)\,dw_1(s) \,\Big|\, \mathfrak{F}_t\right) = 0, \tag{5}$$

and $h_N \to h$ as $N \to \infty$. Since

$$\left|\int_t^{t+h_N} \beta(s,\omega)\,dw_1(s)\right| \leq 2 \sup_u \left|\int_0^u \beta(s,\omega)\,dw_1(s)\right|$$

and the expression to the right of the inequality sign is square integrable in measure P, one can approach the limit in (5) as $N \to \infty$. Let ζ_N be as in Lemma 1. The fact that $\xi(t)$ is a martingale implies the equality

$$\mathsf{E}\xi^2(T) = \mathsf{E}[\xi(T) - \xi(\zeta_N)]^2 + \mathsf{E}\xi^2(\zeta_N).$$

Hence

$$\mathsf{E} \int_0^{\zeta_N} \beta^2(s, \omega)\, ds \leq \mathsf{E}\xi^2(T).$$

Approaching the limit as $N \to \infty$ we complete the proof of the lemma. □

Theorem 3. *If an Itô process defined by formula* (1) *is a Wiener process in* $\mathscr{R}^m$ *with respect to the current* $\{\mathfrak{F}_t\}$, *then*

$$\eta_0 = 0, \qquad a(s, \omega) = 0, \qquad B(s, \omega)B^*(s, \omega) = \boldsymbol{I}$$

for almost all s and ω *(here* $\boldsymbol{I}$ *is the unit operator).*

Proof. Let the conditions of the theorem be satisfied. Then $\eta_0 = 0$ and for any $z \in \mathscr{R}^m$

(6) $$(\eta(t), z) = \int_0^t (a(s, \omega), z)\, ds + \int_0^t |B(s, \omega)z|\, dw_1(s),$$

where

$$w_1(t) = \int_0^t \left(\frac{B(s, \omega)z}{|B(s, \omega)z|}, dw(s) \right)$$

is, in view of Corollary 1 to Theorem 1, a Wiener process. Since $(\eta(t), z)$ is a martingale with respect to $\{\mathfrak{F}_t\}$ and $\mathsf{E}(\eta(t), z)^2 < \infty$, it follows from Lemma 1 that $(a(s, \omega), z) = 0$ for almost all s and ω. Since for a Wiener process

$$\mathsf{E} \sup_{t \leq T} (\eta(t), z)^2 < \infty,$$

it follows that

$$\mathsf{E} \int_0^t |B(s, \omega)z|^2\, ds < \infty$$

for all z.

Let $b(s, \omega)$ be a bounded function with values in $\mathscr{R}^m$. Then

$$\int_0^t (b(s, \omega), d\eta(s)) = \int_0^t (B(s, \omega)b(s, \omega), dw(s)).$$

Consequently,

$$t = \mathsf{E} \int_0^t |B(s, \omega)b(s, \omega)|^2\, ds = \mathsf{E} \int_0^t |b(s, \omega)|^2\, ds$$

and for any $b(s, \omega)$,

(7) $$\mathsf{E}|B(s, \omega)b(s, \omega)|^2 = \mathsf{E}|b(s, \omega)|^2.$$

Let $b_1(s, \omega)$ coincide with the z for which the minimum of $|B(s, \omega)z|$ for $|z|=1$ is attained and let $b_2(s, \omega)$ coincide with the z for which the expression attains its maximum. In this case

$$\mathsf{E}(|B(s, \omega)b_2(s, \omega)|^2 - |B(s, \omega)b_1(s, \omega)|^2) = 0,$$

and hence $|B(s, \omega)b_1(s, \omega)| = |B(s, \omega)b_2(s, \omega)|$. Therefore for all z such that $|z|=1$ we have

$$|B(s, \omega)b_1(s, \omega)| = |B(s, \omega)z| = |B(s, \omega)b_2(s, \omega)|.$$

Consequently, for any vector $z_0 \in \mathcal{R}^m$ the operator

$$\frac{1}{|B(s, \omega)z_0|} B(s, \omega) = U(s, \omega)$$

is a unitary operator. Now let

$$\frac{1}{|B(s, \omega)z_0|} = \alpha(s, \omega).$$

We then obtain from (7) that

$$\mathsf{E}|b(s, \omega)|^2(1-\alpha^2(s, \omega)) = 0.$$

Since $|b(s, \omega)|$ is arbitrary this implies that $\alpha^2(s, \omega) = 1$ for almost all s and ω. We have thus shown that $B(s, \omega)$ is a unitary operator. □

Remark. If a Wiener process $\eta(t)$ is an Itô process with respect to $(w(t), \mathfrak{F}_t)$, then $w(t)$ is also an Itô process with respect to $(\eta(t), \mathfrak{F}_t)$; in other words,

$$w(t) = \int_0^t B^*(s, \omega)\, d\eta(s),$$

provided

$$\eta(t) = \int_0^t B(s, \omega)\, dw(s).$$

This follows from equality $B^*(s, \omega)B(s, \omega) = \boldsymbol{I}$.

Corollary. *If $\eta(t)$ is a Wiener process which is an Itô process with respect to $(w(t), \mathfrak{F}_t)$, then any Itô process with respect to $(\eta(t), \mathfrak{F}_t)$ is an Itô process with respect to $(w(t), \mathfrak{F}_t)$, and any Itô process with respect to $(w(t), \mathfrak{F}_t)$ is an Itô process with respect to $(\eta(t), \mathfrak{F}_t)$.*

Before settling the question concerning the expressions for functions $a(s, \omega)$ and $B(s, \omega)$ in terms of the process $\eta(t)$ defined by formula (1) we shall study processes expressible in terms of stochastic integrals.

The Itô space. Consider the set of processes $I_T(w(t), \mathfrak{F}_t)$ consisting of processes $\eta(t)$ of the form

$$\eta(t)=\int_0^t (b(s,\omega), dw(s)), \qquad t\in[0,T],$$

where $b(s,\omega)$ is a measurable function with values in $\mathcal{R}^m$ such that

$$\int_0^T |b(s,\omega)|^2\, ds<\infty.$$

$I_T(w(t), \mathfrak{F}_t)$ is a linear space. We shall call it the *Itô space*. An important property of this space is its completeness with respect to uniform convergence in probability.

Theorem 4. *If $\eta_n(t)$ is a sequence of processes in $I_T(w(t), \mathfrak{F}_t)$ such that $\sup_{t\le T}|\eta_n(t)-\eta_m(t)|\to 0$ in probability as $n\to\infty$ and $m\to\infty$, then there exists a process $\eta_0(t)\subset I_T(w(t), \mathfrak{F}_t)$ such that $\sup_{t\le T}|\eta_n(t)-\eta_0(t)|\to 0$ as $n\to\infty$ in probability.*

The following lemma will be needed for the proof of the theorem.

Lemma 3. *If*

$$\eta_n(t)=\int_0^t (b_n(s,\omega), dw(s))$$

is a sequence of processes belonging to $I_T(w(t), \mathfrak{F}_t)$ and $\sup_{t\le T}|\eta_n(t)|\to 0$ in probability as $n\to\infty$, then

$$\int_0^T |b_n(s,\omega)|^2\, ds\to 0$$

in probability as $n\to\infty$.

Proof. We may assume without loss of generality that

$$\lim_{n\to\infty}\sup_{t\le T}|\eta_n(t)|=0,$$

with probability 1 (this can always be achieved by resorting to a subsequence—to prove that a sequence converges to zero in probability it is sufficient to show that any subsequence of this sequence contains a subsequence convergent to zero in probability).

Set

$$\tau_N=\sup\,[t\colon t\le T, \sup_{s\le t,\, n>N}|\eta_n(s)|<\varepsilon].$$

Observe that $\tau_N=T$ for N sufficiently large; moreover, τ_N is Markov time. Let $\eta_n^N(t)=\eta_n(t)$ for $t\le\tau_N$ and $\eta_n^N(t)=\eta_n(\tau_N)$ for $t\ge\tau_N$. Then $|\eta_n^N(t)|\le\varepsilon$ for $n>N$

and $\eta_n^N(T)\to 0$ in probability. Hence

$$\mathsf{E}|\eta_n^N(T)|^2=\mathsf{E}\int_0^{\tau_N}|b_n(s,\omega)|^2\,ds\to 0$$

as $n\to\infty$. However,

$$\mathsf{P}\{\int_0^T|b_n(s,\omega)|^2\,ds>\delta\}\le \mathsf{P}\{\tau_N<T\}+\mathsf{P}\{\int_0^{\tau_N}|b_n(s,\omega)|^2\,ds>\delta\}$$

$$\le \mathsf{P}\{\tau_N<T\}+\frac{1}{\delta}\mathsf{E}\int_0^{\tau_N}|b_n(s,\omega)|^2\,ds.$$

Thus

$$\varlimsup_{n\to\infty}\mathsf{P}\{\int_0^T|b_n(s,\omega)|^2\,ds>\delta\}\le \mathsf{P}\{\tau_N<T\}.$$

Approaching the limit in this relation as $N\to\infty$ we obtain the required assertion. □

Proof of the theorem. It follows from the conditions of the theorem that $\eta_n(0)$ converges in probability to a limit. Therefore we can assume without loss of generality that $\eta_n(0)=0$ for all n.

Since

$$\sup_{t\le T}|\int_0^t(b_n(s,\omega)-b_m(s,\omega),dw(s))|\to 0$$

in probability as n and $m\to\infty$, we have in view of Lemma 3

(8) $$\int_0^T|b_n(s,\omega)-b_m(s,\omega)|^2\,ds\to 0$$

in probability as n and $m\to\infty$. Choose a subsequence $\{n_k\}$ such that

$$\mathsf{P}\{\lim_{k,l\to\infty}\int_0^T|b_{n_k}(s,\omega)-b_{n_l}(s,\omega)|^2\,ds=0\}=1.$$

Then there exists a function $b_0(s,\omega)$ such that

(9) $$\mathsf{P}\{\lim_{k\to\infty}\int_0^T|b_{n_k}(s,\omega)-b_0(s,\omega)|^2\,ds=0\}=1,$$

and, moreover, $b_0(s,\omega)$ is measurable and also $\mathfrak{F}_s$-measurable* for all s as a function of ω. It follows from (8) and (9) that

$$\int_0^T|b_n(s,\omega)-b_0(s,\omega)|^2\,ds\to 0$$

in probability as $n\to\infty$. In this case, however, in view of Lemma 2 in Section 2 of

* To achieve this it is necessary first to select from $b_{n_k}(s,\omega)$ a subsequence convergent almost everywhere with respect to the product of Lebesgue measure on the line and measure P and to define $b_0(s,\omega)$ as the limit of this subsequence wherever it exists.

Chapter I we have

$$\mathsf{P}\{\sup_{t\le T} |\int_0^t (b_n(s,\omega), dw(s)) - \int_0^t (b_0(s,\omega), dw(s))| > \varepsilon\}$$

$$\le \mathsf{P}\{\int_0^T |b_n(s,\omega) - b_0(s,\omega)|^2 \, ds > \delta\} + \frac{\delta}{\varepsilon^2}.$$

Approaching the limit as $n\to\infty$ and then letting $\delta\to 0$ we verify that the sequence of processes $\eta_n(t)$ converges uniformly in probability to the process

$$\eta_0(t) = \int_0^t (b_0(s,\omega), dw(s))$$

belonging to $I_T(w(t), \mathfrak{F}_t)$.

The theorem is proved. □

Processes in $I_t(w(t), \mathfrak{F}_t)$ are local martingales and one can find a reducing sequence of random times $\tau_N \uparrow T$ such that the process

$$\eta_N(t) = \eta(\tau_N \wedge t)$$

will be a square integrable martingale. One may choose as τ_N

$$\tau_N = \sup[t: t\le T, \int_0^t |b(s,\omega)|^2 \, ds < N]$$

provided

(10) $$\eta(t) = \int_0^t (b(s,\omega), dw(s)).$$

Since the expression

$$\eta^2(t) - \int_0^t |b(s,\omega)|^2 \, ds$$

is a local martingale as well (for which τ_N is also a reducing sequence of random times), it follows that

$$\langle \eta, \eta\rangle_t = \int_0^t |b(s,\omega)|^2 \, ds.$$

In view of Theorem 22 in Section 1 of Chapter I we have

(11) $$\int_0^t |b(s,\omega)|^2 \, ds = \lim_{\lambda\to 0} \sum_{k=0}^{n-1} (\eta(t_{k+1}) - \eta(t_k))^2$$

in the sense of convergence in probability provided $0 = t_0 < t_1 < \cdots < t_n = t$, $\lambda = \max(t_{k+1} - t_k)$. The expression

(12) $$\langle \eta_1, \eta_2\rangle_t = \lim_{\lambda\to 0} \sum_{k=0}^{n-1} (\eta_1(t_{k+1}) - \eta_1(t_k))(\eta_2(t_{k+1}) - \eta_2(t_k))$$

can be defined for each pair of processes $\eta_1(t)$ and $\eta_2(t)$ belonging to $I_T(w(t), \mathfrak{F}_t)$; the limit in (12) is taken in the sense of convergence in probability and λ and t_k, $k=1, \ldots, n$, are as defined above (cf. Chapter I, Section 1, the corollary to Theorem 22).

If $\eta_k(t)$, $k=1, 2$, are defined by

$$\eta_k(t)=\int_0^t (b_k(s, \omega), dw(s)),$$

then it follows from (11) that

$$\langle \eta_1, \eta_2 \rangle_t = \int_0^t (b_1(s, \omega), b_2(s, \omega))\, ds. \tag{13}$$

Formula (13) allows us to reconstruct the function $b(s, \omega)$ from the process $\eta(t)$ defined in equation (10).

Indeed, let $\zeta_z(t)$, $z \in \mathscr{R}^m$, be defined by

$$\zeta_z(t)=(z, w(t))-\int_0^t (z, dw(s)).$$

Then

$$\langle \eta, \zeta_z \rangle_t = \int_0^t (b(s, \omega), z)\, ds,$$

and hence

$$(b(t, w), z)=\frac{d}{dt}\langle \eta, \zeta_z \rangle_t \tag{14}$$

for almost all t. Clearly, to determine $b(t, \omega)$ it is sufficient to have values of $(b(t, \omega), z)$ only for a z belonging to a basis in $\mathscr{R}^m$.

This result allows us to determine functions $a(s, \omega)$ and $B(s, \omega)$ from the Itô process $\eta(t)$ defined by equation (1).

Observe that for any process $\gamma(t)$ in $\mathscr{R}^1$ of a bounded variation with probability 1 and for a Wiener process $w_1(t)$ we have

$$\lim_{\lambda \to 0} \sum_{k=0}^{n-1} [\gamma(t_{k+1})-\gamma(t_k)][w_1(t_{k+1})-w_1(t_k)]=0$$

(here λ and t_k are as defined above). This is because the sum appearing on the right of the limit sign is bounded above by

$$\operatorname{Var} \gamma(\cdot) \sup_{|s_1-s_2| \leq \lambda} |w_1(s_1)-w_1(s_2)|$$

and $w_1(t)$ is a continuous process.

Let

$$\gamma_z(t)=\int_0^t (a(s, \omega), z)\, ds.$$

Clearly the variation of $\gamma_z(t)$ does not exceed

$$\int_0^T |(a(s,\omega), z)|\, ds.$$

Hence for any z and $x \in \mathcal{R}^m$ we have

$$\lim_{\lambda\to 0} \sum_{k=0}^{n-1} [\gamma_z(t_{k+1}) - \gamma_z(t_k)][\zeta_x(t_{k+1}) - \zeta_x(t_k)] = 0.$$

Furthermore,

$$\lim_{\lambda\to 0} \sum_{k=0}^{n-1} [\zeta_x(t_{k+1}) - \zeta_x(t_k)][\xi_z(t_{k+1}) - \xi_z(t_k)] = \langle \zeta_x, \xi_z \rangle_t = \int_0^t (x, B^*(s,\omega) z)\, ds,$$

provided

$$\xi_z(t) = (z, \int_0^t B(s,\omega)\, dw(s)) = \int_0^t (B^*(s,\omega) z, dw(s)).$$

Thus if $\eta(t)$ is defined by formula (1) we have for all x and $z \in \mathcal{R}^m$

$$\int_0^t (B(s,\omega)x, z)\, ds = \lim_{\lambda\to 0} \sum_{k=1}^{n-1} [\zeta_x(t_{k+1}) - \zeta_x(t_k)][(\eta(t_{k+1}), z) - (\eta(t_k), z)]$$

in the sense of convergence in probability. Consequently,

$$(15) \qquad (B(t,\omega)x, z) = \frac{d}{dt} \lim_{\lambda\to 0} \sum_{k=0}^{n-1} [\zeta_x(t_{k+1}) - \zeta_x(t_k)](\eta(t_{k+1}), z) - (\eta(t_k), z)$$

for almost all t. Formula (15) defines $B(t,\omega)$ for almost all t and ω. If $B(t,\omega)$ is defined, then

$$(16) \qquad (a(t,\omega), z) = \frac{d}{dt} [(\eta(t) - \eta(0), z) - \int_0^t (B^*(s,\omega) z, dw(s))]$$

for almost all t and ω.

Thus we have proved

Theorem 5. *An Itô process $\eta(t)$ defined by* (1) *determines the values of functions $a(t,\omega)$ and $B(t,\omega)$ for almost all t and ω.*

Denote by $I_T(\mathfrak{F}_t)$ the collection of processes $\eta(t)$ for which there exists a measurable numerical function $\beta(s,\omega)$, $\beta^2(s,\omega) \in \mathfrak{M}_1[0,T]$, and a Wiener process $w_1(t)$ with respect to the current $\{\mathfrak{F}_t\}$ such that

$$(17) \qquad \eta(t) = \int_0^t \beta(s,\omega)\, dw_1(s).$$

A set of processes in $I_T(\mathfrak{F}_t)$ can now be described by means of σ-algebras $\mathfrak{F}_t$ only. It is easy to verify that $I_T(w(t), \mathfrak{F}_t) \subset I_T(\mathfrak{F}_t)$ for any Wiener process $w(t)$ with respect to $\{\mathfrak{F}_t\}$. Indeed, if

$$\eta(t) = \int_0^t (b(s, \omega), dw(s)),$$

then

$$\eta(t) = \int_0^t |b(s, \omega)| \, dw_1(s),$$

where

$$w_1(t) = \int_0^t \left(\frac{b(s, \omega)}{|b(s, \omega)|}, dw(s) \right)$$

is, in view of Corollary 1 to Theorem 1, a Wiener process (if $|b(s, \omega)| = 0$ we set $b(s, \omega)/|b(s, \omega)| = z$, where z is a fixed vector in $\mathcal{R}^m$). Clearly processes $\eta(t)$ belonging to $I_T(\mathfrak{F}_t)$ possess the following properties:

1) $\eta(t)$ is a continuous local martingale with respect to the current $\{\mathfrak{F}_t\}$.

2) The monotone process $\langle \eta, \eta \rangle_t$ is absolutely continuous with respect to the Lebesgue measure on the line, i.e., there exists a nonnegative measurable function $\gamma(s, \omega)$ such that

$$\langle \eta, \eta \rangle_t = \int_0^t \gamma(s, \omega) \, ds.$$

It turns out that conditions 1) and 2) assure, under quite general assumptions on the current $\{\mathfrak{F}_t\}$, that $\eta(t)$ belongs to $I_T(\mathfrak{F}_t)$.

Definition. A current $\{\mathfrak{F}_t\}$ is called *nondegenerate* provided there exists at least one process $w_1(t)$ which is a Wiener process with respect to this current.

Theorem 6. *Let a current* $\{\mathfrak{F}_t, 0 \leq t \leq T\}$ *be nondegenerate. Then any process* $\eta(t)$ *satisfying conditions* 1) *and* 2) *belongs to* $I_T(\mathfrak{F}_t)$.

Proof. Let

$$\langle \eta, \eta \rangle_t = \int_0^t \gamma(s, \omega) \, ds$$

and $w(t)$ be a Wiener process with respect to $\{\mathfrak{F}_t\}$ (since $\{\mathfrak{F}_t\}$ is nondegenerate such a process exists). Let

$$g_1(s, \omega) = \begin{cases} 0, & \gamma(s, \omega) = 0 \\ \dfrac{1}{\sqrt{\gamma(s, \omega)}}, & \gamma(s, \omega) > 0, \end{cases}$$

$$g_2(s, \omega) = \begin{cases} 1, & \gamma(s, \omega) = 0 \\ 0, & \gamma(s, \omega) > 0. \end{cases}$$

Set

$$\tilde{w}(t)=\int_0^t g_1(s,\omega)\,d\eta(s)+\int_0^t g_2(s,\omega)\,dw(s)$$

(the definition of stochastic integrals over local martingales is given in Section 2 of Chapter I).

As it follows from Theorem 3 in Section 2 of Chapter I, a stochastic integral over a local martingale is a local martingale. Moreover, since by the definitions of functions g_1 and g_2, $g_1 \cdot g_2=0$ and $g_1^2\gamma+g_2^2=1$, we have

$$\begin{aligned}\langle\tilde{w},\tilde{w}\rangle_t&=\int_0^t g_1^2(s,\omega)\,d\langle\eta,\eta\rangle_s+2\int_0^t g_1(s,\omega)g_2(s,\omega)\,d\langle\eta,w\rangle_s+\int_0^t g_2^2(s,\omega)\,ds\\&=\int_0^t[g_1^2(s,\omega)\gamma(s,\omega)+g_2^2(s,\omega)]\,ds=t.\end{aligned}$$

Hence $\tilde{w}(t)$ is a local $\mathfrak{F}_t$-martingale such that $\langle\tilde{w},\tilde{w}\rangle_t=t$. However, this implies, in view of Theorem 3 in Section 3 of Chapter I, that $\tilde{w}(t)$ is a Wiener process and

$$\begin{aligned}\int_0^t\sqrt{\gamma(s,\omega)}\,d\tilde{w}(s)&=\int_0^t\sqrt{\gamma(s,\omega)}\,g_1(s,\omega)\,d\eta(s)+\int_0^t\sqrt{\gamma(s,\omega)}\,g_2(s,\omega)\,dw(s)\\&=\int_0^t\sqrt{\gamma(s,\omega)}\,g_1(s,\omega)\,d\eta(s).\end{aligned}$$

If

$$\zeta(t)=\int_0^t\sqrt{\gamma(s,\omega)}\,g_1(s,\omega)\,d\eta(s),$$

then

$$\begin{aligned}\langle\eta,\zeta\rangle_t&=\int_0^t\sqrt{\gamma(s,\omega)}\,g_1(s,\omega)\,d\langle\eta,\eta\rangle_s=\int_0^t\gamma(s,\omega)\,ds=\langle\eta,\eta\rangle_t,\\\langle\zeta,\zeta\rangle_t&=\int_0^t\gamma(s,\omega)\,ds=\langle\eta,\eta\rangle_t.\end{aligned}$$

Therefore

$$\langle\eta-\zeta,\eta-\zeta\rangle_t=\langle\eta,\eta\rangle_t-2\langle\eta,\zeta\rangle_t+\langle\zeta,\zeta\rangle_t=0.$$

Hence

$$\mathsf{E}(\eta(t)-\zeta(t))^2=\mathsf{E}\langle\eta-\zeta,\eta-\zeta\rangle_t=0\quad\text{and}\quad\eta(t)=\zeta(t)$$

with probability 1, i.e.,

$$\eta(t)=\int_0^t\sqrt{\gamma(s,\omega)}\,d\tilde{w}(s).$$

The theorem is proved. □

In what follows it will always be assumed that the current $\{\mathfrak{F}_t\}$ is nondegenerate.

We shall now show that the space $I_T(\mathfrak{F}_t)$ is also a complete space.

Theorem 7. *Let $\eta_n(t)$ be a sequence of processes in $I_T(\mathfrak{F}_t)$ such that a process $\eta_0(t)$ exists satisfying*

$$\lim_{n\to\infty} \sup_{t\leq T} |\eta_n(t)-\eta_0(t)|=0$$

in the sense of convergence in probability. Then $\eta_0(t)$ is stochastically equivalent to a process in $I_T(\mathfrak{F}_t)$.

Proof. It is easy to verify that $\eta_0(t)$ is a local $\mathfrak{F}_t$-martingale. We need only show that $\langle\eta_0, \eta_0\rangle_t$ is absolutely continuous with respect to the Lebesgue measure.

We may assume without loss of generality that $\eta_n(t)$ converges to $\eta_0(t)$ uniformly with probability 1.

Let τ_N be determined by the relation

$$\tau_N=\sup\,[t\colon t\leq T, \sup_{n>N} \sup_{s\leq t} |\eta_n(s)-\eta_0(s)|<\varepsilon];$$

set $\eta_n^N(t)=\eta_n(t\wedge\tau_N)$. Then $|\eta_n^N(t)-\eta_0^N(t)|<\varepsilon$ for $n>N$ and hence

$$\lim_{n\to\infty} \mathsf{E}|\eta_n^N(t)-\eta_0^N(t)|^2=0.$$

However,

$$\mathsf{E}\langle\eta_n^N-\eta_0^N, \eta_n^N-\eta_0^N\rangle=\mathsf{E}|\eta_n^N(t)-\eta_0^N(t)|^2.$$

Consequently,

$$\lim_{n\to\infty} \mathsf{E}\langle\eta_n^N-\eta_0^N, \eta_n^N-\eta_0^N\rangle_t=0.$$

But

$$\mathsf{P}\{\langle\eta_n-\eta_0, \eta_n-\eta_0\rangle_T>\delta\}\leq\mathsf{P}\{\tau_N<T\}+\frac{1}{\delta}\mathsf{E}\langle\eta_n^N-\eta_0^N, \eta_n^N-\eta_0^N\rangle_T.$$

Since $\tau_N=T$ for sufficiently large N, it follows that

$$\langle\eta_n-\eta_0, \eta_n-\eta_0\rangle_T\to 0$$

in probability. Without loss of generality we may assume that the last relation is valid with probability 1. Since for $t_1<t_2$

$$\begin{aligned}\langle\eta_0, \eta_0\rangle_{t_2}-\langle\eta_0, \eta_0\rangle_{t_1}\leq 2[&\langle\eta_n, \eta_n\rangle_{t_2}-\langle\eta_n, \eta_n\rangle_{t_1}\\ &+\langle\eta_n-\eta_0, \eta_n-\eta_0\rangle_{t_2}-\langle\eta_n-\eta_0, \eta_n-\eta_0\rangle_{t_1}],\end{aligned}$$

we have for an arbitrary Borel set Λ on $[0, T]$

$$\int_\Lambda d\langle\eta_0, \eta_0\rangle_t \leq 2\int_\Lambda d\langle\eta_n, \eta_n\rangle_t + 2\int_\Lambda d\langle\eta_n - \eta_0, \eta_n - \eta_0\rangle \\ \leq 2\int_\Lambda d\langle\eta_n, \eta_n\rangle_t + 2\langle\eta_n - \eta_0, \eta_n - \eta_0\rangle_T. \tag{18}$$

Let Λ have the Lebesgue measure 0. Utilizing the absolute continuity of $\langle\eta_n, \eta_n\rangle_t$ we obtain

$$\int_\Lambda d\langle\eta_0, \eta_0\rangle_t \leq 2\langle\eta_n - \eta_0, \eta_n - \eta_0\rangle_T.$$

Approaching the limit as $n\to\infty$ we verify that $\langle\eta_0, \eta_0\rangle_T$ is indeed absolutely continuous with respect to the Lebesgue measure. □

Remark. Analogously to formula (18) one can show that for any Borel set $\Lambda \subset [0, T]$ and a pair of processes $\eta_1(t), \eta_2(t) \in I_T(\mathfrak{F}_t)$ the inequality

$$\left|\int_\Lambda d\langle\eta_1, \eta_2\rangle_t\right| \leq \sqrt{\int_\Lambda d\langle\eta_1, \eta_1\rangle_t}\sqrt{\int_\Lambda d\langle\eta_2, \eta_2\rangle_t}$$

is valid. This inequality implies the absolute continuity of $\langle\eta_1, \eta_2\rangle_t$ with respect to the Lebesgue measure. Moreover, if

$$\langle\eta_i, \eta_k\rangle_t = \int_0^t \varphi_{ik}(s, \omega)\, ds, \qquad k = 1, 2,$$

then

$$\varphi_{12}(s, \omega) \leq \sqrt{\varphi_{11}(s, \omega)\varphi_{22}(s, \omega)}.$$

It is possible to introduce the notion of orthogonality of processes on the space $I_T(\mathfrak{F}_t)$. Two processes $\eta_1(t)$ and $\eta_2(t)$ are said to be *orthogonal* if for all $t \in [0, T]$

$$\langle\eta_1, \eta_2\rangle_t = 0.$$

We say that the process $\eta(t)$ is a linear combination of processes $\xi_1(t), \ldots, \xi_n(t)$ if functions $\alpha_1(t, \omega), \ldots, \alpha_n(t, \omega)$ exist such that

$$\sum_{i=1}^n \alpha_i^2(t, \omega) \in \mathfrak{M}_1[0, T] \quad \text{and} \quad \eta(t) = \sum_{i=1}^n \int_0^t \alpha_i(s, \omega)\, d\xi_i(s).$$

Let $\eta_1(t), \ldots, \eta_n(t)$ be a collection of processes in $I_T(\mathfrak{F}_t)$. Assume that they are linearly independent, i.e., not one of them is a linear combination of the remaining ones in the sense indicated above. Then one can construct processes $\xi_1(t), \ldots, \xi_n(t)$ which are pairwise orthogonal and which are linear combinations of $\eta_1(t), \ldots, \eta_n(t)$ and, moreover, $\eta_1(t), \ldots, \eta_n(t)$ are in turn linear combinations of $\xi_1(t), \ldots, \xi_n(t)$. These processes can be constructed using the following formulas:

$$\text{(19)}\qquad \begin{aligned} &\xi_1(t)=\eta_1(t),\\ &\xi_k(t)=\eta_k(t)-\sum_{i=1}^{k-1}\int_0^t \alpha_{ki}(s,\omega)\,d\xi_i(s), \qquad k=2,\dots,n, \end{aligned}$$

where

$$\alpha_{ki}(s,\omega)=\frac{\varphi_{ki}(s,\omega)}{g_{ii}(s,\omega)},$$

if

$$\langle\xi_i,\eta_k\rangle_t=\int_0^t \varphi_{ki}(s,\omega)\,ds, \qquad \langle\xi_i,\xi_i\rangle_t=\int_0^t g_{ii}(s,\omega)\,ds$$

(if $g_{ii}(s,\omega)=0$ and $\varphi_{ki}(s,\omega)=0$ the ratio is also assumed to be 0; cf. the remark following Theorem 7). Formula (19) easily implies that η_k are linear combinations of $\xi_1,\dots,\xi_k$ and that ξ_k are also linear combinations of $\eta_1,\dots,\eta_k$. For $j<k$ we have

$$\langle\xi_k,\xi_j\rangle_t=\langle\eta_k,\xi_j\rangle_t-\sum_{i=1}^{k-1}\int_0^t \alpha_{ki}(s,\omega)\,d\langle\xi_i,\xi_j\rangle_s.$$

Assuming that the pairwise orthogonality of $\xi_1,\dots,\xi_{k-1}$ is verified, we have

$$\langle\xi_k,\xi_j\rangle_t=\int_0^t \varphi_{kj}(s,\omega)\,ds-\int_0^t \alpha_{kj}(s,\omega)g_{jj}(s,\omega)\,ds=0.$$

Utilizing induction we can now deduce the orthogonality of $\xi_1,\dots,\xi_n$.

Let $\{\xi_k(t), k=1,2,\dots\}$ be a sequence of pairwise orthogonal processes in $I_T(\mathfrak{F}_t)$ and let $\eta(t)\in I_T(\mathfrak{F}_t)$.

Assume that

$$\begin{aligned} &\langle\xi_k,\xi_k\rangle_t=\int_0^t g_k(s,\omega)\,ds,\\ &\langle\eta,\xi_k\rangle_t=\int_0^t \varphi_k(s,\omega)\,ds, \qquad \langle\eta,\eta\rangle_t=\int_0^t \varphi(s,\omega)\,ds. \end{aligned}$$

From the remark following Theorem 7 we have that $\varphi_k(s,\omega)$ vanishes almost everywhere where g_k vanishes. Let

$$\alpha_\eta^k(s,\omega)=\frac{\varphi_k(s,\omega)}{g_k(s,\omega)},$$

provided the denominator is not zero; otherwise we let $\alpha_\eta^k(s,\omega)=0$. In this case

$$\eta(t)=\sum_{k=1}^{n}\int_0^t \alpha_\eta^k(s,\omega)\,d\xi_k(s)+\eta_n(t),$$

where $\eta_n(t)$ is orthogonal to $\xi_1(t), \ldots, \xi_n(t)$ and hence also to the process

$$\int_0^t \alpha_\eta^k(s, \omega)\, d\xi_k(s).$$

Therefore

$$\langle \eta, \eta \rangle_t = \int_0^t \varphi(s, \omega)\, ds = \sum_{k=1}^n \int_0^t [\alpha_\eta^k(s, \omega)]^2 \varphi_k(s, \omega)\, ds + \langle \eta_n, \eta_n \rangle_t.$$

This implies that

$$\sum_{k=1}^n [\alpha_\eta^k(t, \omega)]^2 \varphi_k(t, \omega) \leqslant \varphi(t, \omega)$$

for almost all (t, ω) (here we have used the fact that $d\langle \eta_n, \eta_n \rangle / dt \geqslant 0$). Hence,

$$\sum_{k=1}^\infty [\alpha_\eta^k(t, \omega)]^2 \varphi_k(t, \omega) \leqslant \varphi(t, \omega)$$

(the series to the left of the inequality sign is convergent since all its terms are nonnegative). Therefore, setting

$$\zeta_n(t) = \sum_{k=1}^n \int_0^t \alpha_\eta^k(s, \omega)\, d\xi_k(s),$$

we have

(20) $$\langle \zeta_n - \zeta_m, \zeta_n - \zeta_m \rangle_T = \int_0^T \sum_{k=n+1}^m [\alpha_\eta^k(s, \omega)]^2 \varphi_k(s, \omega)\, ds \to 0$$

as n and $m \to \infty$ $(n < m)$. Observe that the inequality

(21) $$\mathsf{P}\{\sup_{t \leqslant T} |\zeta(t)| > c\} \leqslant \frac{N}{c^2} + \mathsf{P}\{\langle \zeta, \zeta \rangle_T > N\}$$

holds for any process $\zeta(t) \in I_T(\mathfrak{F}_t)$. This follows from the fact that $\zeta(t)$ can be expressed as a stochastic integral over a Wiener process and properties of stochastic integrals (cf. Chapter I, Section 3, Lemma 2). We therefore conclude from (20) that

$$\sup_{t \leqslant T} |\zeta_n(t) - \zeta_m(t)| \to 0$$

in probability, and hence in view of Theorem 7 there exists a process $\zeta_0(t) \in I_T(\mathfrak{F}_t)$ such that

$$\sup_{t \leqslant T} |\zeta_n(t) - \zeta_0(t)| \to 0$$

in probability. We have thus shown that the series

$$\sum_{k=1}^{\infty} \int_0^t \alpha_\eta^k(s, \omega)\, d\xi_k(s)$$

converges in probability and that its sum belongs to $I_T(\mathfrak{F}_t)$. Hence

$$\eta(t) = \sum_{k=1}^{\infty} \int_0^t \alpha_\eta^k(s, \omega)\, d\xi_k(s) + \eta_0(t),$$

where $\eta_0(t) \in I_T(\mathfrak{F}_t)$. It is easy to verify that $\eta_0(t)$ is orthogonal to all the processes $\xi_k(t)$, $k = 1, 2, \ldots$. If for all $\eta(t) \in I_T(\mathfrak{F}_t)$ the equality

$$\eta(t) = \sum_{k=1}^{\infty} \int_0^t \alpha_\eta^k(s, \omega)\, d\xi_k(s) \tag{22}$$

is fulfilled, then the sequence $\{\xi_k(t), k = 1, 2, \ldots\}$ is called an *orthogonal basis* in $I_T(\mathfrak{F}_t)$. It follows from the above that in order for a sequence of pairwise orthogonal functionals $\{\xi_k(t), k = 1, 2, \ldots\}$ to be a basis, it is necessary and sufficient that there exist no processes $\eta(t) \in I_T(\mathfrak{F}_t)$ different from zero which are orthogonal to all $\xi_k(t)$.

In the case when $\varphi_k(t, \omega)$ are almost everywhere positive, where

$$\int_0^t \varphi_k(s, \omega)\, ds = \langle \xi_k, \xi_k \rangle_t$$

and $\{\xi_k(t), k = 1, 2, \ldots\}$ form a basis, one can choose the processes

$$w_k(t) = \int_0^t \frac{1}{\sqrt{\varphi_k(s, \omega)}}\, d\xi_k(s)$$

satisfying

$$\langle w_k, w_k \rangle_t = t$$

as a new basis. Processes $w_k(t)$ are Wiener processes and, moreover, are independent. Indeed, if $w^{(m)}(t)$ is a process in $\mathscr{R}^m$ with coordinates $(w_1(t), \ldots, w_m(t))$, then it is a martingale and for $z \in \mathscr{R}^m$ we have

$$\mathsf{E}[(w^{(m)}(t+h) - w^{(m)}(t), z)^2 \mid \mathfrak{F}_t]$$

$$= \mathsf{E}\Big(\sum_{k,j=1}^{m} [\langle w_k, w_j \rangle_{t+h} - \langle w_k, w_j \rangle_t] z_k z_j \mid \mathfrak{F}_t \Big) = h|z|^2$$

(here z_k denotes the coordinates of z). Hence, in view of Corollary 3 to Theorem 1 this process is a Wiener process in $\mathscr{R}^m$.

We shall now investigate the problem when $I_T(\mathfrak{F}_t)$ possesses a finite base in Wiener processes.

Theorem 8. *The space $I_t(\mathfrak{F}_t)$ possesses a finite basis in m Wiener processes provided there exists a Wiener process $w^{(m)}(t)$ in $\mathcal{R}^m$ such that*

$$\sigma[w^{(m)}(s), s \leq t] \subset \mathfrak{F}_t \subset \overline{\sigma[w^{(m)}(s), s \leq t]},$$

where $\sigma[w^{(m)}(s), s \leq t]$ is a σ-algebra generated by the variables $w^{(m)}(s)$, $s \leq t$, and $\overline{\sigma[\cdot]}$ is a completion of this σ-algebra.

Proof. Let $w_1(t), \ldots, w_m(t)$ be the coordinates of $w^{(m)}(t)$. Clearly they belong to $I_T(\mathfrak{F}_t)$ and in view of independence they are pairwise orthogonal. To show that they form a basis in $I_T(\mathfrak{F}_t)$ it is sufficient to verify that conditions

$$\langle \xi, w_k \rangle_t = 0, \qquad k = 1, \ldots, m,$$

where $\xi(t) \in I_T(\mathfrak{F}_t)$ imply the equality $\langle \xi, \xi \rangle_T = 0$. Set

$$\langle \xi, \xi \rangle_t = \int_0^t \gamma(s, \omega)\, ds.$$

Choose a one-dimensional Wiener process $\tilde{w}(t)$ independent of $w^{(m)}(t)$. Let $\alpha(s, \omega) = 1$ if $\gamma(s, \omega) > 0$ and $\alpha(s, \omega) = 0$ if $\gamma(s, \omega) = 0$. Then the process

$$w_{m+1}(t) = \int_0^t \frac{\alpha(s, \omega)}{\sqrt{\gamma(s, \omega)}}\, d\xi(s) + \int_0^t (1 - \alpha(s, \omega))\, d\tilde{w}(s)$$

belongs to $I_T(\hat{\mathfrak{F}}_t)$, where $\hat{\mathfrak{F}}_t = \sigma[\mathfrak{F}_t \cup \sigma[\tilde{w}(s), s \leq t]]$. Moreover, $I_T(\mathfrak{F}_t) \subset I_T(\hat{\mathfrak{F}}_t)$ and

$$\langle w_{m+1}, w_{m+1} \rangle_t = \int_0^t \frac{\alpha^2(s, \omega)}{\gamma(s, \omega)} \gamma(s, \omega)\, ds + \int_0^t (1 - \alpha(s, \omega))^2\, ds = t,$$

since $\tilde{w}(t)$ is orthogonal to all the processes in $I_T(\mathfrak{F}_t)$ because it is independent from each one of the $\{\mathfrak{F}_t\}$-adopted processes. Utilizing the orthogonality of ξ and w_k we verify that

$$\langle w_k, w_{m+1} \rangle_t = 0, \qquad k = 1, \ldots, m.$$

Thus $(w_1(t), \ldots, w_m(t), w_{m+1}(t))$ form an $(m+1)$-dimensional Wiener process and hence $w_{m+1}(t)$ does not depend on $w^{(m)}(t)$, i.e., on the σ-algebra $\mathfrak{F}_t$ as well. Therefore

$$\mathsf{E}(w_{m+1}(T) | \mathfrak{F}_t) = 0.$$

Utilizing the independence of $\tilde{w}(t)$ from $\mathfrak{F}_T$, we obtain

$$\mathsf{E}\left(\int_0^t (1-\alpha(s,\omega))\, d\tilde{w}(s) \mid \mathfrak{F}_T\right)=0.$$

Hence

$$\mathsf{E}\left(\int_0^t \frac{\alpha(s,\omega)}{\sqrt{\gamma(s,\omega)}}\, d\xi(s) \mid \mathfrak{F}_T\right)=0.$$

However, the quantity to the right of the sign of the conditional mathematical expectation in the last equation is $\mathfrak{F}_T$ measurable, i.e.,

$$\zeta(t)=\int_0^t \frac{\alpha(s,\omega)}{\sqrt{\gamma(s,\omega)}}\, d\xi(s)=0$$

for all t. Therefore

$$\langle\zeta,\zeta\rangle_T=\int_0^T \alpha^2(s,\omega)\, ds=0$$

for almost all ω; thus $\alpha(s,\omega)=0$ for almost all (s,ω) and also $\gamma(s,\omega)=0$ for almost all (s,ω). Consequently,

$$\langle\xi,\xi\rangle_T=\int_0^T \gamma(s,\omega)\, ds=0.$$

The theorem is proved. $\square$

In the case when σ-algebras $\mathfrak{F}_t$ satisfy the conditions of the preceding theorem the structure of the space $I_T(\mathfrak{F}_t)$ is quite simple. Observe that in the case when $\eta(t)\in I_T(\mathfrak{F}_t)$ and $\mathsf{E}\eta^2(T)<\infty$ we have

$$\eta(t)=\mathsf{E}(\eta(T)\mid\mathfrak{F}_t)$$

since $\eta(t)$ is a martingale.

The subset $I_T^2(\mathfrak{F}_t)$ of processes $\eta(t)\in I_T(\mathfrak{F}_t)$ satisfying $\mathsf{E}\eta^2(T)<\infty$ is everywhere dense in $I_T(\mathfrak{F}_t)$ in the sense of uniform convergence in probability:

$$\eta(t)=\lim_{N\to\infty}\eta_N(t),$$

provided

$$\eta(t)=\int_0^t \gamma(s,\omega)\, dw(s), \qquad \eta_N(t)=\int_0^t \gamma_N(s,\omega)\, dw(s),$$

and

$$\int_0^T (\gamma_N(s,\omega))^2\, ds\le N \quad\text{and}\quad \int_0^T |\gamma_N(s,\omega)-\gamma(s,\omega)|^2\, ds\to 0,$$

where $w(t)$ is a Wiener process.

Therefore to determine $I_T(\mathfrak{F}_t)$ it is sufficient to describe the everywhere dense set $I_T^2(\mathfrak{F}_t)$. The following theorem provides such a description.

Theorem 9. *If a current $\{\mathfrak{F}_t\}$ satisfies the conditions of Theorem 8 then every process $\eta(t)$ belonging to $I_T^2(\mathfrak{F}_t)$ is of the form*

$$\eta(t)=\mathsf{E}(\eta\,|\,\mathfrak{F}_t), \tag{23}$$

where η is an $\mathfrak{F}_T$-measurable quantity such that $\mathsf{E}\eta=0$ and $\mathsf{E}\eta^2<\infty$; if η satisfies these conditions then the process $\eta(t)$ defined by formula (23) *belongs to $I_T^2(\mathfrak{F}_t)$.*

Proof. For processes belonging to $I_T^2(\mathfrak{F}_t)$, $\eta(T)$ should be taken for η.

Let η be $\mathfrak{F}_t$-measurable with $\mathsf{E}\eta=0$ and $\mathsf{E}\eta^2<\infty$. Then one can find a continuous bounded function $f_\varepsilon(x_1,\dots,x_n)$ $(x_i\in\mathcal{R}^m)$ with bounded first two derivatives and a collection $0\leq t_1<\dots<t_n\leq T$ such that

$$\mathsf{E}(\eta-f_\varepsilon(w(t_1),\dots,w(t_n)))^2\leq\varepsilon.$$

(Here $w(t)$ denotes a process generating the current $\{\mathfrak{F}_t\}$.) Set

$$\eta_\varepsilon(t)=\mathsf{E}(f_\varepsilon(w(t_1),\dots,w(t_n))\,|\,\mathfrak{F}_t).$$

Clearly $\eta_\varepsilon(t)$ is a martingale and $\mathsf{E}\eta_\varepsilon(T)^2<\infty$. We now show that $\eta_\varepsilon(t)\in I_T(\mathfrak{F}_t)$.

Let

$$f_\varepsilon^{(k)}(x_1,\dots,x_k)=\mathsf{E}f_\varepsilon(x_1,\dots,x_k,x_k+w(t_{k+1})-w(t_k),\dots,x_k+w(t_n)-w(t_k)).$$

Then on the interval $[t_{k-1},t_k]$ we have

$$\eta_\varepsilon(t)=\int f_\varepsilon^{(k)}(w(t_1),\dots,w(t_{k-1}),w(t)+y)p(t_k-t,y)\,dy,$$

where $p(t,y)$ is the probability density of $w(t)$. Hence

$$\eta_\varepsilon(t)=\Phi_\varepsilon^{(k)}(w(t_1),\dots,w(t_{k-1}),t,w(t)),\qquad t\in[t_{k-1},t_k],$$

where $\Phi_\varepsilon^{(k)}(w(t_1),\dots,w(t_{k-1}),t,x)$ is a twice continuously differentiable function in t and x.

Applying Itô's formula we have

$$\begin{aligned}\eta_\varepsilon(t)-\eta_\varepsilon(t_{k-1})=\int_{t_{k-1}}^{t}\Big[&\frac{\partial}{\partial s}\Phi_\varepsilon^{(k)}(w(t_1),\dots,w(t_{k-1}),s,w(s))\\ &+\frac{1}{2}\,\mathrm{Sp}\,\frac{d^2}{dx^2}\Phi_\varepsilon^{(k)}(w(t_1),\dots,w(t_{k-1}),s,w(s))\Big]\,ds\\ &+\int_{t_{k-1}}^{t_k}\Big(\frac{d}{dx}\Phi_\varepsilon^{(k)}(w(t_1),\dots,w(t_{k-1}),s,w(s)),dw(s)\Big)\end{aligned}\tag{24}$$

(note that $(d/dx)\Phi_\varepsilon^{(k)}$ takes on values in $\mathscr{R}^m$ and $(d^2/dx^2)\Phi_\varepsilon^{(k)}$ takes on values in $\mathfrak{L}(\mathscr{R}^m)$). Since $\eta_\varepsilon(t)$ is a martingale the first integral on the right-hand side of (24) vanishes in view of Lemma 1. Hence

$$\eta_\varepsilon(t)=\eta_\varepsilon(t_{k-1})+\int_{t_{k-1}}^t\left(\frac{d}{dx}\Phi_\varepsilon^{(k)}(w(t_1),\ldots,w(t_{k-1}),s,w(s)),dw(s)\right).$$

Since $\xi_\varepsilon(0)=\mathsf{E}\eta_\varepsilon=0$ we have verified the existence of a function $b_\varepsilon(s,\omega)$ such that

$$\eta_\varepsilon(t)=\int_0^t(b_\varepsilon(s,\omega),dw(s))$$

and the random variable

$$b_\varepsilon(s,\omega)=\frac{d}{dx}\Phi_\varepsilon^{(k)}(w(t_1),\ldots,w(t_{k-1}),s,w(s))$$

is $\mathfrak{F}_s$-measurable and bounded for a fixed s, $t_{k-1}\le s\le t_k$. Furthermore,

$$\mathsf{P}\{\sup_{0\le t\le T}|\eta_\varepsilon(t)-\eta(t)|>c\}\le\frac{1}{c^2}\mathsf{E}|\eta_\varepsilon-\eta|^2\le\frac{\varepsilon^2}{c^2}.$$

Hence $\eta(t)$ being a uniform limit in probability of the processes in $I_T(\mathfrak{F}_t)$ also belongs to $I_T(\mathfrak{F}_T)$.

Corollary. *Let ξ be an arbitrary $\mathfrak{F}_T$-measurable variable. If $\mathsf{E}\xi^2<\infty$, then*

$$\xi=\mathsf{E}\xi+\int_0^T(b(s,\omega),dw(s)),$$

where $b^2(s,\omega)\in\mathfrak{M}_1[0,T]$ and is $\mathfrak{F}_s$-measurable for each s.

To prove this assertion all that is needed is to observe that the quantity $\xi-\mathsf{E}\xi$ satisfies the conditions of Theorem 9 and hence $\xi-\mathsf{E}\xi=\xi(T)$, where $\xi(t)\in I_T^2(\mathfrak{F}_t)$.

We present an example of a current $\{\mathfrak{F}_t\}$ for which the set $I_T(\mathfrak{F}_t)$ has a basis consisting of a one-dimensional process $w(t)$, while $\mathfrak{F}_t$ is substantially larger than $\sigma[w(s),s\le t]$. Let $w(t)$ be a one-dimensional Wiener process, $\mathfrak{F}_t^{(1)}$ be a current of σ-algebras generated by this process, and τ be a variable independent of $w(t)$ with a continuous distribution and taking on values in $[0,T]$. Set

$$\mathfrak{F}_t^{(2)}=\sigma[\{\tau>s\},s\le t],\qquad \mathfrak{F}_t=\mathfrak{F}_t^{(1)}\cup\mathfrak{F}_t^{(2)}.$$

We show that any process $\eta(t)$ belonging to $I_T(\mathfrak{F}_t)$ is linearly expressible in terms of $w(t)$. It is sufficient to establish this for a process belonging to $I_T^2(\mathfrak{F}_t)$. If

$$\psi(\tau)=\eta(T)$$

then

$$\eta(t)=\mathsf{E}(\eta(T)|\mathfrak{F}_t)=\mathsf{E}(\psi(\tau)|\mathfrak{F}_t^{(2)})=\psi(\tau)\chi_{\{\tau\le t\}}$$

$$+[1-\chi_{\{\tau\le t\}}]\frac{1}{\mathsf{P}\{\tau>t\}}\int_t^T\psi(u)\,dF(u)$$

$$=\frac{1}{\mathsf{P}\{\tau>t\}}\Big[\int_t^T\psi(u)\,dF(u)+\chi_{\{\tau\le t\}}\int_t^T(\psi(\tau)-\psi(u))\,dF(u)\Big],$$

where $F(u)=\mathsf{P}(\tau<u)$ and χ_A is the indicator of the set A. On the other hand $\mathsf{E}(\psi(\tau)|\mathfrak{F}_t)$ is continuous since $\eta(t)$ is such. In this case, however, the function

$$\chi_{\{\tau<t\}}\int_t^T[\psi(\tau)-\psi(u)]\,dF(u)$$

is continuous in t, which is possible only if

$$\int_\tau^T[\psi(\tau)-\psi(u)]\,dF(u)=0$$

for almost all τ. This condition implies that $\psi(\tau)$ is constant for almost all τ, and, since $\mathsf{E}\psi(\tau)=0$, $\psi(\tau)=0$ for almost all τ.

In the general case one can construct a sequence of functions $f_n(s,x,\ldots,x)$ such that

$$\mathsf{E}(\eta(T)-f_n(\tau,w(t_1),\ldots,w(t_n)))^2\to0$$

and

$$\mathsf{E}f_n(s,w(t_1),\ldots,w(t_n))=0$$

(here $\{t_k\}$ is a dense set in $[0,T]$; for example, one can choose

$$f_n(\tau,w(t_1),\ldots,w(t_n))=\mathsf{E}(\eta(T)|\tau,w(t_1),\ldots,w(t_n)).$$

In the same manner as in the proof of Theorem 9, it can be shown that

$$\mathsf{E}(f_n(\tau,w(t_1),\ldots,w(t_n))|\mathfrak{F}_t)=\int_0^t g_n(s,\tau,\omega)\,dw(s),$$

where $g_n(s,\tau,\omega)$ is an $\mathfrak{F}_s$-measurable function for a fixed s. Approaching the limit we verify that $\eta(t)\in I_T(\mathfrak{F}_t,w(t))$.

Itô processes and processes of the diffusion type. Let $\xi(t)$ be a continuous Markov process on $[0,T]$ with values in $\mathcal{R}^m$, $P(t,x,s,dy)$ be its transition probability. Such a process is called a *diffusion process* (cf. Volume II, Chapter I, Section 1) provided there exist functions $a(t, x)$ with values in $\mathscr{R}^m$ and $B(t, x)$ with values in $\mathfrak{L}(\mathscr{R}^m)$ defined on $[0, T]\times\mathscr{R}^m$ such that for any $\varepsilon>0$ the following conditions are satisfied:

(I) $\int_{|y-x|>\varepsilon} P(t, x, s, dy)=o(s-t)$,

(II) $\int_{|y-x|\leq\varepsilon} (y-x)P(t, x, s, dy)=a(t, x)(s-t)+o(s-t)$,

(III) $\forall z\in\mathscr{R}^m \quad \int_{|y-x|\leq\varepsilon} (y-x, z)^2P(t, x, s, dy)=(B(t, x)z, z)(s-t)+o(s-t)$.

We show that under certain additional restrictions the process $\xi(t)$ will be an Itô process with respect to a (certain) Wiener process $w(t)$. As a preliminary, we prove the following lemma.

Lemma 4. *Let $\eta(t)$ be a continuous process and $\{\mathfrak{F}_t\}$ be a current of σ-algebras generated by this process. If for all t there exists a variable ζ_t such that*

$$\sup_{h>0}\left|\frac{1}{h}(\mathsf{E}(\eta(t+h)-\eta(t)|\mathfrak{F}_t)\right|\leq\zeta_t \quad\text{and}\quad \mathsf{E}\zeta_t<\infty$$

and if, moreover,

$$\lim_{h\downarrow 0}\frac{1}{h}\mathsf{E}(\eta(t+h)-\eta(t)|\mathfrak{F}_t)=0,$$

then $\eta(t)$ is a martingale.

Proof. For $s\geq t$ let

$$\psi(s)=\mathsf{E}(\eta(s)|\mathfrak{F}_t).$$

Then

$$\begin{aligned}\lim_{h\downarrow 0}\frac{\psi(s+h)-\psi(s)}{h}&=\lim_{h\downarrow 0}\mathsf{E}\left(\frac{\eta(s+h)-\eta(s)}{h}\middle|\mathfrak{F}_t\right)\\&=\lim_{h\downarrow 0}\mathsf{E}\left(\mathsf{E}\left(\frac{\eta(s+h)-\eta(s)}{h}\middle|\mathfrak{F}_s\right)\middle|\mathfrak{F}_t\right)\\&=\mathsf{E}\left(\lim_{h\downarrow 0}\mathsf{E}\left(\frac{\eta(s+h)-\eta(s)}{h}\middle|\mathfrak{F}_s\right)\middle|\mathfrak{F}_t\right)=0.\end{aligned}$$

The limit transition under the sign of the mathematical expectation is justified in view of the existence of the dominant ζ_s. Moreover since $\eta(s)$ is continuous, so is $\psi(s)$. Therefore $\psi(s)$ is a continuous function which possesses a vanishing right-hand derivative at each point.

Hence $\psi(s)$ is constant for $s \geq t$. Since $\psi(t) = \eta(t)$ it follows that

$$\mathsf{E}(\eta(s) \mid \mathfrak{F}_t) = \eta(t).$$

The lemma is thus proved. □

Theorem 10. *Let condition* (I) *be satisfied uniformly for* $0 \leq t < s \leq T$ *and let* $x \in K$ *for any compact set* $K \subset \mathcal{R}^m$. *Let conditions* (II) *and* (III) *be satisfied. Moreover, let functions* $a(s, x)$ *and* $B(s, x)$ *be continuous and, for each compact* K, *let there exist constants* l *and* c *such that*

1) *for* $x \in K$

$$\left|\int_{|y-x| \leq \varepsilon} (y-x) P(t, x, s, dy)\right| + \int_{|y-x| \leq \varepsilon} |y-x|^2 P(t, x, s, dy) \leq l(s-t),$$

2) $\sup_{|x|>c} P(t, x, s, K) \leq l(s-t)$.

Then a Wiener process $w(t)$ *exists with values in* $\mathcal{R}^m$ *such that* $\xi(t)$ *is an Itô process with respect to* $w(t)$, *and, moreover,*

$$\xi(t) = \xi(0) + \int_0^t a(s, \xi(s))\, ds + \int_0^t B^{1/2}(s, \xi(s))\, dw(s), \tag{25}$$

where $B^{1/2}$ *is the nonnegative-definite square root of operator* B.

Proof. We set

$$\eta(t) = \left(\xi(t) - \xi(0) - \int_0^t a(s, \xi(s))\, ds, z\right)$$

and show that $\eta(t)$ is a local martingale for any $z \in \mathcal{R}^m$.

Let τ_N be the time of the first exit of process $\xi(t)$ out of the set $K_N = \{x : |x| \leq N\}$. Denote by $f_N(x)$ a twice continuously differentiable function satisfying $f_N(x) = (x, z)$ for $|x| \leq N$ and $f_N(x) = 0$ for $|x| \geq N+1$. We now bound the expression

$$\begin{aligned}
&\left|\int [f_N(y) - f_N(x)] P(t, x, s, dy)\right| \\
&\leq \left|\int_{|y-x|>\varepsilon} [f_N(y) - f_N(x)] P(t, x, s, dy)\right| \\
&\quad + \left|\int_{|y-x| \leq \varepsilon} (f_N'(x), y-x) P(t, x, s, dy)\right| \\
&\quad + \tfrac{1}{2}\left|\int_{|y-x| \leq \varepsilon} (f_N''(x + \Theta(y-x))(y-x), y-x) P(t, x, s, dy)\right|
\end{aligned}$$

(here $0 < \Theta < 1$ and Taylor's formula was utilized). For any $C > N+1+\varepsilon$ we have in view of conditions 1) and 2)

$$\begin{aligned}
&\left|\int_{|y-x|>\varepsilon} [f_N(y) - f_N(x)] P(t, x, s, dy)\right| \\
&\leq 2 \sup_z |f_N(z)| \Big[\sup_{|x| \leq C} \int_{|x-y|>\varepsilon} P(t, x, s, dy) \\
&\quad + \sup_{|x| \geq C} P(t, x, s, K_{N+1})\Big] = O(s-t).
\end{aligned}$$

Furthermore, in view of condition 1),

$$\begin{aligned}&\left|\int_{|y-x|\leqslant\varepsilon}(f'_N(x), y-x)P(t, x, s, dy)\right|\\ &\leqslant \sup_{|x|\leqslant N+1}|f'_N(x)|\left|\int_{|y-x|\leqslant\varepsilon}(y-x)P(t, x, s, dy)\right|\\ &=O(s-t).\end{aligned}$$

In the same manner we verify that

$$\int_{|y-x|\leqslant\varepsilon}(f''_N(x+\Theta(y-x))(y-x), y-x)P(t, x, s, dy)=O(s-t)$$

uniformly in s and t. Thus there exists a constant l_1 such that

$$\left|\int[f_N(y)-f_N(x)]P(t, x, s, dy)\right|\leqslant l_1(s-t).$$

Set

$$\begin{aligned}\eta_N(t)=f_N(\xi(t))-f_N(\xi(0))-\int_0^t[(a(s, \xi(s)), f'_N(\xi(s)))\\ +\tfrac{1}{2}\operatorname{Sp} f''_N(\xi(s))B(s, \xi(s))]\,ds.\end{aligned}$$

Since the expression

$$(a(s, x), f'_N(x))+\tfrac{1}{2}\operatorname{Sp} f''_N(x)B(s, x)$$

is bounded, denoting by $\{\mathfrak{F}_t\}$ the current of σ-algebras generated by the process $\xi(t)$ we have

$$\left|\mathsf{E}\left(\frac{\eta_N(s)-\eta_N(t)}{s-t}\,\middle|\,\mathfrak{F}_t\right)\right|\leqslant l_2,$$

where l_2 is a constant. Moreover, it is easy to verify that

$$\lim_{s\downarrow t}\frac{1}{s-t}\int[f_N(y)-f_N(x)]P(t, x, s, dy)=(a(t, x), f'_N(x))+\tfrac{1}{2}\operatorname{Sp} f''_N(x)B(t, x)$$

(this was established, for example, in the course of the proof of Theorem 6 in Volume II, Chapter I, Section 1). Therefore

$$\lim_{s\downarrow t}\mathsf{E}\left(\frac{\eta_N(s)-\eta_N(t)}{s-t}\,\middle|\,\mathfrak{F}_t\right)=0.$$

Hence in view of Lemma 4 $\eta_N(t)$ is a martingale. However, $\eta_N(t)=\eta(t)$ for $t\leqslant\tau_N$. We have thus shown that $\eta(t)$ is a local martingale.

We now show that

$$\zeta(t)=\eta^2(t)-\int_0^t(B(s, \xi(s))z, z)\,ds$$

is also a local martingale. To do this we first observe that for $t \leqslant \tau_N$ the process $\zeta(t)$ coincides with the process

$$\zeta_N(t) = \eta_N^2(t) - \int_0^t (B(s, \xi(s)) f'_N(\xi(s)), f'_N(\xi(s)))\, ds.$$

The process $\zeta_N(t)$ is bounded and

$$\begin{aligned}\left|\mathsf{E}\left(\frac{\zeta_N(s)-\zeta_N(t)}{s-t}\,\middle|\, \mathfrak{F}_t\right)\right| &\leqslant \mathsf{E}\left(\frac{[\eta_N(s)-\eta_N(t)]^2}{s-t}\,\middle|\, \mathfrak{F}_t\right) \\ &\quad + \sup_{x,u} (B(u, x) f'_N(x), f'_N(x)) \\ &\leqslant 2\mathsf{E}\left(\frac{[f_N(\xi(s)) - f_N(\xi(t))]^2}{s-t}\,\middle|\, \mathfrak{F}_t\right) + l_3,\end{aligned}$$

where l_3 is a constant. However, for $C > N + 1 + \varepsilon$ we have

$$\begin{aligned}\mathsf{E}([f_N(\xi(s)) - f_N(\xi(t))]^2 \mid \mathfrak{F}_t) &\leqslant \sup_x \int [f_N(y) - f_N(x)]^2 P(t, x, s, dy) \\ &\leqslant \sup_x \int_{|x-y|\leqslant\varepsilon} [f_N(y) - f_N(x)]^2 P(t, x, s, dy) \\ &\quad + O\Big(\sup_{|x|\leqslant C} \int_{|y-x|>\varepsilon} P(t, x, s, dy) \\ &\quad + \sup_{|x|\geqslant C} P(t, x, s, K_{N+1})\Big) \\ &= O\Big(\sup_{|x|\leqslant C} \int_{|x-y|\leqslant\varepsilon} |x-y|^2 P(t, x, s, dy) + (s-t)\Big).\end{aligned}$$

Hence

$$\left|\mathsf{E}\left(\frac{\zeta_N(s)-\zeta_N(t)}{s-t}\,\middle|\, \mathfrak{F}_t\right)\right| \leqslant l_4$$

for some constant l_4. Furthermore,

$$\begin{aligned}\lim_{s \downarrow t} \mathsf{E}\left(\frac{\zeta_N(s)-\zeta_N(t)}{s-t}\,\middle|\, \mathfrak{F}_t\right) &= \lim_{s\downarrow t} \mathsf{E}\left(\frac{[\eta_N(s)-\eta_N(t)]^2}{s-t}\,\middle|\, \mathfrak{F}_t\right) \\ &\quad - (B(t, \xi(t)) f'_N(\xi(t)), f'_N(\xi(t))) = 0.\end{aligned}$$

Applying Lemma 4 we verify that $\zeta_N(t)$ is a martingale; this shows that $\zeta(t)$ is a local martingale. Hence $\eta(t)$ is a local martingale satisfying

$$\langle \eta, \eta \rangle_t = \int_0^t (B(s, \xi(s)) z, z)\, ds.$$

Let $\tilde{w}(t)$ be a Wiener process in $\mathcal{R}^m$ independent of $\xi(t)$. Denote by $P_1(s, x)$ the projection operator on the range of values of operator $B(s, x)$ and by $P_2(s, x)$ the projection operator on the null space of operator $B(s, x)$. Operators P_1 and P_2 are orthogonal and $P_1 + P_2 = \boldsymbol{I}$. Set

$$\xi_1(t) = \xi(t) - \int_0^t a(s, \xi(s))\, ds,$$
$$w(t) = \int_0^t B^{-1/2}(s, \xi(s))P_1(s, \xi(s))\, d\xi_1(s) + \int_0^t P_2(s, \xi(s))\, d\tilde{w}(s)$$

(the integral over ξ_1 is defined as an integral over a local martingale; $B^{-1/2}P_1 z$ denotes a vector z' belonging to the range of values of B such that $B^{1/2}z' = P_1 z$). Since $\xi_1(t)$ and $\tilde{w}(t)$ are independent it follows that

$$\langle\langle(\xi_1, z), (\tilde{w}, z)\rangle\rangle_t = 0.$$

Therefore $(w(t), z)$ is a continuous local martingale satisfying

$$\langle\langle(w, z), (w, z)\rangle\rangle_t = \int_0^t (P_1(s, \xi(s))z, z)\, ds + \int_0^t (P_2(s, \xi(s))z, z)\, ds = t(z, z).$$

Hence $w(t)$ is a Wiener process in $\mathcal{R}^m$. Furthermore, since $B^{1/2}P_2 = 0$ we have

$$\int_0^t B^{1/2}(s, \xi(s))\, dw(s) = \int_0^t P_1(s, \xi(s))\, d\xi_1(s) + \int_0^t B^{1/2}(s, \xi(s))P_2(s, \xi(s))\, d\tilde{w}(s)$$
$$= \int_0^t P_1(s, \xi(s))\, d\xi_1(s).$$

Also, since for all $z \in \mathcal{R}^m$

$$\left\langle \int_0^t (P_2(s, \xi(s))z, d\xi_1(s)), \int_0^t (P_2(s, \xi(s))z, d\xi_1(s)) \right\rangle$$
$$= \int_0^t (B(s, \xi(s))P_2(s, \xi(s))z, P_2(s, \xi(s))z)\, ds = 0,$$

we have

$$\int_0^t (P_2(s, \xi(s))z, d\xi_1(s)) = 0.$$

Therefore

$$\int_0^t P_1(s, \xi(s))\, d\xi_1(s) = \int_0^t [P_1(s, \xi(s)) + P_2(s, \xi(s))]\, d\xi_1(s) = \xi_1(t) - \xi_1(0).$$

Hence

$$\xi(t) - \xi(0) - \int_0^t a(s, \xi(s))\, ds = \int_0^t B^{1/2}(s, \xi(s))\, dw(s).$$

The theorem is thus proved. □

Remark 1. If one assumes that

$$\sup_{x \in \mathcal{R}^m} \int_{|y-x|>\varepsilon} P(t, x, s, dy) = O(s - t),$$

then condition 2) of the theorem is superfluous. (This condition is utilized to obtain the bound

$$\left|\int_{|y-x|>\varepsilon} (f_N(y)-f_N(x))P(t, x, s, dy)\right| + \int_{|y-x|>\varepsilon} (f_N(y)-f_N(x))^2 P(t, x, s, dy) = O(s-t),$$

which is valid under this additional assumption since the function f_N is bounded.)

Remark 2. Assume that for all $\varepsilon>0$ a continuous process $\xi(t)$ in $\mathcal{R}^m$, $t\in[0, T]$, satisfies for $s>t$ the conditions

(I) $\mathsf{P}\{|\xi(s)-\xi(t)|>\varepsilon\,|\,\mathfrak{F}_t\}=o(s-t)$,

(II) $\mathsf{E}((\xi(s)-\xi(t), z)\psi_\varepsilon(\xi(s)-\xi(t))|\mathfrak{F}_t)=(a(t, \xi(\cdot)), z)(s-t)+o(s-t)$,

(III) $\mathsf{E}((\xi(s)-\xi(t), z)^2\psi_\varepsilon(\xi(s)-\xi(t))|\mathfrak{F}_t)=(B(t, \xi(\cdot))z, z)(s-t)+o(s-t)$,

where $z\in\mathcal{R}^m$,

$$\psi_\varepsilon(x)=1, \quad \text{for } |x|\leq\varepsilon, \qquad \psi_\varepsilon(x)=0 \quad \text{for } |x|>\varepsilon,$$

and $a(t, x(\cdot))$ and $B(t, x(\cdot))$ are functions defined on $[0, T]\times\mathscr{C}^m_{[0,T]}$ ($\mathscr{C}^m_{[0,T]}$ is the set of functions continuous on $[0, T]$ with values in $\mathcal{R}^m$).

If $a(t, x(\cdot))$ and $B(t, x(\cdot))$ are continuous and a constant l exists such that

$$\frac{1}{s-t}\mathsf{P}\{|\xi(s)-\xi(t)|>\varepsilon\,|\,\mathfrak{F}_t\}\leq l, \qquad s>t,$$

then a Wiener process $w(t)$ can be found such that

$$\xi(t)=\xi(0)+\int_0^t a(s, \xi(\cdot))\,ds+\int_0^t B^{1/2}(s, \xi(\cdot))\,dw(s). \tag{26}$$

The proof of this assertion is completely analogous to the proof of Theorem 10.

An absolutely continuous measure substitution. Let $\{\Omega, \mathfrak{S}, \mathsf{P}\}$ be the original probability space, $\mathfrak{F}_t\subset\mathfrak{S}$ be a current of σ-algebras, and $w(s)$ be a Wiener process with respect to this current. If $\rho_T(\omega)$ is a nonnegative functional measurable with respect to $\mathfrak{S}$ such that $\mathsf{E}\rho_T(\omega)=1$, then one can define a new probability measure

$$\tilde{\mathsf{P}}(A)=\int_A \rho_T(\omega)\mathsf{P}(d\omega) \tag{27}$$

on $\{\Omega, \mathfrak{S}\}$. In general the process $w(t)$ is not a Wiener process on the space $\{\Omega, \mathfrak{S}, \tilde{\mathsf{P}}\}$.

However, it turns out that for functionals $\rho_T(\omega)$ of a special type, the classes of Itô processes on probability spaces $\{\Omega, \mathfrak{S}, \mathsf{P}\}$ and $\{\Omega, \mathfrak{S}, \tilde{\mathsf{P}}\}$ do coincide. This fact is a corollary of the following important theorem due to I. V. Girsanov.

Theorem 11. *Let $b(t, \omega)$ be an $\{\mathfrak{F}_t\}$-adopted function on $[0, T]\times\Omega$ with values in $\mathscr{R}^m$ and $|b(t, \omega)|^2 \in \mathfrak{M}_1[0, T]$. Set*

$$\rho_T(\omega) = \exp\{-\int_0^T (b(s, \omega), dw(s)) - \tfrac{1}{2}\int_0^T |b(s, \omega)|^2\, ds\}. \tag{28}$$

Then if $\mathsf{E}\rho_T(\omega) = 1$ the process

$$\tilde{w}(t) = \int_0^t b(s, \omega)\, ds + w(t) \tag{29}$$

is a Wiener process on the probability space $\{\Omega, \mathfrak{S}, \tilde{\mathsf{P}}\}$ with respect to the current of σ-algebras $(\mathfrak{F}_t\}$.

As a preliminary to the proof of the theorem we shall establish several auxiliary assertions.

Lemma 5. *If $b(s, \omega)$ is a bounded step function in t such that $|b(s, \omega)| \leq N$, then*

$$\mathsf{E}(\exp\{\int_t^T (b(s, \omega), dw(s))\} \mid \mathfrak{F}_t) \leq e^{N^2(T-t)/2} \tag{30}$$

for $t \in [0, T]$.

Proof. We have for $s > t$

$$\mathsf{E}(\exp\{(b(t, \omega), w(s) - w(t))\} \mid \mathfrak{F}_t) = \exp\{\tfrac{1}{2}|b(t, \omega)|^2 (s-t)\} \leq e^{N^2(s-t)/2}.$$

Let $t = t_0 < t_1 < \cdots < t_n = T$ and $b(s, \omega) = b(t_k, \omega)$ for $t_k \leq s < t_{k+1}$. Then

$$\begin{aligned} &\mathsf{E}\left(\exp\left\{\sum_{k=0}^{n-1} (b(t_k, \omega), w(t_{k-1}) - w(t_k))\right\} \middle| \mathfrak{F}_t\right) \\ &\quad = \mathsf{E}(\cdots \mathsf{E}(\mathsf{E}(\exp\{(b(t_{n-1}, \omega), w(t_n) - w(t_{n-1}))\} \mid \mathfrak{F}_{t_{n-1}}) \\ &\qquad \times \exp\{(b(t_{n-2}, \omega), w(t_{n-1}) - w(t_{n-2}))\} \mid \mathfrak{F}_{t_{n-2}}) \times \cdots \\ &\qquad \times \exp\{(b(t_0, \omega), w(t_1) - w(t_0))\} \mid \mathfrak{F}_{t_0}) \\ &\quad \leq \prod_{k=0}^{n-1} \exp\{\tfrac{1}{2}N^2(t_{k+1} - t_k)\}. \quad \square \end{aligned}$$

Corollary. *Inequality* (30) *is valid for any $\{\mathfrak{F}_t\}$-adopted function $b(t, \omega)$ such that $|b(t, \omega)| \leq N$.*

This corollary is obtained by means of a limit transition.

Lemma 6. *If $b(t, \omega)$* is an $\{\mathfrak{F}_t\}$-adopted function such that $|b(t, \omega)| \leq N$ then

$$\mathsf{E}(\exp\{-\int_t^T (b(s, \omega), dw(s)) - \tfrac{1}{2}\int_t^T |b(s, \omega)|^2\, ds\} \mid \mathfrak{F}_t) = 1.$$

Proof. Let

$$\rho_{t_1,t_2}(\omega)=\exp\{-\int_{t_1}^{t_2}(b(s,\omega),dw(s))-\tfrac{1}{2}\int_{t_1}^{t_2}|b(s,\omega)|^2\,ds\}. \tag{31}$$

Utilizing Itô's formula we obtain

$$\rho_{t_1,t_2}(\omega)=1+\int_{t_1}^{t_2}\rho_{t_1,s}(\omega)(b(s,\omega),dw(s)).$$

Since in view of the corollary to Lemma 5

$$\mathsf{E}|\rho_{t,s}(\omega)|^2|b(s,\omega)|^2\leqslant N^2e^{2N^2(s-t)},$$

it follows that

$$\mathsf{E}(\int_{t_1}^{t_2}\rho_{t_1,s}(\omega)(b(s,\omega),dw(s))|\mathfrak{F}_{t_1})=0$$

This completes the proof of the lemma. □

Corollary 1. *For any $\{\mathfrak{F}_t\}$-adopted function $b(t,\omega)$ such that $|b(t,\omega)|^2\in\mathfrak{M}_1[0,T]$ we have*

$$\mathsf{E}(\exp\{-\int_t^T(b(s,\omega),dw(s))-\tfrac{1}{2}\int_t^T|b(s,\omega)|^2\,ds\}|\mathfrak{F}_t)\leqslant 1.$$

This inequality is a corollary of Fatou's theorem.

Corollary 2. *If $\mathsf{E}\rho_T(\omega)=1$, then also for $0\leqslant t_1<t_2\leqslant T$*

$$\mathsf{E}(\rho_{t_1,t_2}(\omega)|\mathfrak{F}_{t_1})=1.$$

Indeed,

$$1=\mathsf{E}\rho_{0,t_1}(\omega)\mathsf{E}(\rho_{t_1,t_2}(\omega)\mathsf{E}(\rho_{t_2,T}(\omega)|\mathfrak{F}_{t_2})|\mathfrak{F}_{t_1}). \tag{32}$$

If the inequality

$$\mathsf{E}(\rho_{t_1,t_2}(\omega)|\mathfrak{F}_{t_1})<1$$

were valid with a positive probability, the expression on the right-hand side of (32) would have been less than 1. We now proceed with the proof of Theorem 11.

Proof of Theorem 11. Denote by $\tilde{\mathsf{E}}$ the mathematical expectation with respect to measure $\tilde{\mathsf{P}}$.

It is sufficient to verify that

$$\tilde{\mathsf{E}}(\exp\{i(z,\tilde{w}(t_2)-\tilde{w}(t_1))\}|\mathfrak{F}_{t_1})=\exp\{-\tfrac{1}{2}|z|^2(t_2-t_1)\}$$

for $t_1 < t_2$, i.e., that for any bounded $\mathfrak{F}_{t_1}$-measurable variable η

(33) $$\tilde{\mathsf{E}}\eta \exp\{i(z, \tilde{w}(t_2) - \tilde{w}(t_1))\} = \exp\{-\tfrac{1}{2}|z|^2(t_2 - t_1)\}\tilde{\mathsf{E}}\eta.$$

It follows from Corollary 2 to Lemma 6 that for any $\mathfrak{F}_t$-measurable variable ξ

(34) $$\tilde{\mathsf{E}}\xi = \mathsf{E}\xi\rho_{0,t}(\omega)\rho_{t,T}(\omega) = \mathsf{E}\xi\rho_{0,t}(\omega)\mathsf{E}(\rho_{t,T}(\omega)\,|\,\mathfrak{F}_t) = \mathsf{E}\xi\rho_{0,t}(\omega).$$

Hence (33) is equivalent to the equality

(35) $$\mathsf{E}\eta' \exp\{i(z, \tilde{w}(t_2) - \tilde{w}(t_1))\}\rho_{t_1,t_2}(\omega) = \exp\{-\tfrac{1}{2}|z|^2(t_2 - t_1)\}\mathsf{E}\eta',$$

where $\eta' = \eta\rho_{0,t_1}(\omega)$ is a variable such that $\mathsf{E}|\eta'| < \infty$.

From Itô's formula we have

$$\begin{aligned} &d \exp\{i(z, \tilde{w}(t) - \tilde{w}(t_1))\}\rho_{t_1,t}(\omega) \\ &= \exp\{i(z, \tilde{w}(t) - \tilde{w}(t_1))\}\rho_{t_1,t}(\omega)[-(b(t,\omega), dw(t)) \\ &\quad + i(b(t,\omega), z)\,dt + i(z, dw(t)) - \tfrac{1}{2}|z|^2\,dt - i(b(t,\omega), z)\,dt]. \end{aligned}$$

Therefore

$$\begin{aligned} \exp\{i(z, \tilde{w}(t_2) - \tilde{w}(t_1))\}\rho_{t_1,t_2}(\omega) = 1 + \int_{t_1}^{t_2} &\exp\{i(z, \tilde{w}(t) - \tilde{w}(t_1))\}\rho_{t_1,t}(\omega) \\ &\times [i(z, dw(t)) - (b(t,\omega), dw(t))] \\ &- \tfrac{1}{2}|z|^2 \int_{t_1}^{t_2} \exp\{i(z, \tilde{w}(t) - \tilde{w}(t_1))\}\rho_{t_1,t}(\omega)\,dt. \end{aligned}$$

Assume that $|b(t,\omega)| \le N$. Since

$$[\rho_{t_1,t}(\omega)(|z| + |b(t.\,\omega)|)]^2 \le (N + |z|)^2 \exp\{-2\int_{t_1}^{t} (b(s,\omega), dw(s))\},$$

we have in view of Lemma 5 that

$$\mathsf{E}\int_{t_1}^{t_2} [\rho(t_1, t)(|z| + |b(t,\omega)|)]^2\,dt < \infty.$$

Hence

$$\mathsf{E}\eta' \int_{t_1}^{t_2} \exp\{i(z, \tilde{w}(t) - \tilde{w}(t_1))\}\rho_{t_1,t}(\omega)[-(b(t,\omega), dw(t)) + i(z, dw(t))] = 0$$

and

$$\begin{aligned} &\mathsf{E}\eta' \exp\{i(z, \tilde{w}(t_2) - \tilde{w}(t_1))\}\rho_{t_1,t_2}(\omega) \\ &= \mathsf{E}\eta' - \frac{|z|^2}{2}\int_{t_1}^{t_2} \mathsf{E}\eta' \exp\{i(z, \tilde{w}(t) - \tilde{w}(t_1))\}\rho_{t_1,t}(\omega)\,dt. \end{aligned}$$

Viewing this relation as an equation in

$$\mathsf{E}\eta' \exp\{i(z, \tilde{w}(t_2)-\tilde{w}(t_1))\}\rho_{t_1,t_2}(\omega)$$

for $t_2 \in [t, T]$ and solving this equation we verify that (35) is valid for bounded $b(t, \omega)$.

Now let $b_N(s, \omega)$ be a sequence of bounded functions such that

$$\int_0^T |b_N(s, \omega)-b(s, \omega)|^2\, ds \to 0$$

in probability. Then setting

$$\begin{aligned}(36)\qquad &\tilde{w}_N(t)=\int_0^T b_N(s, \omega)\, ds + w(t),\\ &\rho_{t,s}^N(\omega)=\exp\{-\textstyle\int_t^s (b_N(u, \omega), dw(u))-\frac{1}{2}\int_t^s |b_N(u, \omega)|^2\, du\},\end{aligned}$$

we have

$$\begin{aligned}(37)\qquad &\mathsf{E}\eta\rho_{0,t_2}^N(\omega)\exp\{i(z, \tilde{w}_N(t_2)-\tilde{w}_N(t_1))\}\\ &=\mathsf{E}\eta\rho_{0,t_1}^N(\omega)\exp\left\{-\frac{|z|^2}{2}(t_2-t_1)\right\}.\end{aligned}$$

Since $\tilde{w}_N(t)\to\tilde{w}(t)$ as $N\to\infty$ in probability it follows that

$$\begin{aligned}(38)\qquad &\lim_{N\to\infty}\mathsf{E}\eta\rho_{0,t_2}(\omega)\exp\{i(z, \tilde{w}_N(t_2)-\tilde{w}_N(t_1))\}\\ &=\mathsf{E}\eta\rho_{0,t_2}(\omega)\exp\{i(z, \tilde{w}(t_2)-\tilde{w}(t_1))\}.\end{aligned}$$

Furthermore,

$$\begin{aligned}(39)\qquad &|\mathsf{E}\eta\rho_{0,t_2}^N(\omega)\exp\{i(z, \tilde{w}_N(t_2)-\tilde{w}_N(t_1))\}\\ &-\mathsf{E}\eta\rho_{0,t_2}(\omega)\exp\{i(z, \tilde{w}_N(t_2)-\tilde{w}_N(t_1))\}|\\ &\leq C\mathsf{E}|\rho_{0,t_2}^N(\omega)-\rho_{0,t_2}(\omega)|,\end{aligned}$$

where C is such that $|\eta|\leq C$.

In the same manner

$$|\mathsf{E}\eta\rho_{0,t_1}^N(\omega)-\mathsf{E}\eta\rho_{0,t_1}(\omega)|\leq C\mathsf{E}|\rho_{0,t_1}^N(\omega)-\rho_{0,t_1}(\omega)|.$$

We show that

$$(40)\qquad \lim_{N\to\infty}\mathsf{E}|\rho_{0,t}^N(\omega)-\rho_{0,t}(\omega)|=0$$

for all $t\in[0, T]$. We have

$$\mathsf{E}|\rho_{0,t}^N(\omega)-\rho_{0,t}(\omega)|=\mathsf{E}(|\rho_{0,t}^N(\omega)-\rho_{0,t}(\omega)|+\rho_{0,t}(\omega)-\rho_{0,t}^N(\omega)),$$

since $\mathsf{E}\rho_{0,t}(\omega)=\mathsf{E}\rho_{0,t}^N(\omega)=1$. On the other hand,

$$|\rho_{0,t}^N(\omega)-\rho_{0,t}(\omega)|+\rho_{0,t}(\omega)-\rho_{0,t}^N(\omega)\leq 2\rho_{0,t}(\omega).$$

Since the expression of the left-hand side of the preceding equality tends to zero in probability, an application of Lebesgue's theorem yields (40).

Utilizing (38) and (39) we verify that

$$\lim_{N\to\infty} \mathsf{E}\eta\rho_{0,t_2}^N(\omega)\exp\{i(z,\tilde{w}_N(t_2)-\tilde{w}_N(t_1))\}$$
$$=\mathsf{E}\eta\rho_{0,t_2}(\omega)\exp\{i(z,\tilde{w}(t_2)-\tilde{w}(t_1))\}.$$

Also (40) implies that

$$\lim_{N\to\infty} \mathsf{E}\eta\rho_{0,t_1}^N=\mathsf{E}\eta\rho_{0,t_1}.$$

Relationship (35) is thus established and the theorem is proved. □

Remark. Assume that the process $w_1(t)$ with values in $\mathcal{R}^k$ is independent of $w(t)$ and of $\rho_T(\omega)$. Then $w_1(t)$ is a Wiener process also on $\{\Omega, \mathfrak{S}, \tilde{\mathsf{P}}\}$.

Indeed consider the compound process $w^*(t)=\{w(t); w_1(t)\}$ in the space $\mathcal{R}^{m+k}$. This is a Wiener process. If $b^*(s,\omega)$ in $\mathcal{R}^{m+k}$ is defined as $\{b(s,\omega); 0\}$, then

$$\int_0^T (b^*(s,\omega), dw^*(s))=\int_0^T (b(s,\omega), dw(s)).$$

In view of Theorem 11 the process $\tilde{w}^*(t)=\int_0^t b^*(s,\omega)\,ds+w^*(t)$ is a Wiener process on $\{\Omega, \mathfrak{S}, \tilde{\mathsf{P}}\}$. Therefore both components of the compound process

$$\{w(t)+\int_0^t b(s,\omega)\,ds; w_1(t)\}$$

will be Wiener processes and, moreover, these components are mutually independent.

This remark allows us to construct a mapping of the set of Itô's processes on $\{\Omega, \mathfrak{S}, \mathsf{P}\}$ into the set of Itô's processes on the space $\{\Omega, \mathfrak{S}, \tilde{\mathsf{P}}\}$.

Indeed, let $\bar{w}(t)$ be an arbitrary one-dimensional Wiener process with respect to $\{\mathfrak{F}_t\}$. We shall assume that a basis consisting of Wiener processes in $I_T(\mathfrak{F}_t)$ exists. Processes of $w_1(t),\ldots,w_m(t)$ which are the coordinates of $w(t)$ in $\mathcal{R}^m$ may be assumed to be elements of this basis. In this case

$$\bar{w}(t)=\int_0^t (c(s,\omega), dw(s))+\int_0^t \alpha(s,\omega)\,dw^*(s), \tag{41}$$

where $c(s,\omega)$ is a function with values in $\mathcal{R}^m$ and $\alpha(s,\omega)$ is a function with values in $\mathcal{R}^1$ and $w^*(t)$ is a Wiener process independent of $w(t)$. We obtain this representation if, in a representation of $\bar{w}(t)$ in terms of the elements of the basis, we collect separately integrals over $w_1,\ldots,w_m$ and then separately over the remaining Wiener processes. The process $\bar{w}(t)$ is a Wiener process on $\{\Omega, \mathfrak{S}, \mathsf{P}\}$ if and only if in representation (41)

$$|c(s, \omega)|^2 + \alpha^2(s, \omega) = 1$$

for almost all s and ω. Correspond to it the process

$$\tilde{\bar{w}}(t) = \int_0^t (c(s, \omega).\, d\tilde{w}(s)) + \int_0^t \alpha(s, \omega) dw^*(s). \tag{42}$$

This is a process in $I_T(\mathfrak{F}_t)$ on the probability space $\{\Omega, \mathfrak{S}, \mathsf{P}\}$. It is easy to verify that $\langle \tilde{\bar{w}}, \tilde{\bar{w}} \rangle_t = t$; hence $\tilde{\bar{w}}(t)$ is a Wiener process on $\{\Omega, \mathfrak{S}, \tilde{\mathsf{P}}\}$. If $\eta(t)$ is a numerical Itô process on the probability space $\{\Omega, \mathfrak{S}, \mathsf{P}\}$ it means that a Wiener process $\bar{w}(t)$ and functions $\beta(s, \omega)$ and $\gamma(s, \omega)$ exist such that

$$\eta(t) = \eta_0 + \int_0^t \gamma(s, \omega)\, ds + \int_0^t \beta(s, \omega)\, d\bar{w}(s).$$

We correspond to it the Itô process on $\{\Omega, \mathfrak{S}, \tilde{\mathsf{P}}\}$

$$\tilde{\eta}(t) = \eta_0 + \int_0^t \gamma(s, \omega)\, ds + \int_0^t \beta(s, \omega)\, d\tilde{\bar{w}}(s). \tag{43}$$

The mapping constructed in such a manner is an isomorphism, in a certain sense, between the spaces of Itô processes on $\{\Omega, \mathfrak{S}, \mathsf{P}\}$ and $\{\Omega, \mathfrak{S}, \tilde{\mathsf{P}}\}$. It is easy to see that the mapping is invertible, linear, and homogeneous, and commutes with the operations of stochastic integration. Utilizing formulas (41)–(43) we obtain

$$\tilde{\eta}(t) = \eta_0 + \int_0^t [\gamma(s, \omega) + \beta(s, \omega)(c(s, \omega), b(s, \omega))]\, ds + \int_0^t \beta(s, \omega)\, d\bar{w}(s).$$

This, however, implies that the spaces of Itô processes on the probability spaces $\{\Omega, \mathfrak{S}, \mathsf{P}\}$ and $\{\Omega, \mathfrak{S}, \tilde{\mathsf{P}}\}$ coincide.

We now present yet another theorem which gives sufficient conditions for $\mathsf{E}\rho_T(\omega) = 1$. Clearly it is sufficient to consider the case $m = 1$.

Theorem 12. *If an $\{\mathfrak{F}_t\}$-adopted function $b(t, \omega)$ satisfies the condition*

$$\mathsf{E} \exp\{\tfrac{1}{2} \int_0^T b^2(t, \omega)\, dt\} < \infty,$$

then

$$\mathsf{E} \exp\{\int_0^T b(t, \omega)\, dw(t) - \tfrac{1}{2} \int_0^T b^2(t, \omega)\, dt\} = 1.$$

Proof. Let $b(s, \omega) \neq 0$ and the variables ζ_t be defined by the equality

$$t = \int_0^{\zeta_t} b^2(s, \omega)\, ds,$$

$$\tilde{\mathfrak{F}}_t = \mathfrak{F}_{\zeta_t}, \qquad \tilde{w}(t) = w(\zeta_t).$$

As follows from Corollary 4 to Theorem 1, the process $\tilde{w}(t)$ is a Wiener process with respect to the current of σ-algebras $\{\tilde{\mathfrak{F}}_t\}$. The quantity $\int_0^T b^2(s, \omega) = \tau$ is a Markov time with respect to the current $\{\tilde{\mathfrak{F}}_t\}$. To prove the theorem it is sufficient to show that for any Markov time τ such that $\mathsf{E}\, e^{(1/2)\tau} < \infty$,

$$\mathsf{E} \exp\{\tilde{w}(\tau) - \tfrac{1}{2}\tau\} = 1.$$

First let τ_a be a Markov time of a special kind: the time of the first arrival of the process $\tilde{w}(t)$ at the line $t - a$ $(a > 0)$. It is easy to see that the process

$$\eta(t) = \exp\{\tilde{w}(t) - \tfrac{1}{2}t\}$$

is a martingale. Since τ_a coincides with the time of the first arrival of the continuous process with independent increments $-\tilde{w}(t) + t$ at the level a, in view of formulas (68), (70), and (71) in Volume II, Chapter IV, Section 2, we have

$$\mathsf{E}\, e^{-\lambda \tau_a} = e^{aB(\lambda)},$$

where $B(\lambda)$ satisfies the relation

$$B(\lambda) + \tfrac{1}{2}B(\lambda)^2 = \lambda,$$

and since $B(0) = 0$, it follows that

$$B(\lambda) = 1 - \sqrt{1 + 2\lambda}.$$

Thus

$$\mathsf{E}\, e^{-\lambda \tau_a} = \exp\{a(1 - \sqrt{1 + 2\lambda})\}. \tag{44}$$

Although the results stated above are valid only for $\operatorname{Re} \lambda \geq 0$ it is easy to deduce from the analyticity of the right-hand side of (44) for $\operatorname{Re} \lambda > -\frac{1}{2}$ and its continuity for $\operatorname{Re} \lambda \geq -\frac{1}{2}$ that formula (44) is valid also for $\operatorname{Re} \lambda \geq -\frac{1}{2}$. In particular,

$$\mathsf{E}\, e^{\tau_a/2} = e^a. \tag{45}$$

Since $\eta(\tau_a) = \exp\{\tilde{w}(\tau_a) - \frac{1}{2}\tau_a\} = e^{(1/2)\tau_a - a}$, it follows from formula (45) that $\mathsf{E}\eta(\tau_a) = 1$. Since $\eta(t)$ is a martingale and in view of the strong Markov property of $\tilde{w}(t)$ we have that for any pair of Markov times ζ_1 and ζ_2 such that $\zeta_1 \leq \zeta_2$

$$\mathsf{E}(\eta(\zeta_2) \mid \tilde{\mathfrak{F}}_{\zeta_1}) \leq 1.$$

Therefore for any Markov time $\zeta \leq \tau_a$

$$\eta(\zeta) \geq \mathsf{E}(\eta(\tau_a) \mid \tilde{\mathfrak{F}}_\zeta)$$

and $\mathsf{E}\eta(\zeta) \geq 1$, i.e., $\mathsf{E}\eta(\zeta) = 1$. Clearly $\tau_a \wedge \tau \leq \tau_a$, and

$$1 = \mathsf{E}\eta(\tau_a \wedge \tau) = \mathsf{E}\eta(\tau_a)\chi_{\{\tau_a \leq \tau\}} + \mathsf{E}\eta(\tau)\chi_{\{\tau \leq \tau_a\}}. \tag{46}$$

However,

$$\eta(\tau_a)\chi_{\{\tau_a\le\tau\}}=e^{-a+(\tau_a/2)}\chi_{\{\tau_a\le\tau\}}\le e^{-a}\chi_{\{\tau_a\le\tau\}}e^{\tau/2}.$$

Hence

$$\lim_{a\to\infty}\mathsf{E}\eta(\tau_a)\chi_{\{\tau_a\le\tau\}}=0$$

since the quantity to the right of the sign of the mathematical expectation is dominated by an integrable function $e^{\tau/2}$ and approaches zero as $a\to\infty$. Taking into account that $\chi_{\{\tau\le\tau_a\}}\uparrow 1$ as $a\uparrow\infty$ and approaching the limit in (46) we obtain that $\mathsf{E}_\eta(\tau)=1$. In the general case the proof follows from:

$$E\exp\{\textstyle\int_0^T\mathscr{C}(s,\omega)dw(s)-\frac{1}{2}\int_0^T\mathscr{C}^2(s,\omega)ds\}$$
$$=E\exp\{\textstyle\int_0^T\mathscr{C}_1(s,\omega)d\bar{w}(s)-\frac{1}{2}\int_0^T\mathscr{C}_1^2(s,\omega)\,ds\},$$

where

$$\mathscr{C}_1^2(s,\omega)=1+\mathscr{C}^2(s,\omega),\ \bar{w}(t)=\textstyle\int_0^t\mathscr{C}(s,\omega)\,(\mathscr{C}_1(s,\omega))^{-1}\,dw(s)$$
$$+\textstyle\int_0^t(\mathscr{C}_1(s,\omega))^{-1}\,dw_1(s)$$

and $w_1(t)$ is independent of $\mathfrak{F}_t$. □

§ 2. Stochastic Differential Equations for Processes of Diffusion Type

In this section *processes of diffusion type*, i.e., processes satisfying the stochastic differential equation

$$(1)\qquad d\xi(t)=a(t,\xi(\cdot))\,dt+B(t,\xi(\cdot))\,dw(t),$$

where $\xi(t)$ is the process under consideration, $w(t)$ is a Wiener process, and $\xi(t)$ and $w(t)$ take on values in $\mathscr{R}^m$ are considered. The functions $a(t,x(\cdot))$ and $B(t,x(\cdot))$ are defined on $[0,T]\times\mathscr{C}^m_{[0,T]}$ and take on values in $\mathscr{R}^m$ and $\mathfrak{L}(\mathscr{R}^m)$, respectively. A solution of equation (1) is sought on the interval $[0,T]$; the initial condition is always assumed to be $\xi(0)=0$. In order that the right-hand side of (1) can be viewed as a stochastic differential we shall assume that the following condition is fulfilled:

1) Functions $a(t,x(\cdot))$ and $B(t,x(\cdot))$ are measurable jointly in the variables and for all $t\in[0,T]$ as functions of $x(\cdot)$ are measurable with respect to the σ-algebra $\mathfrak{S}_t$ generated by cylinders in $\mathscr{C}^m_{[0,T]}$ with bases over $[0,t]$. The last requirement is equivalent to:

$$a(t,x(\cdot))=a(t,x_1(\cdot)),\qquad B(t,x(\cdot))=B(t,x_1(\cdot))$$

provided $x_1(s)=x(s)$ for $s\le t$.

A solution of (1) is defined to be a process $\xi(t)$ such that the process $w_t(s)=w(t+s)-w(t)$ does not depend on $\mathfrak{F}^\xi_t$, which is the σ-algebra generated by process $\xi(\cdot)$ up to time t.

Let $\{\mathfrak{F}_t\}$ be a current of σ-algebras generated by the process $w(t)$. If $\mathfrak{F}^\xi_t\subset\mathfrak{F}_t$, i.e., $\xi(t)$ is $\mathfrak{F}_t$-measurable for each t, then such a solution $\xi(t)$ of equation (1) is called a *strong* solution. Other solutions—in cases in which it must be emphasized that they are not necessarily strong—are called *weak*. When weak solutions of (1) are considered, the probability space is not often stipulated: $\xi(t)$ is a weak solution of (1) if $\xi(t)$ is defined on some probability space on which a Wiener process $w(t)$ is defined such that (1) is satisfied. As a probability space we often choose the measurable space $\{\mathcal{C}^m_{[0,T]}, \mathfrak{S}_T\}$ with different probability measures (i.e., a measure associated with process $w(t)$ or one associated with $\xi(t)$).

Sometimes in addition to condition 1) the following condition is imposed:

2) $a(t, x(\cdot))$ and $B(t, x(\cdot))$ are continuous jointly in the variables.

Equations of type (1) were studied in Chapter II but under the following more restrictive conditions:

3) For any compact set $K \subset \mathcal{C}^m_{[0,T]}$ a constant l_K exists such that for $x(\cdot) \in K$ and $y(\cdot) \in K$ the inequality

$$|a(t, x(\cdot)) - a(t, y(\cdot))| + \|B(t, x(\cdot)) - B(t, y(\cdot))\| \leq l_K \|x - y\|_m$$

is satisfied (where $\|\cdot\|_m$ is the norm in $\mathcal{C}^m_{[0,T]}$ and $\|\cdot\|$ is the norm in $\mathfrak{L}(\mathcal{R}^m)$.)

4) An l exists such that for all $x(\cdot)$

$$|a(t, x(\cdot))| + \|B(t, x(\cdot))\| \leq l(1 + \|x\|_m).$$

Under these conditions it was proved in Chapter II that a solution of (1) exists, is unique and is a *strong* solution.

Measures associated with solutions of equation (1). A general construction of a measure associated with a random process was presented in Volume I, Chapter V, Section 1.

Since we are discussing only continuous solutions of (1), the measures associated with solutions of $\xi(t)$ should naturally be considered on $\mathcal{C}^m_{[0,T]}$. Let $\xi(t)$ be a solution of (1) and μ_ξ be an associated measure on $\mathcal{C}^m_{[0,T]} = \Omega$. First we shall assume that $B(t, x(\cdot))$ is a nondegenerate operator for all $t \in [0, T]$, $x(\cdot) \in \mathcal{C}^m_{[0,T]}$. In this case one can assert that:

a) the process

$$y(t) = x(t) - \int_0^t a(s, x(\cdot))\, ds \tag{2}$$

is a local martingale on the probability space $\{\Omega, \mathfrak{S}, \mu_\xi\}$;

b) the process

$$z(t) = \int_0^t B^{-1}(s, x(\cdot))\, dy(s) \tag{3}$$

is a Wiener process on this space.

Both processes are local martingales with respect to the current of σ-algebras $\{\mathfrak{Y}_t\}$ on $\mathcal{C}^m_{[0,T]}$.

Clearly, in the case when the process $z(t)$ on $\{\Omega, \mathfrak{S}_T, \mu_\xi\}$ is defined by formulas (2) and (3) ($z(t)$ is a measurable function of the point $\omega = x(\cdot) \in \mathcal{C}^m_{[0,T]} = \Omega$), we have

$$x(t)=\int_0^t a(s, x(\cdot))\,ds+\int_0^t B(s, x(\cdot))\,dz(s). \tag{4}$$

Thus, if a measure μ_ξ is such that conditions a) and b) are satisfied then it is associated with a solution of equation (1).

Observe that the function $a(s, x(\cdot))$ for which $\{y(t), \mathfrak{S}_t\}$ is a local martingale is uniquely defined for almost all s, $(x(\cdot))$ with respect to the product of the Lebesgue measure on the line and the measure μ_ξ. Indeed, if $\{y_1(t), \mathfrak{S}_t\}$ is a local martingale on the same probability space, so is $\{y(t)-y_1(t), \mathfrak{S}_t\}$ where $y_1(t)=x(t)-\int_0^t a_1(s, x(\cdot))\,ds$. However, it is easy to verify that $\langle y-y_1, y-y_1\rangle_t=0$ and hence

$$\mathsf{E}\left(\int_0^t [a_1(s, x(\cdot))-a(s, x(\cdot))]^2\,ds\right)^2=0,$$

which implies that $a_1(s, x(\cdot))=a(s, x(\cdot))$ for almost all s and for almost all $x(\cdot)$ with respect to measure μ_ξ.

If the operator $B(t, x(\cdot))$ degenerates, the process $y(t)$ defined by equation (2) will be a local martingale. The process $z(t)$ in equation (3) may also be defined as follows,

$$z(t)=\lim_{\varepsilon\to\infty}\int_0^t (B(s, x(\cdot))+d\boldsymbol{I})^{-1}\,\varepsilon y(s),$$

assuming that $B(s, x(\cdot))$ is a nonnegative symmetric operator. This process will also be a local square integrable martingale for which the equality

$$\langle (z, u), (z, u)\rangle_t=\int_0^t |P(s, x(\cdot))u|^2\,ds$$

is valid for all $u\in\mathcal{R}^m$; here $P(s, x(\cdot))$ is the projection operator on the range of values of operator $B(s, x(\cdot))$. The last assertion is a corollary of the equality

$$\lim_{\varepsilon\to 0} |(B(s, x(\cdot))+\varepsilon\boldsymbol{I})^{-1}B(s, x(\cdot))u|^2=|P(s, x(\cdot))u|^2.$$

Now let $Q(s, x(\cdot))$ be the projection operator on the subspace orthogonal to the range of values of $B(s, x(\cdot))$ (we shall always assume here that $B(s, x(\cdot))$ is symmetric and nonnegative). Let $w_1(t)$ be a Wiener process in $\mathcal{R}^m$ independent of $z(t)$. Then the process

$$w(t)=z(t)+\int_0^t Q(s, x(\cdot))\,dw_1(s) \tag{5}$$

is a Wiener process since is it a square integrable martingale and, moreover, for all $u\in\mathcal{R}^m$

$$\langle (w, u), (w, u)\rangle_t=t.$$

Clearly the process $x(t)$ satisfies the equation

$$x(t)=\int_0^t a(s, x(\cdot))\, ds+\int_0^t B(s, x(\cdot))\, dw(s).$$

If the measure μ_ξ is given, then a nonnegative symmetric operator $B(s, x(\cdot))$ is uniquely determined from the relation: for all $u \in \mathcal{R}^m$

$$\langle\langle(y, u),(y, u)\rangle\rangle_t=\int_0^t|B(s, x(\cdot)) u|^2\, ds \tag{6}$$

almost everywhere with respect to measure μ_ξ.

Thus the following theorem is valid.

Theorem 1. *If a measure μ_ξ is such that the process $\{y_t, \mathfrak{F}_t\}$ defined by equation* (2) *is a local martingale on $\{\mathscr{C}_{[0, T]}^m, \mathfrak{S}_T, \mu_\xi\}$ for which relation* (6) *is fulfilled, and the functions $a(t, x(\cdot))$ and $B(t, x(\cdot))$ satisfy condition* (1), *then the measure μ_ξ is associated with a weak solution of equation* (1).

Next we shall consider measures which are absolutely continuous with respect to measure μ_ξ associated with a solution of equation (1). Let μ be such a measure and let

$$\begin{aligned} \rho_T(x(\cdot)) &= \frac{d\mu}{d\mu_\xi}(x(\cdot)) \\ &= \exp\left\{\int_0^T(b(s, x(\cdot)), dz(s))-\tfrac{1}{2}\int_0^T|b(s, x(\cdot))|^2\, ds\right\}, \end{aligned} \tag{7}$$

where $b(t, x(\cdot))$ is a function defined on $[0, T] \times \mathscr{C}_{[0, T]}^m$ with values in $\mathcal{R}^m$ satisfying the same conditions as those imposed on $a(t, x(\cdot))$; $z(t)$ is determined by equations (2) and (3) and is a function of $x(\cdot)$. Assume that

$$\int \rho_T(x(\cdot)) \mu_\xi(dx(\cdot))=1. \tag{8}$$

Let $\rho_T(x(\cdot))$ be defined by formula (7) when T is replaced by t. Then from Itô's formula we have for $u \in \mathcal{R}^m$

$$\begin{aligned} (u, \xi(t)) \rho_t(\xi(\cdot)) = & \int_0^t(u, \xi(s)) \rho_s(\xi(\cdot))(b(s, \xi(\cdot)), dw(s)) \\ & +\int_0^t \rho_s(\xi(\cdot))(B^*(s, \xi(\cdot)) u, dw(s))+\int_0^t[(a(s, \xi(\cdot)), u) \\ & +(B^*(s, \xi(\cdot)) u, b(s, \xi(\cdot))) \rho_s(\xi(\cdot))]\, ds. \end{aligned}$$

Using this equality and formula (34) in Section 1 it is easy to verify that the process $\{y(t), \mathfrak{S}_t\}$, where

$$y(t)=x(t)-\int_0^t(a(s, x(\cdot))+B(s, x(\cdot)) b(s, x(\cdot)))\, ds,$$

is a local martingale on the space $\{\mathscr{C}^m_{[0,T]}, \mathfrak{S}_T, \mu\}$. Simple calculations yield that

$$\langle (y, u), (y, u)\rangle_t = \int_0^t (B(s, x(\cdot))u, B(s, x(\cdot))u)\, ds.$$

Therefore the process $x(t)$ on $\{\mathscr{C}^m_{[0,T]}, \mathfrak{S}_T, \mu\}$ satisfies condition b) stipulated above. Hence, in view of Theorem 1, the measure μ is associated with a solution of a stochastic differential equation (1). Thus the following theorem is valid.

Theorem 2. *Let $\xi(t)$ be a solution of equation* (1) *and let the function $b(t, x(\cdot))$ defined on $[0, T]\times\mathscr{C}^m_{[0,T]}$ with values in $\mathscr{R}^m$ satisfy condition* 1). *If $\rho_T(x(\cdot))$ is defined by equality* (7) *and condition* (8) *is satisfied, then a solution of equation*

$$d\xi_1(t) = a_1(t, \xi_1(\cdot))\, dt + B(t, \xi_1(\cdot))\, dw(t), \tag{9}$$

exists, where

$$a_1(t, x(\cdot)) = a(t, x(\cdot)) + B(t, x(\cdot))b(t, x(\cdot)) \tag{10}$$

such that the measure associated with solution $\xi_1(t)$ is absolutely continuous with respect to measure μ_ξ. Moreover,

$$\frac{d\mu_{\xi_1}}{d\mu_\xi}(x(\cdot)) = \rho_T(x(\cdot)). \tag{11}$$

Corollary. *If equation* (1) *possesses a solution $\xi(t)$ then equation* (9) *also possesses a solution $\xi_1(t)$ for all $a_1(t, x(\cdot))$ for which there exists $b(t, x(\cdot))$, such that $a_1(t, x(\cdot))$ is determined by equation* (10) *and the function $\rho_T(x(\cdot))$ defined by* (7) *satisfies relation* (8). *In particular, if $B(t, x(\cdot))$ possesses a uniformly bounded inverse operator and equation* (1) *possesses a solution for some bounded $a(t, x(\cdot))$, then equation* (9) *possesses a solution for every bounded $a(t, x(\cdot))$.*

Remark. Since $\rho_T(x(\cdot))$ is everywhere positive, so is $d\mu_{\xi_1}/d\mu_\xi$ and hence the measures μ_ξ and μ_{ξ_1} are equivalent.

We shall now discuss separately the case when $B(t, x(\cdot)) = \boldsymbol{I}$ ($\boldsymbol{I}$ is the unit operator). If $\xi(t)$ is a solution of the stochastic equation

$$d\xi(t) = a(t, \xi(\cdot))\, dt + dw(t), \tag{12}$$

$$\rho_T(x(\cdot)) = \exp\{-\int_0^T (a(s, x(\cdot)), dx(s)) + \tfrac{1}{2}\int_0^T |a(s, x(\cdot))|^2\, ds\} \tag{13}$$

and condition (8) is fulfilled, then the measure μ_ξ associated with the process $\xi(\cdot)$ is absolutely continuous with respect to the measure μ_w associated with the process $w(t)$ and, furthermore,

$$\frac{d\mu_\xi}{d\mu_w}(x(\cdot)) = (\rho_T(x(\cdot)))^{-1}. \tag{14}$$

Indeed, in view of Theorem 2 a solution of equation

$$d\xi_1(t) = a_1(t, \xi_1(\cdot))\,dt + dw(t)$$

exists such that $\mu_\xi \sim \mu_{\xi_1}$. Moreover, $a_1(t, x(\cdot)) = a(t, x(\cdot)) - \boldsymbol{I}a(t, x(\cdot)) = 0$. Consequently, $\xi_1(t) = w(t)$.

Formula (14) follows from formula (11).

Assume now that $a(s, x(\cdot))$ is such that equation (8) may not be satisfied for ρ_T. Set

$$a_N(t, x(\cdot)) = \begin{cases} a(t, x(\cdot)) & \text{if } \int_0^t |a(s, x(\cdot))|^2\,ds < N \\ 0 & \text{if } \int_0^t |a(s, x(\cdot))|^2\,ds \geq N. \end{cases}$$

Let

$$\xi_N(t) = \int_0^t a_N(s, \xi(\cdot))\,ds + w(t),$$
$$\tau_N = \sup\,[t\colon \int_0^t |a(s, \xi(\cdot))|^2\,ds < N].$$

Then

$$\xi_N(t) = \xi(t) \quad \text{for } t < \tau_N \quad \text{and} \quad a_N(t, \xi(\cdot)) = a_N(t, \xi_N(\cdot)) = 0 \quad \text{for } t > \tau_N,$$

since

$$\int_0^t |a(s, \xi_N(\cdot))|^2\,ds \geq \int_0^{\tau_N} |a(s, \xi_N(\cdot))|^2\,ds = \int_0^{\tau_N} |a(s, \xi(\cdot))|^2\,ds = N.$$

Hence

$$\xi_N(t) = \int_0^t a_N(s, \xi_N(\cdot))\,ds + w(t),$$

and in view of that shown above

$$(15)\qquad \frac{d\mu_{\xi_N}(\cdot)}{d\mu_w}(x(\cdot)) = \exp\{\int_0^T (a_N(s, x(\cdot)), dx(s)) - \tfrac{1}{2}\int_0^T |a_N(s, x(\cdot))|^2\,ds\}.$$

For any measurable set $E \subset \mathscr{C}^m_{[0,T]}$ we have

$$\mu_{\xi_N}(E \cap \{x(\cdot)\colon \int_0^T |a(s, x(\cdot))|^2\,ds < N\}) = \mu_\xi(E \cap \{x(\cdot)\colon \int_0^T |a(s, x(\cdot))|^2\,ds < N\});$$

furthermore,

$$\exp\{\int_0^T (a_N(s, x(\cdot)), dx(s)) - \tfrac{1}{2}\int_0^T |a_N(s, x(\cdot))|^2\,ds\} = \exp\{\int_0^T (a(s, x(\cdot)), dx(s)) - \tfrac{1}{2}\int_0^T |a(s, x(\cdot))|^2\,ds\},$$

provided

$$\int_0^T |a(s, x(\cdot))|^2 \, ds < N.$$

Therefore formula (14) is valid for all $x(\cdot)$ such that $\int_0^T |a(s, x(\cdot))|^2 \, ds < N$. Since N can be chosen arbitrarily it follows that formula (14) is valid for all $x(\cdot)$ such that

$$\int_0^T |a(s, x(\cdot))|^2 \, ds < \infty.$$

Thus for any measurable set E we have

$$\begin{aligned}\mu_\xi(E) &= \lim_{N\to\infty} \mu_\xi(E \cap \{x(\cdot) : \int_0^T |a(s, x(\cdot))|^2 \, ds < N\}) \\ &= \lim_{N\to\infty} \int_{E\cap\{x(\cdot):\int_0^T |a(s,x(\cdot))|^2 ds < N\}} \hat{\rho}_T(x(\cdot))\mu_w(dx) \\ &= \int_E \hat{\rho}_T(x(\cdot))\mu_w(dx),\end{aligned}$$

where

$$\hat{\rho}_T(x(\cdot)) = \exp\{\int_0^T (a(t, x(\cdot)), dx(t)) - \tfrac{1}{2}\int_0^T |a(t, x(\cdot))|^2 \, dt\}, \tag{16}$$

provided

$$\mathsf{P}\{\int_0^T |a(t, w(\cdot))|^2 \, dt < \infty\} = 1,$$
$$\mathsf{P}\{\int_0^T |a(t, \xi(\cdot))|^2 \, dt < \infty\} = 1.$$

This means that

$$\frac{d\mu_\xi}{d\mu_w}(x(\cdot)) = \hat{\rho}_T(x(\cdot)).$$

Since the right-hand side of (16) is positive, $d\mu_w/d\mu_\xi$ also exists and

$$\frac{d\mu_w}{d\mu_\xi}(x(\cdot)) = (\hat{\rho}_T(x(\cdot)))^{-1}.$$

Assume now that

$$\mu_\xi(\{x(\cdot) : \int_0^T |a(s, x(\cdot))|^2 \, ds < \infty\}) = 1. \tag{17}$$

Then it can be shown in exactly the same manner as above that

$$\frac{d\mu_\xi}{d\mu_w}(x(\cdot)) = \hat{\rho}_T(x(\cdot)),$$

if we assume that for all $x(\cdot)$ satisfying the condition $\int_0^T |a(s, x(\cdot)|^2\, ds = \infty$,

(18) $$\exp\{\int_0^T (a(t, x(\cdot)), dx(t)) - \tfrac{1}{2}\int_0^T |a(t, x(\cdot))|^2\, dt\} = 0$$

is valid (since the measure μ_ξ is concentrated on $x(\cdot)$ satisfying $\int_0^T |a(s, x(\cdot)|^2\, ds < \infty$).* Equality (18) is quite natural because

$$\int_0^T (a(t, w(\cdot)), dw(t)) - \tfrac{1}{2}\int_0^T |a(t, w(\cdot))|^2\, dt$$

for $\int_0^T |a(t, w(\cdot))|^2\, dt < \infty$ has the same distributions as $w_1(\zeta) - \frac{1}{2}\zeta$, where ζ is distributed as $\int_0^T |a(t, w(\cdot))|^2 dt$ and $w_1(s)$ is a one-dimensional Wiener process; $w_1(\zeta) - \frac{1}{2}\zeta \to -\infty$ as $\zeta \to \infty$.

If, however,

$$\mu_w(\{x(\cdot): \int_0^T |a(s, x(\cdot))|^2\, ds < \infty\}) = 1,$$

then

$$\begin{aligned} \mu_w(E) &= \lim_{N\to\infty} \mu_w(E \cap \{x(\cdot): \int_0^T |a(s, x(\cdot))|^2\, ds < N\}) \\ &= \lim_{N\to\infty} \int_{E\cap\{x(\cdot): \int_0^T |a(s, x(\cdot))|^2\, ds < N\}} \hat{\rho}_T(x(\cdot))^{-1} \mu_\xi(dx) \\ &= \int_E \hat{\rho}_T^{-1}(x(\cdot))\mu_\xi(dx), \end{aligned}$$

provided $\hat{\rho}_T^{-1}(x(\cdot)) = 0$ for $x(\cdot)$ such that $\int_0^T |a(s, x(\cdot))|^2\, ds = \infty$. Hence also

$$\frac{d\mu_w}{d\mu_\xi}(x(\cdot)) = \hat{\rho}_T^{-1}(x(\cdot))$$

(in view of the assumptions on $a(t, x(\cdot))$ the right-hand side of the last equality is always well defined).

Thus the following theorem is valid:

Theorem 3. *Let $a(s, x(\cdot))$ be a function defined for $s \in [0, T]$, $x(\cdot) \in \mathscr{C}^m_{[0,T]}$, with values in $\mathscr{R}^m$ satisfying condition* 1). *If*

$$\xi(t) = \int_0^t a(s, \xi(\cdot))\, ds + w(t),$$

* The Itô integral $\int_0^T (f(s), dx(s))$ is defined here as the limit

$$\lim_{N\to\infty} \int_0^T (f_N(s), dx(s)),$$

where

$$f_N(t) = \chi_{[0,N]}(\int_0^t |f(s)|^2\, ds) f(t),$$

which exists almost everywhere on the set $\{x(\cdot): \int_0^T |f(s)|^2\, ds < \infty\}$.

where $w(t)$ is a Wiener process in $\mathcal{R}^m$ with respect to $\{\mathfrak{F}_t^\xi\}$, then

a) $\mathsf{P}\{\int_0^T |a(s, \xi(\cdot))|^2\, ds < \infty\} = 1 \Rightarrow \mu_\xi \ll \mu_w$,
b) $\mathsf{P}\{\int_0^T |a(s, w(\cdot))|^2\, ds < \infty\} = 1 \Rightarrow \mu_w \ll \mu_\xi$.

Moreover, in case a)

$$\frac{d\mu_\xi}{d\mu_w}(x(\cdot)) = \hat{\rho}_T(x(\cdot))$$

(*taking stipulation* (18) *into account*), *and in case* b)

$$\frac{d\mu_w}{d\mu_\xi} = \hat{\rho}_T^{-1}(x(\cdot)).$$

The existence of solutions of stochastic differential equations. The aim of this subsection is to prove the following theorem.

Theorem 4. *Let coefficients $a(t, x(\cdot))$ and $B(t, x(\cdot))$ satisfy conditions* 1) *and* 2) *as well as condition* 4) *stated in the beginning of this section. Then equation* (1) *possesses a solution.*

Thus condition 3) is not necessary for the existence of a solution of equation (1). To prove the theorem several auxiliary assertions are required.

Lemma 1. *Let condition* 4) *be satisfied. Then a solution $\xi(t)$ of equation* (1) *satisfies inequality*

$$\mathsf{E}(\sup_{t\le T} |\xi(t)|^2) \le C,$$

where C is a constant which depends on T and l only.

Proof. We have

$$\sup_{s\le t} |\xi(s)|^2 \le 2T \int_0^t |a(s, \xi(\cdot))|^2\, ds + 2 \sup_{s\le t} \left|\int_0^s B(u, \xi(\cdot))\, dw(u)\right|^2.$$

Hence,

$$\begin{aligned}\mathsf{E} \sup_{s\le t} |\xi(s)|^2 &\le 2T \int_0^t \mathsf{E}|a(s, \xi(\cdot))|^2\, ds + 8\mathsf{E}\left|\int_0^t B(s, \xi(\cdot))\, dw(s)\right|^2 \\ &= \mathsf{E} \int_0^t [2T|a(s, \xi(\cdot))|^2 + 8\,\mathrm{Sp}\, B(s, \xi(\cdot))B^*(s, \xi(\cdot))]\, ds \\ &\le K\Big(1 + \int_0^t \mathsf{E} \sup_{u\le s} |\xi(u)|^2\, ds\Big),\end{aligned}$$

where K is a constant which depends on T and l only. The last inequality yields the required result. □

Remark. Analogously one shows that for each n there exists a constant C_n which depends on T and l only such that

$$\mathsf{E}\sup_{t\leq T}|\xi(t)|^n\leq C_n.$$

Lemma 2. *Under the conditions of Lemma 1 there exists a constant K which depends on T and l only such that*

$$\mathsf{E}|\xi(t+h)-\xi(t)|^4\leq Kh^2.$$

Proof. Utilizing Itô's formula we write

$$\begin{aligned}|\xi(t+h)-\xi(t)|^4&=(\xi(t+h)-\xi(t),\,\xi(t+h)-\xi(t))^2\\&=4\int_t^{t+h}|\xi(s)-\xi(t)|^2(B^*(s,\xi(\cdot))[\xi(s)-\xi(t)],\,dw(s))\\&\quad+2\int_t^{t+h}\{|\xi(s)-\xi(t)|^2[2(\xi(s)-\xi(t),\,a(s,\xi(\cdot)))\\&\quad+\operatorname{Sp}B(s,\xi(\cdot))B^*(s,\xi(\cdot))]+|B^*(s,\xi(\cdot))[\xi(s)-\xi(t)]|^2\}\,ds.\end{aligned}$$

Hence

$$\begin{aligned}\mathsf{E}|\xi(t+h)-\xi(t)|^4&\leq K_1\mathsf{E}\int_t^{t+h}|\xi(s)-\xi(t)|^3(1+\sup_{u\leq s}|\xi(u)|)\,ds\\&\quad+K_2\mathsf{E}\int_t^{t+h}|\xi(s)-\xi(t)|^2(1+\sup_{u\leq s}|\xi(u)|^2)\,ds.\end{aligned}$$

Now applying Hölder's inequality and Lemma 1 we obtain

$$\begin{aligned}\mathsf{E}|\xi(t+h)-\xi(t)|^4&\leq K_3\Big(\Big[\int_t^{t+h}\mathsf{E}|\xi(s)-\xi(t)|^4\,ds\Big]^{3/4}h^{1/4}\\&\quad+\Big[\int_t^{t+h}\mathsf{E}|\xi(s)-\xi(t)|^4\,ds\Big]^{1/2}h^{1/2}\Big).\end{aligned}\tag{19}$$

From this inequality, utilizing the fact that in view of the remark following Lemma 1

$$\mathsf{E}|\xi(s)-\xi(t)|^4\leq 16C_4,$$

we obtain that

$$\mathsf{E}|\xi(s)-\xi(t)|^4\leq K_4h$$

for some K_4.

Substituting $K_4(s-t)$ in equation (19) in place of $\mathsf{E}|\xi(s)-\xi(t)|^4$ we arrive at the required result. □

Corollary. *Let $\mathcal{M}_l$ be the set of measures μ_ξ associated with solutions $\xi(t)$ of equation (1) for different $a(s,x(\cdot))$ and $B(s,x(\cdot))$, satisfying conditions 1) and 2) and condition 4) with the same constant l. Then $\mathcal{M}_l$ is weakly compact.*

This follows from Lemma 5 and Theorem 2 in Volume I, Chapter VI, Section 4. We now proceed to prove Theorem 4.

Proof of Theorem 4. We construct sequences $a_n(t, x(\cdot))$ and $B_n(t, x(\cdot))$ satisfying the following conditions:

(I) for some l'

$$|a_n(t, x(\cdot))| + \|B_n(t, x(\cdot))\| \leq l'(1 + \|x(\cdot)\|_m);$$

(II) a constant H_n exists such that

$$|a_n(t, x(\cdot)) - a_n(t, y(\cdot))| + \|B_n(t, x(\cdot)) - B_n(t, y(\cdot))\| \leq H_n \|x(\cdot) - y(\cdot)\|_m;$$

(III) $$|a(t, x(\cdot)) - a_n(t, x(\cdot))| + \|B_n(t, x(\cdot)) - B(t, x(\cdot))\| \to 0$$

uniformly on any compact set in $\mathscr{C}^m_{[0,T]}$.

Such sequences of functions can be constructed as follows (we shall discuss the sequences of functions $a_n(t, x(\cdot))$ only).

Let $\Gamma_n(x)$ be piecewise linear functions which coincide with $x(\cdot)$ at the points $(k/n)T \in [0, T]$. Denote by $\gamma_n(t, x_0, \ldots, x_n)$ a piecewise linear function in $\mathscr{C}^m_{[0,T]}$ taking on value x_k at the point $t = (k/n)T$. Define, furthermore,

$$a_n(t, x(\cdot)) = \int \cdots \int a(t, \gamma_n(\cdot, x_0, \ldots, x_n)) \exp\left\{-\frac{\varepsilon_n}{n} \sum_{k=0}^{n} x_k^2\right\} \times \prod_{k=0}^{n} \left(g\left(\frac{x(kT/n) - x_k}{\varepsilon_n}\right) \frac{dx_k}{\varepsilon_n}\right),$$

where the function $g(z)$ is defined for $z \in \mathscr{R}^m$, is nonnegative, does not vanish only for $|z| \leq 1$, possesses a bounded derivative, $\int g(z)\, dz = 1$, and $\varepsilon_n \to 0$. Then

$$\begin{aligned} |a_n(t, x(\cdot))| &\leq \sup_{|x_i| \leq \varepsilon_n} |a(t, \gamma_n(\cdot, x(0) + x_0, \ldots, x(T) + x_n))| \\ &\leq l(1 + \sup_t |\gamma(\cdot, x(0) + x_0, \ldots, x(T) + x_n)| \\ &\leq l(1 + \varepsilon_n + \sup_s |x(s)|), \end{aligned}$$

and condition (I) is valid.

Condition (II) follows from the fact that $a_n(t, x(\cdot))$ is a function of $x((k/n)T)$ and possesses bounded derivatives with respect to these variables.

Finally,

$$|a_n(t, x(\cdot)) - a(t, x(\cdot))| \le \int \cdots \int |a(t, \gamma_n(\cdot, x(0)+\bar{x}_0, \ldots, x(T)+\bar{x}_n)) - a(t, \gamma_n(\cdot, x(0), \ldots, x(T))| \prod_{k=0}^{n} \left(g\left(\frac{\bar{x}_k}{\varepsilon_n}\right)\frac{d\bar{x}_k}{\varepsilon_n}\right)$$
$$+ |a(t, x(\cdot)) - a(t, \gamma_n(\cdot, x(0), \ldots, x(T)))|$$
$$+ \int \cdots \int |a(t, \gamma_n(\cdot, x_0, \ldots, x_n))| \times \left(1 - \exp\left\{-\varepsilon_n \frac{1}{n} \sum_{k=0}^{n} x_k^2\right\}\right) \times \prod_{k=0}^{n} \left(g\left(\frac{x(kT/n) - x_k}{\varepsilon_n}\right)\frac{dx_k}{\varepsilon_n}\right).$$

The first summand tends to zero uniformly on every compact K since for an arbitrary compact $K \subset \mathscr{C}^m_{[0,T]}$ one can find a compact K^1 such that $\gamma(\cdot, x(0), \ldots, x(T)) \in K^1$ for all $x \in K$ and, moreover,

$$\sup_t |\gamma_n(t, x(0), \ldots, x(T)) - \gamma_n(t, x(0)+z_0, \ldots, x(T)+z_n)| \le \varepsilon_n$$

for $|z_k| \le \varepsilon_n$. The second summand tends to zero uniformly on every compact since

$$\sup_t |x(t) - \Gamma_n x(t)| \to 0$$

uniformly on every compact. Finally, the third summand is bounded by

$$l(1 + \varepsilon_n + \sup_t |x(t)|)\varepsilon_n \sup_t (|x(t)| + \varepsilon_n)^2$$

and hence tends to zero uniformly as $n \to \infty$ on every compact.

Now let $\xi_n(t)$ be a solution of the stochastic equation

$$\xi_n(t) = \int_0^t a_n(s, \xi_n(s))\, ds + \int_0^t B_n(s, \xi_n(s))\, dw(s). \tag{20}$$

Denote by μ_n a measure on $\mathscr{C}^m_{[0,T]}$ associated with the solution $\xi_n(t)$ of equation (20). Since the coefficients of this equation satisfy conditions 3) and 4) the solution of equation (20) exists and is unique. In view of the preceding corollary the set of measures μ_n is compact. Therefore we can assume without loss of generality that μ_n converges weakly to a measure μ. Let $g_t(x(\cdot))$ be a continuous $\mathfrak{S}_t$-measurable function on $\mathscr{C}^m_{[0,T]}$. Then we have for all $u \in \mathscr{R}^m$

$$\begin{aligned} &\int [(x(t+h) - x(t), u) - \int_t^{t+h} (a_n(s, x(\cdot)), u)\, ds] g_t(x(\cdot))\mu_n(dx) \\ &= \mathsf{E}(\xi_n(t+h) - \xi_n(t) - \int_t^{t+h} a_n(s, \xi_n(\cdot))\, ds, u) g_t(\xi_n(\cdot)) = 0. \end{aligned} \tag{21}$$

Furthermore,

$$\overline{\lim_{n\to\infty}} \int_t^{t+h} |(a(s, x(\cdot)) - a_n(s, x(\cdot)), u)| g_t(x(\cdot))\mu_n(dx)$$

$$= \overline{\lim_{n\to\infty}} \int_t^{t+h} \frac{|a(s, x(\cdot)) - a_n(s, x(\cdot))| \cdot |u|}{1 + \|x(\cdot)\|_m}$$

$$\times g_t(x(\cdot))(1 + \|x(\cdot)\|_m)\mu_n(dx).$$

If $\nu_n(dx) = (1 + \|x(\cdot)\|_m)\mu_n(dx)$, the measures ν_n are uniformly bounded and are weakly convergent. Therefore for any $\varepsilon > 0$ there exists a compact K_ε such that

$$\overline{\lim_{n\to\infty}} \nu_n(\mathscr{C}^m_{[0,T]} - K_\varepsilon) \leq \varepsilon.$$

Consequently,

$$\overline{\lim_{n\to\infty}} \int_t^{t+h} |(a(s, x(\cdot)) - a_n(s, x(\cdot)), u)| \, |g_t(x(\cdot))| \mu_n(dx)$$

$$= O(\overline{\lim_{n\to\infty}} \nu_n(\mathscr{C}^m_{[0,T]} - K_\varepsilon)) = O(\varepsilon)$$

(here we utilized conditions (I) and (III), which are satisfied for $a_n(s, x(\cdot))$). Since $\varepsilon > 0$ is arbitrary

$$\lim_{n\to\infty} \int \int_t^{t+h} |a(s, x(\cdot)) - a_n(s, x(\cdot))| \, |g_t(x(\cdot))| \, ds \, \mu_n(dx) = 0. \tag{22}$$

It follows from the weak convergence of measures μ_n that

$$\begin{aligned} &\lim_{n\to\infty} \int (x(t+h) - x(t), u) g_t(x(\cdot))\mu_n(dx) \\ &\quad = \int (x(t+h) - x(t), u) g_t(x(\cdot))\mu(dx), \\ &\lim_{n\to\infty} \int \int_t^{t+h} (a(s, x(\cdot)), u) g_t(x(\cdot)) \, ds \, \mu_n(dx) \\ &\quad = \int \int_t^{t+h} (a(s, x(\cdot)), u) g_t(x(\cdot)) \, ds \, \mu(dx). \end{aligned} \tag{23}$$

Therefore approaching the limit in (21) as $n \to \infty$ and taking (22) and (23) into account we obtain

$$\int [(x(t+h) - x(t), u) - \int_t^{t+h} (a(s, x(\cdot)), u) \, ds] g_t(x(\cdot))\mu(dx) = 0. \tag{24}$$

Hence the process

$$y(t) = x(t) - \int_0^t a(s, x(\cdot)) \, ds$$

is a martingale on the probability space $\{\mathscr{C}^m_{[0,T]}, \mathfrak{S}_T, \mu\}$.

Utilizing the equality

$$\int [(x(t-h)-x(t)-\int_0^{t+h} a_n(s, x(\cdot))\,ds, u)^2 - \int_t^{t+h} (B_n^*(s, x(\cdot))u, B_n^*(s, x(\cdot))u)\,ds] g_t(x(\cdot))\mu_n(dx)=0,$$

one can prove in the same manner as above that the process

$$(y(t), u)^2 - \int_0^t (B^*(s, x(\cdot))u, B^*(s, x(\cdot))u)\,ds$$

is a martingale on $\{\mathscr{C}^m_{[0,T]}, \mathfrak{S}_T, \mu\}$. Hence

$$\langle (y(\cdot), u), (y(\cdot), u)\rangle_t = \int_0^t (B^*(s, x(\cdot))u, B^*(s, x(\cdot))u)\,ds.$$

Therefore in view of Theorem 1, the measure μ is associated with the process $\xi(t)$ which is a solution of equation (1). The existence of a solution has thus been proved. □

Remark. Let τ be a finite Markov time with respect to the current $\{\mathfrak{S}_t\}$. Utilizing the independence of the process $w(t+\tau)-w(\tau)$ from σ-algebra $\mathfrak{S}_\tau$ one can prove, in the same manner as in Theorem 4, the existence of the processes $\xi(s)$ on $[\tau, T]$ satisfying relation

$$\xi(t)-\xi(\tau)=\int_\tau^t a(s, \xi(\cdot))\,ds + \int_\tau^t B(s, \xi(\cdot))\,dw(s), \qquad \tau \leq t \leq T,$$

provided $\xi(s)$ is defined on $[0, \tau]$ and $\xi(s)$ is $\mathfrak{S}_\tau$-measurable for $s \leq \tau$.

Uniqueness of a solution. The following fact established by I. V. Girsanov plays an important part in the study of uniqueness of solutions of stochastic equations.

Lemma 3 (I. V. Girsanov). *Let $(X, \mathfrak{B}, \mu)$ be a probability space and $(Y, \mathfrak{L})$ be a measurable space. If a measurable mapping $x=f(y)$ exists of the space $(Y, \mathfrak{L})$ into $(X, \mathfrak{B})$ and measurable mappings $g_1(x)$ and $g_2(x)$ of the space $(X, \mathfrak{B})$ into $(Y, \mathfrak{L})$ satisfying, for almost all x with respect to measure μ, the relation*

$$f(g_1(x)) = f(g_2(x)) = x,$$

and the measures ν_i on $(Y, \mathfrak{L})$ defined by equation $\nu_i(C) = \mu(g_i^{-1}(C))$ are such that $\nu_2 \ll \nu_1$, then for almost all x with respect to measure μ

$$g_1(x) = g_2(x).$$

Proof. Let $Y_i = g_i(X)$, $y \in Y_1 \cap Y_2$. Then $y = g_1(x_1) = g_2(x_2)$. Hence

$$x_1 = f(g_1(x_1)) = f(g_2(x_2)) = x_2$$

and $g_1(x_1) = g_2(x_2)$. This relation is valid for all $x \in f(Y_1 \cap Y_2)$. Observe that

$$\nu_i(Y_i) = \mu(X) = 1, \qquad \nu_i(Y - Y_i) = 0.$$

Thus $\nu_2(Y - Y_2) = 0$ and $\nu_2(Y - Y_1) = 0$ since $\nu_1(Y - Y_1) = 0$ and $\nu_2 \ll \nu_1$. This implies that $\nu_2(Y_1 \cap Y_2) = 1$. Therefore

$$1 = \nu_2(Y_1 \cap Y_2) = \mu(g_2^{-1}(Y_1 \cap Y_2)) = \mu(f(Y_1 \cap Y_2)).$$

The lemma is thus proved. □

Now consider the stochastic differential equation (1). Assume that the operator $B(t, x(\cdot))$ is invertible. Then

$$w(t) = -\int_0^t B^{-1}(s, \xi(\cdot))\, a(s, \xi(\cdot))\, ds + \int_0^t B^{-1}(s, \xi(\cdot))\, d\xi(s); \tag{25}$$

the stochastic integral is well defined for each solution $\xi(\cdot)$ of equation (1) since the process $\xi(t)$ is an Itô process. Equation (25) determines a single-valued measurable mapping of the measurable space $\{\mathscr{C}^m_{[0,T]}, \mathfrak{S}_T\}$ into probability space $\{\mathscr{C}^m_{[0,T]}, \mathfrak{S}_T, \mu\}$, where μ is the measure associated with the process $w(t)$; this mapping corresponds to mapping f in Girsanov's lemma. Various $\{\mathfrak{F}_t\}$-adopted solutions of equation (1) correspond to mappings $g_i(x)$ in this lemma. Thus the following theorem is valid.

Theorem 5. *Let the coefficients of equation* (1) *satisfy conditions* 1) *and* 2) *and moreover, let the operator $B(t, x(\cdot))$ be invertible for all $t \in [0, T]$ and $x(\cdot) \in \mathscr{C}^m_{[0,T]}$. If $\xi_1(t)$ and $\xi_2(t)$ are two strong solutions of equation* (1) *and if $\mu_{\xi_2} \ll \mu_{\xi_1}$, then $\xi_1(t) = \xi_2(t)$ with probability* 1.

Remark. We say that a solution of equation (1) is *weakly unique* if measures μ_{ξ_1} and μ_{ξ_2}, associated with two solutions $\xi_1(t)$ and $\xi_2(t)$ of equation (1), coincide.

Theorem 5 asserts that if an operator $B(t, x(\cdot))$ is nondegenerate, weak uniqueness implies uniqueness of the strong solution. This fact will often be utilized below.

Corollary 1. *Let $a(s, x(\cdot))$ be such that $\int_0^T |a(s, x(\cdot))|^2\, ds < \infty$ for all $x(\cdot) \in \mathscr{C}^m_{[0,T]}$. Then the equation*

$$\xi(t) = w(t) + \int_0^t a(s, \xi(\cdot))\, ds \tag{26}$$

possesses at most one strong solution.

Indeed, in view of Theorem 3 the measure associated with process $\xi(t)$ is equivalent to the Wiener measure. Hence measures associated with two arbitrary strong solutions of equation (26) are equivalent and by Theorem 5 these solutions coincide with probability 1.

Corollary 2. *Let operator $B(s, x(\cdot))$ be invertible for all*

$$s \in [0, T], \qquad x(\cdot) \in \mathscr{C}^m_{[0,T]}$$

and

$$b(s, x(\cdot)) = B^{-1}(s, x(\cdot))a(s, x(\cdot))$$

be a bounded function. Then a solution of equation (1) *will be strongly unique provided the solution of the equation*

$$d\xi_1(t) = B(t, \xi_1(\cdot))\, dw(t) \tag{27}$$

is weakly unique.

Proof. Indeed, Theorem 2 implies that for any solution $\xi(t)$ of equation (1) one can find a solution $\xi_1(t)$ of equation (27) such that measures μ_{ξ_1} and μ_ξ associated with processes $\xi_1(t)$ and $\xi(t)$ are equivalent (note that condition (8) is fulfilled since $b(s, x(\cdot)$ is bounded). If a solution of equation (27) is weakly unique, then the measures associated with solutions of equation (1) are equivalent since they are equivalent to the measure associated with the weakly unique solution of the first equation. It remains now to apply Theorem 5. □

Corollary 2 allows us to reduce the problem of weak uniqueness of a solution of equation (1) to the identical problem for equation (27).

Indeed, assume that equation (1) possesses a weakly unique solution for all $a(s, x(\cdot))$ such that $B^{-1}(s, x(\cdot))a(s, x(\cdot))$ is bounded. For some $a_1(s, x(\cdot))$ satisfying conditions 1) and 2), let equation (1) possess two solutions $\xi_i(t)$ $(i = 1, 2)$ and let $\mu_{\xi_1} \neq \mu_{\xi_2}$. Set

$$\tau_N = \sup\,[t \leq T\colon \sup_{s \leq t} |B^{-1}(s, \xi_1(\cdot))a(s, \xi_1(\cdot))| \leq N,$$
$$\sup_{s \leq t} |B^{-1}(s, \xi_2(\cdot))a(s, \xi_2(\cdot))| \leq N].$$

Since $|B^{-1}(s, x(\cdot))a(s, x(\cdot))|$ is bounded on each compact set and for any $\varepsilon > 0$ one can find a compact set K_ε such that

$$\mathsf{P}\{\xi_1(\cdot) \in K_\varepsilon\} \geq 1 - \varepsilon, \qquad \mathsf{P}\{\xi_2(\cdot) \in K_\varepsilon\} \geq 1 - \varepsilon,$$

it follows that $\mathsf{P}\{\tau_N = T\} \to 1$ as $N \to \infty$. Let

$$a_1^N(t, x(\cdot)) = a(t, x(\cdot)),$$

provided

$$\sup_{s \leq t} |B^{-1}(s, x(\cdot))a(s, x(\cdot))| \leq N,$$

and

$$a_1^N(t, x(\cdot)) = a(t_N, x(\cdot)) \quad \text{for } t_N < t,$$

provided

$$|B^{-1}(s, x(\cdot))a(s, x(\cdot))| < N \quad \text{for } s < t_N,$$

and

$$|B^{-1}(t_N, x(\cdot))a(t_N, x(\cdot))| = N.$$

Clearly $a_1^N(t, x(\cdot))$ satisfies conditions 1) and 2) and $|B^{-1}(t, x(\cdot))a_1^N(t, x(\cdot))| \leq N$. Let $\tilde{\xi}_i(t) = \xi_i(t)$ for $t \leq \tau_N$ and

$$\tilde{\xi}_i(t) - \tilde{\xi}_i(\tau_N) = \int_{\tau_N}^t a_1^N(s, \tilde{\xi}_i(\cdot))\, ds + \int_{\tau_N}^t B(s, \tilde{\xi}_i(\cdot))\, dw(s)$$

for $t > \tau_N$ (the existence of such a $\tilde{\xi}_i(t)$ is assured by the remark following Theorem 4). Clearly, for $s < \tau_N$ we have

$$a_1^N(s, \tilde{\xi}_i(\cdot)) = a_1(s, \tilde{\xi}_i(\cdot));$$

hence $\tilde{\xi}_i(\cdot)$ is a solution of the equation

$$d\tilde{\xi}_i(t) = a_1^N(t, \tilde{\xi}_i(\cdot))\, dt + B(t, \tilde{\xi}_i(\cdot))\, dw(t).$$

If $\mu_{\xi_1} \neq \mu_{\xi_2}$, then also $\mu_{\tilde{\xi}_1} \neq \mu_{\tilde{\xi}_2}$ for N sufficiently large. Thus the weak nonuniqueness of a solution of equation (1) for some $a_1(s, x(\cdot))$ results in the weak nonuniqueness of a solution for $a(s, x(\cdot))$ such that $B^{-1}(s, x(\cdot))a(s, x(\cdot))$ is bounded, and hence in the weak nonuniqueness of a solution of equation (27). In what follows we shall thus consider the problem of uniqueness of the solution of equation (27). First we shall discuss the one-dimensional case.

A function $B(t, x(\cdot))$ defined and measurable on $[0, T] \times \mathscr{C}_{[0,T]}^m$ and measurable with respect to $\mathfrak{S}_t$ for each t is said to be *invariantly time dependent* if for any function $x(\cdot) \in \mathscr{C}_{[0,T]}^m$ and a continuous function $\lambda(t)$ which maps $[0, T]$ onto $[0, T]$ in a one-to-one manner ($\lambda(0) = 0$) the relation

$$B(t, x_1(\cdot)) = B(\lambda(t), x(\cdot))$$

is satisfied where $x_1(t) = x(\lambda(t))$. Furthermore, we define the functional

$$\lambda_0(t, \xi(\cdot)) = \int_0^t [d\xi(t)]^2;$$

if $\xi(t)$ is an Itô process then this functional coincides with $\langle \xi, \xi \rangle_t$. Set

$$\lambda_1(t, \xi(\cdot)) = \int_0^t \frac{d_s\lambda_0(s, \xi(\cdot))}{B^2(s, \xi(\cdot))} = \int_0^t B^{-2}(s, \xi(\cdot))\, d\langle \xi, \xi \rangle_s. \tag{28}$$

Functional $\lambda_0(t, \xi(\cdot))$ is invariantly dependent on t. The same is valid for functional $\lambda_1(t, \xi(\cdot))$, provided $B(t, \xi(\cdot))$ is invariantly dependent on t.

Let $\xi(\cdot)$ be a solution of equation (27). In this case $\lambda_1^{\cdot}(t, \xi(\cdot))=t$. We define quantities τ_t by the equality

(29) $$t=\lambda_0(\tau_t, \xi(\cdot)).$$

Set $w_1(t)=\xi(\tau_t)$. Clearly $w_1(t)$ is a local martingale. Moreover,

$$\langle w_1, w_1\rangle_t=\lambda_0(\tau_t, \xi(\cdot))=t.$$

Hence $w_1(t)$ is a Wiener process. Denote by φ_t the inverse function for τ_t: $\tau_{\varphi_t}=t$. Then $\xi(t)=w_1(\varphi_t)$.

Function φ_t can be defined by means of the process $w_1(\cdot)$. Indeed, utilizing the relations

$$t=\lambda_1(t, \xi(\cdot))=\lambda_0(t, w_1(\cdot)),$$
$$\lambda_1(t, \xi(\cdot))=\lambda_1(\varphi_t, w_1(\cdot)),$$

we observe that φ_t is defined by the equality

(30) $$\lambda_1(\varphi_t, w_1(\cdot))=t.$$

Thus if $\xi(t)$ is a solution of equation (27), then a Wiener process $w_1(\cdot)$ exists such that $\xi(t)=w_1(\varphi_t)$, where φ_t is uniquely determined by means of process w_1 via the relation (30) and $\lambda_1(t, \xi)$ is defined for all Itô processes by means of equality (28). Relation (30) is equivalent to the equality

(31) $$t=\int_0^{\varphi_t} B^{-2}(s, w_1(\cdot))\, ds.$$

If $\xi_1(t)$ and $\xi_2(t)$ are two solutions of equations (27), and since under the stated conditions $\xi_1(t)$ and $\xi_2(t)$ are the same functions of Wiener processes $w_1(t)$ and $w_2(t)$, it follows that the measures associated with processes $\xi_1(t)$ and $\xi_2(t)$ coincide. Applying Theorem 5 and Corollary 2 we obtain the following result.

Theorem 6. *Let the coefficients of equation*

(32) $$d\xi(t)=a(t, \xi(\cdot))\, dt+B(t, \xi(\cdot))\, dw(t)$$

in $\mathcal{R}^1$ *satisfy conditions* 1) *and* 2) *and, moreover, let* $B(t, x(\cdot))>0$ *and be invariantly dependent on* t.

Then the solution of equation (32) *is weakly unique and a strong solution of* (32) *is unique.*

Remark. It may happen that although function $B(t, x(\cdot))$ does not depend invariantly on t, it can be rewritten in a form in which it does depend invariantly on

t. This is due to the fact that it may be possible to construct a function $B_1(t, x(\cdot))$ invariantly dependent on t and such that

$$B(t, \xi(\cdot)) = B_1(t, \xi(\cdot))$$

with probability 1. In order to obtain such a representation of $B(t, \xi(\cdot))$ it is necessary to eliminate the explicit dependence of $B(t, \xi(\cdot))$ on t, expressing t in the form of a function $\varphi(t, \xi(\cdot))$ invariantly dependent on t.

We now show how this can be accomplished. Let $\lambda_0(t, \xi(\cdot)) = \langle \xi, \xi \rangle_t$ (clearly this function does not depend on $a(\cdot, \cdot)$ for $\xi(\cdot)$ which is a solution of equation (32).) Next set

$$F(t, \xi(\cdot)) = \int_0^t B^2(s, \xi(\cdot))\, ds$$

and define $\varphi(t, \xi(\cdot))$ by the relation

$$F(\varphi(t, \xi(\cdot)), \xi(\cdot)) = \lambda_0(t, \xi(\cdot)).$$

Since $\lambda_0(t, \xi(\cdot)) = F(t, \xi(\cdot))$ it is easy to verify that $\varphi(t, \xi(\cdot)) = t$ with probability 1. Furthermore, $\varphi(t, \xi(\cdot))$ is determined by the behavior of $\xi(\cdot)$ up to time t since both $\lambda_0(t, \xi(\cdot))$ and $F(t, \xi(\cdot))$ possess this property. We now determine the conditions under which $\varphi(t, \xi(\cdot))$ depends invariantly on t.

First we shall assume that $B(t, \xi(\cdot)) = B_k(t, \xi(\cdot))$ for $t_k \leqslant t < t_{k+1}$ where $0 = t_0 < t_1 < \cdots < t_n = T$ and $B_k(t, \xi(\cdot))$ is invariantly dependent on t. In this case for $t_k \leqslant t \leqslant t_{k+1}$

$$\varphi(t, \xi(\cdot)) = \varphi(t_k, \xi(\cdot)) + \int_{t_k}^t \frac{d\lambda_0(s, \xi(\cdot))}{B_k^2(s, \xi(\cdot))}.$$

Since both $\lambda_0(t, \xi(\cdot))$ and $B_k^2(t, \xi(\cdot))$ are invariantly dependent on t so is $\varphi(t, \xi(\cdot))$. Therefore the stated assertion is valid for functions $B(t, x(\cdot))$, which are the limits of step functions in t, invariantly dependent on t on the intervals of constancy.

The function $B(s, x)$ jointly continuous in the variables may serve as an example of such a function: since the function $B_k(x)$ is clearly invariantly dependent on t, $B(s, x)$ can be approximated by means of the step functions in t.

Now let $\xi(t)$ be a solution of equation (27) in $\mathcal{R}^m$, operator $B(s, x(\cdot))$ be invariantly dependent on t, and $g(t, x(\cdot))$ be a positive function on $[0, T] \times \mathscr{C}^m_{[0,T]}$ also invariantly dependent on t. We define the functions τ_t by means of equality

$$t = \int_0^{\tau_t} g(s, \xi(\cdot))\, ds.$$

These functions are continuous and increase in t, and, for each t, τ_t is a Markov time for the Wiener process $w(t)$.

Set $\xi_1(t) = \xi(\tau_t)$,.

$$w_1(t) = \int_0^{\tau_t} \sqrt{g(s, \xi(\cdot))}\, dw(s).$$

The process $w_1(t)$ is also a Wiener process in $\mathcal{R}^m$. Since

$$\xi_1(t)=\int_0^{\tau_t} B(s,\xi(\cdot))\,dw(s)=\int_0^t B(\tau_s,\xi_1(\cdot))\,dw(\tau_s)$$

$$=\int_0^t \frac{1}{\sqrt{g(s,\xi_1(\cdot))}} B(s,\xi_1(\cdot))\,d\int_0^{\tau_s}\sqrt{g(u,\xi(\cdot))}\,dw(u)$$

$$=\int_0^t \frac{1}{\sqrt{g(s,\xi_1(\cdot))}} B(s,\xi_1(\cdot))\,dw_1(s),$$

$\xi_1(t)$ satisfies the stochastic differential equation

$$d\xi_1(t)=\frac{1}{\sqrt{g(s,\xi_1(\cdot))}} B(s,\xi_1(\cdot))\,dw_1(s), \tag{33}$$

where $w_1(t)$ is a Wiener process. Assume that a function $\lambda_1(t,\xi(\cdot))$ exists, invariantly dependent on t such that $\lambda_1(t,\xi(\cdot))=t$ with probability 1. Then defining φ_t to be the inverse function for τ_t, $t=\tau_{\varphi_t}$, we obtain

$$t=\lambda_1(t,\xi(\cdot))=\lambda_1(\varphi_t,\xi_1(\cdot)). \tag{34}$$

Hence $\xi(t)=\xi_1(\varphi_t)$, where φ_t is determined from $\xi_1(\cdot)$ by means of equality (34) and satisfies equation (27). Using this fact we arrive at the following theorem.

Theorem 7. *Let $B(s,x(\cdot))$ be an operator function with values in $\mathfrak{L}(\mathcal{R}^m)$ and $g(s,x(\cdot))$ be a function with values in $\mathcal{R}^1$. Moreover, let both functions be defined and measurable on $[0,T]\times\mathscr{C}^m_{[0,T]}$ and be invariantly dependent on t. If a solution of equation* (33) *is weakly unique, then the strong solution of equation* (27) *is unique.*

Proof. If $\xi(t)$ and $\xi'(t)$ are two solutions of equation (27), then it follows from that proven above that these solutions are of the form

$$\xi(t)=\xi_1(\varphi_t) \quad\text{and}\quad \xi'(t)=\xi_1'(\varphi_t'),$$

where $\xi_1(t)$ satisfies equation (33) and $\xi'(t)$ satisfies the equation

$$d\xi_1'(t)=\frac{1}{\sqrt{g(s,\xi_1'(\cdot))}} B(s,\xi_1'(\cdot))\,dw_1'(s),$$

where $w_1'(\cdot)$ is a Wiener process and φ_t and φ_t' are defined by the equalities

$$t=\lambda_1(\varphi_t,\xi_1(\cdot))=\lambda_1(\varphi_t',\xi_1'(\cdot)),$$

$$\lambda_1(t,x(\cdot))=\langle z_u(\cdot),z_u(\cdot)\rangle_t,$$

$$z_u(t)=\int_0^t (u,B^{-1}(s,\xi(\cdot))\,d\xi(s),$$

$$u\in\mathcal{R}^m,\qquad |u|=1.$$

Hence $\xi(t)$ and $\xi'(t)$ are obtained from Wiener processes $w_1(t)$ and $w_1'(t)$ by means of the same transformations; therefore the measures associated with these processes coincide. To complete the proof it remains only to apply Theorem 5. □

Itô processes and stochastic differential equations. Let $w(t)$ be a Wiener process in $\mathcal{R}^m$, and $\{\mathfrak{F}_t\}$ be a current of σ-algebras generated by this process. Consider the Itô process

$$\xi(t)=\int_0^t a(s,\omega)\,ds+\int_0^t B(s,\omega)\,dw(s), \tag{35}$$

where $a(t,\omega)$ and $B(t,\omega)$ are $\{\mathfrak{F}_t\}$-adopted functions taking on values in $\mathcal{R}^m$ and $\mathfrak{L}(\mathcal{R}^m)$ respectively.

Below we state conditions under which $\xi(t)$ is a solution of a stochastic differential equation of form (1).

Denote by $\{\mathfrak{F}_t^\xi\}$ the current of σ-algebras generated by process $\xi(t)$.

Theorem 8. *Let the following conditions be satisfied:*

1) *$a(s,\omega)$ and $B(s,\omega)$ are continuous in s for almost all ω;*
2) $\int_0^T \mathsf{E}|a(s,\omega)|^2\,ds<\infty$;
3) *$B(s,\omega)$ is a positive symmetric operator.*

Then there exist measurable functions $a(s,x(\cdot))$ and $B(s,x(\cdot))$ defined for $s\in[0,T]$, $x(\cdot)\in\mathcal{C}^m_{[0,T]}$, and taking on values in $\mathcal{R}^m$ and $\mathfrak{L}(\mathcal{R}^m)$, respectively, and satisfying condition 1) *stated in the beginning of this section (following equation* (1)), *and a Wiener process $\bar{w}(t)$ with respect to $\{\mathfrak{F}_t^\xi\}$ such that $\xi(t)$ satisfies the relation*

$$\xi(t)=\int_0^t a(s,\xi(\cdot))\,ds+\int_0^t B(s,\xi(\cdot))\,d\bar{w}(s). \tag{36}$$

Proof. Set

$$\bar{a}(s,\omega)=\mathsf{E}(a(s,\omega)\,|\,\mathfrak{F}_s^\xi),\qquad \eta(t)=\xi(t)-\int_0^t \bar{a}(s,\omega)\,ds.$$

Clearly $\eta(t)$ is an $\mathfrak{F}_t^\xi$-measurable function. We show that $\eta(t)$ is a local martingale with respect to the current $\{\mathfrak{F}_t^\xi\}$.

Let

$$\tau_N=\sup\Big[t\le T,\ \sup_{s\le t}|\eta(s)|\le N\Big],\qquad \eta_N(t)=\eta(t\wedge\tau_N).$$

Then

$$\eta_N(t)=\int_0^{t\wedge\tau_N}[a(s,\omega)-\bar{a}(s,\omega)]\,ds+\int_0^{t\wedge\tau_N}B(s,\omega)\,dw(s).$$

Noting that $\eta_N(t)$ is bounded and utilizing condition 2) of the theorem we obtain

$$\mathsf{E}\Big|\int_0^{t\wedge\tau_N}B(s,\omega)\,dw(s)\Big|^2<\infty.$$

Hence since $\mathfrak{F}_{t_1} \supset \mathfrak{F}^{\xi}_{t_1}$ we have for $t_1 < t_2$

$$\mathsf{E}\left(\int_{t_1\wedge\tau_N}^{t_2\wedge\tau_N} B(s, \omega)\, dw(s) \,\middle|\, \mathfrak{F}^{\xi}_{t_1}\right) = \mathsf{E}\left(\int_{t_1\wedge\tau_N}^{t_2\wedge\tau_N} B(s, \omega)\, dw(s) \,\middle|\, \mathfrak{F}_{t_1}\right) = 0.$$

Furthermore, the process

$$\bar{\eta}(t) = \int_0^t [a(s, \omega) - \bar{a}(s, \omega)]\, ds$$

satisfies for $t_1 < t_2$ the condition

$$\mathsf{E}\left(\int_{t_1}^{t_2} [a(s, \omega) - \bar{a}(s, \omega)]\, ds \,\middle|\, \mathfrak{F}^{\xi}_{t_1}\right) = \mathsf{E}\left(\int_{t_1}^{t_2} \mathsf{E}(a(s, \omega) - \bar{a}(s, \omega) \mid \mathfrak{F}^{\xi}_s)\, ds \,\middle|\, \mathfrak{F}^{\xi}_{t_1}\right) = 0.$$

This implies that for any Markov times τ_N with respect to the current $\{\mathfrak{F}^{\xi}_t\}$ we also have

$$\mathsf{E}\left(\int_{t_1\wedge\tau_N}^{t_2\wedge\tau_N} [a(s, \omega) - \bar{a}(s, \omega)]\, ds \,\middle|\, \mathfrak{F}^{\xi}_{t_1}\right) = 0.$$

We have thus shown that $\eta_N(t)$ is a martingale, whence $\eta(t)$ is a local martingale.

Since

$$\eta(t) = \int_0^t [a(s, \omega) - \bar{a}(s, \omega)]\, ds + \int_0^t B(s, \omega)\, dw(s),$$

it follows that for any $z \in \mathcal{R}^m$

$$\langle\langle (z, \eta), (z, \eta) \rangle\rangle_t = \int_0^t |B(s, \omega) z|^2\, ds.$$

The quantity on the left-hand side of the last equality is $\mathfrak{F}^{\xi}_t$-measurable, therefore so is $|B(t, \omega) z|$. Hence $B(t, \omega)$ is an $\{\mathfrak{F}^{\xi}_t\}$-adopted function (the positive symmetric operator B is determined by the values of $(B^2 z, z) = |Bz|^2$). Set

$$\bar{w}(t) = \int_0^t B^{-1}(s, \omega)\, d\eta(t).$$

This process is $\{\mathfrak{F}^{\xi}_t\}$-adopted, is a local martingale with respect to $\{\mathfrak{F}^{\xi}_t\}$, and satisfies for all $z \in \mathcal{R}^m$

$$\langle\langle (z, \bar{w}), (z, \bar{w}) \rangle\rangle_t = |z|^2 t,$$

whence it is a Wiener process in $\mathcal{R}^m$ with respect to the current $\{\mathfrak{F}^{\xi}_t\}$. Clearly,

$$\begin{aligned} &\eta(t) = \int_0^t B(s, \omega)\, d\bar{w}(t), \\ &\xi(t) = \int_0^t \bar{a}(s, \omega)\, ds + \int_0^t B(s, \omega)\, d\bar{w}(t). \end{aligned} \tag{37}$$

It remains only to observe that since $\bar{a}(t, \omega)$ and $B(t, \omega)$ are $\{\mathfrak{F}^{\xi}_t\}$-adopted there exist functions $a(t, x(\cdot))$ and $B(t, x(\cdot))$ such that

$$\bar{a}(t, \omega) = a(t, \xi(\cdot)), \qquad B(t, \omega) = B(t, \xi(\cdot))$$

with probability 1. Substituting these values into (37) we obtain (36). $\square$

§ 3. Diffusion Processes in $\mathcal{R}^m$

These processes were discussed before in Chapter I of Volume II. In this Section a diffusion process is viewed as a solution $\xi(t)$ of the following differential equation,

$$d\xi(t)=a(t,\xi(t))\,dt+B(t,\xi(t))\,dw(t),$$

where $w(t)$ is a Wiener process, both processes $\xi(t)$ and $w(t)$ take on values in $\mathcal{R}^m$, and $a(t,x)$ and $B(t,x)$ are measurable functions on $[0,T]\times\mathcal{R}^m$ taking on values in $\mathcal{R}^m$ and $\mathfrak{L}(\mathcal{R}^m)$, respectively. Equation (1) is to be solved under the initial condition $\xi(0)$ ($\xi(s)$), where $\xi(0)$ ($\xi(s)$) is a given random variable independent of $w(t)$ ($w(t+s)-w(s)$).

Equations of the form (1) were studied in Section 2 of Chapter II. Under the assumption that the functions $a(t,x)$ and $B(t,x)$ satisfy the local Lipschitz condition, i.e., for every N there exists $l_{N,T}$ such that for $t\leq T$, $|x|\leq N$ and $|y|\leq N$

(2) $$|a(t,x)-a(t,y)|+\|B(t,x)-B(t,y)\|\leq l_{N,T}|x-y|,$$

it was established therein that the solution of equation (1) is unique. Moreover, if the condition

(3) $$|a(t,x)|+\|B(t,x)\|\leq K_T(1+|x|)$$

is satisfied then a solution of (1) necessarily exists. This solution will also necessarily be adopted to the current of σ-algebras $\mathfrak{F}_t$ generated by the variable $\xi(0)$ ($\xi(s)$) and by $w(u)$ ($w(u+s)-w(s)$), $u\leq t$.

In this section we do not require that condition 2) be satisfied. However, we also do not assume that solutions of (1) are $\{\mathfrak{F}_t\}$-adopted (i.e., are strong). We shall obtain more general conditions here for the existence of a weak solution and also conditions for the weak uniqueness and hence for the uniqueness of the strong solution.

Absolute continuity of measures associated with diffusion processes. In this subsection a formula is obtained for the density of a measure associated with a diffusion process with respect to another measure. Also some results concerning one-dimensional distributions of a solution of equation (1) are presented. Throughout this subsection the following assumption is made (condition **A**):

A. Coefficients $a(t,x)$ and $B(t,x)$ satisfy conditions (2) and (3) and $B(t,x)$ is a nondegenerate continuous operator for all $t\in[0,\infty)$ and $x\in\mathcal{R}^m$.

Theorem 1. *Let $a_1(t,x)$, $a_2(t,x)$, and $B(t,x)$ satisfy condition* A *and $\xi_i(t)$ be solutions of the equation*

(4) $$d\xi_i(t)=a_i(t,\xi_i(t))\,dt+B(t,\xi_i(t))\,dw(t)$$

on $[s, \infty)$ with the initial values $\xi_i(s)$. Let $\mu_i^{s,T}$ denote the measure associated with the process $\xi_i(t)$ on $[s, T]$ and $\nu_s^i(dx)$ be the distribution of $\xi_i(s)$. Then the condition $\nu_s^1(\cdot)\sim\nu_s^2(\cdot)$ implies the equivalence of the measures $\mu_1^{s,T}$ and $\mu_2^{s,T}$; moreover,

$$\frac{d\mu_2^{s,T}}{d\mu_1^{s,T}}(\xi_1(\cdot))=\frac{d\nu_s^2}{d\nu_s^1}(\xi_1(s))$$

$$\times\exp\{\textstyle\int_s^T (B^{-1}(t,\xi_1(t))(a_2(t,\xi_1(t))-a_1(t,\xi_1(t))), dw(t)) \tag{5}$$

$$-\tfrac{1}{2}\textstyle\int_s^T |B^{-1}(t,\xi_1(t))(a_2(t,\xi_1(t))-a_1(t,\xi_1(t)))|^2\,dt\}.$$

Proof. It follows from the results in Section 2 of Chapter II that $\xi_i(t)$ are Markov processes. Utilizing general properties of densities of Markov measures, one can verify that it is sufficient to prove the theorem for the case when $\xi_1(s)=\xi_2(s)=x$ with probability 1 (cf. Volume I, Chapter VII, Section 6, equation (6)). We shall prove the theorem under this assumption.

If the function $B^{-1}(t,x)[a_2(t,x)-a_1(t,x)]$ is bounded, the assertion of the theorem follows from Theorem 2 in Section 2 and the uniqueness of the solution of equation (4).

Let $a_n(t,x)$ satisfy condition **A** for all $n>2$ with the same constants $l_{N,T}$ and K_T:

$$a_n(t,x)=\begin{cases} a_2(t,x), & |x|\leqslant n,\\ a_1(t,x), & |x|\geqslant 2n.\end{cases}$$

Denote by $\xi_n(t)$ the solution of equation (4) on $[s,\infty)$ if $a_2(t,x)$ is replaced by $a_n(t,x)$ and $\xi_n(s)=x$. Then

$$B^{-1}(t,x)[a_n(t,x)-a_1(t,x)]$$

is bounded and hence

$$\frac{d\mu_n^{s,T}}{d\mu_1^{s,T}}(\xi_1(\cdot))=\exp\{\textstyle\int_s^T\big(B^{-1}(t,\xi(t))[a_n(t,\xi_1(t))-a_1(t,\xi_1(t))], dw(t)\big) \tag{6}$$

$$-\tfrac{1}{2}\textstyle\int_s^T |B^{-1}(t,\xi_1(t))[a_n(t,\xi_1(t))-a_1(t,\xi_1(t))]|^2\,dt\}.$$

Let $\tau^n=\max[t\leqslant T:\sup_{s\leqslant t}|\xi_2(s)|\leqslant n]$. Since the solution of equation (4) is unique, processes $\xi_2(t)$ and $\xi_n(t)$ coincide on $[s,\tau^n]$. Therefore we have for all Borel sets $A\subset\mathscr{C}_{[s,T]}^m$

$$\mu_n^{s,T}(A\cap S_r)=\mu_2^{s,T}(A\cap S_r),$$

where $S_r=\{x(\cdot): \sup_{s\leqslant t\leqslant T}|x(t)|<r\}$. Consequently for $n>r$ we have on S_r

$$\frac{d\mu_n^{s,T}}{d\mu_1^{s,T}}(x(\cdot))=\frac{d\mu_2^{s,T}}{d\mu_1^{s,T}}(x(\cdot)).$$

However, for $\xi_1(\cdot)\in S_r$, $n>r$, the right-hand side of (6) is the same as for $n=2$, i.e., it coincides with the right-hand side of (5) (since $(dv_s^2/dv_s^1)(\cdot)=1$). We thus have shown that measures $\mu_2^{s,T}$ and $\mu_1^{s,T}$ are equivalent on $\bigcup_r S_r$ and since $\bigcup_{r=1}^{\infty} S_r=\mathscr{C}_{[0,T]}^m$ they are equivalent in general. Formula (5) is also verified and the theorem is proved. □

Denote by $R_\lambda(s, x, E)$ the quantity

$$\mathsf{E}_{s,x}\int_s^\infty e^{-\lambda(t-s)}\chi_E(\xi(t))\,dt, \tag{7}$$

where $\lambda>0$, E is a Borel set in $\mathcal{R}^m$, and $\xi(t)$ is a solution of equation (1) on $[s,\infty)$ with the initial condition $\xi(s)=x$. As was established in Section 2 of Chapter II, a solution of (1) is a Markov process with the transition probability given by

$$P(s, x, t, E)=\mathsf{P}_{s,x}\{\xi(t)\in E\}, \tag{8}$$

where $\xi(t)$ is the same solution as in (7), $\mathsf{E}_{s,x}$ is the mathematical expectation with respect to a measure associated with this process $\xi(t)$, and $R_\lambda(s, x, E)$ is a measure on E. We are interested in the existence of a density for this measure with respect to the Lebesgue measure. In addition to condition **A** we shall also assume the following condition:

B. For any N there exists $C_N>0$ such that

$$\operatorname{Sp}(\boldsymbol{I}-B(t,x)B^*(t,x))^2\leq 1-C_N$$

for $|x|\leq N$, $t\leq N$.

Below the following two lemmas will be required.
Let $f(s,x)$ be a measurable bounded function; set

$$G_\lambda f(s,x)=\mathsf{E}\int_s^\infty e^{-\lambda(t-s)}f(t, x+w(t)-w(s))\,dt,$$

where $w(t)$ is a Wiener process in $\mathcal{R}^m$.

Lemma 1. *Let*

$$Lg(s,x)=(b(s,x), g'_x)+\tfrac{1}{2}\operatorname{Sp} C(s,x)g''_{xx}$$

be a differential operator with coefficients $b(s,x)$ and $C(s,x)$ which are defined on $[0,\infty)\times\mathcal{R}^m$, taking on values in $\mathcal{R}^m$ and $\mathfrak{L}(\mathcal{R}^m)$, respectively, are measurable, and satisfy the inequalities

$$|b(s,x)|\leq\delta,\qquad \operatorname{Sp} C(s,x)C^*(s,x)\leq\theta^2.$$

Then for each $\varepsilon>0$ and all sufficiently large $\lambda>0$

$$\iint[LG_\lambda g(s,x)]^2\,ds\,dx\leq(\theta^2+\varepsilon)\iint g^2(s,x)\,ds\,dx. \tag{9}$$

Proof. Let

$$g(s,x)=\iint \exp\{i\alpha s+i(x,y)\}\tilde{g}(\alpha,y)\,d\alpha\,dy.$$

Then

$$\begin{aligned}G_\lambda g(s,x)&=\mathsf{E}\int_s^\infty\iint \exp\{i\alpha s+i(x,y)+i(w(t)-w(s),y)-\lambda(t-s)\}\\&\qquad\times\tilde{g}(\alpha,y)\,d\alpha\,dy\\&=\iint \exp\{i\alpha s+i(x,y)\}\frac{1}{\lambda-i\alpha+\frac{1}{2}|y|^2}\tilde{g}(\alpha,y)\,d\alpha\,dy.\end{aligned}\tag{10}$$

Therefore in view of the Parseval equality

$$\iint\left|\frac{\partial}{\partial x^k}G_\lambda g(s,x)\right|^2 ds\,dx=(2\pi)^m\iint\frac{|y^k|^2}{(\lambda+\frac{1}{2}|y|^2)^2+\alpha^2}|\tilde{g}(\alpha,y)|^2\,d\alpha\,dy,$$

$$\iint\left|\frac{\partial^2}{\partial x^k\,\partial x^j}G_\lambda g(s,x)\right|^2 ds\,dx=(2\pi)^m\iint\frac{|y^ky^j|^2}{(\lambda+\frac{1}{2}|y|^2)^2+\alpha^2}|\tilde{g}(\alpha,y)|^2\,d\alpha\,dy.$$

Hence for any $\varepsilon_1>0$

$$\begin{aligned}\iint|L[G_\lambda g(s,x)]|^2\,ds\,dx&\leqslant(1+\varepsilon_1)\frac{\theta^2}{4}\iint\sum_{k,j}\left|\frac{\partial^2}{\partial x^k\,\partial x^j}G_\lambda g(s,x)\right|^2 ds\,dx\\&\quad+\left(1+\frac{1}{\varepsilon_1}\right)\delta\iint\sum_k\left(\frac{\partial}{\partial x^k}G_\lambda g(s,x)\right)^2 ds\,dx\\&=(2\pi)^m\iint\frac{\frac{1}{4}(1+\varepsilon_1)\theta^2|y|^4+(1+(1/\varepsilon_1))\delta|y|^2}{(\lambda+\frac{1}{2}|y|^2)^2+\alpha^2}\\&\quad\times|\tilde{g}(\alpha,y)|^2\,d\alpha\,dy\leqslant(2\pi)^m(1+\varepsilon_1)\theta^2\iint|\tilde{g}(\alpha,y)|^2\,d\alpha\,dy\\&=(1+\varepsilon_1)\theta^2\iint g^2(s,x)\,ds\,dx,\end{aligned}$$

provided $\lambda>(1+1/\varepsilon_1)\delta/(1+\varepsilon_1)\theta^2$.

The lemma is thus proved. □

Corollary. *For $\theta<1$ and all $f\in\mathcal{L}_2([0,\infty)\times\mathcal{R}^m)$ the equation*

$$f=g+LG_\lambda g\tag{11}$$

possesses a solution in $\mathcal{L}_2([0,\infty)\times\mathcal{R}^m)$. Let $\|\cdot\|_2$ be the norm in $\mathcal{L}_2([0,\infty)\times\mathcal{R}^m)$. Then

$$\|g\|_2\leqslant\frac{1}{1-\theta_1}\|f\|_2,\qquad 0<\theta_1<1.$$

We shall denote the solution of (11) *by*

$$g=(\boldsymbol{I}+LG_\lambda)^{-1}f;$$

here $(\boldsymbol{I}+LG_\lambda)^{-1}$ *is an operator from* $\mathscr{L}_2([0,\infty)\times\mathscr{R}^m)$ into $\mathscr{L}_2([0,\infty)\times\mathscr{R}^m)$.

For all λ *sufficiently large the inequality*

$$\|(\boldsymbol{I}+LG_\lambda)^{-1}\|_2\leq\frac{1}{1-\theta_1}$$

is valid.

Remark 1. It follows from formula (10) that G_λ is also an operator from $\mathscr{L}_2([0,\infty)\times\mathscr{R}^m)$ into $\mathscr{L}_2([0,\infty)\times\mathscr{R}^m)$ and, moreover, $\|G_\lambda\|\leq 1/\lambda$.

Next we set

$$G_\lambda^\xi f(s,x)=\mathsf{E}_{s,x}\int_s^\infty e^{-\lambda(t-s)}f(t,\xi(t))\,dt, \tag{12}$$

where $\xi(t)$ is the same solution of (1) as in (7). If $f(t,x)$ possesses continuous bounded derivatives f'_t, f'_x, and f''_{xx}, then using Itô's formula we can write the expression

$$\begin{aligned}\mathsf{E}_{s,x}f(t,\xi(t))=f(s,x)+\mathsf{E}_{s,x}\int_s^t[f'_u(u,\xi(u))\\+(a(u,\xi(u)),f'_u(u,\xi(u)))\\+\tfrac{1}{2}\operatorname{Sp}B(u,\xi(u))B^*(u,\xi(u))f''_{xx}(u,\xi(u))]\,du.\end{aligned}$$

Substituting this expression into (10) we obtain

$$G_\lambda^\xi f(s,x)=\frac{1}{\lambda}f(s,x)+\frac{1}{\lambda}\mathsf{E}_{s,x}\int_s^\infty e^{-\lambda(t-s)}L_\xi f(t,\xi(t))\,dt,$$

where

$$L_\xi u(t,x)=\frac{\partial u}{\partial t}(t,x)+\left(a(t,x),\frac{\partial}{\partial x}u(t,x)\right)+\tfrac{1}{2}\operatorname{Sp}B(t,x)B^*(t,x)u''_{xx}(t,x),$$

or

$$f(s,x)=G_\lambda^\xi[\lambda f-L_\xi f](s,x). \tag{13}$$

Lemma 2. *If there exists* $c>0$ *such that for all* $x\in\mathscr{R}^m$ *and* $s\in[0,\infty)$ *the inequalities*

$$\operatorname{Sp}(B(s,x)B^*(s,x)-\boldsymbol{I})^2\leq 1-c,\qquad |a(s,x)|\leq\frac{1}{c},$$

are fulfilled, then for $\lambda > 0$ sufficiently large,

$$G_\lambda^\xi f(s, x) = G_\lambda(\boldsymbol{I} - L_1 G_\lambda)^{-1} f(s, x),$$

where

$$L_1 u(t, x) = (a(t, x), u_x'(t, x)) + \tfrac{1}{2} \operatorname{Sp}(B(t, x)B^*(t, x) - \boldsymbol{I}) u_{xx}''.$$

There exist λ_0 and H depending on c only such that for $\lambda > \lambda_0$,

$$\|G_\lambda^\xi\|_2 \leq H.$$

Proof. Substitute $f = G_\lambda g$ into (13), where $g \in \mathscr{L}_2([0, \infty) \times \mathscr{R}^m)$. We obtain

$$G_\lambda g = G_\lambda^\xi[\lambda G_\lambda g - L_\xi G_\lambda g].$$

Utilizing the form of $G_\lambda g$

$$G_\lambda g(s, x) = \int_0^\infty e^{-\lambda t} g(t + s, x + w(t))\, dt,$$

we verify that

$$\frac{\partial}{\partial s} G_\lambda g(s, x) + \tfrac{1}{2} \Delta G_\lambda g(s, x) = \mathsf{E} \int_0^\infty e^{-\lambda t}\left[\frac{\partial g}{\partial t} + \tfrac{1}{2} \Delta g(t + s, x + w(t))\right] dt$$

$$= G_\lambda\left[\frac{\partial g}{\partial t} + \tfrac{1}{2} \Delta g\right] = -g + \lambda G_\lambda g,$$

where $\Delta u = \operatorname{Sp} u_{xx}''$. The last equality is obtained by means of Ito's formula in exactly the same manner as formula (13) was obtained. Hence

$$\lambda G_\lambda g - L_\xi G_\lambda g = g - L_1 G_\lambda g.$$

Thus

$$G_\lambda g = G_\lambda^\xi(g - L_1 G_\lambda g). \tag{14}$$

Let g be a solution of the equation

$$g - L_1 G_\lambda g = f.$$

If $f \in \mathscr{L}_2([0, \infty) \times \mathscr{R}^m)$ then under the conditions of the lemma this solution exists, i.e., $g = (\boldsymbol{I} - L_1 G_\lambda)^{-1} f$. Substituting this g into (14) we obtain the required result. □

Theorem 2. *If the conditions of Lemma 2 are satisfied, then for any integrable and square integrable function $\varphi(x)$ and for all $\lambda > \lambda_0$, where λ_0 depends only on c, the quantity*

$$\int \varphi(x) R_\lambda(s, x, E)\, dx$$

is absolutely continuous as a function of E with respect to the Lebesgue measure $m(E)$. If

$$R_\lambda(s, \varphi, E) = \int \varphi(x) R_\lambda(s, x, E)\, ds,$$

then the function

$$r_s(y) = \frac{dR_\lambda(s, \varphi, \cdot)}{dm}(y)$$

is integrable in the $(2-\varepsilon)$th power for any ε, $0<\varepsilon<1$, and for each s_0 there exists a constant H_{s_0} which depends on c, ε, and s_0 only such that

$$\int (r(y))^{2-\varepsilon}\, dy \le H_{s_0}\left(\int \varphi^2(x)\, dx + \int |\varphi(x)|\, dx\right)$$

for $\lambda > \lambda_0$ and $s \le s_0$.

Proof. Let $f(s, x) = \chi_E(x)\, e^{-\delta s}$. Then

$$G_\lambda^\xi f(s, x) = e^{-\delta s} R_{\lambda+\delta}(s, x, E).$$

Consequently,

$$R_{\lambda+\delta}(s, x, E) = e^{\delta s} G_\lambda^\xi f(s, x).$$

Utilizing Lemma 2 we can write:

$$R_{\lambda+\delta}(s, x, E) = e^{\delta s} G_\lambda (\boldsymbol{I} - L_1 G_\lambda)^{-1} f(s, x).$$

Denote

$$(\boldsymbol{I} - L_1 G_\lambda)^{-1} f = g.$$

Then

$$\begin{aligned} R_{\lambda+\delta}(s, x, E) &= e^{\delta s} \int_0^\infty \int \frac{1}{(2\pi t)^{m/2}} \exp\left\{-\frac{|x-y|^2}{2t} - \lambda t\right\} g(t, y)\, dt\, dy, \\ R_{\lambda+\delta}(s, \varphi, E) &= e^{\delta s} \int_0^\infty \int \varphi(t, y)\, e^{-\lambda t} g(t, y)\, dt\, dy, \end{aligned} \tag{15}$$

where

$$\varphi(t, y) = (2\pi t)^{-m/2} \int \varphi(x) \exp\left\{-\frac{|x-y|^2}{2t}\right\} dx.$$

If

$$\varphi(x) = \int e^{i(z,x)} \tilde{\varphi}(z)\, dz,$$

then

$$\varphi(t, y) = \int \tilde{\varphi}(z)\, e^{-t|z|^2/2}\, e^{i(z,y)}\, dz.$$

Therefore

$$\int |\varphi(t, y)|^2\, dy = (2\pi)^m \int |\tilde{\varphi}(z)|^2 e^{-t|z|^2}\, dz < (2\pi)^m \int |\tilde{\varphi}(z)|^2\, dz = \int |\varphi(x)|^2\, dx.$$

Hence

$$\begin{aligned} R_{\lambda+\delta}(s, \varphi, E) &\leq e^{\delta s} \sqrt{\frac{1}{2\lambda} \int |\varphi(x)|^2\, dx}\ \sqrt{\int_0^\infty \int g^2(t, y)\, dt\, dy} \\ &\leq e^{\delta s} H_1 \sqrt{\int \varphi^2(x)\, dx \int_0^\infty \int \chi_E(x)\, e^{-2\delta s}\, dx\, ds}; \end{aligned}$$

here $H_1 = (1/\sqrt{2\lambda}) \|(\boldsymbol{I} - L_1 G_\lambda)^{-1}\|$. Finally we obtain

$$R_{\lambda+\delta}(s, \varphi, E) \leq H_1 \frac{e^{\delta s}}{\sqrt{2\delta}} \sqrt{\int \varphi^2(x)\, dx \cdot m(E)}.$$

The existence of the density of $R_{\lambda+\delta}(s, \varphi, E)$ with respect to the Lebesgue measure follows from this inequality. We note that

$$\int_E r_s(y)\, dy \leq H_2 \sqrt{m(E)}$$

for some H_2. Let $E_\alpha = \{y : r_s(y) > \alpha\}$. Then $\alpha m(E_\alpha) \leq H_2 (m(E_\alpha))^{1/2}$. Hence $m(E_\alpha) \leq \alpha^{-2} H_2^2$. Therefore

$$\begin{aligned} \int r_s^{2-\varepsilon}(y)\, dy &= \int_{r_s(y) \leq 1} r_s^{2-\varepsilon}(y)\, dy + \sum_{n=0}^\infty \int_{2^n \leq r_s(y) < 2^{n+1}} r_s^{2-\varepsilon}(y)\, dy \\ &\leq \int r_s(y)\, dy + \sum_{n=0}^\infty H_2^2 \frac{2^{2-\varepsilon}}{2^{n\varepsilon}}. \end{aligned}$$

To complete the proof it remains only to note that

$$\left|\int r_s(y)\, dy\right| \leq \int |\varphi(x)| R_\lambda(s, x, \mathscr{R}^m)\, dx = \frac{1}{\lambda} \int |\varphi(x)|\, dx. \quad \square$$

Remark 1. Assume that $g(t, y)$ appearing in formula (15) belongs to $\mathscr{L}_p([0, \infty)\times\mathscr{R}^m)$, where p is such that

$$\int_0^\infty \int \left[(2\pi t)^{-m/2} \exp\left\{-\frac{|x-y|^2}{2t} - \lambda t\right\}\right]^q dt\, dy < \infty, \qquad \frac{1}{p}+\frac{1}{q}=1. \tag{16}$$

Then

$$R_{\lambda+\delta}(s, x, E) \leq C\|g\|_p,$$

where $\|\cdot\|_p$ is the norm in $\mathscr{L}_p$. If, moreover, $\|g\|_p \leq (m(E))^\alpha$, where $\alpha > 0$, then

$$R_{\lambda+\delta}(s, x, E) \leq C_1(m(E))^\alpha.$$

One can deduce from this inequality, in the same manner as in Theorem 2, the existence of the density of $R_{\lambda+\delta}(s, x, E)$ with respect to the measure $m(E)$ as well as the integrability of this density in the

$$\left[\frac{1}{1-\alpha} - \varepsilon\right]\text{th}$$

power for any $\varepsilon > 0$. In order that (16) be valid it is necessary that

$$\frac{m}{2}(q-1) < 1, \qquad q < \frac{m+2}{m}, \qquad p > \frac{m+2}{2}.$$

If $m = 1$ one can choose $q = p = 2$.

Corollary. *If $m = 1$, under the conditions of Lemma 2 $R_\lambda(s, x, E)$ is absolutely continuous with respect to the Lebesgue measure, and if*

$$r_\lambda(s, x, y) = \frac{dR_\lambda(s, x, \cdot)}{dm}(y), \tag{17}$$

then there exists for all ε, $o < \varepsilon < 1$, and s_0 a constant C_2 which also depends on c such that

$$\int r_\lambda^{2-\varepsilon}(s, x, y)\, dy \leq C_2 \tag{18}$$

for $\lambda > \lambda_0$ and $s \leq s_0$.

Remark 2. Assume that the coefficients $a(t, x)$ and $B(t, x)$ in equation (1) do not depend on t and are equal to $a(x)$ and $B(x)$. Set

$$G_\lambda^\xi f(x) = \mathsf{E}_{0,x} \int_0^\infty e^{-\lambda t} f(\xi(t))\, dt,$$
$$G_\lambda f(x) = \mathsf{E} \int_0^\infty e^{-\lambda t} f(x + w(t))\, dt,$$

where $\xi(t)$ is a solution of equation (1) on $[0, \infty)$ under the initial condition $\xi(0)=x$. One can verify in the same manner as Lemmas 1 and 2 were proved that

$$G_\lambda^\xi f = G_\lambda[\boldsymbol{I}-L_1G_\lambda]^{-1}f;$$

however, in this case all the operators are considered in $\mathscr{L}_2(\mathscr{R}^m)$. Moreover, the norm $\|(\boldsymbol{I}-L_1G_\lambda)^{-1}\|_2$ is finite. Therefore

$$G_\lambda^\xi\chi_E(x)=\int\left[\int_0^\infty \exp\left\{-\frac{|x-y|^2}{2t}-\lambda t\right\}(2\pi t)^{-m/2}\,dt\right]g(y)\,dt,$$

where $g=[\boldsymbol{I}-L_1G_\lambda]^{-1}$. It is easy to see that the function

$$\Gamma_\lambda(|x-y|)=\int_0^\infty (2\pi t)^{-m/2}\exp\left\{-\frac{|x-y|^2}{2t}-\lambda t\right\}dt$$

is integrable in y in qth power, provided $q<m/(m-2)$. In particular, for $m\le 3$ this function is square integrable. Hence if the coefficients $a(\cdot)$ and $B(\cdot)$ satisfying the conditions of the lemma are independent of t, relations (17) and (18) are valid for $m\le 3$ and the constant C_2 does not depend on s.

We now prove a lemma which asserts that certain measures associated with a solution of equation (1) possess densities with respect to the Lebesgue measure.

Lemma 3. *Let $\xi(t)$ satisfy equation* (1) *with the coefficients satisfying the condition*

$$|a(t,x)|+\|B(t,x)\|+\|B^{-1}(t,x)\|\le C,$$

where C is a constant. If $f(t,x)$ is a twice continuously differentiable numerical function such that the derivatives f'_t, f'_x and f''_{xx} satisfy the condition

$$|f'_t|+|f'_x|+|f'_x|^{-1}+\|f''_{xx}\|\le C_1,$$

then there exists for each T and ε, $0<\varepsilon<1$, a constant $C_{T,\varepsilon}$ which depends on C only and C_1 such that the set function of a set $E\subset\mathscr{R}^1$ given by

$$\mathsf{E}\int_0^T \chi_E(f(t,\xi(t))\,dt$$

is absolutely continuous with respect to the Lebesgue measure in $\mathscr{R}^1$ and, if $p_T(y)$ is the density of this measure with respect to the Lebesgue measure, then

$$\int (p_T(y))^{2-\varepsilon}\,dy\le C_{T,\varepsilon}.$$

Proof. Set

$$\eta_t=f(t,\xi(t))=f(0,\xi(0))+\int_0^t \alpha(s,\xi(s))\,ds+\int_0^t \beta(s,\xi(s))\,d\bar w(s),$$

where

$$\alpha(s,x)=f'_s(s,x)+(a(s,x),f'_x(s,x))+\tfrac{1}{2}\operatorname{Sp}B(s,x)B^*(s,x)f''_{xx}(s,x),$$

$$\beta(s,x)=|B^*(s,x)f'_x(s,x)|,$$

$$\bar{w}(t)=\int_0^t\left(\frac{B^*(s,\xi(s))f'_x(s,\xi(s))}{|B^*(s,\xi(s))f'_x(s,\xi(s))|},\ dw(s)\right);$$

$\bar{w}(t)$ is a one-dimensional Wiener process. Let τ_t be defined by the equality

$$t=\int_0^{\tau_t}\beta^2(s,\xi(s))\,ds.$$

Then

$$\mathsf{E}\int_0^T\chi_E(\eta_t)\,dt=\mathsf{E}\int_0^{\tau_T}\chi_E(\eta_{\tau_s})\,d\tau_s\leq\frac{1}{\delta}\mathsf{E}\int_0^{T/\delta}\chi_E(\eta_{\tau_s})\,ds,$$

where $\delta>0$ is such that $\delta\leq\beta^2(s,x)$. Furthermore,

$$\begin{aligned}\eta_{\tau_t}&=\int_0^{\tau_t}\beta(s,\xi(s))\,d\bar{w}(s)+\int_0^{\tau_t}\alpha(s,\xi(s))\,ds+f(0,\xi(0))\\&=\tilde{w}(t)+\int_0^t\gamma(s)\,ds+f(0,\xi(0)),\end{aligned}$$

where $\gamma(s)$ is a bounded function. The process η_{τ_t} will be a Wiener process if the original measure is replaced by a measure absolutely continuous with respect to the original one possessing the density

$$\rho_T=\exp\{-\int_0^T\gamma(s)\,d\tilde{w}(s)-\tfrac{1}{2}\int_0^T\gamma^2(s)\,ds\}$$

(cf. Theorem 11 in Section 1). Therefore using the Cauchy–Schwarz inequality we obtain

$$\begin{aligned}\mathsf{E}\int_0^{T_1}\chi_E(\tilde{w}(t))\,dt&=\mathsf{E}\int_0^{T_1}\chi_E(\eta_{\tau_t})\rho_{T_1}\,dt\\&\geq\frac{\left(\int_0^{T_1}\mathsf{E}\chi_E(\eta_{\tau_t})\,dt\right)^2}{\int_0^{T_1}\mathsf{E}\rho_{T_1}^{-1}\,dt}\geq\frac{\left(\int_0^{T_1}\mathsf{E}\chi_E(\eta_{\tau_t})\,dt\right)^2}{T_1e^{C_2T_1}}\end{aligned}$$

(here we also utilized Lemma 5 in Section 1), where the constant C_2 depends only on the maximum of $|\gamma(t)|$ so that it can be chosen to depend on C and C_1 only. We have thus shown that since

$$\mathsf{E}\int_0^{T_1}\chi_E(\tilde{w}(t))\,dt\leq\frac{1}{\sqrt{2\pi}}m(E)\int_0^{T_1}t^{-1/2}\,dt=\sqrt{\frac{T_1}{8\pi}}m(E),$$

(19) $$\mathsf{E}\int_0^T\chi_E(\eta_t)\,dt\leq H\sqrt{m(E)}$$

is valid for some H. Now from inequality (19) the assertion of the lemma is deduced in the same manner as that of Theorem 2. □

Remark. If $m=1$, choosing $f(t,x)=x$ we can verify that under the conditions of Lemma 3 for the solution of equation (1), there exists a function $g(y)$ integrable in the $(2-\varepsilon)$th power (for any ε, $0<\varepsilon<1$) such that

$$\int_0^T \mathsf{P}\{\xi(t)\in E\}\,dt=\int_E g(y)\,dy.$$

The existence of a solution. It follows from Theorem 3 in Section 2 that equation (1) possesses for any continuous $a(t,x)$ and $B(t,x)$ satisfying condition (3) a weak solution. If one utilizes Theorem 2 in Section 2 then under the assumption that $B(t,x)$ is nondegenerate the requirement of continuity of $a(t,x)$ may be dropped.

Theorem 3. *Let $a(t,x)$ and $B(t,x)$ satisfy condition* (3), *$a(t,x)$ be measurable, and $B(t,x)$ be continuous and nondegenerate for all $t\in[0,\infty)$ and $x\in\mathcal{R}^m$. Then equation* (1) *possesses a (weak) solution on* $[0,T]$ *obeying the initial condition* $\xi(0)$.

Proof. Equation

$$d\xi_0(t)=B(t,\xi_0(t))\,dw(t) \tag{20}$$

possesses a solution under the initial condition $\xi_0(0)=\xi(0)$. Therefore in view of Theorem 2 in Section 2 equation (1) possesses a solution for all $a(t,x)$ such that the function

$$B^{-1}(t,x)\,a(t,x)$$

is bounded.

Set

$$a_N(t,x)=\begin{cases} a(t,x), & |x|\le N\\ 0, & |x|>N.\end{cases}$$

In this case the function $B^{-1}(t,x)a_N(t,x)$ is bounded and hence in view of Lemma 6 in Section 1,

$$\mathsf{E}\exp\{\textstyle\int_0^T (B^{-1}(t,\xi_0(t))a_N(t,\xi_0(t)),dw(t))-\frac{1}{2}\int_0^T|B^{-1}(t,\xi_0(t))a_N(t,\xi_0(t))|^2\,dt\}=1.$$

Let μ_N be a measure on $\mathscr{C}^m_{[0,T]}$ defined by the equality

$$\mu_N(A)=\mathsf{E}\rho_N^T(\xi_0(\cdot))\chi_A(\xi_0(\cdot)),$$

where A is a Borel set in $\mathscr{C}^m_{[0,T]}$, and

$$\rho_N^T=\exp\{\textstyle\int_0^T (B^{-1}(t,\xi_0(t))a_N(t,\xi_0(t)),dw(t))-\frac{1}{2}\int_0^T|B^{-1}(t,\xi_0(t))a_N(t,\xi_0(t))|^2\,dt\}.$$

As follows from Theorem 2 in Section 2, the measure μ_N is associated with the solution of the equation

$$d\xi_N(t)=a_N(t,\xi_N(t))\,dt+B(t,\xi_N(t))\,dw(t).$$

Let $\tau_N = \max[t \leq T, \sup_{s \leq t} |\xi_N(s) \leq N]$. Then $\xi_N(t)$ for $t < \tau_N$ is a solution of (1). Clearly $\mu_N(A \cap S_r)$ does not depend on N for $N > r$ $(S_r = \{x(\cdot): \sup_t |x(t)| \leq r\})$. Hence there exists the limit

$$\lim_{N\to\infty} \mu_N(A \cap S_r) = \mu(A \cap S_r)$$

for all A and r. In particular,

$$\mu(S_r) = \lim_{N\to\infty} \mu_N(S_r).$$

However,

$$\mu(S_r) = \mathsf{E} \exp\{\textstyle\int_0^T (B^{-1}(t, \xi_0(t))a(t, \xi_0(t)), dw(t)) \\ -\frac{1}{2}\int_0^T |B^{-1}(t, \xi_0(t))\, a(t, \xi_0(t))|^2\, dt\} \chi_{S_r}(\xi_0(\cdot)).$$

It follows from inequality (3) that there exists a constant K which does not depend on N such that

$$\mathsf{E}(\sup_{0 \leq t \leq T} |\xi_N(t)|^2 \,|\, \xi(0)) \leq K(1 + |\xi(0)|^2),$$

(cf., for instance, Theorem 9 in Section 1 of Chapter II). Therefore

$$\lim_{r\to\infty} \varliminf_{N\to\infty} \mu_N(S_r) = 1.$$

Hence,

$$\mathsf{E} \exp\{\textstyle\int_0^T (B^{-1}(t, \xi_0(t))a(t, \xi_0(t)), dw(t)) - \frac{1}{2}\int_0^T |B^{-1}(t, \xi_0(t))a(t, \xi_0(t))|^2\, dt\} \geq 1.$$

However, in such a case, in view of Corollary 1 to Lemma 6 in Section 1 only the equality may hold in the last relation. Applying once more Theorem 2 in Section 2 we verify the existence of a solution of (1); moreover, $\mu(A)$ is the measure associated with the solution. The theorem is proved. □

We now utilize the results of the preceding subsection to prove the existence theorem in the case of discontinuous $B(t, x)$.

Theorem 4. *Let the coefficients of equation* (1) *satisfy inequality* (3), $B(t, x)$ *satisfy condition* **B**, *and let* $B(t, x)$ *be continuous in* t *uniformly with respect to* x *for* $|x| \leq N$ *for each* N. *Then equation* (1) *possesses a solution on* $[0, T]$ *provided only* $\xi(0)$ *possesses a square integrable probability density* $\varphi_0(x)$.

Proof. If we prove the existence of a solution of equation (20) under the initial condition $\xi_0(0) = \xi(0)$ then the same arguments as in the proof of Theorem 3 will yield the existence of a solution of equation (1). We therefore proceed to prove the existence of a solution of equation (20).

In view of the conditions imposed on $B(t, x)$ one can choose a sequence of functions $B_n(t, x)$ such that

1) for each N

$$\lim_{n\to\infty} \int_{|x|\le N} \sup_{0\le t\le T} \|B_n(t, x) - B(t, x)\|\, dx = 0,$$

2) $B_n(t, x)$ satisfies condition (3), all with the same constant K_T,
3) $B_n(t, x)$ satisfies condition (2), but the constants $l_{T,N}$ may depend on n,
4) $B_n(t, x)$ satisfies condition **B**, all with the same constants c_N.

The following function can be chosen as $B_n(t, x)$:

$$B_n(t, x) = (2\pi\delta_n)^{-m} \int \exp\left\{-\frac{|y|^2}{2\delta_n}\right\} B(t, x+y)\, dy,$$

where $\delta_n \to 0$. Clearly conditions 2)–4) are satisfied for this function. It is easy to see that $B_n(t, x)$ are also uniformly continuous in t for $|x| \le N$. Therefore for any $\varepsilon > 0$ one can find t_i, $0 = t_0 < t_1 < \cdots < t_k = T$, such that

$$\sup_{0\le t\le T} \|B_n(t, x) - B(t, x)\| \le \varepsilon + \sup_i \|B_n(t_i, x) - B(t_i, x)\|.$$

The last inequality implies condition 1). The equation

$$d\xi_n(t) = B_n(t, \xi_n(t))\, dw(t), \qquad \xi_n(0) = \xi(0),$$

possesses a solution for each n and, moreover, this solution is unique. Utilizing Lemma 2 in Section 2 one can verify that the sequence of measures μ_n associated with the processes $\xi_n(t)$ in $\mathscr{C}^m_{[0,T]}$ is compact. Without loss of generality it may be assumed that μ_n is weakly convergent to a measure μ. We now show that this measure is associated with a solution of equation (20) under the initial condition $\xi(0)$. The proof is analogous to the proof of Theorem 3 in Section 2. Since $\xi_n(t)$ is a local martingale, it follows that the process $\xi_0(t)$ with which the measure μ is associated will also be a local martingale. Clearly the distribution of $\xi_0(0)$ will coincide with the limit distribution for $\xi_n(0)$, i.e., with $\xi(0)$. To prove the theorem it is sufficient to show (in view of Theorem 1 in Section 2) that for each $z \in \mathscr{R}^m$

$$\eta_0(t) = (\xi_0(t), z)^2 - \int_0^t |B^*(s, \xi_0(s))z|^2\, ds$$

is also a local martingale. For each n

$$\eta_n(t) = (\xi_n(t), z)^2 - \int_0^t |B_n^*(s, \xi_n(s))z|^2\, ds$$

is a local martingale because

$$\xi_n(t) = \xi(0) + \int_0^t B_n(s, \xi_n(s))\, dw(s),$$

i.e., $\xi_n(t)$ is an Itô process. Therefore it is sufficient to show that finite-dimensional distributions of the process $\eta_n(t)$ converge to the finite-dimensional distributions of the process $\eta_0(t)$ (in that case also the measures associated with the processes $\eta_n(t)$ on $\mathscr{C}_{[0,T]}$ will converge weakly to the measure associated with the process $\eta_0(t)$). Let $\eta_n^{(k)}(t)$ be defined by the equalities

$$\eta_n^{(k)}(t)=(\xi_n(t), z)^2-\int_0^t |B_k^*(s, \xi_n(s))z|^2\, ds, \qquad k=0, 1, \ldots,$$

where $B_0(s, x)=B(s, x)$. Since $B_k(s, x)$ are continuous, in view of the weak convergence of measures μ_n to μ, the finite-dimensional distributions of $\eta_n^{(k)}(t)$ $(k>0)$ converge to the finite-dimensional distributions of $\eta_0^{(k)}(t)$. Therefore the theorem will be established if we verify that for all $\varepsilon>0$, $t\in[0, T]$,

$$\overline{\lim_{k\to\infty}}\ \overline{\lim_{n\to\infty}}\ \mathsf{P}\{|\eta_n^{(k)}(t)-\eta_n(t)|>\delta\}=0, \tag{21}$$

and that this relation is valid for $n=0$. Since

$$\sup_n \mathsf{P}\{\sup_{0\le t\le T} |\xi_n(t)|>r\}$$

approaches zero as $r\to\infty$, (21) will be verified if we show that

$$\overline{\lim_{k\to\infty}}\ \overline{\lim_{n\to\infty}}\ \mathsf{P}\{|\eta_n^{(k)}(t)-\eta_n(t)|>\delta, \qquad \sup_{0\le s\le T} |\xi_n(s)|<r\}=0 \tag{22}$$

for all r and that this relation is also valid for $n=0$.

Assume that $\tilde{B}_n(t, x)=B_n(t, x)$ for $|x|\le r$, $\tilde{B}_n(t, x)$ satisfies condition (2) and that for some $c>0$ (which depends on r but not on n)

$$\mathrm{Sp}\,(\tilde{B}_n(t, x)\tilde{B}_n^*(t, x)-\boldsymbol{I})^2\le 1-c.$$

Denote by $\tilde{\xi}_n(t)$ the solution of equation (20) (for $n>0$) with $B(t, x)$ replaced by $\tilde{B}_n(t, x)$. Since this solution is unique, $\tilde{\xi}_n(t)=\xi_n(t)$, provided $\sup_{0\le t\le T}|\xi_n(t)|<r$. Therefore

$$\begin{aligned}\int_0^T \mathsf{P}\{\xi_n(t)\in E, \sup_{0\le s\le T} |\xi_n(s)|<r\}\, dt &= \int_0^T \mathsf{P}\{\tilde{\xi}_n(t)\in E, \sup_{0\le s\le T} |\tilde{\xi}_n(s)|<r\}\, dt\\ &\le e^{\lambda T}\int_0^\infty \mathsf{P}\{\tilde{\xi}_n(t)\in E\}\, e^{-\lambda t}\, dt\\ &= e^{\lambda T}\int \varphi_0(x)\, R_\lambda^{\tilde{\xi}_n}(0, x, E)\, dx,\end{aligned}$$

where

$$R_\lambda^{\tilde{\xi}_n}(0, x, E)=\mathsf{E}(\int_0^\infty \chi_E(\tilde{\xi}_n(t))\, e^{-\lambda t}\, dt|\tilde{\xi}_n(0)=x).$$

In view of Theorem 2 there exists a function $g_n(y)$ such that

$$\int_0^T \mathsf{P}\{\xi_n(t)\in E, \sup_{0\le s\le T} |\xi_n(s)|<r\}\, dt=\int_E g_n(y)\, dy; \tag{23}$$

moreover,

$$\sup_n \int_{|y|\le r} (g_n(y))^{2-\varepsilon}\, dy < \infty. \tag{24}$$

Utilizing the weak convergence of measures μ_n to μ we verify that (23) and (24) are valid for $n=0$ as well. Therefore

$$\begin{aligned}
&\mathsf{P}\{|\eta_n^{(k)}(t) - \eta_n(t)| > \delta, \sup_{0\le s\le T} |\xi_n(s)| < r\} \\
&\le \frac{1}{\delta} \int \int_0^t \big| |B_n^*(s, y)z|^2 - |B_k^*(s, y)z|^2 \big| \mathsf{P}\{\xi_n(s) \in dy, \sup_{0\le t\le T} |\xi_n(t)| < r\}\, ds \\
&\le \frac{1}{\delta} \int_{|y|\le r} \sup_{0\le s\le T} \big| |B_n^*(s, y)z|^2 - |B_k^*(s, y)z|^2 \big| \int_0^T \mathsf{P}\{\xi_n(s) \in dy, \sup_{0\le t\le T} |\xi_n(t)| < r\}\, dy \\
&\le \frac{1}{\delta} \int_{|y|\le r} \sup_{0\le s\le T} \big| |B_n^*(s, y)z|^2 - |B_k^*(s, y)z|^2 \big| g_n(y)\, dy \\
&\le \frac{1}{\delta} \Big(\int_{|y|\le r} \sup_{0\le s\le T} \big| |B_n^*(s, y)z|^2 \\
&\quad - |B_k^*(s, y)z|^2 \big|^{(2-\varepsilon)/(1-\varepsilon)}\, dy \Big)^{(1-\varepsilon)/(2-\varepsilon)} \Big(\int (g_n(y))^{2-\varepsilon}\, dy \Big)^{1/(2-\varepsilon)}.
\end{aligned}$$

The last inequality together with condition 1) imply (22). The theorem is thus proved. □

Remark 1. If $B(t, x)$ does not depend on t and $m \le 3$, then a solution of (1) exists under any initial condition.

To verify this one must utilize Remark 2 following Theorem 2, which implies that (22) is fulfilled for any initial distribution of $\xi(0)$.

We now present yet additional conditions which assure the existence of a solution of (1) under an arbitrary initial condition. We shall assume that $B(t, x)$ satisfies the following condition:

C. For any r there exist twice continuously differentiable functions $f_1(t, x), \ldots, f_k(t, x)$ such that

$$\left|\frac{\partial f_k}{\partial t}\right| + \left|\frac{\partial f_k}{\partial x}\right| + \left|\frac{\partial f_k}{\partial x}\right|^{-1} + \left\|\frac{\partial^2 f_k}{\partial x^2}\right\| \le C,$$

and Borel sets $\Lambda_1, \ldots, \Lambda_k$ in $\mathscr{R}^1$ of Lebesgue measure 0 such that $B(t, x)$ is continuous in x for $f_1(t, x) \notin \Lambda_1, \ldots, f_k(t, x) \notin \Lambda_k$, $t \in [0, T]$, $|x| \le r$.

Theorem 5. *If $a(t, x)$ and $B(t, x)$ satisfy condition* (3) *and $B(t, x)$ satisfies condition* **C** *and, moreover, $\|B^{-1}(t, x)\|$ is locally bounded, then equation* (1) *possesses a solution for any initial value of $\xi(0)$.*

Proof. As in Theorems 3 and 4 it is sufficient to consider the case when $a(t, x) = 0$.

Let $B_n(t, x)$ be as in Theorem 4. Then $B_n(t, x) \to B(t, x)$ at all points of continuity of $B(t, x)$. Denote by $G_1, \ldots, G_k$ open sets in $\mathscr{R}^1$ such that $G_i \supset \Lambda_i$, where $f_1, \ldots, f_k$ and $\Lambda_1, \ldots, \Lambda_k$ are as stated in condition **C**. Denote by $F_r(G_1, \ldots, G_k)$ the set of pairs $(t; x) \in [0, T] \times \mathscr{R}^m$ such that $|x| \leq r$ and $f_i(t, x) \notin G_i$ for all $i = 1, \ldots, k$. The set $F_r(G_1, \ldots, G_k)$ is closed and $B_n(t, x)$ tends to $B(t, x)$ uniformly on $F_r(G_1, \ldots, G_k)$ since the function $B(t, x)$ is uniformly continuous in x on this set. As in the proof of Theorem 4 it may be assumed that measures μ_n associated with processes $\xi_n(t)$, which are constructed in the same manner as in Theorem 4, converge to a limiting measure μ. To prove the theorem we must show (similarly to the proof of Theorem 4) that finite-dimensional distributions of processes $\eta_n(t)$ converge to the finite-dimensional distributions of the process $\eta_0(t)$ where $\eta_n(t)$ are as in Theorem 4. To show this we shall verify that for all continuous $x(s)$ the functional

$$\int_0^T |B^*(s, x(s))z|^2\, ds \tag{25}$$

is continuous almost everywhere with respect to measure μ. Clearly this functional is continuous for all $x(\cdot)$ in $\mathscr{C}^m_{[0,T]}$ such that $B^*(s, x)$ is continuous in x for almost all s at the point $(s; x(s))$. Denote by Γ the set of points of discontinuity of $B^*(s, x)$ in x in the set $[0, T] \times \mathscr{R}^m$. The functional (25) will be continuous for $x(\cdot)$ such that

$$\int_0^T \chi_\Gamma(s, x(s))\, ds = 0.$$

It follows from condition **C** that

$$\int_0^T \chi_{\Gamma \cap S_r}(s, x(s))\, ds \leq \sum_{j=1}^k \int_0^T \chi_{G_j}(f_j(s, x(s)))\, ds.$$

Moreover, the weak convergence of μ_n to μ and the fact that G_i are open imply that

$$\varliminf_{n\to\infty} \mathsf{E}\chi_{G_j}(f_j(s, \xi_n(s)))\chi\{\sup_{0\leq t\leq T} |\xi_n(t)| > r\} \geq \mathsf{E}\chi_{G_j}(f_j(s, \xi_0(s)))\chi\{\sup_{0\leq t\leq T} |\xi_0(t)| < r\},$$

where $\xi_0(\cdot)$ is the process with which measure μ is associated. Consequently,

$$\int_0^T \mathsf{E}\chi_{G_j}(f_j(s, \xi_0(s)))\chi\{\sup_{0\leq t\leq T} |\xi_0(t)| < r\}\, ds$$

$$\leq \varliminf_{n\to\infty} \int_0^T \mathsf{E}\chi_{G_j}(f_j(s, \xi_n(s)))\chi\{\sup_{0\leq t\leq T} |\xi_n(t)| < r\} ds$$

$$\leq \varliminf_{n\to\infty} \int_0^T \mathsf{E}\chi_{G_j}(f_j(s, \tilde{\xi}_n)(s)))\, ds,$$

where $\tilde{\xi}_n(t)$ are as in the proof of Theorem 4. Utilizing (19) we verify that

$$\mathsf{E}\int_0^T \chi_r(s, \xi_0(s))\, ds \cdot \chi\{\sup_{0\le t\le T} |\xi_0(t)|<r\} \le H \sum_{j=1}^k \sqrt{m(G_j)}.$$

Since $m(\Lambda_j)=0$ and $G_j \supset \Lambda_j$ is an arbitrary open set, the measure $m(G_j)$ may be made arbitrarily small. Hence

$$\mathsf{E}\int_0^T \chi_r(s, \xi_0(s))\, ds \cdot \chi\{\sup_{0\le t\le T} |\xi_0(t)|<r\}=0.$$

Approaching the limit as $r\to\infty$ we obtain

$$\mathsf{E}\int_0^T \chi_r(s, \xi_0(s))\, ds=0,$$

i.e., $\mu(\{\chi(\cdot): \int_0^T \chi_r(s, x(s))\, ds=0\})=1$. We have thus shown that the functional (25) is continuous almost everywhere with respect to measure μ. This implies that finite-dimensional distributions of the processes

$$(\xi_n(t), z)^2-\int_0^t |B^*(s, \xi_n(s))z|^2\, ds$$

converge to the finite-dimensional distributions of $\eta_0(t)$. To complete the proof it is required to show that for any $\delta>0$

$$\text{(26)}\qquad \overline{\lim_{n\to\infty}}\, \mathsf{P}\{|\int_0^t |B^*(s, \xi_n(s))z|^2\, ds-\int_0^t |B_n^*(s, \xi_n(s))z|^2\, ds|>\delta\}=0.$$

Equality (26) follows from the following relations:

$$\begin{aligned}
&\mathsf{P}\{\int_0^T ||B^*(s, \xi_n(s))z|^2-|B_n^*(s, \xi_n(s))z|^2|\, ds>\delta, \sup_{0\le t\le T} |\xi_n(t)|<r\} \\
&\quad = \mathsf{P}\{\int_0^T ||B^*(s, \tilde{\xi}_n(s))z|^2-|B_n^*(s, \tilde{\xi}_n(s))z|^2|\, ds>\delta, \sup_{0\le t\le T} |\tilde{\xi}_n(t)|<r\} \\
&\quad \le \mathsf{P}\Big\{2|z|^2K_T^2(1+r^2)\int_0^T \sum_{j=1}^k \chi_{G_j}(f_j(t, \tilde{\xi}_n(t)))\, dt>\frac{\delta}{2}\Big\} \\
&\qquad + \mathsf{P}\Big\{\int_0^T ||B^*(s, \tilde{\xi}_n(s))z|^2-B_n^*(s, \tilde{\xi}_n(s))z|^2| \\
&\qquad \times \Big|\prod_{j=1}^k (1-\chi_{G_j}(f_j(s, \tilde{\xi}_n(s))))\, ds>\frac{\delta}{2};\ \sup_{0\le t\le T} |\tilde{\xi}_n(t)|<r\Big\}.
\end{aligned}$$

Since

$$(|B^*(s, x)z|^2-|B_n^*(s, x)z|^2)\prod_{j=1}^k (1-\chi_{G_j}(f_j(s, x)))\to 0$$

for $|x|\le r$, the second summand in the last inequality tends to 0. By choosing appropriate G_j the first summand becomes small simultaneously for all n; this is due to the fact that

$$\mathsf{E}\int_0^T \chi_{G_j}(f_j(t, \xi_n(t)))\,dt \leq H\sqrt{m(G_j)}.$$

Since r is arbitrary, equality (26) follows and the theorem is proved. □

Remark 1. Under the conditions of Theorem 3, the Remark following Theorem 4, and Theorem 5 a solution of equation (1) exists on the interval $[\tau, T]$ under the initial condition $\xi(\tau)$ provided only τ and $\xi(\tau)$ do not depend on $w(s+\tau)-w(\tau)$ for $s>0$, i.e., if τ is a Markov time for the process $w(t)$.

To verify this one must rewrite equation (1) in the form

$$d\xi'(t) = a(t+\tau, \xi'(t))\,dt + B(t+\tau, \xi'(t))\,dw'(t), \tag{27}$$

where $\xi'(t)=\xi(t+\tau)$ and $w'(t)=w(t+\tau)-w(\tau)$. All the arguments presented in Theorems 3–5 are applicable to equation (27); it is only necessary to replace probabilities and mathematical expectations in the proofs from the beginning by conditional probabilities and mathematical expectations for a fixed τ. In particular, under the conditions of Theorem 4 a solution of equation (27) exists provided the conditional distribution of $\xi(\tau)$ for a fixed τ possesses for almost all τ a square integrable probability density.

Remark 2. If the coefficients in equation (1) satisfy the conditions of Theorems 3 and 5 or of the Remark following Theorem 4 on every finite interval $[0, T]$, then this equation possesses a solution under a given initial condition on $[0, \infty)$.

The solution can be constructed in the following manner. Choose a sequence $T_n \uparrow \infty$ and let $\xi_n(t)$ be a solution of equation (1) on the interval $[T_n, T_{n+1}]$ under the initial condition $\xi_n(T_n)=\xi_{n-1}(T_n)$ and $\xi_0(t)$ be a solution of (1) on $[0, T_1]$ under the initial condition $\xi(0)$. Then the process $\xi(t)=\xi_n(t)$ for $t\in[T_n, T_{n+1}]$ is the required solution of (1).

The uniqueness of the solution. We shall consider solutions of equation (1) on $[0, \infty)$ and prove the weak uniqueness of the solution on this interval. Note that from the uniqueness of a solution of (1) on $[0, \infty)$ one can obtain its weak uniqueness on the interval $[0, T]$ as well. For this purpose it is sufficient to extend the coefficients $a(t, x)$ and $B(t, x)$ to $[T, \infty)$ in such a manner that equation (1) will have a solution on this interval under any initial condition. Then one may extend the solution of equation (1) on $[0, T]$ to the solution on the interval $[0, \infty)$ and utilize the weak uniqueness of the solution on $[0, \infty)$.

Below the following condition will be used:

D. $a(s, x)$ and $B(s, x)$ are defined and measurable on $[0, \infty)\times\mathscr{R}^m$ and satisfy condition (3) for any T; equation (1) possesses a weak solution on $[\tau, \infty)$ for any Markov time τ for $w(t)$ and $\xi(\tau)$ which is independent of $w(t+\tau)-w(\tau)$, and for all $(s_0; x_0) \in [0, \infty) \times \mathcal{R}^m$ we have

$$(28) \quad \overline{\lim_{s \to s_0, x \to x_0}} \operatorname{Sp} (B(s, x)B^*(s, x) - B(s_0, x_0)B^*(s, x))^2 \leq \|B^{-1}(s_0, x_0)B^{*-1}(s_0, x_0)\|^{-2}.$$

Theorem 6. *If condition* **D** *is fulfilled then a solution of equation* (1) *on* $[0, \infty)$ *under the initial condition* $\xi(0)$ *is weakly unique for any* $\xi(0)$.

Proof. Observe that condition **D** implies the existence (for each point $(s_0; x) \in [0, \infty) \times \mathcal{R}^m$) of a $\rho > 0$ such that for $|s - s_0| < \rho$ and $|x - x_0| < \rho$

$$(29) \quad \operatorname{Sp} (B(s, x)B^*(s, x) - B(s_0, x_0)B^*(s_0, x_0))^2 \leq (1-\rho)\|B^{-1}(s_0, x_0)B^{*-1}(s_0, x_0)\|^{-2}.$$

Let $\bar{B}(s, x) = B^{-1}(s_0, x_0)B(s, x)$ for $|s - s_0| < \rho$, $|x - x_0| < \rho$, $\bar{B}(s, x) = \boldsymbol{I}$ otherwise; $\bar{a}(s, x) = B^{-1}(s_0, x_0)a(s, x)$ for $|x - x_0| < \rho$, $|s - s_0| < \rho$, $\bar{a}(s, x) = 0$ otherwise. If $\xi(t)$ is a solution of (1) on $[s_0, \infty)$ and τ is the time of the first departure from the set $\{x: |x - x_0| < \rho\}$ and $\tilde{\xi}(t) = B^{-1}(s_0, x_0)\xi(t)$, then $\tilde{\xi}(t)$ for $t \in [s_0, \tau \wedge (s_0 + \rho)]$ will be a solution of equation

$$(30) \quad d\tilde{\xi}(t) = \bar{a}(t, \tilde{\xi}(t))\, dt + \bar{B}(t, \xi(t))\, dw(t).$$

In view of the choice of ρ the inequality

$$\operatorname{Sp} (\bar{B}(s, x)\bar{B}^*(s, x) - \boldsymbol{I})^2 \leq 1 - \rho$$

is satisfied. Using the fact that a solution of (1) on any finite interval may be obtained by putting together solutions of equations of the form (30) (since every compact set in $[0, \infty) \times \mathcal{R}^m$ may be covered by a finite number of regions of the form

$$\{(s; x): |s - s_0| < \rho, |x - x_0| < \rho\},$$

and a solution of (1) is continuous and thus has for any $\varepsilon > 0$ a finite number of ε-oscillations on a finite interval) one can verify that it is sufficient to prove the weak uniqueness for a solution of equation (30). Therefore we shall proceed with the proof under the assumption that instead of (28) inequality

$$(31) \quad \sup_{s,x} \operatorname{Sp} (B(s, x)B^*(s, x) - \boldsymbol{I})^2 < 1$$

is fulfilled and that $|a(s, x)| \leq C$ for some $C < \infty$. Denote by $\xi_{s,x}(t)$ a solution of equation (1) on $[s, \infty)$ under the initial condition $\xi_{s,x}(s) = x$. Utilizing Lemma 2 one may assert that a function $Q_{s,x}(t, E)$ exists such that

$$Q_{s,x}(t, E) = \mathsf{P}\{\xi_{s,x}(t) \in E\},$$

and that this function does not depend on the choice of the solution. Indeed, for any solution $\xi_{s,x}(t)$ the function $Q_{s,x}(t, E)$ is determined by the corresponding Laplace transform

$$\int_s^\infty e^{-\lambda(t-s)} Q_{s,x}(t, E)\, dt = e^{s\delta} G^\xi_{\lambda-\delta}[e^{-\delta s}\chi_E(x)], \tag{32}$$

where $G^\xi_\lambda = G_\lambda(I - L_1 G_\lambda)^{-1}$ in view of Lemma 2, is uniquely determined by the coefficients of equation (1). Analogously relation (13) may be written as

$$f(s, \xi(s)) = \mathsf{E}[\int_0^\infty e^{-\lambda(t-s)} g(t, \xi(t))\, dt \,|\, \mathfrak{F}^\xi_s],$$

where $\{\mathfrak{F}^\xi_s\}$ is a current of σ-algebras generated by the process $\xi(\cdot)$ (here we have utilized the fact that $w(t+s) - w(t)$ is independent of $\mathfrak{F}^\xi_s$), $f(s, x) = G_\lambda(I - L_1 G_\lambda)^{-1} g(s, x)$. Hence

$$\mathsf{E}(\chi_E(\xi(t)) \,|\, \mathfrak{F}^\xi_s) \quad \text{and} \quad Q_{s,\xi(s)}(t, E)$$

have the same Laplace transforms. Therefore

$$\mathsf{E}(\chi_E(\xi(t)) \,|\, \mathfrak{F}^\xi_s) = Q_{s,\xi(s)}(t, E) \tag{33}$$

with probability 1. However, (33) implies that $\xi(t)$ is a Markov process with the transition probability $Q_{s,x}(t, E)$ defined by the Laplace transform

$$\int_0^\infty e^{-\lambda t} Q_{s,x}(t, E)\, dt = e^{\delta s} G_{\lambda-\delta}[I - L_1 G_{\lambda-\delta}]^{-1} g_{\delta,E}(s, x), \tag{34}$$

where $g_{\delta,E}(s, x) = e^{-\delta s}\chi_E(x)$. Thus each solution of (1) is a Markov process with a given transition probability which depends only on the coefficients of (1). Consequently the measure associated with $\xi(t)$ on $\mathscr{C}^m_{[0,T]}$ for any T is uniquely determined by the distribution of $\xi(0)$. Thus the very same measure is associated with two solutions under the initial condition $\xi(0)$. The theorem is proved. □

Remark. Since under the conditions of the theorem the operator $B(t, x)$ is nondegenerate, in view of Girsanov's lemma in Section 2 the weak uniqueness of a solution implies the uniqueness of a strong solution.

Continuous dependence of solutions on a parameter. Here we shall be concerned mainly with the dependence of a solution on the initial conditions. However, we shall first prove a general theorem from which assertions concerning a continuous dependence on other parameters may also be obtained.

Theorem 7. *Let $\xi^n(t)$, $n = 0, 1, \ldots$, be solutions of equations*

$$\xi^n(t) = x_n + \int_0^t a_n(s, \xi^n(s))\, ds + \int_0^t B_n(s, \xi^n(s))\, dw(s) \tag{35}$$

and let the following conditions be satisfied:

1) *A constant K exists such that for* $0\leq s\leq T$

$$|a_n(s, x)|+\|B_n(s, x)\|\leq K(1+|x|).$$

2) *For each n the condition of Theorem 6 is fulfilled.*
3) *For each r there exist functions* $f_1(t, x,), \ldots, f_k(t, x)$ *such that*

$$\sup_{t,x}\left(\left|\frac{\partial}{\partial t} f_j(t, x)\right|+\left|\frac{\partial}{\partial x} f_j(t, x)\right|+\left|\frac{\partial}{\partial x} f_j(t, x)\right|^{-1}+\left\|\frac{\partial^2}{\partial x^2} f_j(t, x)\right\|\right)<\infty$$

and Borel sets $\Lambda_1, \ldots, \Lambda_k$ *in* $\mathcal{R}^1$ *of Lebesgue measure* 0 *such that for* $(t; x)\in [0, T]\times\{x: |x|<r\}-\bigcup_{j=1}^k \{(t; x); f_j(t, x)\in \Lambda_j\}$

$$\lim a_n(t, x)=a_0(t, x) \quad \text{and} \quad \lim B_n(t, x)=B_0(t, x).$$

Then measures μ_n *on* $\mathscr{C}^m_{[0,T]}$ *associated with processes* $\xi^n(t)$ *converge weakly to the measure* μ_0 *which is associated with the process* $\xi^0(t)$ *provided only* $x_n\to x_0$.

Proof. (This proof is very similar to the proof of Theorem 5.) The sequence of measures μ_n is compact. If μ is a limit point of this sequence then by using condition 3) one verifies, in the same manner as in Theorem 5, that finite-dimensional distributions of processes

$$\eta_n(t)=\xi^n(t)-\int_0^t a_n(s, \xi^n(s))\, ds$$

and processes

$$(\eta_n(t), z)^2-\int_0^t |B_n^*(s, \xi^n(s))z|^2\, ds$$

converge respectively to the finite-dimensional distributions of the processes

$$\eta_0(t)=\bar{\xi}(t)-\int_0^t a_0(s, \bar{\xi}(s))\, ds,$$
$$(\eta_0(t), z)^2-\int_0^t |B_0^*(s, \bar{\xi}(s))z|^2\, ds,$$

where $\bar{\xi}(t)$ is a process with which measure μ is associated. Therefore in view of Theorem 1 in Section 1 $\bar{\xi}(t)$ is a solution of (35) for $n=0$. (In Theorem 5 the uniqueness of a solution for $n>0$ was assumed; actually, in the course of the proof only the weak uniqueness was utilized since only measures associated with solutions were considered.) By assumption, solution (35) is weakly unique. Therefore the compact sequence μ_n possesses a unique limit joint. The theorem is proved. □

Below, the following condition will be used:

E. For any $r>0$ there exist functions $f_1(t, x), \ldots, f_k(t, x)$ such that

$$\sup_{t,x}\left(\left|\frac{\partial}{\partial t} f_j(t, x)\right|+\left|\frac{\partial}{\partial x} f_j(t, x)\right|+\left|\frac{\partial}{\partial x} f_j(t, x)\right|^{-1}+\left\|\frac{\partial^2}{\partial x^2} f_j(t, x)\right\|\right)<\infty$$

and $a(t, x)$ and $B(t, x)$ are continuous on the set

$$[0, T]\times\{x: |x|<r\}-\bigcup_{j=1}^{k}\{(t, x): f_j(t, x)\in \Lambda_j\},$$

where Λ_j are Borel sets of Lebesgue measure 0; moreover, $a(t, x)$ and $B(t, x)$ satisfy the condition of Theorem 6.

Corollary 1. *Let $\xi(t)$ be a solution of equation (1) for which condition* **E** *is satisfied. Set*

$$\mathsf{T}_t^s f(x)=\int f(y)P(s, x, t, dy).$$

For any bounded continuous function f on $\mathcal{R}^m$ the function $\mathsf{T}_t^s f(x)$ is jointly continuous in the variables.

Proof. Let $s_n\to s_0$, $t_n\to t_0$, and $x_n\to x_0$. Denote by $\xi_n(u)$ a solution of the equation

$$\xi_n(u)=x_n+\int_0^u a(s_n+v, \xi_n(v))\,dv+\int_0^u B(s_n+v, \xi_n(v))\,dw(v).$$

Condition 3) of the theorem is fulfilled for the coefficients

$$a_n(v, x)=a(s_n+v, x),\qquad B_n(v, x)=B(s_n+v, x),\qquad n=0, 1, \ldots.$$

Conditions 1) and 2) of the theorem are also satisfied. Hence

$$\lim_{n\to\infty}\mathsf{E}f(\xi_n(u))=\mathsf{E}f(\xi_0(u)).$$

However, $\mathsf{E}f(\xi_n))=\mathsf{T}_{s_n+u}^{s_n}$, so that

$$\lim_{n\to\infty}\mathsf{T}_{s_n+u}^{s_n}f(x_n)=\mathsf{T}_{s_0+u}^{s_0}f(x_0).$$

Next, using Lemma 2 in Section 2 we verify that there exists a constant C such that uniformly in n

$$\mathsf{E}|\xi_n(u)-\xi_n(u+h)|^2\le Ch.$$

Therefore

$$\begin{aligned}\overline{\lim_{n\to\infty}}\,|\mathsf{T}_{t_n}^{s_n}f(x_n)-\mathsf{T}_{t_0}^{s_0}f(x_0)|&\le\overline{\lim_{n\to\infty}}\,|\mathsf{T}_{t_n}^{s_n}f(x_n)-\mathsf{T}_{s_n+(t_0-s_0)}^{s_n}f(x_n)|\\&=\overline{\lim_{n\to\infty}}\,\mathsf{E}|f(\xi_n(t_n-s_n))-f(\xi_n(t_0-s_0))|.\end{aligned}$$

The last expression is 0 since the quantity to the right of the sign of the mathematical expectation is bounded and approaches 0 in probability. Indeed, setting $u_n = t_n - s_n$ we have

$$\begin{aligned}(36)\quad &\overline{\lim_{n\to\infty}}\, \mathsf{P}\{|f(\xi_n(u_n)) - f(\xi_n(u_0))| > \varepsilon\}\\ &\leqslant \overline{\lim_{n\to\infty}}\, \mathsf{P}\{|\xi_n(u_0) > N\} + \overline{\lim_{n\to\infty}}\, \mathsf{P}\{|\xi_n(u_n) - \xi_n(u_0)| > \delta_N\},\end{aligned}$$

where δ_N is such that $|f(y)-f(x)| \leqslant \varepsilon$ for $|y| \leqslant N$ and $|x-y| \leqslant \delta_N$. Since $|u_n - u_0| \to 0$, the second summand in the right-hand side of (36) equals 0. The first one can be made arbitrarily small by an appropriate choice of ε. The assertion is thus proved. □

Corollary 2. *If condition* **E** *is satisfied and $\mu_{s,x}$ denotes the measure associated with the process $\xi(t-s)$ where $\xi(t)$ is a solution of equation* (1) *on $[s, \infty)$ under the initial condition $\xi(s) = x$, then $\mu_{s,x}$ is weakly continuous in s and x.*

The proof of this assertion was obtained in Corollary 1.

If $F_s(x(\cdot))$ is a family of bounded functionals continuous in s and $x(\cdot)$ on every compact set in $[0, T] \times \mathscr{C}^m_{[0,T]}$ then

$$\int F_s(x(\cdot))\mu_{s,y}(dx)$$

is also continuous in s and y, since

$$\begin{aligned}&\overline{\lim_{s\to s_0,\, y\to y_0}} \left|\int F_s(x(\cdot))\mu_{s,y}(dx) - \int F_{s_0}(x(\cdot))\mu_{x_0,y_0}(dx)\right|\\ &\leqslant \lim_{s\to s_0,\, y\to y_0} \int |F_s(x(\cdot)) - F_{s_0}(x(\cdot))|\mu_{s,y}(dx) = 0.\end{aligned}$$

(The last step follows from the fact that the integrand tends to zero on every compact set and for each $\varepsilon > 0$ one can choose a compact K such that

$$\underline{\lim_{s\to s_0,\, y\to y_0}}\, \mu_{s,y}(K) \geqslant 1 - \varepsilon.)$$

Corollary 3. *Let $f_n(x)$, $n = 0, 1, \ldots$ be a sequence of continuous and bounded functions on $\mathcal{R}^m$ such that for any N*

$$\lim_{n\to\infty} \sup_{|x|\leqslant N} |f_n(x) - f_0(x)| = 0.$$

Then under the conditions of Theorem 7 we have for every N

$$(37)\qquad \lim_{n\to\infty} \sup_{|x|\leqslant N} |\mathsf{E} f_n(\xi_x^{(n)}(t)) - \mathsf{E} f_0(\xi_x^{(0)}(t))| = 0,$$

where $\xi_x^{(n)}$ is a solution of (35) *under the initial condition x.*

Proof. To prove (37) it is sufficient to show that for any convergent sequence $\{x_n\}$ belonging to $\mathscr{R}^m$ with $\lim x_n = x_0$,

$$\lim_{n\to\infty} |\mathsf{E}f_n(\xi_{x_n}^{(n)}(t)) - \mathsf{E}f_0(\xi_{x_n}^{(0)}(t))| = 0,$$

which is equivalent to the equality

$$\lim_{n\to\infty} |\mathsf{E}f_n(\xi_{x_n}^{(n)}(t)) - \mathsf{E}f_0(\xi_{x_n}^{(n)}(t))| = 0,$$

since Theorem 7 and Corollary 1 imply that

$$\lim_{n\to\infty} \mathsf{E}f_0(\xi_{x_n}^{(n)}(t)) = \mathsf{E}f_0(\xi_{x_0}^{(0)}(t)),$$

$$\lim_{n\to\infty} \mathsf{E}f_0(\xi_{x_n}^{(0)}(t)) = \mathsf{E}f_0(\xi_{x_0}^{(0)}(t)).$$

However,

$$\overline{\lim_{n\to\infty}}\, \mathsf{E}|f_n(\xi_{x_n}^{(n)}(t)) - f_0(\xi_{x_n}^{(n)}(t))| \le \overline{\lim_{n\to\infty}} \sup_{|x|\le N} |f_n(x) - f_0(x)|$$

$$+ \overline{\lim_{n\to\infty}}\, \mathsf{P}\{|\xi_{x_n}^{(n)}(t)| > N\} \sup_x |f_n(x) - f_0(x)|.$$

The first summand on the right-hand side equals 0 and the second can be made arbitrarily small by choosing N sufficiently large. Formula (37) is proved. □

Corollary 4. *Let the functions $a_n(s, x)$ and $B_n(s, x)$, $n \neq 0$, in Theorem 7 be twice continuously differentiable in x and jointly continuous in the variables; let the function $f_n(x)$ be continuous and bounded with bounded derivatives up to the second order. If $P_n(s, y, t, E)$ is the transition probability for the process which is a solution of (36), then, as follows from Theorem 6 in Section 2 of Chapter II, the function*

$$u_n(s, x) = \int P_n(s, x, t, dy) f_n(y)$$

for $s \le t$ is the unique bounded solution of the Cauchy problem

$$\frac{\partial}{\partial s} u_n(s, x) + \left(a_n(s, x), \frac{\partial}{\partial x} u_n(s, x)\right) + \tfrac{1}{2}\operatorname{Sp} B_n^2(s, x) \frac{\partial^2}{\partial x^2} u_n(s, x) = 0,$$

(38)

$$u_n(t, x) = f_n(x).$$

In view of Corollary 3

$$u_n(s, x) \to u_0(s, x),$$

where

$$u_0(s, x) = \int P_0(s, x, t, dy) f_0(y),$$

provided only $\sup_{|x| \le N} |f_n(x) - f_0(x)| \to 0$ for all N and $f_n(x)$ are bounded and continuous. Therefore $u_0(s, x)$ is a generalized solution of equation (38) in the following sense: for any twice continuously differentiable functions $a_n(s, x)$ and $B_n(s, x)$ satisfying the conditions of Theorem 7 and any sequence of continuous jointly bounded functions $f_n(x)$ convergent to $f_0(x)$, the bounded solutions of Cauchy's problem (38) converge to the same continuous function $u_0(s, x)$.

Consider equation (1) with coefficients satisfying condition **E**. Denote by $\xi_{s,x}(t)$ the solution of equation (1) on $[s, \infty)$ under the initial condition $\xi_{s,x}(s) = x$.

Let G be a domain in $[0, \infty) \times \mathscr{R}^m$. Denote

$$\tau_{s,x}(G) = \inf [t \colon \xi_{s,x}(t) \notin G]$$

($\tau_{s,x}(G)$ may equal $+\infty$). Consider a continuous bounded function $f(t, y)$ vanishing for t sufficiently large (simultaneously for all y). Set

$$\varphi_G(s, x) = \mathsf{E}(\tau_{s,x}(G), \xi_{s,x}(\tau_{s,x}(G))). \tag{39}$$

Note that

$$f(\tau_{s,x}(G), \xi_{s,x}(\tau_{s,x}(G))) = g_s(\eta_{s,x}(\cdot)),$$

where $\eta_{s,x}(t) = \xi_{s,x}(t - s)$ and $g_s(x(\cdot))$ is a functional on $\mathscr{C}^m_{[0,T]}$, provided T is such that $f(t, y) = 0$ for $t \ge T$. This functional $g_s(x(\cdot))$ is defined in the following manner: if $\bar{t}$ is the time when $x(\cdot)$ hits the boundary of G for the first time, then

$$g_s(x(\cdot)) = f(s + \bar{t}, x(\bar{t})) \chi_{\{s + \bar{t} \le T\}}.$$

Clearly $g_s(x(\cdot))$ is continuous in s uniformly in $x(\cdot)$ on each bounded set in $\mathscr{C}^m_{[0,T]}$. Furthermore, for a fixed s, $g_s(x(\cdot))$ is continuous at point $\bar{x}(\cdot)$, provided there exists a sequence $t_k \downarrow \bar{t}$ such that $\bar{x}(t_k)$ does not belong to the closure of G and $\bar{x}(t) \in G$ for $t < \bar{t}$.

Thus in view of Corollary 2, in order that the function $\varphi_G(s, x)$ be continuous at the point $\bar{s}$, $\bar{x}$ it is sufficient that a sequence $\varepsilon_k \downarrow 0$ exist with probability 1 for the process $\xi_{\bar{s},\bar{x}}(\cdot)$ such that

$$\xi_{\bar{s},\bar{x}}(\tau_{\bar{s},\bar{x}}(G) + \varepsilon_k) \notin [G],$$

where $[G]$ is the closure of G.

Assume now that the boundary of G is smooth. Let $\vec{n}$ be the normal in $(-\infty, \infty) \times \mathscr{R}^m$ to the boundary of G at the point $(\tau_{\bar{s},\bar{x}}(G); \xi_{\bar{s},\bar{x}}(\tau_{\bar{s},\bar{x}}(G)))$. Clearly $\vec{n}$ is an $\mathfrak{F}_{\tau_{\bar{s},\bar{x}}(G)}$-measurable variable, where $\{\mathfrak{F}_t\}$ is the current of σ-algebras generated by $\xi_{\bar{s},\bar{x}}(t)$. Let n_t and n_x be projections of $\vec{n}$ on $(-\infty, \infty)$ and $\mathscr{R}^m$, respectively. Consider for $\varepsilon>0$ the process

$$\zeta(\varepsilon)=n_t\varepsilon+(n_x, \xi_{\bar{s},\bar{x}}(\tau_{\bar{s},\bar{x}}(G)+\varepsilon)-\xi_{\bar{s},\bar{x}}(\tau_{\bar{s},\bar{x}}(G))).$$

For sufficiently small ε it follows from the relation

$$\frac{\zeta(\varepsilon)}{\varepsilon+|\xi_{\bar{s},\bar{x}}(\tau_{\bar{s},\bar{x}}(G)+\varepsilon)-\xi_{\bar{s},\bar{x}}(\tau_{\bar{s},\bar{x}}(G))|}\geqslant\delta, \tag{40}$$

where $\delta>0$ is a fixed arbitrarily small number, that

$$(\tau_{\bar{s},\bar{x}}(G)+\varepsilon;\ \xi_{\bar{s},\bar{x}}(\tau_{\bar{s},\bar{x}}(G)+\varepsilon))\notin[G],$$

since this point is located at the interior of a cone whose axis is an exterior normal line to G', the boundary of G.

Insofar as

$$\zeta(\varepsilon)=O(\varepsilon)+\int_0^\varepsilon(g(s,\omega), dw(s)),$$

where $|g(s,\omega)|\geqslant\rho>0$, ρ being a nonrandom number, utilizing Corollary 4 to Theorem 1 in Section 1 and the law of the iterated logarithm for a Wiener process (cf. Theorem 4 in Volume II, Chapter IV, Section 3), we verify the existence of a sequence $\{\varepsilon_k\}$ such that

$$\lim_{k\to\infty}\left(\zeta(\varepsilon_k)\Big/\sqrt{\varepsilon_k\ln\ln\frac{1}{\varepsilon_k}}\right)>\rho_1,$$

where $\rho_1>0$ is a nonrandom constant. Analogous considerations show that

$$\overline{\lim_{k\to\infty}}\left(|\xi_{\bar{s},\bar{x}}(\tau_{\bar{s},\bar{x}}(G)+\varepsilon_k)-\xi_{\bar{s},\bar{x}}(\tau_{\bar{s},\bar{x}}(G))|\Big/\sqrt{\varepsilon_k\ln\ln\frac{1}{\varepsilon_k}}\right)<\rho_2.$$

Hence if $\delta=\rho_1/\rho_2$, relation (40) is fulfilled for $\varepsilon=\varepsilon_k$ for k sufficiently large.

Thus the following theorem is valid.

Theorem 8. *If condition* **E** *is satisfied and the domain G in $[0,\infty)\times\mathscr{R}^m$ possesses a smooth boundary such that the normal line $\vec{n}$ to the boundary has at each point a nonzero projection on $\mathscr{R}^m$ and the function is continuous and bounded on $[0,\infty)\times\mathscr{R}^m$ with $f(t,y)=0$ for $t>T$ for some $T>0$, then the function*

$$\mathsf{E}f(\tau_{s,x}(G), \xi_{s,x}(\tau_{s,x}(G))) \tag{41}$$

is jointly continuous in the variables.

Corollary. *Let the function $f(t,y)$ be continuous and bounded on $[0,\infty)\times\mathscr{R}^m$ and $|f(t,y)|\to0$ uniformly in y as $t\to\infty$. If the remaining conditions of Theorem 8 are satisfied, then function (41) is also continuous in s and x.*

Proof. Indeed, the function $f(t, y)$ can be represented as the series

$$f(t, y) = \sum_{k=1}^{\infty} f_k(t, y),$$

where f_k satisfy the conditions of Theorem 8 and $|f_k(t, y)| \le \alpha_k$, $\sum \alpha_k < \infty$. It is clear that in this case

$$\mathsf{E} f(\tau_{s,x}(G), \xi_{s,x}(\tau_{s,x}(G))) = \sum_{k=1}^{\infty} \mathsf{E} f_k(\tau_{s,x}(G), \xi_{s,x}(\tau_{s,x}(G))), \tag{42}$$

and the series in (42) is uniformly convergent. Since the summands in the right-hand side of (42) are continuous, so is the sum of the series. □

Homogeneous diffusion processes. In this subsection some basic facts (which to a large extent follow from preceding theorems) related to homogeneous diffusion processes, i.e., processes which are solutions of equation

$$d\xi(t) = a(\xi(t))\, dt + B(\xi(t))\, dw(t) \tag{43}$$

are presented.

It is assumed throughout this subsection that the following conditions are fulfilled.

I. $a(x)$ and $B(x)$ are measurable functions on $\mathscr{R}^m$ with values in $\mathscr{R}^m$ and $\mathfrak{L}(\mathscr{R}^m)$, respectively, and $B^{-1}(x)$ is locally bounded.

II. The inequality $|a(x)| + \|B(x)\| \le K(1 + |x|)$ is satisfied for some K.

III. For all $x_0 \in \mathscr{R}^m$

$$\overline{\lim_{x \to x_0}} \operatorname{Sp}(B(x)B^*(x) - B(x_0)B^*(x_0))^2 < \|B^{-1}(x_0)B^{*-1}(x_0)\|^{-2},$$

and for any $r > 0$ one can find functions $f_1(x), \ldots, f_k(x)$ such that for some $C > 0$ and $i = 1, \ldots, k$,

$$|f_i'(x)| + |f_i'(x)|^{-1} + \|f_i''(x)\| \le C$$

and $a(x)$ and $B(x)$ are continuous for

$$x \in \{x: |x| < r\} - \bigcup_{j=1}^{k} \{x: f_j(x) \in \Lambda_j\},$$

where Λ_j are Borel sets in $\mathscr{R}^1$ of Lebesgue measure 0.

Then the following assertions are valid.

1. *A solution of equation* (43) *exists, is weakly unique, and is a homogeneous Markov process.*

Homogeneity follows from the weak uniqueness and the fact that $\xi(t+h)$, as a function of t, is a solution of equation (43) if the process $w(t)$ appearing in this equation is replaced by the Wiener process $w_h(t) = w(t+h) - w(h)$. Note that the transition probability $P(t, x, E)$ of the process $\xi(t)$ is defined by equality

$$P(t, x, E) = \mathsf{P}\{\xi_x(t) \in E\},$$

where $\xi_x(t)$ is a solution of (43) under the initial condition $\xi_x(0) = x$.

2. *If*

$$(44) \qquad \mathsf{T}_t f(x) = \int f(y) P(t, x, dy),$$

then $\mathsf{T}_t f(x)$ *is jointly continuous in the variables. This follows from Corollary* 1 *to Theorem* 7.

Thus the process $\xi(t)$ *is a homogeneous stochastically continuous Feller process. Therefore it is strong Markovian.*

An operator $\tilde{\mathsf{A}}$ is called a *quasi-generating operator* of a Markov process $\xi(t)$ if it is defined on a set $\mathcal{D}_{\mathsf{A}} \subset \mathscr{C}_{\mathcal{R}^m}$, $\tilde{\mathsf{A}} f$ is a locally bounded Borel function, and for all $t > 0$ and $x \in \mathcal{R}^m$

$$(45) \qquad \mathsf{E}_x f(\xi(t)) - f(x) = \mathsf{E}_x \int_0^t \tilde{\mathsf{A}} f(\xi(s))\, ds.$$

If formula (45) is valid for $t = \tau$, where τ is a Markov moment such that $\tau \le \tau_U$ and U is a neighborhood of point x, then the operator $\tilde{\mathsf{A}}$ is called a *quasi-characteristic operator.*

3. *A quasi-generating operator of a process which is a solution of* (43) *is defined on all functions* $f \in \mathscr{C}^2_{\mathcal{R}^m}$ *where* $\mathscr{C}^2_{\mathcal{R}^m}$ *is the space of twice continuously differentiable functions such that*

$$\sup_x (|f(x)| + |f'_x(x)| + \|f''_{xx}(x)\|) < \infty;$$

moreover,

$$(46) \qquad \tilde{\mathsf{A}} f(x) = (a(x), f'_x(x)) + \tfrac{1}{2} \operatorname{Sp} B(x) B^*(x) f''_{xx}(x).$$

This relationship follows from equality

$$\mathsf{E}_x[f(\xi(T)) - f(\xi(0))] = \mathsf{E}_x \int_0^T ((a(\xi(t)), f'_x(\xi(t))) + \tfrac{1}{2} \operatorname{Sp} B(\xi(t)) B^*(\xi(t)) f''_{xx}(\xi(t)))\, dt,$$

which is a corollary of Itô's formula.

Utilizing equality (45) one can verify, in the same manner as in Lemma 2 proved in Volume II, Chapter II, Section 5, the equality

$$(47) \qquad \mathsf{E}_x f(\xi(\tau)) - f(x) = \mathsf{E}_x \int_0^\tau \tilde{\mathsf{A}} f(\xi(s))\, ds$$

for any Markov time τ such that $\mathsf{E}_x \tau < \infty$.

Lemma 4. *If G is a bounded domain and ζ_G is the time of the first exit out of G, then $\mathsf{E}_x\zeta_G$ is uniformly bounded in x.*

Proof. Let $f(x)=\exp\{(z, x)\}$ for $x\in G$ and for $x\notin G$ the function $f(x)$ is extended in such a manner that it belongs to $\mathscr{C}^2_{\mathscr{R}^m}$. Then for $x\in G$

$$\tilde{A}f(x)=[(a(x), z)+\tfrac{1}{2}|B^*(x)z|^2]\exp\{(z, x)\}.$$

In view of the conditions imposed on $a(x)$ and $B(x)$ one can choose $z\in\mathscr{R}^m$ such that $\tilde{A}f(x)>\delta>0$, $x\in G$. Substituting this function into (47) and replacing τ by the Markov time $\zeta_G\wedge T$ we have

$$\mathsf{E}_x f(\xi(\zeta_G\wedge T))-f(x)=\mathsf{E}_x\int_0^{\zeta_G\wedge T}\tilde{A}f(\xi(t))\,dt>\mathsf{E}_x[\zeta_G\wedge T]\delta.$$

Approaching the limit as $T\to\infty$ we obtain from the last equality that

$$\mathsf{E}_x\zeta_G\leq(2/\delta)\sup_{x\in G}e^{(x,z)}. \tag{48}$$

The lemma is proved. □

We shall now prove a theorem which will allow us to express generalized solutions of the boundary problem for an elliptic (partial) differential equation by means of solutions of a stochastic differential equation.

Theorem 9. *Let G be a bounded domain with a smooth boundary, $\varphi(x)$ be continuous in G' (the boundary of G), and $a(x)$ and $B(x)$ satisfy conditions* I–III. *Furthermore, let $a_n(x)$ and $B_n(x)$ be twice continuously differentiable functions, condition* II *be satisfied for these functions with the same constant, $\|B_n^{-1}\|$ be bounded in G and $a_n(x)\to a(x)$, $B_n(x)\to B(x)$ at all points of continuity of $a(x)$ and $B(x)$, and φ_n be a sequence of continuous functions uniformly convergent to φ on G'. Then the sequence of functions $u_n(x)$ which are solutions of the equations*

$$(a_n(x), u_n'(x))+\tfrac{1}{2}\operatorname{Sp}B_n(x)B^*(x)u_n''(x)=0 \tag{49}$$

under the boundary condition $u_n(x)=\varphi_n(x)|_{x\in G'}$ converges to the function

$$u(x)=\mathsf{E}_x\varphi(\xi(\zeta_G)). \tag{50}$$

Proof. We shall assume that $a_n(x)=a(x)$, $B_n(x)=B(x)$ for $x\notin G$. Let $\xi^n(t)$ be a solution of the equation

$$d\xi^n(t)=a_n(\xi^n(t))\,dt+B_n(\xi^n(t))\,dw(t),$$

and u_n satisfy (49). Utilizing relations (46) and (47) we obtain

$$u_n(x)=\mathsf{E}^n_x\varphi(\xi^n(\zeta_G)),$$

where E_x^n is the mathematical expectation of the Markov process ξ^n. Theorem 7 implies that finite-dimensional distributions of processes $\xi^n(t)$ converge to the finite-dimensional distributions of the process $\xi(t)$.

We show that

$$\lim_{n\to\infty} \mathsf{E}_x^n\varphi(\xi^n(\zeta_G)) = \mathsf{E}_x\varphi(\xi(\zeta_G)).$$

Combining Theorems 7 and 8 we verify that for all $\lambda > 0$

$$\lim_{n\to\infty} \mathsf{E}_x^n e^{-\lambda\zeta_G}\varphi(\xi^n(\zeta_G)) = \mathsf{E}_x e^{-\lambda\zeta_G}\varphi(\xi(\zeta_G)). \tag{51}$$

It remains to show that

$$\lim_{N\to\infty} \overline{\lim_{n\to\infty}} \mathsf{P}_x^n\{\zeta_G > N\} = 0.$$

This assertion follows from the fact that in view of (51) we have for almost all N

$$\lim_{n\to\infty} \mathsf{P}_x^n\{\zeta_G > N\} = \mathsf{P}_x\{\zeta_G > N\},$$

and that ζ_G is finite (the latter follows, for instance, from (48)). □

Thus if the function $u(x)$ of the form (50) is smooth, it is then a solution of equation

$$(a(x), u'(x)) + \tfrac{1}{2}\operatorname{Sp} B(x)B^*(x)u''(x) = 0$$

in G under the boundary condition φ. In the general case this function can be viewed as a generalized solution of this equation.

Homogeneous processes with integrable kernel of a potential. We shall consider solution of equation (43) such that in addition to conditions I–III the following condition is satisfied:

IV. For any bounded Borel set E

$$G(x, E) = \int_0^\infty P(t, x, E)\, dt < \infty, \qquad \lim_{t\to\infty} \sup_x P(t, x, E) = 0, \tag{52}$$

the potential $G(x, E)$ is absolutely continuous with respect to the Lebesgue measure m, and the kernel of the potential

$$g(x, y) = \frac{dG(x, \cdot)}{dm}(y)$$

satisfies the condition

$$\sup_x \int (g(x, y))^\alpha\, dy < \infty \tag{53}$$

for some $\alpha>1$. Note that a study of solutions of equation (43) can always be reduced to the case when the potential $G(x, E)$ is defined. To achieve this, it is required to consider instead of process $\xi(t)$ the process $(\xi(t); w_3(t))$ in $\mathscr{R}^{m+3}$, where $\xi(t)$ is a solution of (43) in $\mathscr{R}^m$ and $w_3(t)$ is a Wiener process in $\mathscr{R}^3$ independent of $\xi(t)$. It is easy to see that the composite process $(\xi(t); w_3(t))$ also satisfies an equation of the form (43); for this process a potential is defined and (52) is satisfied, since

$$\int_0^\infty \mathsf{P}\{w_3(t)\in E_3\}\,dt<\infty, \qquad \lim_{t\to\infty}\sup_{x\in\mathscr{R}^3}\mathsf{P}\{w_3(t)+x\in E_3\}=0$$

for any bounded Borel set $E_3\subset\mathscr{R}^3$. The existence of $g(x, y)$ under certain assumptions follows from Remark 2 to Theorem 2. For example, inequality (53) is satisfied if $a(x)$, $B(x)$, and $B^{-1}(x)$ are bounded and satisfy a Hölder condition for $m\geqslant 3$ since under these assumptions $P(t, x, E)$ possesses density $\rho_t(x, y)$ such that for some c_1 and c_2

$$\rho_t(x, y)\leqslant c_1 t^{-m/2}\exp\{-c_2|x-y|^2\}$$

(the last assertion follows from the properties of fundamental solutions of parabolic differential equations, see, e.g., A. Friedman [1], Chapter I, § 5–6).

If (53) is fulfilled and $\beta=\alpha/(1-\alpha)$, then for all $f\in\mathfrak{L}_\beta(\mathscr{R}^m)$ the operator

$$\mathsf{G}f(x)=\mathsf{E}_x\int_0^\infty f(\xi(t))\,dt=\int g(x, y)f(y)\,dy$$

is defined; moreover, $\sup_x|\mathsf{G}f(x)|\leqslant K\|f\|_\beta$, where

$$K=\sup_x\left(\int(g(x, y))^\alpha\,dy\right)^{1/\alpha}.$$

Utilizing an approximation of functions in $\mathfrak{L}_\beta(\mathscr{R}^m)$ in terms of functions in $\mathfrak{L}_\beta(\mathscr{R}^m)\cap\mathscr{C}_{\mathscr{R}^m}$, continuity in x of the function

$$\mathsf{E}_x\int_0^T f(\xi(t))\,dt, \qquad f\in\mathscr{C}_{\mathscr{R}^m},$$

for all $T>0$ and the second condition (52) we verify that $\mathsf{G}f(x)\in\mathscr{C}_{\mathscr{R}^m}$ for all $f\in\mathfrak{L}_\beta(\mathscr{R}^m)$.

Let $f\in\mathfrak{L}_\beta(\mathscr{R}^m)$; then

$$\mathsf{E}_x\mathsf{G}f(\xi(t))-\mathsf{G}f(x)=-\mathsf{E}_x\int_0^t f(\xi(s))\,ds.$$

Therefore for all locally bounded functions $f\in\mathfrak{L}_\beta(\mathscr{R}^m)$ $\mathsf{G}f\in\mathscr{D}_{\tilde{\mathsf{A}}}$ and $\tilde{\mathsf{A}}\mathsf{G}f=-f$. Consider the expression

$$\mathsf{G}f(\xi(t))-\mathsf{G}f(\xi(0))+\int_0^t f(\xi(s))\,ds;$$

this expression is a martingale. Assume that the function $\mathsf{G}f$ is twice continuously differentiable with respect to x.

Denote by $\tilde{\mathsf{A}}_0$ the differential operator

$$\tilde{\mathsf{A}}_0 g=(a(x), g'(x))+\tfrac{1}{2}\operatorname{Sp} B(x)B^*(x)g''_{xx}(x),$$

defined on the space $\mathscr{C}^2_{\mathscr{R}^m}$ of twice continuously differentiable functions.

From Itô's formula

$$\mathsf{G}f(\xi(t))-\mathsf{G}f(\xi(0))=\int_0^t \tilde{\mathsf{A}}_0\mathsf{G}f(\xi(s))\,ds+\int_0^t (B(\xi(s))(\mathsf{G}f)'_x(\xi(s)), dw(s)).$$

Since $\int_0^t [f(\xi(s))-\tilde{\mathsf{A}}_0\mathsf{G}f(\xi(s))]\,ds$ is a martingale, it follows that

$$\int_0^t [f(\xi(s))-\tilde{\mathsf{A}}_0\mathsf{G}f(\xi(s))]\,ds=0$$

with probability 1; hence

$$\mathsf{G}f(\xi(t))-\mathsf{G}f(\xi(0))+\int_0^t f(\xi(s))\,ds=\int_0^t (b(\xi(s)), dw(s)), \tag{54}$$

where $b(x)=B(x)\cdot(\mathsf{G}f)'_x(x)$. We show that representation (54) is valid for all $f\in\mathfrak{L}_\beta(\mathscr{R}^m)$ under some additional assumptions.

Denote by $\hat{\mathscr{C}}_{\mathscr{R}^m}$ the set of functions $f\in\mathscr{C}_{\mathscr{R}^m}$ such that

$$\lim_{|x|\to\infty} f(x)=0, \qquad \hat{\mathscr{C}}^2_{\mathscr{R}^m}=\hat{\mathscr{C}}_{\mathscr{R}^m}\cap\mathscr{C}^2_{\mathscr{R}^m}.$$

Theorem 10. *Let the following conditions be satisfied:*

1) $\mathsf{G}f\in\hat{\mathscr{C}}_{\mathscr{R}^m}$, *where* $f\in\mathfrak{L}_\beta(\mathscr{R}^m)$;

2) $\tilde{\mathsf{A}}_0[\hat{\mathscr{C}}^2_{\mathscr{R}^m}]\cap\mathfrak{L}_\beta(\mathscr{R}^m)$ *is dense in* $\mathfrak{L}_\beta(\mathscr{R}^m)$ *in the norm* $\|\cdot\|_\beta$.

Then there exists for all $f\in\mathfrak{L}_\beta(\mathscr{R}^m)$ *a function* $b(x)$ *with values in* $\mathscr{R}^m$ *satisfying*

$$\sup_x \int |b(y)|^2 g(x, y)\,dy<\infty \tag{55}$$

such that equality (54) *is fulfilled.*

Proof. Let $\varphi_n\in\hat{\mathscr{C}}^2_{\mathscr{R}^m}$ be a sequence such that $\|\tilde{\mathsf{A}}_0\varphi_n-f\|_\beta\to 0$. Then

$$\sup_x |\mathsf{G}\tilde{\mathsf{A}}_0\varphi_n(x)-\mathsf{G}f(x)|\to 0.$$

Note that $\varphi_n+\mathsf{G}\tilde{\mathsf{A}}_0\varphi_n$ satisfies equality

$$\begin{aligned}&\mathsf{E}_x[\varphi_n(\xi(t))+\mathsf{G}\tilde{\mathsf{A}}_0\varphi_n(\xi(t))]-\varphi_n(x)-\mathsf{G}\tilde{\mathsf{A}}_0\varphi_n(x)\\&=\mathsf{E}_x\int_0^t \tilde{\mathsf{A}}_0\varphi_n(\xi(s))\,ds-\mathsf{E}_x\int_0^t \tilde{\mathsf{A}}_0\varphi_n(\xi(s))\,ds=0.\end{aligned}$$

Since

$$\lim_{t\to\infty} |\mathsf{E}_x \mathsf{G}\tilde{\mathsf{A}}_0\varphi_n(\xi(t))| = \lim_{t\to\infty} |\mathsf{E}_x \int_t^\infty \tilde{\mathsf{A}}_0\varphi_n(\xi(s)), ds|$$

$$= \mathsf{E}_x \lim_{t\to\infty} \int_t^\infty |\tilde{\mathsf{A}}_0\varphi_n(\xi(s))| \, ds = 0,$$

because $\tilde{\mathsf{A}}_0\varphi_n \in \mathfrak{L}_\beta(\mathscr{R}^m)$ and for any N $\overline{\lim}_{t\to\infty} |\mathsf{E}_x\varphi_n(\xi(t))| \leq \sup_{|y|\geq N} |\varphi_n(y)|$, in view of the second of conditions (52) and $\varphi_n \in \hat{\mathscr{C}}_{\mathscr{R}^m}$ it follows that $\varphi_n + \mathsf{G}\tilde{\mathsf{A}}_0\varphi_n = 0$. Hence

$$\sup_x |\varphi_n(x) - \mathsf{G}f(x)| \to 0.$$

Denote by $b_n(x)$ the function $B(x)\varphi_n'(x)$. Then

$$\varphi_n(\xi(t)) - \varphi_n(\xi(0)) - \int_0^t \tilde{\mathsf{A}}_0\varphi_n(\xi(s)) \, ds = \int_0^t (b_n(\xi(s)), dw(s)). \tag{56}$$

It follows from the uniform convergence of φ_n to $\mathsf{G}f$ and the inequality

$$\sup_x \mathsf{E}_x \left(\int_0^t |\tilde{\mathsf{A}}_0\varphi_n(\xi(s)) - \tilde{\mathsf{A}}_0\varphi_l(\xi(s))| \, ds\right)^2$$

$$\leq 2\left(\sup_x \mathsf{E}_x \int_0^t |\tilde{\mathsf{A}}_0\varphi_n(\xi(s)) - \tilde{\mathsf{A}}_0\varphi_l(\xi(s))| \, ds\right)^2$$

$$\leq 2K^2 \|\tilde{\mathsf{A}}_0\varphi_n - \tilde{\mathsf{A}}_0\varphi_l\|_\beta^2,$$

which is a corollary to Lemma 3 in Volume II, Chapter II, Section 6, that

$$\lim_{n,l\to\infty} \sup_x \mathsf{E}_x \int_0^t |b_n(\xi(s)) - b_l(\xi(s))|^2 \, ds = 0$$

uniformly in t. Hence

$$\lim_{n,l\to\infty} \sup_x \mathsf{E}_x \int_0^\infty |b_n(\xi(s)) - b_l(\xi(s))|^2 \, ds$$

$$= \lim_{n,l\to\infty} \sup_x \int |b_n(y) - b_l(y)|^2 g(x, y) \, dy = 0.$$

Set

$$b(y) = \sum_{k=1}^\infty (b_{n_{k+1}}(y) - b_{n_k}(y)) + b_{n_1}(y),$$

where n_k is chosen in such a manner that

$$K_1 = \sum_{k=1}^\infty \sqrt{\sup_x \int |b_{n_{k+1}}(y) - b_{n_k}(y)|^2 g(x, y) \, dy} < \infty.$$

Then

$$\sqrt{\int |b(y)|^2 g(x, y)\, dy} \leq \sqrt{\int |b_{n_1}(y)|^2 g(x, y)\, dy} + K_1,$$

and the condition of the theorem is fulfilled for $b(y)$. Since

$$\lim_{n\to\infty} \sup_x \int |b(y) - b_n(y)|^2 g(x, y)\, dy = 0,$$

it follows that

(57) $$\int_0^t (b_n(\xi(t)), dw(t)) \to \int_0^t (b(\xi(t)), dw(t))$$

in probability P_x for any x. Approaching the limit in (56) as $n \to \infty$ we complete the proof of the theorem. □

Remark. In the course of the proof the following assertion was established:

If $b_n(y)$ is a sequence of functions with values in $\mathscr{R}^m$ satisfying

$$\sup_x \int |b_n(y)|^2 g(x, y)\, dy < \infty$$

and

$$\lim_{l,n\to\infty} \sup_x \int |b_n(y) - b_l(y)|^2 g(x, y)\, dy = 0,$$

then there exists a function $b(y)$ such that (57) *is satisfied in the sense of convergence in probability P_x for any x.*

We shall now investigate the form of W-functionals on diffusion processes satisfying the conditions of Theorem 10 (the definition of a W-functional is presented in Volume II, Chapter II, Section 6).

First we shall verify some properties of functionals of the integral type.

Lemma 5. *Let $f(x)$ be an arbitrary measurable nonnegative function such that $f(x) = 0$ for $|x| \geq N$ and for some t_0*

$$\sup_x \mathsf{E}_x \int_0^{t_0} f(\xi(t))\, dt < \alpha < \infty.$$

Then for any open set G containing $\{x: |x| \leq N\}$ there exists a constant C such that

(58) $$\mathsf{E}_x \int_0^\infty f(\xi(t))\, dt \leq C \mathsf{E}_x \int_0^\infty \chi_G(\xi(t))\, dt;$$

moreover, C depends on α, t_0, and G only.

Proof. Let τ be an arbitrary Markov time. It follows from the inequality

$$\int_0^\tau f(\xi(t))\,dt \leqslant \sum_{k=0}^{\infty} \chi_{\{\tau > kt_0\}} \int_{kt_0}^{kt_0+t_0} f(\xi(s))\,ds$$

that

$$\mathsf{E}_x \int_0^\tau f(\xi(t))\,dt \leqslant \alpha \sum_{k=0}^{\infty} \mathsf{E}_x \chi_{\{\tau > kt_0\}} \leqslant \alpha(\mathsf{E}_x \tau + t_0).$$

Setting $\tau = \tau_G$ we verify that

$$\sup_x \mathsf{E}_x \int_0^{\tau_G} f(\xi(t))\,dt \leqslant \alpha\Big(\sup_x \mathsf{E}_x \tau_G + t_0\Big) \leqslant \bar{\alpha} < \infty.$$

Furthermore, we have

$$\inf_{|x|=N} \mathsf{E}_x \int_0^{\tau_G} \chi_G(\xi(t))\,dt \geqslant s_0 \cdot \inf_{|x|=N} \mathsf{P}_x\{\sup_{0\leqslant s\leqslant s_0} |\xi(s)-\xi(0)| < \delta\}, \tag{59}$$

where δ is the distance between $\{x: |x| = N\}$ and G'. The right-hand side of (59) is positive for s_0 sufficiently small since under our assumptions the process is uniformly stochastically continuous. Let

$$\inf_{|x|\leqslant N} \mathsf{E}_x \int_0^{\tau_G} \chi_G(\xi(t))\,dt = d_0.$$

Then we have for all x, $|x| = N$,

$$\mathsf{E}_x \int_0^{\tau_G} \chi_G(\xi(t))\,dt \geqslant d_0 \geqslant \frac{d_0}{\bar{\alpha}} \mathsf{E}_x \int_0^{\tau_G} f(\xi(t))\,dt. \tag{60}$$

Clearly inequality (60) remains valid also for $|x| \leqslant N$. Let ζ_1 be the first time after time τ_G at which $\{x: |x| = N\}$ was reached, ζ_1' be the first time after ζ_1 at which G' was reached, ζ_2 be the first time after ζ_1' at which $\{x: |x| = N\}$ was reached, ζ_2' be the first time after ζ_2 at which G' was reached, and so on. Then

$$\int_0^\infty f(\xi(t))\,dt = \int_0^{\tau_G} f(\xi(t))\,dt + \sum_{k=1}^{\infty} \int_{\zeta_k}^{\zeta_k'} f(\xi(t))\,dt.$$

However, since $|\xi(\zeta_k)| = N$,

$$\begin{aligned}\mathsf{E}_x \int_{\zeta_k}^{\zeta_k'} f(\xi(t))\,dt &\leqslant \mathsf{E}_x \mathsf{E}_{\xi(\zeta_k)} \int_0^{\tau_G} f(\xi(s))\,ds \\ &\leqslant \frac{\bar{\alpha}}{d_0} \mathsf{E}_x \mathsf{E}_{\xi(\zeta_k)} \int_0^{\tau_G} \chi_G(\xi(s))\,ds \\ &= \frac{\bar{\alpha}}{d_0} \mathsf{E}_x \int_{\zeta_k}^{\zeta_k'} \chi_G(\xi(t))\,dt.\end{aligned}$$

If $|x| \leqslant N$, then

$$\mathsf{E}_x \int_0^{\tau_G} f(\xi(t))\,dt \leqslant \frac{\bar{\alpha}}{d_0} \mathsf{E}_x \int_0^{\tau_G} \chi_G(\xi(t))\,dt.$$

If, however, $|x| > N$, then, denoting by τ the first time at which the set $\{x: |x| - N\}$ was reached, we have

$$\begin{aligned}\mathsf{E}_x \int_0^{\tau_G} f(\xi(t))\,dt &= \mathsf{E}_x \mathsf{E}_{\xi(\tau)} \int_0^{\tau_G} f(\xi(t))\,dt \\ &\leqslant \frac{\bar{\alpha}}{d_0} \mathsf{E}_x \mathsf{E}_{\xi(\tau)} \int_0^{\tau_G} \chi_G(\xi(t))\,dt \leqslant \frac{\bar{\alpha}}{d_0} \mathsf{E}_x \int_0^{\tau_G} \chi_G(\xi(t))\,dt.\end{aligned}$$

Thus

$$\begin{aligned}\mathsf{E}_x \int_0^{\infty} f(\xi(t))\,dt &\leqslant \frac{\bar{\alpha}}{d_0} \mathsf{E}_x \left[\int_0^{\tau_G} \chi_G(\xi(t))\,dt + \sum_{k=1}^{\infty} \int_{\zeta_k}^{\zeta_k''} \chi_G(\xi(t))\,dt\right] \\ &\leqslant \frac{\bar{\alpha}}{d_0} \mathsf{E}_x \int_0^{\infty} \chi_G(\xi(t))\,dt.\end{aligned}$$

The lemma is proved. □

Remark. Utilizing the fact that W-functionals can be approximated by functionals of an integral type we verify the following result.

If W-functional α_t possesses a bounded support F (F is a closed set), then there exists

$$\lim_{t \to \infty} \alpha_t = \alpha_\infty.$$

If $G \supset F$, where G is an open set, then there exists a constant C dependent on G, F, t_0, and $\sup_x \mathsf{E}_x \alpha_{t_0}$ such that

$$\mathsf{E}_x \alpha_\infty \leqslant C \mathsf{E}_x \int_0^{\infty} \chi_G(\xi(t))\,dt.$$

Let α_t be a W-functional with a bounded support F. Construct a sequence of measurable nonnegative functions $\varphi_1^{(n)}(x)$ such that the following conditions be satisfied:

a) $\varphi_1^{(n)}(x)$ are nonvanishing in a bounded domain only;
b) $\sup_t \sup_x \mathsf{E}_x \int_0^t \varphi_1^{(n)}(\xi(t))\,dt < \infty$;
c) $\int_0^t \varphi_1^{(n)}(\xi(t))\,dt \to \alpha_t$ in measure P_x for all x and t.

As function $\varphi_1^{(n)}(x)$ one can choose

$$n(1 - \mathsf{E}_x \exp\{-\alpha_{1/n}\})\chi_G(x).$$

Condition c) is satisfied in view of Theorem 1 in Volume II, Chapter II, Section 6.

Condition b) follows from the inequality

$$\mathsf{E}_x \int_0^t \varphi_1^{(n)}(\xi(t))\,dt \le \mathsf{E}_x n \int_0^t \mathsf{E}_{\xi(s)}\alpha_{1/n}\,ds$$
$$= \mathsf{E}_x n \int_0^t [\alpha_{s+(1/n)} - \alpha_s]\,ds = \mathsf{E}_x n \int_t^{t+1/n} \alpha_s\,ds \le \mathsf{E}_x \alpha_{t+1/n}.$$

Since in view of Lemma 3 in Volume II, Chapter II, Section 6, condition b) implies uniform boundedness of all the moments of

$$\int_0^t \varphi_1^{(n)}(\xi(s))\,ds$$

(in n and x), it follows from condition c) that for all β

$$\lim_{n\to\infty} \sup_x \mathsf{E}_x \left|\int_0^t \varphi_1^{(n)}(\xi(s))\,ds - \alpha_t\right|^\beta = 0.$$

Theorem 10 yields the existence of $\varphi_n(x)$ and $b_n(x)$ such that $\tilde{\mathsf{A}}\varphi_n(s) = \varphi_1^{(n)}(x)$ and (56) is fulfilled when $\tilde{\mathsf{A}}_0\varphi_n$ is replaced by $\tilde{\mathsf{A}}\varphi_n$. Therefore in view of Lemma 5

$$\mathsf{E}_x \int_0^t \varphi_1^{(n)}(\xi(s))\,ds = \mathsf{E}_x\varphi_n(\xi(t)) - \varphi_n(x)$$

and

$$\varphi_n(x) = -\int_0^\infty \mathsf{E}_x \varphi_1^{(n)}(\xi(t))\,dt \tag{61}$$

are uniformly bounded. Utilizing the uniform in n convergence of the integrals in the right-hand side of (61) we obtain that

$$\lim_{n\to\infty} \varphi_n(x) = -\mathsf{E}_x\alpha_\infty$$

uniformly in x. Therefore setting $\varphi(x) = -\mathsf{E}_x\alpha_\infty$ we obtain

$$\lim_{n\to\infty} \sup_x \mathsf{E}_x[\varphi_n(\xi(t)) - \varphi_n(\xi(0)) - \int_0^t \varphi_1^{(n)}(\xi(s))\,ds - (\varphi(\xi(t)) - \varphi(\xi(0)) - \alpha_t]^2 = 0.$$

Hence in view of the remark following Theorem 10 there exists a function $b(x)$ such that

$$\mathsf{E}_x[\int_0^t (b_n(\xi(s)), dw(s)) - \int_0^t (b(\xi(s)), dw(s))]^2 = 0.$$

Using the fact that for any W-functional α_t there exists a sequence of W-functionals $\alpha_t^{(n)}$ with bounded supports and a monotonically increasing sequence of domains G_n, $\bigcup G_n = \mathscr{R}^m$, such that $\alpha_t^{(n)} = \alpha_t$ for $t \le \tau_{G_n}$, we verify the following theorem:

Theorem 11. *Under the conditions of Theorem* 10 *there exist for each W-functional α_t measurable functions $\varphi(x)$ with values in $\mathscr{R}^1$ and $b(x)$ with values in $\mathscr{R}^m$ such* *that*

(62) $$\alpha_t = \varphi(\xi(t)) - \varphi(\xi(0)) + \int_0^t (b(\xi(s)), dw(s)).$$

§ 4. Continuous Homogeneous Markov Processes in $\mathscr{R}^m$

In this section we shall investigate the local structure of continuous Markov processes in $\mathscr{R}^m$. We shall therefore assume that process x_t is defined on an open set G possessing a compact closure and is cut off at the time ζ of the first exit out of G, and that there exists the limit

$$x_\zeta = \lim_{t \uparrow \zeta} x_t$$

which belongs to the boundary Γ of the domain G. Moreover, we shall assume that $\sup_x \mathsf{E}_x \zeta < \infty$ and the process is uniformly stochastically continuous, i.e., that

$$\lim_{t \downarrow 0} \sup_{x \in G} P(t, x, V_\varepsilon(x)) = 0,$$

where $V_\varepsilon(x) = \{y: |x-y| > \varepsilon\}$.

Since we study local properties, G can be chosen arbitrarily small and our assumptions are thus nonrestrictive.

Should we be able to describe generating operators of the processes under consideration we would at the same time describe characteristic operators of continuous processes in $\mathscr{R}^m$. Following that and utilizing the results of Volume II, Chapter II, Section 5, we would be able to investigate the form of generating operators of continuous processes in $\mathscr{R}^m$ as well.

Let P_x and E_x be the probability and the mathematical expectation corresponding to the process, $P(t, x, E)$ be the transition probability, $\mathsf{T}_t f(x) = \mathsf{E}_x f(x_t)$, and A be the quasi-generating operator of the process, i.e., the operator defined on the continuous functions f in G for which $\mathsf{A}f$ is bounded and

(1) $$\mathsf{T}_t f(x) = f(x) + \int_0^t \mathsf{T}_s \mathsf{A} f(x)\, ds.$$

It is easy to see that the expression

(2) $$\hat{f}_t = f(x_t) - f(x_0) - \int_0^t \mathsf{A} f(x_s)\, ds$$

for $f \in \mathscr{D}_\mathsf{A}$ (where $\mathscr{D}_\mathsf{A}$ is the domain of definition of A) is a martingale with respect to measure P_x for any $x \in G$. Moreover, $\hat{f}_t$ is a homogeneous additive continuous functional on process x_t (cf. Volume II, Chapter II, Section 6). This indicates that for the study of operator A is is quite important to describe those continuous, additive homogeneous functionals of the process which are also martingales. This is done in the next subsection.

The *M*-functionals. Before presenting a definition of the class of functionals under consideration we shall establish a property of the functional defined by (2).

Lemma 1. *If f is continuous on $G \cup \Gamma$, then there exists for a functional $\hat{f}_t$ defined by formula* (2) *a W-functional φ_t such that for all $x \in G$*

$$\mathsf{E}_x \hat{f}_t^2 = \mathsf{E}_x \varphi_t. \tag{3}$$

Proof. Denote

$$v(t, x) = \mathsf{E}_x \hat{f}_t^2.$$

Since $\hat{f}_t$ is a martingale

$$v(t+h, x) = \mathsf{E}_x \hat{f}_{t+h}^2 = \mathsf{E}_x \hat{f}_t^2 + \mathsf{E}_x [\hat{f}_{t+h} - \hat{f}_t]^2 = v(t, x) + \mathsf{E}_x v(h, x_t).$$

Thus $v(t, x)$ is a W-function (cf. Volume II, Chapter II, Section 6, p. 171). Furthermore,

$$\begin{aligned} v(t, x) &\leq 2\mathsf{E}_x[(f(x_t) - f(x_0))^2 + (\textstyle\int_0^t \mathsf{A} f(x_s)\, ds)^2] \\ &= 2\mathsf{E}_x[f^2(x_t) - f^2(x_0) + 2f(x_0)(f(x_0) - f(x_t)) + (\textstyle\int_0^t \mathsf{A} f(x_s)\, ds)^2] \\ &= 2\mathsf{T}_t f^2(x) - 2f^2(x) + O(t + t^2). \end{aligned}$$

In view of the uniform stochastic continuity of the process

$$\mathsf{T}_t f^2(x) \to f^2(x)$$

uniformly in x, we thus have

$$\lim_{t \downarrow 0} \sup_x v(t, x) = 0.$$

Therefore in accordance with Theorem 3 in Volume II, Chapter II, Section 6, there exists a unique W-functional φ_t such that relation (3) is satisfied. □

Definition. A continuous homogeneous additive functional α_t is called an *M-functional* if

a) $\mathsf{E}_x \alpha_t = 0$ for all x and t,

b) there exists a W-functional φ_t such that for all $t > 0$

$$\mathsf{E}_x \alpha_t^2 = \mathsf{E}_x \varphi_t.$$

Condition a) implies that α_t is a martingale.

Lemma 2. *If α_t and β_t are two M-functionals so is the sum $\alpha_t + \beta_t$.*

To prove this assertion it is sufficient to verify only condition b) of the definition of an M-functional. Since

$$\mathsf{E}_x(\alpha_t+\beta_t)^2 \leq 2\mathsf{E}_x\alpha_t^2+2\mathsf{E}_x\beta_t^2$$

and there exist W-functionals φ_t and ψ_t such that

$$\mathsf{E}_x\alpha_t^2=\mathsf{E}_x\varphi_t, \qquad \mathsf{E}_x\beta_t^2=\mathsf{E}_x\psi_t,$$

in view of Theorem 3 in Volume II, Chapter II, Section 6, the relation

$$\lim_{h\downarrow 0,\Delta\downarrow 0} \mathsf{E}_x\frac{1}{h}\int_0^t v(h, x_s)v(\Delta, x_s)\,ds \leq 2 \lim_{h\downarrow 0,\Delta\downarrow 0}\left(\mathsf{E}_x\frac{1}{h}\int_0^t \mathsf{E}_{x_s}\varphi_h\mathsf{E}_{x_s}\varphi_\Delta\,ds\right.$$
$$\left.+\mathsf{E}_x\frac{1}{h}\int_0^t \mathsf{E}_{x_s}\psi_h\mathsf{E}_{x_s}\varphi_\Delta\,ds\right)=0$$

is satisfied for the W-functional $v(t,x)=\mathsf{E}_x(\alpha_t+\beta_t)^2$. In this case, however, in view of the same theorem there exists a W-functional χ_t such that

$$v(t,x)=\mathsf{E}_x\chi_t.$$

In what follows we shall denote a W-functional φ_t for which $\mathsf{E}_x\varphi_t=\mathsf{E}_x\alpha_t^2$, where α_t is an M-functional, by $\langle\alpha,\alpha\rangle_t$; thus

$$\mathsf{E}_x\alpha_t^2=\mathsf{E}_x\langle\alpha,\alpha\rangle_t.$$

Remark. The function $\langle\alpha,\alpha\rangle_t$ is such that $\alpha_t^2-\langle\alpha,\alpha\rangle_t$ is a martingale with respect to measure P_x. The existence of such a function for each x follows from Theorem 9 in Section 1 of Chapter I. However, in the definition of an M-functional it is required that this function be independent of x and, moreover, be a W-functional.

Let α_t and β_t be two M-functionals. In view of Lemma 2, so is $\alpha_t+\beta_t$. Therefore the process

$$\tfrac{1}{2}[\langle\alpha+\beta,\alpha+\beta\rangle_t-\langle\alpha,\alpha\rangle_t-\langle\beta,\beta\rangle_t]=\langle\alpha,\beta\rangle_t \tag{4}$$

is also a continuous additive functional representable as a difference of two W-functionals. In what follows such functionals are called *$\tilde{W}$-functionals.*

Denote the set of M-functionals by Φ_M and introduce the following convergence on the space Φ_M: a sequence $\alpha_t^{(n)}$ is called *convergent to* α_t if for all t and x

$$\lim_{n\to\infty}\mathsf{E}_x(\alpha_t^{(n)}-\alpha_t)^2=0 \quad\text{and}\quad \sup_{n,x}\mathsf{E}_x(\alpha_t^{(n)})^2<\infty.$$

We now show that Φ_M becomes a complete space in the sense of this convergence. Indeed, assume that

$$\alpha_t^{(n)} \in \Phi_M, \qquad \lim_{n,m\to\infty} \mathsf{E}_x(\alpha_t^{(n)} - \alpha_t^{(m)})^2 = 0.$$

Let $n_k(x)$ be defined by relation

$$\sup_{n,m \geq n_k(x)} \mathsf{E}_x(\alpha_1^{(n)} - \alpha_1^{(m)})^2 \leq 2^{-k},$$

and

$$\alpha_t(x) = \lim_{k\to\infty} \alpha_t^{(n_k(x))}.$$

This limit exists for every x with probability $\mathsf{P}_x = 1$. Set $\alpha_t = \alpha_t(x_0)$. Since $\alpha_t = \alpha_t(x)$ with probability $\mathsf{P}_x = 1$ we have for all x

$$\text{(5)} \qquad \lim_{n\to\infty} \mathsf{E}_x(\alpha_t^{(n)} - \alpha_t)^2 = 0.$$

Hence $\mathsf{E}_x\alpha_t = 0$. We now show that α_t is an additive almost homogeneous functional. It is sufficient to verify that

$$\text{(6)} \qquad \theta_h\alpha_t = \alpha_{t+h} - \alpha_h$$

with probability $\mathsf{P}_x = 1$ for any x. Clearly

$$\theta_h\alpha_t = \lim_{k\to\infty} [\alpha_{t+h}^{(n_k(x_h))} - \alpha_h^{(n_k(x_h))}];$$

however,

$$\mathsf{E}_x[\alpha_{t+h}^{(n_k(x_h))} - \alpha_h^{(n_k(x_h))} - \alpha_{t+h}^{(n_k(x))} + \alpha_h^{(n_k(x))}]^2 = \mathsf{E}_x\mathsf{E}([\alpha_t^{(n_k(x_h))} - \alpha_t^{(n_k(x))}]^2/x_h) \to 0$$

as $k\to\infty$ since $n_k(x_h)$ and $n_k(x)\to\infty$.

Moreover, $\alpha_{t+h}^{(n_k(x))} - \alpha_t^{(n_k(x))} \to \alpha_{t+h} - \alpha_t$ in measure P_x. Thus assertion (6) is verified.

It follows from (5) that

$$\mathsf{P}_x\{\sup_{s\leq t} |\alpha_s^{(n)} - \alpha_s| > \varepsilon\} \leq \frac{1}{\varepsilon^2}\mathsf{E}_x(\alpha_s^{(n)} - \alpha_s)^2 \to 0;$$

hence α_t is a continuous functional. To prove the existence of $\langle\alpha, \alpha\rangle_t$ the following lemma, which is of interest in its own right, will be required.

Lemma 3. *Let* $0 = t_{n0} < t_{n1} < \cdots < t_{nn} = t$ *and* $\max_k (t_{n_k} - t_{n_{k-1}}) \to 0$ *as* $n \to \infty$. *Then*

(7) $$\langle \alpha, \alpha \rangle_t = \lim_{n\to\infty} \sum_{k=0}^{n-1} (\alpha_{t_{nk+1}} - \alpha_{t_{nk}})^2$$

in the sense of convergence in probability P_x *for any* $x \in G$.

Proof. We shall prove the stronger assertion: for all $x \in G$

(8) $$\lim_{n\to\infty} \mathsf{E}_x\left(\langle \alpha, \alpha \rangle_t - \sum_{k=0}^{n-1} (\alpha_{t_{nk+1}} - \alpha_{t_{nk}})^2\right)^2 = 0.$$

If

$$\eta_{nk} = (\alpha_{t_{nk+1}} - \alpha_{t_{nk}})^2 - \langle \alpha, \alpha \rangle_{t_{nk+1}} + \langle \alpha, \alpha \rangle_{t_{nk}},$$

then

$$\mathsf{E}_x(\eta_{nk} \mid \mathfrak{F}_{t_{nk}}) = \mathsf{E}_{x_{t_{nk}}} \alpha^2_{t_{nk+1} - t_{nk}} - \mathsf{E}_{x_{t_{nk}}} \langle \alpha, \alpha \rangle_{t_{nk+1} - t_{nk}} = 0,$$

where $\mathfrak{F}_t$ is the σ-algebra generated by x_s, $s \leq t$.

Therefore

$$\mathsf{E}_x\left(\sum_{k=0}^{n-1} \eta_{nk}\right)^2 = \mathsf{E}_x \sum_{k=0}^{n-1} \eta^2_{nk}.$$

To prove (8) it is sufficient to verify that

(9) $$\mathsf{E}_x \sum_{k=0}^{n-1} \eta^2_{nk} \to 0.$$

The continuity of α_t and that of $\langle \alpha, \alpha \rangle_t$ imply that

$$\sup_k |\eta_{nk}| \to 0$$

as $n \to \infty$. Furthermore, the variable

$$\sum_k |\eta_{nk}| \leq \langle \alpha, \alpha \rangle_t + \sum_{k=0}^{n-1} (\alpha_{t_{nk+1}} - \alpha_{t_{nk}})^2$$

is bounded in probability since

$$\mathsf{E}_x \sum_{k=0}^{n-1} |\eta_{nk}| \leq 2\mathsf{E}_x \langle \alpha, \alpha \rangle_t.$$

Thus (9) will follow once the uniform integrability of

$$\sum_{k=0}^{n-1} \eta_{nk}^2$$

is established. To verify the latter we bound the expectation

$$\mathsf{E}_x\left(\sum_{k=0}^{n-1} \eta_{nk}^2\right)^2.$$

Clearly,

$$\begin{aligned}\mathsf{E}_x\left(\sum_{k=0}^{n-1} \eta_{nk}^2\right)^2 &= \mathsf{E}_x\left(\sum_{k=0}^{n-1} |\eta_{nk}|\right)^4 \\ &\leq \mathsf{E}_x\left(\langle \alpha, \alpha\rangle_t + \sum_{k=0}^{n-1} (\alpha_{t_{nk+1}} - \alpha_{t_{nk}})^2\right)^4 \\ &\leq 8\mathsf{E}_x\langle \alpha, \alpha\rangle_t^4 + 8\mathsf{E}_x\left(\sum_{k=0}^{n-1} (\alpha_{t_{nk+1}} - \alpha_{t_{nk}})^2\right)^4.\end{aligned}$$

Lemma 3 in Volume II, Chapter II, Section 6, assures us that $\mathsf{E}_x\langle \alpha, \alpha\rangle_t^4$ is finite. To show that the second summand is bounded we estimate the probability

$$Q_x(\lambda) = \mathsf{P}_x\left\{\sum_{k=0}^{n-1} (\alpha_{t_{nk+1}} - \alpha_{t_{nk}})^2 > \lambda\right\}.$$

Let

$$\xi_k = \begin{cases} (\alpha_{t_{nk+1}} - \alpha_{t_{nk}})^2 & \text{if } |\alpha_{t_{nk+1}} - \alpha_{t_{nk}}| \leq \sqrt{c}, \\ 0 & \text{if } |\alpha_{t_{nk+1}} - \alpha_{t_{nk}}| > \sqrt{c}. \end{cases}$$

Then

$$Q_x(2rc) \leq \mathsf{P}_x\{\sup_k |\alpha_{t_{nk+1}} - \alpha_{t_{nk}}| > \sqrt{c}\} + \mathsf{P}_x\left\{\sum_{k=0}^{n-1} \xi_k > 2rc\right\}.$$

However,

$$\begin{aligned}\mathsf{P}_x\left\{\sum_{k=0}^{n-1} \xi_k > 2rc\right\} &\leq \sum_l \mathsf{P}_x\left\{\sum_0^l \xi_k \leq 2(r-1)c,\ \sum_0^{l+1} \xi_k > 2(r-1)c,\ \sum_{l+2}^{n-1} \xi_k > c\right\} \\ &\leq \sum_l \mathsf{P}_x\left\{\sum_0^l \xi_k \leq 2(r-1)c,\ \sum_0^{l+1} \xi_k > 2(r-1)c\right\} \\ &\quad \times \sup_{y \in G} \mathsf{P}_y\left\{\sum_{l+2}^{n-1} \xi_k > c\right\} \\ &\leq \sup_{y \in G} \mathsf{P}_y\left\{\sum_0^{n-1} \xi_k > c\right\} \mathsf{P}_x\left\{\sum_{k=0}^{n-1} \xi_k > 2(r-1)c\right\}.\end{aligned}$$

Therefore

$$\mathsf{P}_x\left\{\sum_{k=0}^{n-1} \xi_k > 2rc\right\} \leqslant \left(\sup_{y\in G} \mathsf{P}_y\left\{\sum_{k=0}^{n-1} \xi_k > c\right\}\right)^r \leqslant \frac{1}{c^r}(\sup_{y\in G} \mathsf{E}_y\langle\alpha, \alpha\rangle_t)^r.$$

Furthermore,

$$\sup_k |\alpha_{t_{nk+1}} - \alpha_{t_{nk}}| \leqslant 2 \sup_{u\leqslant t} |\alpha_u|.$$

Since α_u is a martingale we have in view of formula (18) in Volume II, p. 57, that

$$\mathsf{P}_x\{2 \sup_{u\leqslant t} |\alpha_u| > \sqrt{c}\} \leqslant \frac{2^{2r}\mathsf{E}_x\alpha_t^{2r}}{c^r}.$$

(Actually a continuous analog of formula (18) is used; its validity is discussed in Volume I, p. 185.)

We now show that $\mathsf{E}_x\alpha_t^{2r}$ is finite for all r and is bounded uniformly in x. Let ρ be chosen in such a manner that

$$\sup_x \mathsf{P}_x\{\sup_{u\leqslant t} |\alpha_u| > \rho\} \leqslant \frac{1}{\rho^2} \sup_x \mathsf{E}_x\langle\alpha, \alpha\rangle_t \leqslant \frac{1}{2}.$$

Denoting by ζ the Markov moment for which $\sup_{u\leqslant\zeta} |\alpha_u| = \lambda - \rho$ for the first time we obtain

$$\mathsf{P}_x\{\sup_{u\leqslant t} |\alpha_u| > \lambda\} = \mathsf{P}_x\{\zeta < t, \sup_{u\leqslant t} |\alpha_u - \alpha_\zeta| > \rho\} \leqslant \tfrac{1}{2}\mathsf{P}_x\{\sup_{u\leqslant t} |\alpha_u| > \lambda - \rho\}.$$

Hence

$$\mathsf{P}_x\{\sup_{u\leqslant t} |\alpha_u| > k\rho\} \leqslant \frac{1}{2^k}.$$

This inequality implies the existence of uniformly bounded moments of α_t of all orders. Thus

$$Q_x(2rc) \leqslant c^{-r} \sup_{y\in G} (\mathsf{E}_y\langle\alpha, \alpha\rangle_t)^r - 2^{2r}c^{-r} \sup_{y\in G} \mathsf{E}_y\alpha_t^{2r}.$$

From here we obtain for any r the existence of a constant A_r such that

$$Q_x(c) \leqslant \frac{A_r}{c^r}.$$

Whence

$$\mathsf{E}_x\left(\sum_{k=0}^{n-1} (\alpha_{t_{nk+1}} - \alpha_{t_{nk}})^2\right)^4$$

is uniformly bounded in n. The proof of the lemma is complete. □

Corollary. *If $t_{n0} < t_{n1} < \cdots < t_{nn} = t$ and $\max_k (t_{nk+1} - t_{nk}) \to 0$ then*

$$\langle \alpha, \beta \rangle_t = \lim_{n\to\infty} (\alpha_{t_{nk+1}} - \alpha_{t_{nk}})(\beta_{t_{nk+1}} - \beta_{t_{nk}})$$

in the mean square with respect to measure P_x for all x.

We now show that if α_t is a functional such that (5) is fulfilled with $\alpha_t^{(n)} \in \Phi_M$, then $\alpha_t \in \Phi_M$ also. For this purpose it is sufficient to verify the existence of the W-functional $\langle \alpha, \alpha \rangle_t$. Utilizing the relation

$$\langle \alpha^{(n)}, \alpha^{(n)} \rangle_t = \lim_{m\to\infty} \sum_{k=0}^{m-1} (\alpha_{t_{mk+1}}^{(n)} - \alpha_{t_{mk}}^{(n)})^2$$

in the sense of mean square convergence with respect to measure P_x we may write

$$\begin{aligned}
&\mathsf{E}_x |\langle \alpha^{(n)}, \alpha^{(n)} \rangle_t - \langle \alpha^{(p)}, \alpha^{(p)} \rangle_t| \\
&\leq \lim_{m\to\infty} \mathsf{E}_x \left| \sum_{k=0}^{m-1} ([\alpha_{t_{mk+1}}^{(n)} - \alpha_{t_{mk}}^{(n)}]^2 - [\alpha_{t_{mk+1}}^{(p)} - \alpha_{t_{mk}}^{(p)}]^2) \right| \\
&\leq \lim_{m\to\infty} \Bigg\{ \mathsf{E}_x \sum_{k=0}^{m-1} [\alpha_{t_{mk+1}}^{(n)} - \alpha_{t_{mk}}^{(n)} - \alpha_{t_{mk+1}}^{(p)} - \alpha_{t_{mk}}^{(p)}]^2 \qquad (10) \\
&\qquad \times \mathsf{E}_x \sum_{k=0}^{m-1} [\alpha_{t_{mk+1}}^{(n)} - \alpha_{t_{mk}}^{(n)} + \alpha_{t_{mk+1}}^{(p)} - \alpha_{t_{mk}}^{(p)}]^2 \Bigg\}^{1/2} \\
&= \{\mathsf{E}_x [\alpha_t^{(n)} - \alpha_t^{(p)}]^2 \mathsf{E}_x [\alpha_t^{(n)} + \alpha_t^{(p)}]^2\}^{1/2}.
\end{aligned}$$

Consequently,

$$\lim_{n,p\to\infty} \mathsf{E}_x |\langle \alpha^{(n)}, \alpha^{(n)} \rangle_t - \langle \alpha^{(p)}, \alpha^{(p)} \rangle_t| = 0.$$

Therefore the limit of the quantities $\langle \alpha^{(n)}, \alpha^{(n)} \rangle_t$ exists for each x with respect to measure P_x. It can be shown in the same manner as in the case of α_t that there exists a nonnegative additive almost homogeneous functional $\langle \alpha, \alpha \rangle_t$ such that for all x

$$\lim_{n\to\infty} \mathsf{E}_x |\langle \alpha^{(n)}, \alpha^{(n)} \rangle_t - \langle \alpha, \alpha \rangle_t| = 0.$$

Approaching the limit in relation

$$\mathsf{E}_x \langle \alpha^{(n)}, \alpha^{(n)} \rangle_t = \mathsf{E}_x (\alpha_t^{(n)})^2$$

we verify that

$$\mathsf{E}_x \langle \alpha, \alpha \rangle_t = \mathsf{E}_x \alpha_t^2.$$

It remains to show that $\langle\alpha, \alpha\rangle_t$ is continuous. Let τ be a Markov moment such that $\tau \leq t$. In the same manner as relation (10) was obtained, one can derive the bound

$$\text{(11)} \quad \mathsf{E}_x|\langle\alpha^{(n)}, \alpha^{(n)}\rangle_\tau - \langle\alpha^{(p)}, \alpha^{(p)}\rangle_\tau| \leq \{\mathsf{E}_x|\alpha_t^{(n)} - \alpha_t^{(p)}|^2 |\mathsf{E}_x|\alpha_t^{(n)} - \alpha_t^{(p)}|^2\}^{1/2}.$$

Let

$$\tau = \begin{cases} t & \text{if } \sup\limits_{s \leq t} |\langle\alpha^{(n)}, \alpha^{(n)}\rangle_s - \langle\alpha^{(p)}, \alpha^{(p)}\rangle_s| \leq \varepsilon, \\ \inf\,[s: |\langle\alpha^{(n)}, \alpha^{(n)}\rangle_s - \langle\alpha^{(p)}, \alpha^{(p)}\rangle_s| > \varepsilon] & \text{otherwise.} \end{cases}$$

Then

$$\varepsilon\, \mathsf{P}_x\{\sup_{s \leq t} |\langle\alpha^{(n)}, \alpha^{(n)}\rangle_s - \langle\alpha^{(p)}, \alpha^{(p)}\rangle_s| > \varepsilon\} \leq \mathsf{E}_x|\langle\alpha^{(n)}, \alpha^{(n)}\rangle_\tau - \langle\alpha^{(p)}, \alpha^{(p)}\rangle_\tau|.$$

Hence $\langle\alpha^{(n)}, \alpha^{(n)}\rangle_t$ converges to $\langle\alpha, \alpha\rangle_t$ uniformly with respect to t in probability P_x. This implies that $\langle\alpha, \alpha\rangle_t$ is continuous. Consequently $\langle\alpha, \alpha\rangle_t$ is a W-functional and we have proved

Theorem 1. *The space Φ_M of M-functionals is a linear space which is complete in the sense of convergence: $\alpha_t^{(n)} \to \alpha_t$, provided*

$$\lim_{n\to\infty} \mathsf{E}_x(\alpha_t^{(n)} - \alpha_t)^2 = 0 \quad \text{and} \quad \sup_{x,n} \mathsf{E}_x(\alpha^{(n)})^2 < \infty$$

for all $x \in G$.

Below, we shall consider also stochastic integrals over M-functionals. Utilizing the construction in Section 2 of Chapter I, the integral $\int_0^t f(s)\, d\alpha_s$ can be defined for a function f adopted to the current $\{\mathfrak{F}_s\}$ and satisfying the condition $\int_0^t f^2(s)\, d\langle\alpha, \alpha\rangle_s < \infty$.

We shall now discuss a narrower class of stochastic integrals of the form

$$\text{(12)} \qquad \int_0^t g(x_s)\, d\alpha_s,$$

where $g(x)$ is a Borel function such that

$$\text{(13)} \qquad \sup_x \mathsf{E}_x \int_0^t g^2(x_s)\, d\langle\alpha, \alpha\rangle_s < \infty.$$

Under this assumption the integral (12) can also be viewed as an M-functional (more precisely there exists an M-functional stochastically equivalent to this stochastic integral with respect to any measure P_x for all $x \in G$).

We shall first assume that $g(x)$ is a continuous function in x.

In this case it follows from the definition of an integral that

$$\int_0^t g(x_s)\,d\alpha_s = \lim_{h\downarrow 0} \sum_{kh<t} g(x_{kh})[\alpha^t_{kh+h} - \alpha_{kh}], \qquad \alpha^t_s = \alpha_{t\wedge s},$$

in the mean square with respect to measure P_x for any x. (This is because

$$\lim_{h\to 0} \mathsf{E}_x\Big[\int_0^t g(x_s)\,d\alpha_s - \sum_{kh<t} g(x_{kh})[\alpha^t_{kh+h} - \alpha_{kh}]\Big]^2 = 0.)$$

Since $\langle\alpha, \alpha\rangle_t$ is a W-functional it follows that $\mathsf{E}_x\langle\alpha, \alpha\rangle_t \leqslant c_1 + c_2 t$ for some constants c_1 and c_2. Therefore

$$\mathsf{E}_x\Big[\int_0^t g(x_s)\,d\alpha_s - \sum_{kh<t} g(x_{kh})[\alpha^t_{kh+h} - \alpha_{kh}]\Big]^2 \leqslant 2\|g\|^2(c_1 + c_2 t). \tag{14}$$

There exists a positive $h_l(x)$ such that for all x and l and $h < h_l(x)$

$$\sum_{N=1}^{\infty} \frac{1}{N^4} \mathsf{E}_x\Big[\int_0^N g(x_s)\,d\alpha_s - \sum_{kh<N} g(x_{kh})[\alpha^N_{kh+h} - \alpha_{kh}]\Big]^2 \leqslant 2^{-l}. \tag{15}$$

Relationship (15) implies that with probability $\mathsf{P}_x = 1$ we have uniformly in t on every bounded set

$$\int_0^t g(x_s)\,d\alpha_s = \lim_{l\to\infty} \sum_{kh_l(x)<t} g(x_{kh_l(x)})[\alpha^t_{(k+1)h_l(s)} - \alpha^t_{kh_l(x)}]. \tag{16}$$

Denote by $I_{x_0}(t)$ the limit appearing in the right-hand side of (16). Then one can verify in the same manner as in the proof of Theorem 1 that $I_{x_0}(t)$ is an additive almost homogeneous functional. Since

$$\mathsf{E}_x(I_{x_0}(t))^2 = \mathsf{E}_x\Big(\int_0^t g(x_s)\,d\alpha_s\Big)^2 = \mathsf{E}_x \int_0^t g^2(x_s)\,d\langle\alpha, \alpha\rangle_s$$

and the function

$$\langle I_{x_0}, I_{x_0}\rangle_t = \int_0^t g^2(x_s)\,d\langle\alpha, \alpha\rangle_s$$

is a W-functional while $\mathsf{E}_x I_{x_0}(t) = 0$ it follows that $I_{x_0}(t)$ is an M-functional.

Now let $g(x)$ be a Borel function such that (13) is fulfilled and there exists a sequence of continuous functions $g_n(x)$ such that

$$\begin{aligned} &\lim_{n\to\infty} \mathsf{E}_x \int_0^t [g(x_s) - g_n(x_s)]^2\,d\langle\alpha, \alpha\rangle_s = 0, \\ &\sup_{x,n} \mathsf{E}_x \int_0^t g_n^2(x_s)\,d\langle\alpha, \alpha\rangle_s < \infty. \end{aligned} \tag{17}$$

Then the sequence of M-functionals equivalent to

$$\int_0^t g_n(x_s)\,d\alpha_s$$

converges to an M-functional. In what follows this M-functional is considered to be the integral

$$\int_0^t g(x_s)\, d\alpha_s.$$

We now show that given an arbitrary bounded Borel function $g(x)$ there exists an M-functional $\alpha_t(g)$ such that

$$\alpha_t(g) = \int_0^t g(x_s)\, d\alpha_s$$

with probability $\mathsf{P}_x = 1$ for all x.

Denote by $\mathcal{F}$ the set of functions g for which the corresponding M-functional exists. The set $\mathcal{F}$ contains all continuous functions and given that a bounded convergent sequence of functions g_n belongs to $\mathcal{F}$, its limit also belongs to $\mathcal{F}$ (since the bounded convergence of g_n to g implies that (17) is fulfilled). Thus $\mathcal{F}$ contains all bounded Borel functions.

If g is an arbitrary Borel function satisfying (13), by setting

$$g^N(x) = \begin{cases} g(x) & \text{for } |g(x)| < N \\ N \operatorname{sgn} g(x) & \text{for } |g(x)| \geq N \end{cases}$$

we have

$$\lim_{N\to\infty} \mathsf{E}_x \int_0^t [g(x_s) - g^N(x_s)]^2\, d\langle \alpha, \alpha\rangle_s = 0$$

since $g(x) - g^N(x) \to 0$ for all x and $|g(x) - g^N(x)| \leq |g(x)|$. Therefore there exists in Φ_M the limit

$$\lim_{N\to\infty} \int_0^t g^N(x_s)\, d\alpha_s,$$

i.e., $g \in \mathcal{F}$. We have thus proved

Theorem 2. *Let $\alpha_t \in \Phi_M$. For any Borel function $g(x)$ satisfying condition* (13) *there exists an M-functional $\alpha_t(g)$ such that with probability* $\mathsf{P}_x = 1$

(18) $$\alpha_t(g) = \int_0^t g(x_s)\, d\alpha_s$$

for all $t > 0$.

From now on, this M-functional will be interpreted as the stochastic integral appearing in the right-hand side of (18).

Remark 1.

$$\langle \alpha(g), \alpha(g)\rangle_t = \int_0^t g^2(x_s)\, d\langle \alpha, \alpha\rangle_s,$$
$$\langle \alpha(g_1), \alpha(g_2)\rangle_t = \int_0^t g_1(x_s) g_2(x_s)\, d\langle \alpha, \alpha\rangle_s.$$

Remark 2. If g_1 and g_2 are Borel functions such that

$$\int_0^t g_1^2(x_s) g_2^2(x_s)\, d\langle\alpha, \alpha\rangle_s < \infty,$$

then

$$\int_0^t g_1(x_s)\, d\alpha_s(g_2) = \int_0^t g_1(x_s) g_2(x_s)\, d\alpha_s.$$

In particular, if $g_1 > 0$ then

$$\alpha_t = \int_0^t \frac{1}{g_1(x_s)}\, d\alpha_s(g_1).$$

Differentiation of *M*-functionals. Below, the following auxiliary assertions will be needed.

Lemma 4. *Let φ_t and γ_t be W-functionals. If there exists a measurable function $g(s)$ such that for all x with probability $\mathsf{P}_x = 1$*

$$\gamma_t = \int_0^t g(s)\, d\varphi_s,$$

then there exists a Borel function $\hat{g}(x)$ such that

$$\mathsf{P}_x\{\gamma_t = \int_0^t \hat{g}(x_s)\, d\varphi_s\} = 1$$

for all x.

Proof. Since φ_s and γ_s are continuous monotone functions for all ω, setting

$$\bar{g}(s) = \varliminf_{h \downarrow 0} \frac{\gamma_{s+h} - \gamma_s}{\varphi_{s+h} - \varphi_s}$$

we obtain $\bar{g}(s) = g(s)$ for almost all s (in measure $d\varphi_s$). Hence

$$\gamma_t = \int_0^t \bar{g}(s)\, d\varphi_s.$$

Set

$$\hat{g}(x) = \mathsf{E}_x \varliminf_{h \downarrow 0} \frac{\gamma_h}{\varphi_h} = \mathsf{E}_x \bar{g}(0).$$

Since functionals γ and φ are homogeneous

$$\mathsf{E}(\bar{g}(s) | \mathfrak{F}_s) = \mathsf{E}(\theta_s \bar{g}(0) | \mathfrak{F}_s) = \mathsf{E}_{x_s} \bar{g}(0) = \hat{g}(x_s).$$

The weak measurability of the (underlying) process implies that $\hat{g}(x)$ is Borel measurable.

Finally note that in view of Lemma 2 in Volume II, Chapter II, Section 6,

$$\mathsf{P}_x\{\bar{g}(0)=\hat{g}(x)\}=1.$$

Therefore

$$\mathsf{P}_x\{\bar{g}(s)=\theta_s\bar{g}(0)=\theta_s\bar{g}(x_0)=\hat{g}(x_s)\}=1$$

and the lemma is proved. □

Lemma 5. *If φ_t and γ_t are two W-functionals, then*

$$\gamma_t=\gamma_t^{(1)}+\gamma_t^{(2)},$$

where $\gamma_t^{(1)}$ and $\gamma_t^{(2)}$ are W-functionals and, moreover, there exists a Borel function $\hat{g}(x)$ such that

$$\gamma_t^{(1)}=\int_0^t \hat{g}(x_s)\,d\varphi_s,$$

while $\gamma_t^{(2)}$ is a singular function of t relative to φ_t for almost all ω with respect to measure P_x.

Proof. The function $\hat{g}(x)$ is constructed in the same manner as in Lemma 4. Then the absolutely continuous component of γ_t relative to φ_t will be

$$\gamma_t^{(1)}=\int_0^t \hat{g}(x_s)\,d\varphi_s\leq\gamma_t.$$

Therefore $\gamma_t^{(1)}$ is a W-functional: $\mathsf{E}_x\gamma_t^{(1)}\leq\mathsf{E}_x\gamma_t$. Evidently the singular component $\gamma_t^{(2)}=\gamma_t-\gamma_t^{(1)}$ will be nonnegative and being a difference of two additive homogeneous continuous functionals will also be such a functional. Since $\gamma_t^{(2)}\leq\gamma_t$, $\gamma_t^{(2)}$ is a W-functional as well. Moreover, by construction $\gamma_t^{(2)}$ is singular relative to φ_t. The lemma is thus proved. □

Corollary 1. *If γ_t is a $\tilde{W}$-functional and φ_t is a W-functional then there exists a Borel function $\hat{g}(x_s)$ such that*

$$\gamma_t=\int_0^t \hat{g}(x_s)\,d\varphi_s+\gamma_t^{(2)}\quad \textit{and}\quad \int_0^t|\hat{g}(x_s)|\,d\varphi_s<\infty,$$

where $\gamma_t^{(2)}$ is also a $\tilde{W}$-functional singular relative to φ_t. If, moreover, γ_t is absolutely continuous relative to φ_t, then

$$\gamma_t=\int_0^t \hat{g}(x_s)\,d\varphi_s.$$

Theorem 3. *Let α_t and β_t be M-functionals. Then there exists a Borel function $g(x)$ such that*

$$\int_0^t|g(x_s)|\,d\langle\alpha,\alpha\rangle_s<\infty$$

and

(19) $$\langle \alpha, \beta \rangle_t = \int_0^t g(x_s)\, d\langle \alpha, \alpha \rangle_s.$$

Proof. In view of the corollary to the preceding lemma, to prove the theorem it is sufficient to verify that $\langle \alpha, \beta \rangle_t$ is absolutely continuous relative to $\langle \alpha, \alpha \rangle_t$.

Let $(c_1, d_1), (c_2, d_2), \ldots, (c_k, d_k)$ be a system of disjoint intervals. In view of the corollary to Lemma 3, we have

$$\sum_{j=1}^k |\langle \alpha, \beta \rangle_{d_j} - \langle \alpha, \beta \rangle_{c_j}| = \lim_{n\to\infty} \sum_{j=1}^k \left| \sum_{i=0}^{n-1} (\alpha_{t_{i+1}^{(j)}} + \alpha_{t_i^{(j)}})\beta_{t_{i+1}^{(j)}} - \beta_{t_i^{(j)}}) \right|,$$

where $t_i^{(j)} = c_j + (i/n)(d_j - c_j)$. Therefore

$$\begin{aligned}\sum_{j=1}^k |\langle \alpha, \beta \rangle_{d_j} - \langle \alpha, \beta \rangle_{c_j}| &\leq \lim_{n\to\infty} \sqrt{\sum_{j=1}^k \sum_{i=0}^{n-1} (\alpha_{t_{i+1}^{(j)}} - \alpha_{t_i^{(j)}})^2} \\ &\quad \times \sqrt{\sum_{j=1}^k \sum_{i=0}^{n-1} (\beta_{t_{i+1}^{(j)}} - \beta_{t_i^{(j)}})^2} \\ &\leq \sqrt{\sum_{j=1}^k [\langle \alpha, \alpha \rangle_{d_j} - \langle \alpha, \alpha \rangle_{c_j}] \sum_{j=1}^k [\langle \beta, \beta \rangle_{d_j} - \langle \beta, \beta \rangle_{c_j}]}.\end{aligned}$$

The last inequality yields that for any Borel set Λ on the line

$$\left|\int_\Lambda d\langle \alpha, \beta \rangle_t\right| \leq \sqrt{\int_\Lambda d\langle \alpha, \alpha \rangle_t \int_\Lambda d\langle \beta, \beta \rangle_t}.$$

Thus $\langle \alpha, \beta \rangle_t$ is absolutely continuous relative to $\langle \alpha, \alpha \rangle_t$. The theorem is proved. □

From now on the function $g(x)$ satisfying (19) will be denoted by $(\partial\beta/\partial\alpha)(x)$. Thus

(20) $$\langle \alpha, \beta \rangle_t = \int_0^t \frac{\partial \beta}{\partial \alpha}(x_s)\, d\langle \alpha, \alpha \rangle_s.$$

Two M-functionals α and β are called *orthogonal* if $\langle \alpha, \beta \rangle_t = 0$ with probability $\mathsf{P}_x = 1$ for all x and t.

The equality

$$\mathsf{E}_x \alpha_t \beta_t = \mathsf{E}_x \langle \alpha, \beta \rangle_t$$

implies that the condition $\mathsf{E}_x \alpha_t \beta_t = 0$ is a necessary condition for the orthogonality of α_t and β_t. We now show that it is also a sufficient one. Let the condition be satisfied. Then

$$\mathsf{E}_x(\alpha_t + \beta_t)^2 = \mathsf{E}_x \alpha_t^2 + \mathsf{E}_x \beta_t^2 = \mathsf{E}_x \langle \alpha, \alpha \rangle_t + \mathsf{E}_x \langle \beta, \beta \rangle_t.$$

Hence in view of the uniqueness of the W-functional corresponding to a given M-functional

$$\langle\alpha+\beta, \alpha+\beta\rangle_t = \langle\alpha, \alpha\rangle_t + \langle\beta, \beta\rangle_t;$$

whence $\langle\alpha, \beta\rangle_t = 0.$

Lemma 6. *Let $g_1(x)$ and $g_2(x)$ be two functions for which the stochastic integrals*

$$\alpha_t(g_1) = \int_0^t g_1(x_s)\, d\alpha_s, \qquad \beta_t(g_2) = \int_0^t g_2(x_s)\, d\beta_s$$

exist. Then

$$\langle\alpha(g_1), \beta(g_2)\rangle_t = \int_0^t g_1(x_s)g_2(x_s)\, d\langle\alpha, \beta\rangle_s. \tag{21}$$

Proof. Assume first that α and β are orthogonal, and show that $\alpha_t(g_1)$ and $\beta_t(g_2)$ are also orthogonal. Consider, to begin with, the case of continuous g_1 and g_2. Then

$$\mathsf{E}_x\alpha_t(g_1)\beta_t(g_2) = \lim \mathsf{E}_x \sum g_1(x_{s_k})[\alpha_{s_{k+1}} - \alpha_{s_k}] \sum g_2(x_{s_k})[\beta_{s_{k+1}} - \beta_{s_k}] = 0$$

$$0 = s_0 < s_1 < \cdots < s_n = t, \qquad \max(s_{k+1} - s_k) \to 0),$$

since

$$\mathsf{E}_x([\alpha_{s_{k+1}} - \alpha_{s_k}][\beta_{s_{k+1}} - \beta_{s_k}] \mid \mathfrak{F}_{s_k}) = \mathsf{E}_x \mathsf{E}_{x_{s_k}} \alpha_{s_{k+1}-s_k}\beta_{s_{k+1}-s_k} = 0$$

in view of the orthogonality of α and β.

We verify the equality

$$\mathsf{E}_x\alpha_t(g_1)\beta_t(g_2) = 0$$

by means of a limit transition over continuous functions (as was done in the proof of Theorem 2), provided only that α and β are orthogonal.

Now let $\beta_t = \beta_t^{(1)} + \beta_t^{(2)}$, where $\beta_t^{(2)}$ is orthogonal to α_t and

$$\beta_t^{(1)} = \int_0^t f(x_s)\, d\alpha_s.$$

Then

$$\beta_t(g_2) = \beta_t^{(1)}(g_2) + \beta_t^{(2)}(g_2).$$

Moreover, $\beta_t^{(2)}(g_2)$ is orthogonal to $\alpha_t(g_1)$. Hence $\langle\alpha(g_1), \beta^{(2)}(g_2)\rangle_t = 0$. Therefore

$$\begin{aligned}\langle\alpha(g_1), \beta(g_2)\rangle_t &= \langle\alpha(g_1), \beta^{(1)}(g_2)\rangle = \langle\alpha(g_1), \alpha(fg_2)\rangle_t \\ &= \int_0^t g_1(x_s)f(x_s)g_2(x_s)\, d\langle\alpha, \alpha\rangle_s.\end{aligned}$$

However,

$$\langle \alpha, \beta \rangle_t = \langle \alpha, \alpha(f) \rangle_t = \int_0^t f(x_s)\, d\langle \alpha, \alpha \rangle_s.$$

Formula (21) is thus proved, provided the decomposition of β_t utilized above is valid.

Set $f(x) = (\partial\beta/\partial\alpha)(x)$. Then

$$\beta_t^{(2)} = \beta_t - \int_0^t \frac{\partial \beta}{\partial \alpha}(x_s)\, d\alpha_s \tag{22}$$

and

$$\langle \alpha, \beta^{(2)} \rangle_t = \langle \alpha, \beta \rangle_t - \int_0^t \frac{\partial \beta}{\partial \alpha}(x_s)\, d\langle \alpha, \alpha \rangle_s = 0,$$

i.e., $\beta_t^{(2)}$ is orthogonal to α_t. The lemma is proved. □

Remark 1. Formula (22), which allows us to construct an M-functional orthogonal to α_t starting with β_t, can be generalized in the following manner. Let $\alpha_t^{(1)}, \ldots, \alpha_t^{(n)}$ be a sequence of M-functionals. Then

$$\beta_t^{(k)} = \alpha_t^{(k)} - \sum_{i=1}^{k-1} \int_0^t \frac{\partial \alpha^{(k)}}{\partial \beta^{(i)}}(x_s)\, d\beta_s^{(i)}, \qquad k > 1, \quad \beta_t^{(1)} = \alpha_t^{(1)}, \tag{23}$$

form a sequence of pairwise orthogonal functionals. This can be verified using an induction argument: if $\beta_t^{(1)}, \ldots, \beta_t^{(k-1)}$ are pairwise orthogonal then we have for $j < k$

$$\langle \beta^{(k)}, \beta^{(j)} \rangle_t = \langle \alpha^{(k)}, \beta^{(j)} \rangle_t - \int_0^t \frac{\partial \alpha^{(k)}}{\partial \beta^{(j)}}(x_s)\, d\langle \beta^{(j)}, \beta^{(j)} \rangle_s = 0.$$

It is also easy to verify the inequality

$$\sum_{j=1}^{k-1} \int_0^t \left(\frac{\partial \alpha^{(k)}}{\partial \beta^{(j)}}(x_s) \right)^2 d\langle \beta^{(j)}, \beta^{(j)} \rangle_s \leq \langle \alpha^{(k)}, \alpha^{(k)} \rangle_t. \tag{24}$$

Remark 2. In the case when $\alpha_t \equiv 0$ we shall assume that

$$\frac{\partial \beta}{\partial \alpha} \equiv 0.$$

Definition. M-functional β_t is called *adopted to a functional* α_t if

$$\beta_t = \int_0^t \frac{\partial \beta}{\partial \alpha}(x_s)\, d\alpha_s.$$

Definition. A system of functionals $\beta_t^{(1)}, \beta_t^{(2)}, \ldots$ is called *adopted to a system of functionals* $\alpha_t^{(1)}, \alpha_t^{(2)}, \ldots$ if for each k there exists $n \geq 1$ and functions $g_1(x), \ldots, g_n(x)$ such that

$$\beta_t^{(k)} = \sum_{i=1}^{n} \int_0^t g_i(x_s)\, d\alpha_s^{(i)}. \tag{25}$$

Two systems of functionals $\{\alpha_t^{(i)}, i = 1, 2, \ldots\}$ and $\{\beta_t^{(i)}, i = 1, 2, \ldots\}$ are called *equivalent* if each one is adopted to the other.

Functionals $\beta_t^{(k)}$ defined by equation (23) are adopted to the system $\{\alpha_t^{(i)}, i = 1, 2, \ldots\}$. This can easily be verified using an induction argument. Namely, we show that $\beta_t^{(k)}$ is expressed in terms of $\alpha_t^{(1)}, \ldots, \alpha_t^{(k)}$ via a formula of type (25). This is true for $k = 1$. Assuming that $\beta_t^{(i)}$ for $i \leq k-1$ are expressed in terms of $\alpha_t^{(1)}, \ldots, \alpha_t^{(i)}$ we obtain using formula (23) that $\beta_t^{(k)}$ is expressed in terms of $\alpha_t^{(1)}, \ldots, \alpha_t^{(k)}$. On the other hand,

$$\alpha_t^{(k)} = \beta_t^{(k)} + \sum_{i=1}^{k-1} \int_0^t \frac{\partial \alpha^{(k)}}{\partial \beta^{(i)}}(x_s)\, d\beta_s^{(i)}.$$

Hence the following theorem is valid.

Theorem 4. *For any sequence of M-functionals there exists an equivalent sequence of pairwise orthogonal M-functionals.*

A sequence of pairwise orthogonal M-functionals $\{\alpha_t^{(k)}, k = 1, 2, \ldots\}$ is called a *basis* in Φ_M provided for any M-functional β_t

$$\beta_t = \sum_{k=1}^{\infty} \int_0^t \frac{\partial \beta}{\partial \alpha^{(k)}}(x_s)\, d\alpha_s^{(k)}, \tag{26}$$

where the convergence of the series is in the sense of convergence in Φ_M.

We say that a sequence $\{\beta_t^{(k)}, k = 1, 2, \ldots\}$ is *closed* in Φ_M if there is no M-functional γ_t different from zero which is orthogonal to all $\beta_t^{(k)}, k = 1, 2, \ldots$.

Clearly, if a sequence is closed so is any other sequence equivalent to it. A sequence of pairwise orthogonal M-functionals $\{\alpha_t^{(k)}, k = 1, \ldots\}$ which is closed forms a basis. Indeed, for any M-functional β_t the series

$$\sum_{k=1}^{\infty} \int_0^t \frac{\partial \beta}{\partial \alpha^{(k)}}(x_s)\, d\alpha_s^{(k)}$$

converges in Φ_M since for $n > p$,

$$\left\langle \sum_{k=p}^{n} \int_0^t \frac{\partial \beta}{\partial \alpha^{(k)}}(x_s)\, d\alpha_s^{(k)}, \sum_{k=p}^{n} \int_0^t \frac{\partial \beta}{\partial \alpha^{(k)}}(x_s)\, d\alpha_s^{(k)} \right\rangle = \sum_{k=p}^{n} \int_0^t \left[\frac{\partial \beta}{\partial \alpha^{(k)}}(x_s)\right]^2 d\langle \alpha^{(k)}, \alpha^{(k)} \rangle_s \to 0$$

as n and $p \to \infty$ in view of the inequality

$$\mathsf{E}_x\left(\sum_{k=1}^{n} \int_0^t \frac{\partial \beta}{\partial \alpha^{(k)}}(x_s)\, d\alpha_s^{(k)}\right)^2 = \mathsf{E}_x \sum_{k=1}^{n} \int_0^t \left(\frac{\partial \beta}{\partial \alpha^{(k)}}(x_s)\right)^2 d\langle \alpha^{(k)}, \alpha^{(k)}\rangle_s \le \mathsf{E}_x\langle \beta, \beta\rangle_t$$

(this inequality is analogous to (24)). Consider now the M-functional

$$\beta_t - \sum_{k=1}^{\infty} \int_0^t \frac{\partial \beta}{\partial \alpha^{(k)}}(x_s)\, d\alpha_s^{(k)}.$$

It is easy to verify that it is orthogonal to all $\alpha_t^{(k)}$ and hence equals 0. This implies (26), i.e., $\{\alpha_t^{(k)}, k = 1, 2, \ldots\}$ is a basis in Φ_M. We shall now prove the existence of a closed sequence $\{\beta_t^{(k)}\}$ in Φ_M. This, at the same time, will establish the existence of a basis.

Choose in $\mathscr{D}_{\mathsf{A}}$ a sequence of functions $\{f_k(x), k = 1, 2, \ldots\}$ such that its closure in bounded convergence contains all the continuous functions.

Furthermore, let

$$\varphi_{n,k,r}(x) = \frac{1}{h_n} \int_r^{r+h_n} \mathsf{T}_s f_k(x)\, ds, \qquad h_n = 2^{-n},$$

where r runs over all the nonnegative rational numbers.

Clearly the set of functions $\{\varphi_{n,k,r}\}$ is countable. We shall arrange these functions in a sequence $\{\varphi_1, \varphi_2, \ldots\}$.

Set

$$\beta_t^{(k)} = \varphi_k(x_t) - \varphi_k(x_0) - \int_0^t \mathsf{A}\varphi_k(x_s)\, ds.$$

Let β_t be an M-functional such that

$$\mathsf{E}_x \beta_t \beta_t^{(k)} = 0.$$

Then

$$\mathsf{E}_x \varphi_k(x_t)\beta_t = \mathsf{E}_x \beta_t \int_0^t \mathsf{A}\varphi_k(x_s)\, ds = \mathsf{E}_x \int_0^t \beta_s \mathsf{A}\varphi_k(x_s)\, ds.$$

Analogously for $s < t$

$$\mathsf{E}_x \varphi_k(x_t)[\beta_t - \beta_s] = \mathsf{E}_x \int_s^t (\beta_u - \beta_s)\mathsf{A}\varphi_k(x_u)\, du.$$

Consequently,

$$|\mathsf{E}_x \varphi_k(x_t)[\beta_t - \beta_s]| \le \|\mathsf{A}\varphi_k\| \sqrt{\mathsf{E}_x(\beta_t - \beta_s)^2}\,(t - s).$$

Let $s < t$, $s + h_m < t$. Then for a rational $t - s$

$$|\mathsf{E}_x \varphi_k(x_t)[\beta_{s+h_n} - \beta_s]| = |\mathsf{E}_x \mathsf{T}_{t-s-h_n}\varphi_k(x_s + h_n)[\beta_{s+h_n} - \beta_s]|$$
$$\le \|\mathsf{A}\mathsf{T}_{t-s-h_n}\varphi_k\| \sqrt{\mathsf{E}_x(\beta_{s+h_n} - \beta_s)^2}\, h_n.$$

However $\|\mathsf{T}_{t-s-h_n}\mathsf{A}\varphi\| \leq \|\mathsf{A}\varphi\|$. Hence if $0<\varepsilon_n<h_n$ and $(t-\varepsilon_n)/h_n$ is an integer, then

$$|\mathsf{E}_x\varphi_k(x_t)\beta_t| \leq |\mathsf{E}_x\varphi_k(x_t)\beta_{\varepsilon_n}| + \sum_{\varepsilon_n+lh_n<t} |\mathsf{E}_x\varphi_k(x_t)[\beta_{\varepsilon_n+(l+1)h_n} - \beta_{\varepsilon_n+lh_n}]|$$

$$\leq \|\varphi_k\|\sqrt{\mathsf{E}_x\beta^2_{\varepsilon_n}} + \|\mathsf{A}\varphi_k\| \sum_{\varepsilon_n+lh_n<t} \sqrt{\mathsf{E}_x[\beta_{\varepsilon_n+(l+1)h_n} - \beta_{\varepsilon_n+lh_n}]^2}\, h_n.$$

Since $\mathsf{E}_x[\beta_{s+h_n}-\beta_s]^2$ is bounded and tends to 0, so does the last expression above. Hence we have shown that

$$\mathsf{E}_x\varphi_k(x_t)\beta_t = 0.$$

Applying limiting transitions we verify that for any bounded Borel function $\varphi(x)$ and every x

$$\mathsf{E}_x\varphi(x_t)\beta_t = 0.$$

Let $t_1 < t_2 < \cdots < t_l$. Then for $t_l \leq t$ we have

$$\begin{aligned}\mathsf{E}_x f_1(x_{t_1}) \ldots f_l(x_{t_l})\beta_t &= \mathsf{E}_x f_1(x_{t_1}) \ldots f_l(x_{t_l})\beta_{t_l}\\ &= \mathsf{E}_x f_1(x_{t_1}) \ldots f_{l-1}(x_{t_{l-1}})\mathsf{T}_{t_l-t_{l-1}} f_l(x_{t_{l-1}})\beta_{t_{l-1}}\\ &\quad+ \mathsf{E}_x f_1(x_{t_1}) \ldots f_{l-1}(x_{t_{l-1}})\mathsf{E}_{x_{t_{l-1}}} f_l(x_{t_l-t_{l-1}})\beta_{t_l-t_{l-1}}\\ &= \mathsf{E}_x f_1(x_{t_1}) \ldots f_{l-1}(x_{t_{l-1}})\mathsf{T}_{t_l-t_{l-1}} f_l(x_{t_{l-1}})\beta_{t_{l-1}}.\end{aligned}$$

Hence

$$\mathsf{E}_x f_1(x_{t_1}) \ldots f_l(x_{t_l})\beta_t = 0.$$

However, β_t is an $\{\mathfrak{F}_t\}$-measurable variable. Therefore the last relation implies that $\mathsf{E}_x\beta_t^2 = 0$. We have thus shown that the system under consideration is closed.

Hence the following theorem is valid.

Theorem 5. *There exists a basis in the space Φ_M.*

Maximal functionals. The rank of a process.

Definition. An M-functional α_t is called *maximal* if the condition that α_t is adopted to a functional β_t implies that β_t is adopted to α_t.

Lemma 7. *Let $\{\alpha_t^{(k)}, k=1,2,\ldots\}$ be a basis in Φ_M, $c_k \neq 0$, and the series*

(27) $$\sum_{k=1}^{\infty} c_k\alpha_t^{(k)} = \beta_t$$

converge in Φ_M. Then β_t is a maximal functional.

Proof. Let β_t be adopted to a functional γ_t:

$$\beta_t = \int_0^t \frac{\partial \beta}{\partial \gamma}(x_s)\, d\gamma_s.$$

Then using the representation

$$\gamma_t = \sum_{k=1}^{\infty} \int_0^t \frac{\partial \gamma}{\partial \alpha^{(k)}}(x_s)\, d\alpha_s^{(k)},$$

we obtain

$$\beta_t = \sum_{k=1}^{\infty} \int_0^t \frac{\partial \beta}{\partial \gamma}(x_s) \frac{\partial \gamma}{\partial \alpha^{(k)}}(x_s)\, d\alpha_s^{(k)}. \tag{28}$$

A comparison between (27) and (28) yields

$$\frac{\partial \beta}{\partial \gamma}(x_s) \frac{\partial \gamma}{\partial \alpha^{(k)}}(x_s) = c_k$$

almost everywhere in measure $d\langle \alpha^{(k)}, \alpha^{(k)} \rangle_s$. Therefore

$$\frac{\partial \beta}{\partial \gamma}(x_s) \neq 0, \qquad \frac{\partial \gamma}{\partial \alpha^{(k)}}(x_s) = c_k \left(\frac{\partial \beta}{\partial \gamma}(x_s) \right)^{-1}$$

almost everywhere in measure $d\langle \alpha^{(k)}, \alpha^{(k)} \rangle_s$. Hence

$$\gamma_t = \int_0^t \sum_{k=1}^{\infty} c_k \left(\frac{\partial \beta}{\partial \gamma}(x_s) \right)^{-1} d\alpha_s^{(k)} = \int_0^t \left(\frac{\partial \beta}{\partial \gamma}(x_s) \right)^{-1} d\beta_s.$$

Thus γ_t is adopted to β_t and the lemma is proved. □

Remark. It was established in the process of the proof that if β_t is a maximal functional adopted to γ_t, then

$$\frac{\partial \beta}{\partial \gamma}(x) = \left(\frac{\partial \gamma}{\partial \beta}(x) \right)^{-1}.$$

Clearly, if β_t is a maximal functional adopted to an M-functional γ_t, then γ_t is also a maximal functional, since if γ_t is adopted to a functional δ_t so is β_t; hence δ_t is adopted to both β_t and γ_t. The adoption relation for maximal functionals is thus an equivalence relation.

We shall now investigate the "magnitude" of the class of maximal functionals.

Lemma 8. *If α_t is an M-functional and β_t is a maximal functional, then $\langle \alpha, \alpha \rangle_t$ is absolutely continuous with respect to $\langle \beta, \beta \rangle_t$.*

Proof. Let

$$\langle\alpha,\alpha\rangle_t=\varphi_t^{(1)}+\varphi_t^{(2)},\qquad \langle\beta,\beta\rangle_t=\psi_t^{(1)}+\psi_t^{(2)},$$

where $\varphi_t^{(1)}$ is absolutely continuous and $\varphi_t^{(2)}$ is singular with respect to $\langle\beta,\beta\rangle_t$, and $\psi_t^{(1)}$ is absolutely continuous and $\psi_t^{(2)}$ is singular with respect to $\langle\alpha,\alpha\rangle_t$. Lemma 5 assures that such a representation is possible.

Set $\nu_t=\langle\alpha,\alpha\rangle_t+\langle\beta,\beta\rangle_t$. Since $\varphi_t^{(1)}$ and $\psi_t^{(1)}$ are both absolutely continuous with respect to ν_t, in view of Lemma 4 there exist Borel nonnegative functions $f_i(x)$ and $g_i(x)$ such that

$$\varphi_t^{(i)}=\int_0^t f_i(x_s)\,d\nu_s,\qquad \psi_t^{(i)}=\int_0^t g_i(x_s)\,d\nu_s.$$

Moreover,

$$f_2(x_s)[g_1(x_s)+g_2(x_s)]=0,\qquad g_2(x_s)[f_1(x_s)+f_2(x_s)]=0$$

almost everywhere in measure $d\nu_s$.

Let

$$\bar\beta_t=\beta_t+\int_0^t k(x_s)\,d\alpha_s,$$

where $k(x)=1$ if $f_2(x)>0$ and $k(x)=0$ if $f_2(x)=0$. We have

$$\beta_t=\bar\beta_t-\int_0^t k(x_s)\,d\alpha_s=\bar\beta_t-\int_0^t h(x_s)\,d\bar\beta_s,$$

since

$$\int_0^t k(x_s)\,d\bar\beta_s=\int_0^t k(x_s)\,d\beta_s+\int_0^t k(x_s)\,d\alpha_s=\int_0^t k(x_s)\,d\alpha_s$$

in view of the equality

$$\begin{aligned}\mathsf{E}_x\Big(\int_0^t k(x_s)\,d\beta_s\Big)^2&=\mathsf{E}_x\int_0^t k(x_s)\,d\langle\beta,\beta\rangle_s\\&=\mathsf{E}_x\int_0^t k(x_s)[g_1(x_s)+g_2(x_s)]\,d\nu_s=0.\end{aligned}$$

Therefore β_t is adopted to $\bar\beta_t$. But then $\bar\beta_t$ is also adopted to β_t. Hence

$$\langle\bar\beta,\bar\beta\rangle_t=\int_0^t\Big(\frac{\partial\bar\beta}{\partial\beta}(x_u)\Big)^2\,d\langle\beta,\beta\rangle_u.$$

However, β_t and $\int_0^t k(x_s)\,d\alpha_s$ are orthogonal. Consequently,

$$\langle\bar\beta,\bar\beta\rangle_t=\langle\beta,\beta\rangle_t+\int_0^t k(x_s)\,d\langle\alpha,\alpha\rangle_s=\langle\beta,\beta\rangle_t+\varphi_t^{(2)}.$$

Since $\langle\bar\beta,\bar\beta\rangle_t$ is absolutely continuous with respect to $\langle\beta\beta\rangle_t$, so is $\varphi_t^{(2)}$. Thus $\varphi_t^{(2)}=0$ (because $\varphi_t^{(2)}$ is, by construction, singular with respect to $\langle\beta,\beta\rangle_t$). We have thus verified that $\langle\alpha, \alpha\rangle_t$ is absolutely continuous with respect to $\langle\beta, \beta\rangle_t$. The lemma is proved. □

Corollary 1. *If α_t and β_t are maximal functionals then $\langle\alpha, \alpha\rangle_t$ and $\langle\beta, \beta\rangle_t$ are mutually absolutely continuous.*

Corollary 2. *Let α_t be a maximal functional and β_t be an M-functional such that $\langle\alpha, \alpha\rangle_t$ is absolutely continuous with respect to $\langle\beta, \beta\rangle_t$. Then β_t is also a maximal functional.*

Indeed, if β_t is adopted to γ_t, then $\langle\beta, \beta\rangle_t$ is absolutely continuous with respect to $\langle\gamma, \gamma\rangle_t$, which in view of Lemma 8 is absolutely continuous with respect to $\langle\alpha, \alpha\rangle_t$. Hence $\langle\beta, \beta\rangle_t$ and $\langle\gamma, \gamma\rangle_t$ are mutually absolutely continuous.

In view of the fact that

$$\beta_t = \int_0^t \frac{\partial\beta}{\partial\gamma}(x_s)\, d\gamma_s, \qquad \frac{d\langle\beta, \beta_t\rangle_t}{d\langle\gamma, \gamma\rangle_t} = \left(\frac{\partial\beta}{\partial\gamma}(x_t)\right)^2,$$

$(\partial\beta/\partial\gamma)(x_s)$ is positive everywhere in measure $d\langle\gamma, \gamma\rangle_s$. Therefore

$$\gamma_t = \int_0^t \left(\frac{\partial\beta}{\partial\gamma}(x_s)\right)^{-1} d\beta_s,$$

i.e., the functional γ_t is adopted to β_t. We have thus shown that β_t is maximal.

Definition. A W-functional δ_t is called a *functional of a standard type* if $\delta_t = \langle\alpha, \alpha\rangle_t$, where α_t is a maximal functional.

From Corollaries 1 and 2 follows

Theorem 6. *In order that an M-functional α_t be maximal it is necessary and sufficient that $\langle\alpha, \alpha\rangle_t$ be a functional of a standard type.*

Remark. If δ_t is a functional of a standard type and a W-functional γ_t is of the form

$$\gamma_t = \int_0^t g(x_s)\, d\delta_s,$$

where $g(x_s)$ is positive almost everywhere in measure $d\delta_s$, then γ_t is a functional of a standard type as well.

Indeed, if $\delta_t = \langle\alpha, \alpha\rangle_t$, then $\gamma_t = \langle\beta, \beta\rangle_t$, where

$$\beta_t = \int_0^t \sqrt{g(x_s)}\, d\alpha_s$$

and the maximality of β_t follows from Corollary 2.

Theorem 7. *For any M-functional α_t there exists a maximal functional β_t such that α_t is adopted to β_t.*

Proof. Let $\bar{\beta}_t$ be a maximal functional. Then $\langle\alpha, \alpha\rangle_t$ is absolutely continuous with respect to $\langle\bar{\beta}, \bar{\beta}\rangle_t$. Hence

$$\langle\alpha, \alpha\rangle_t = \int_0^t g(x_s)\, d\langle\bar{\beta}, \bar{\beta}\rangle_s.$$

Let $f(x)\cdot g(x)=1$ if $g>0$ and $f=0$ if $g=0$. Set $\langle\bar{\beta}, \bar{\beta}\rangle_t = \gamma_t^{(1)} + \gamma_t^{(2)}$, where $\gamma_t^{(1)}$ is absolutely continuous with respect to $\langle\alpha, \alpha\rangle_t$ and $\gamma_t^{(2)}$ is singular. Then

$$\langle\bar{\beta}, \bar{\beta}\rangle_t = \int_0^t f(x_s)\, d\langle\alpha, \alpha\rangle_s + \gamma_t^{(2)};$$

$\gamma_t^{(2)}$ is absolutely continuous with respect to $\langle\bar{\beta}, \bar{\beta}\rangle_t$ and

$$\gamma_t^{(2)} = \int_0^t h(x_s)\, d\langle\bar{\beta}, \bar{\beta}\rangle_s,$$

where $h(x)=0$ if $f>0$ and $h(x)=1$ if $f=0$. Now set

$$\beta_t = \alpha_t + \int_0^t h(x_s)\, d\bar{\beta}_s.$$

Observe that

$$\alpha_t = \int_0^t (1-h(x_s))\, d\alpha_s.$$

Indeed, if

$$\tilde{\alpha}_t = \int_0^t h(x_s)\, d\alpha_s,$$

then

$$\langle\tilde{\alpha}, \tilde{\alpha}\rangle_t = \int_0^t h^2(x_s) g(x_s)\, d\langle\bar{\beta}, \bar{\beta}\rangle_s = 0$$

in view of the fact that $hg=0$. Therefore

$$\begin{aligned}\langle\beta, \beta\rangle_t &= \langle\alpha, \alpha\rangle_t + \int_0^t h^2(x_s)\, d\langle\bar{\beta}, \bar{\beta}\rangle_s + 2\int_0^t (1-h(x_s))h(x_s)\, d\langle\alpha, \bar{\beta}\rangle_s \\ &= \int_0^t [g(x_s)+h(x_s)]\, d\langle\bar{\beta}, \bar{\beta}\rangle_s,\end{aligned}$$

since $(1-h(x))h(x)=0$. Because $g(x)+h(x)>0$ it follows that $\langle\beta, \beta\rangle_t$ is absolutely continuous with respect to $\langle\bar{\beta}, \bar{\beta}\rangle_t$ and β_t is therefore a functional of a maximal type. Functional α_t is adopted to β_t since

$$\int_0^t (1-h(x_s))\, d\beta_s = \int_0^t (1-h(x_s))\, d\alpha_s + \int_0^t (1-h(x_s))h(x_s)\, d\bar{\beta}_s = \alpha_t.$$

The theorem is proved. □

Corollary. *There exists a complete system of maximal functionals.*

To construct such a system it is necessary first to choose a complete system of M-functionals $\alpha_t^{(k)}$ and then to find for each $\alpha_t^{(k)}$ a maximal functional $\beta_t^{(k)}$ to which $\alpha_t^{(k)}$ is adopted.

We say that a sequence of M-functionals $\alpha_t^{(1)}, \ldots, \alpha_t^{(n)}, \ldots$ is *nondegenerate* if for each n the functional $\alpha_t^{(n)}$ is *not* adopted to the system $\{\alpha_t^{(1)}, \ldots, \alpha_t^{(n-1)}\}$.

Choose a complete nondegenerate sequence of maximal functionals $\beta_t^{(1)}, \ldots, \beta_t^{(n)}, \ldots$ and consider the matrix

$$D(x) = \left\| \begin{matrix} \frac{\partial \beta^{(1)}}{\partial \beta^{(1)}}(x) \ldots \frac{\partial \beta^{(1)}}{\partial \beta^{(k)}}(x) \ldots \\ \vdots \quad \cdots \quad \vdots \quad \cdots \\ \frac{\partial \beta^{(i)}}{\partial \beta^{(1)}}(x) \ldots \frac{\partial \beta^{(i)}}{\partial \beta^{(k)}}(x) \ldots \\ \ldots\ldots\ldots\ldots\ldots\ldots \end{matrix} \right\| = \left\| \frac{\partial \beta^{(i)}}{\partial \beta^{(k)}}(x) \right\|, \tag{29}$$

which has as many rows and columns as there are functionals in the sequence $\{\beta_t^{(k)}\}$. If $\bar{\beta}_t^{(1)}, \ldots, \bar{\beta}_t^{(n)}, \ldots$ is some other complete nondegenerate sequence of maximal functionals and $\bar{D}(x)$ is the analogue of matrix (29), constructed from this sequence then

$$D(x) = C(x)\bar{D}(x)C_1(x), \tag{30}$$

where

$$C(x) = \left\| \frac{\partial \beta_i}{\partial \bar{\beta}_j}(x) \right\|, \quad C_1(x) = \left\| \frac{\partial \bar{\beta}_i}{\partial \beta_j}(x) \right\|.$$

Analogously

$$\bar{D}(x) = C_1(x)D(x)C(x). \tag{31}$$

Relations (30) and (31) show that the rank of matrix $D(x)$ at point x does not depend on the choice of sequence $\{\beta_t^{(k)}\}$. We shall denote this rank by $r(x)$ and call it *the rank of the process at point x.*

Note that functions

$$\frac{\partial \beta^{(i)}}{\partial \beta^{(k)}}(x_s), \quad \frac{\partial \beta^{(i)}}{\partial \bar{\beta}^{(k)}}(x_s), \quad \frac{\partial \bar{\beta}^{(i)}}{\partial \beta^{(k)}}(x_s), \quad \frac{\partial \bar{\beta}^{(i)}}{\partial \bar{\beta}^{(k)}}(x_s)$$

are defined only up to the sets of measure zero in $d\sigma_s$, where σ_t is an arbitrary functional of the standard type. Therefore $r(x)$ is also defined up to the sets A such that

$$\int_0^t \chi_A(x_s)\, d\sigma_s = 0$$

for all t (χ_A is the indicator of the set A). In particular the rank may not be defined on a set of this kind. In the case when a process has no M-functionals different from zero we shall assume that the rank of the process is 0.

Random time substitution. If δ_t is a positive additive continuous functional and τ_t is defined by equality $\delta_{\tau_t} = t$, then the process $y_t = x_{\tau_t}$ will also be a strong Markov continuous process in G (cf. Volume II, Chapter II, Section 6, pp. 183–184).

Lemma 9. *If γ_t is a W-functional on process x_t then $\hat{\gamma}_t = \gamma_{\tau_t}$ will be a W-functional on process y_t.*

Proof. If $\mathcal{N}_t$ is a σ-algebra generated by variables x_s, $s \leq t$, and $\hat{\mathcal{N}}_t$ is a σ-algebra generated by variables y_s, $s \leq t$, then $\hat{\mathcal{N}}_t \subset \mathcal{N}_{\tau_t}$ and the variable γ_{τ_t} is $\mathcal{N}_{\tau_t}$-measurable, i.e., $\hat{\gamma}_t$ is $\hat{\mathcal{N}}_t$-measurable. Thus γ_{τ_t} is obviously continuous.

Let $\hat{\theta}_h$ be a shift operator for process y_t. Since the variables x_{τ_s} and τ_s are $\hat{\mathcal{N}}_s$-measurable the operator $\hat{\theta}_h$ is defined for these variables. It is easy to see that

$$\hat{\theta}_h x_\tau = x_{\tau_{s+h}}, \qquad \hat{\theta}_h \tau_s = \tau_{s+h} - \tau_h.$$

Assume now that γ_t is of the form

$$\gamma_t = \int_0^t g(x_u)\, du, \tag{32}$$

where g is a continuous function. Then

$$\gamma_{\tau_t} = \int_0^{\tau_t} g(x_u)\, du = \int_0^t g(x_{\tau_s})\, d\tau_s = \lim_{n\to\infty} \sum_{k=1}^n g(x_{\tau_{t_k}})(\tau_{t_k} - \tau_{t_{k-1}}),$$

where $t_k = kt/n$. Therefore

$$\begin{aligned}\hat{\theta}_h \gamma_{\tau_t} &= \lim_{n\to\infty} \sum_{k=1}^n g(x_{\tau_{t_k+h}})(\tau_{t_k+h} - \tau_{t_{k-1}+h})\\ &= \int_h^{t+h} g(x_{\tau_s})\, d\tau_s = \int_{\tau_h}^{\tau_t+h} g(x_u)\, du = \gamma_{\tau_{t+h}} - \gamma_{\tau_h}.\end{aligned}$$

We have verified the formula

$$\hat{\theta}_h \gamma_{\tau_t} = \gamma_{\tau_{t+h}} - \gamma_{\tau_h} \tag{33}$$

for functionals γ_t of the form (32), provided g is a continuous function. Utilizing the convergence of functionals (32) under the bounded convergence of functions g we verify the validity of (33), provided γ_t is of the form (32), where g is a bounded Borel function. However, in view of Theorem 1 in Volume II, Chapter II, Section 6, p. 167, any W-functional is a limit of functionals of the form (32), where g is a bounded Borel function. Thus (33) is verified for an arbitrary W-functional γ_t.

We have shown that $\hat{\gamma}_t$ is a continuous nonnegative additive functional. To verify that it is a W-functional we observe that

$$\mathsf{E}_x\hat{\gamma}_\infty \leq \mathsf{E}_x\gamma_\infty \quad \text{and} \quad \sup_x \mathsf{E}_x\gamma_\infty \leq C(\sup_x \mathsf{E}_x\zeta + 1), \tag{34}$$

where ζ is the cutoff time of the process (i.e., the time of the first exit out of G). Inequality (34) follows from the fact that

$$\mathsf{E}_x\gamma_t \leq C_1(t+1)$$

for some C_1 and by assumption $\sup_x \mathsf{E}_x\zeta < \infty$.

The lemma is thus proved. □

Lemma 10. *If α_t is an M-functional of process x_t, then $\hat{\alpha}_t = \alpha_{\tau_t}$ is an M-functional of the process y_t; moreover,*

$$\langle \hat{\alpha}, \hat{\alpha} \rangle_t = \langle \alpha, \alpha \rangle_{\tau_t}. \tag{35}$$

Proof. $\hat{\mathcal{N}}_t$-measurability and continuity is verified in exactly the same manner as in the preceding lemma. The fact that $\hat{\alpha}_t$ is a martingale and relation (35) follow from Theorem 6 in Section 1 of Chapter I and from Lemma 3.

It follows from Lemma 9 that since $\langle \alpha, \alpha \rangle_t$ is a W-functional so is $\langle \alpha, \alpha \rangle_{\tau_t}$. It remains to show that $\hat{\alpha}_t$ is an additive functional, i.e., to verify relation

$$\hat{\theta}_h \hat{\alpha}_t = \hat{\alpha}_{t+h} - \hat{\alpha}_h, \tag{36}$$

where $\hat{\theta}_h$ is a shift operator for process y_t.

First assume that

$$\alpha_t = f(x_t) - f(x_0) - \textstyle\int_0^t \mathsf{A}f(x_s)\, ds, \tag{37}$$

where f is a function in $\mathscr{D}_\mathsf{A}$. Then

$$\alpha_{\tau_t} = f(x_{\tau_t}) - f(x_0) - \textstyle\int_0^{\tau_t} \mathsf{A}f(x_s)\, ds.$$

Therefore in view of the properties of operator $\hat{\theta}_h$, namely, $\hat{\theta}_h x_{\tau_t} = x_{\tau_{t+h}}$, and formula (33) applied to the functional $\gamma_t = \int_0^t \mathsf{A}f(x_s)\, ds$ we have

$$\hat{\theta}_h \alpha_{\tau_t} = f(x_{\tau_{t+h}}) - f(x_{\tau_h}) - \textstyle\int_{\tau_h}^{\tau_{t+h}} \mathsf{A}f(x_s)\, ds.$$

Formula (36) has thus been established for the functionals of the form (37). It now remains only to observe that functionals of the form given by (37) are complete in Φ_M. □

Corollary 1. *If α_t and β_t are M-functionals on the process x_t,*

$$\hat{\alpha}_t = \alpha_{\tau_t}, \qquad \hat{\beta}_t = \beta_{\tau_t},$$

then

$$\frac{\partial \alpha}{\partial \beta}(x) = \frac{\partial \hat{\alpha}}{\partial \hat{\beta}}(x). \tag{38}$$

Indeed, if $(\partial \alpha / \partial \beta)(x) = g(x)$ then

$$\langle \alpha, \beta \rangle_t = \int_0^t g(x_s)\, d\langle \beta, \beta \rangle_s.$$

Therefore

$$\langle \hat{\alpha}, \hat{\beta} \rangle_t = \int_0^{\tau_t} g(x_s)\, d\langle \beta, \beta \rangle_s = \int_0^t g(x_{\tau_s})\, d\langle \beta, \beta \rangle_{\tau_s} = \int_0^t g(y_s)\, d\langle \hat{\beta}, \hat{\beta} \rangle_s,$$

i.e.,

$$\frac{\partial \hat{\alpha}}{\partial \hat{\beta}}(y_s) = g(y_s).$$

Corollary 2. *Under a random time substitution W-functionals pass into W-functionals and M-functionals into M-functionals, derivatives of M-functionals with respect to each other remain unchanged, orthogonal functionals pass into orthogonal ones, maximal into maximal ones, complete systems into complete systems, and a basis into a basis.*

Now let δ_t be a positive functional such that all the standard functionals are absolutely continuous with respect to it. (A positive standard functional, if it exists, can be chosen for δ_t; otherwise we may set $\delta_t = t + \gamma_t$, where γ_t is a standard functional.) If one carries out the random substitution by means of functional δ_t then every standard functional γ_t passes into a functional $\hat{\gamma}_t$ which is absolutely continuous with respect to the function $\hat{\delta}_t = \delta_{\tau_t} = t$; thus all the standard functionals will be absolutely continuous with respect to the Lebesgue measure, and hence for any M-functional $\hat{\alpha}_t$ on the process y_t there exists a function $g_{\hat{\alpha}}(x)$ such that

$$\langle \hat{\alpha}, \hat{\alpha} \rangle_t = \int_0^t g_{\hat{\alpha}}(y_s)\, ds. \tag{39}$$

A process y_t obtained as a result of such a random time substitution is called a *process with an absolutely continuous standard functional.*

Theorem 8. *Let y_t be a process with an absolutely continuous standard functional. If $\varphi_1, \ldots, \varphi_m \in \mathcal{D}_{\hat{A}}$ where $\hat{A}$ is a quasi-generating operator of process y_t and $F(t_1, \ldots, t_m)$ is a twice continuously differentiable function, then $F(\varphi_1, \ldots, \varphi_m) \in$*

$\mathcal{D}_{\hat{A}}$ and

$$(40)\qquad \hat{A}F(\varphi_1, \ldots, \varphi_m) = \sum_1^m \frac{\partial F}{\partial t_k}(\varphi_1, \ldots, \varphi_m)\hat{A}\varphi_k + \frac{1}{2}\sum_1^m \frac{\partial^2 F}{\partial t_i\,\partial t_k}(\varphi_1, \ldots, \varphi_m) b_{\varphi_i,\varphi_k},$$

where $b_{\varphi_i,\varphi_k}(x)$ is defined by the equality

$$(41)\qquad \langle \alpha^i, \alpha^k\rangle_t = \int_0^t b_{\varphi_i,\varphi_k}(y_s)\,ds,$$

$$(42)\qquad \alpha_t^i = \varphi_i(y_t) - \varphi_i(y_0) - \int_0^t \hat{A}\varphi_i(y_s)\,ds.$$

Proof. Firstly note that the existence of functions b_{φ_i,φ_k} follows from the absolute continuity of $\langle \alpha^i, \alpha^k\rangle_t$ with respect to $\langle \alpha^i, \alpha^i\rangle_t$ and from formula (39). Applying Itô's formula (Theorem 1 in Section 3 of Chapter I) to the function

$$F(\varphi_1(y_t), \ldots, \varphi_m(y_t)) = F\left(\varphi_1(y_0) + \alpha_t^1 + \int_0^t \hat{A}\varphi_1(y_s)\,ds, \ldots, \varphi_m(y_0) + \alpha_t^m + \int_0^t \hat{A}\varphi_m(y_s)\,ds\right),$$

we obtain

$$\begin{aligned} F(\varphi_1(y_t), \ldots, \varphi_m(y_t)) &= F(\varphi_1(y_0), \ldots, \varphi_m(y_0)) \\ &\quad + \int_0^t \sum_{k=1}^m F'_{t_k}(\varphi_1(y_s), \ldots, \varphi_m(y_s))[d\alpha_s^k + \hat{A}\varphi_k(y_s)\,ds] \\ &\quad + \frac{1}{2}\int_0^t \sum_{k,j=1}^m F''_{t_k t_j}(\varphi_1(y_s), \ldots, \varphi_m(y_s))\,d\langle \alpha^k, \alpha^j\rangle_s \\ &= F(\varphi_1(y_0), \ldots, \varphi_m(y_0)) \\ &\quad + \sum_{k=1}^m \int_0^t F'_{t_k}(\varphi_1(y_s), \ldots, \varphi_m(y_s))\,d\alpha_s^k \\ &\quad + \int_0^t \Bigg[\sum_{k=1}^m F'_{t_k}(\varphi_1(y_s), \ldots, \varphi_m(y_s))\hat{A}\varphi_k(y_s) \\ &\quad + \frac{1}{2}\sum_{k,j=1}^m F''_{t_k t_j}(\varphi_1(y_s), \ldots, \varphi_m(y_s)) b_{\varphi_j,\varphi_k}(y_s)\Bigg]\,ds. \end{aligned}$$

Taking the mathematical expectation of both sides we verify the validity of the theorem. □

Assume that the following condition is satisfied:

(A) For any point $x \in G$ one can find a neighborhood G_1 such that in the closure of G_1 coordinates $\varphi_1, \ldots, \varphi_m$ belonging to $\mathcal{D}_{\hat{A}}$ may be introduced.

Consider the process $\hat{y}_t$ obtained from y_t by cutting off at time ζ_1 of the first exit out of the neighborhood of G_1. In view of Itô's formula referred to above we have for any twice continuously differentiable function $F(t_1, \ldots, t_m)$ and time $\tau \leq \zeta_1$

$$F(\varphi_1(y_\tau), \ldots, \varphi_m(y_\tau)) = F(\varphi_1(y_0), \ldots, \varphi_m(y_0)) + \int_0^\tau \sum_{k=1}^m F'_{t_k}(\varphi_1(y_s), \ldots, \varphi_m(y_s))\, d\alpha_s^k + \int_0^\tau \Big[\sum_{k=1}^m F'_{t_k}(\varphi_1(y_s), \ldots, \varphi_m(y_s)) \hat{\mathsf{A}}\varphi_k(y_s) + \frac{1}{2} \int_0^t \sum_{k,j=1}^m F''_{t_k t_j}(\varphi_1(y_s), \ldots, \varphi_m(y_s)) b_{\varphi_j,\varphi_k}(y_s) \Big] ds$$

and hence

$$\mathsf{E}_y F(\varphi_1(y_\tau), \ldots, \varphi_m(y_\tau)) = F(\varphi_1(y), \ldots, \varphi_m(y)) + \mathsf{E}_y \int_0^t L(F(\varphi_1, \ldots, \varphi_m)](y_s)\, ds, \tag{43}$$

and

$$L[F(\varphi_1, \ldots, \varphi_m)](y) = \sum_{k=1}^m F'_{t_k}(\varphi_1(y), \ldots, \varphi_m(y)) \hat{a}_k(y) + \frac{1}{2} \sum_{k,j=1}^m F''_{t_k t_j}(\varphi_1(y), \ldots, \varphi_m(y)) \hat{b}_{kj}(y),$$

$$\hat{a}_k(y) = \hat{\mathsf{A}}\varphi_k(y), \qquad \hat{b}_{kj}(y) = b_{\varphi_k,\varphi_j}(y).$$

Consider the mapping $U(x)$ in $G_1 \subset \mathscr{R}^m$ defined by relation $x \to \{\varphi_1(x), \ldots, \varphi_m(x)\}$. Under this mapping the set G_1 is mapped continuously in a one-to-one manner into a domain $\hat{G}_1$ in $\mathscr{R}^m$. Hence $\hat{x}_t = U(\hat{y}_t)$ is a Markov process; moreover, at the cutoff time $\hat{\zeta}$ the process $\hat{x}_{\zeta-0}$ belongs to the boundary of $\hat{G}_1$.

It follows from (43) that for any twice continuously differentiable function $F(x)$ in $\hat{G}_1$ and a Markov time $\tau \leq \hat{\zeta}$ the relation

$$\hat{\mathsf{E}}_x F(\hat{x}_\tau) = F(x) + \hat{\mathsf{E}}_x \int_0^\tau \hat{L}[F](\hat{x}_s)\, ds \tag{44}$$

is valid, where $\hat{\mathsf{E}}_x$ is the expectation of the process $\hat{x}_t$ and

$$\hat{L}[F](x) = \sum_{i=1}^m a_i(x) \frac{\partial F}{\partial x^i}(x) + \frac{1}{2} \sum_{i,j=1}^m b_{ij}(x) \frac{\partial^2 F}{\partial x^i\, \partial x^j}(x), \tag{45}$$

$$a_i(U(x)) = \hat{a}_i(x), \qquad b_{ij}(U(x)) = \hat{b}_{ij}(x)$$

($x^1, \ldots, x^m$ are the coordinates of x in $\mathscr{R}^m$).

Formulas (44) and (45) show that for the process $\hat{x}_t$ a quasi-characteristic operator is defined on all twice continuously differentiable functions and is a differential operator of the second order, i.e., is of the same form as in the case of diffusion processes.

We now show that the process $\hat{x}_t$ satisfies a stochastic differential equation. For this purpose we first observe that the processes $\hat{y}_t$ and $\hat{x}_t$ possess the same class of W-functionals and that of M-functionals since the σ-algebras generated by these processes coincide.

Relation (42) yields that

$$\hat{x}_t^i - \hat{x}_0^i = \int_0^t a_i(\hat{x}_s)\,ds + \hat{\alpha}_t^i, \tag{46}$$

where $\hat{\alpha}_t^i$ is an M-functional on the process $\hat{x}_t$ (this functional is obtained in a natural manner from functional α_t^i) and $\hat{x}_t^i$ is the ith coordinate of $\hat{x}_t$.

Lemma 11. *If condition* (A) *is fulfilled, then* $\hat{\alpha}_t^i, i = 1, \ldots, m$, *defined by* (46) *form a complete system of M-functionals.*

Proof. Let β_t be an M-functional of the process $\hat{x}_t$ orthogonal to all the functionals $\hat{\alpha}_t^i$. Then for any twice continuously differentiable function $F(x)$ we obtain from the equality

$$F(\hat{x}_t) = F(\hat{x}_0) + \int_0^t \hat{L}[F](x_s)\,ds + \sum_1^m \int_0^t \frac{\partial F}{\partial x^i}(x_s)\,d\hat{\alpha}_s^i$$

that

$$\hat{\mathsf{E}}_x \beta_t F(\hat{x}_t) = \hat{\mathsf{E}}_x \beta_t \int_0^t \hat{L}[F](x_s)\,ds.$$

From this relationship one obtains

$$\hat{\mathsf{E}}_x \beta_t F(\hat{x}_t) = 0$$

in the same manner as in the proof of Theorem 5 for any twice continuously differentiable function $F(x)$ and hence for any continuous function $F(x)$. This implies (in the same manner as in Theorem 5) that $\mathsf{E}_x \beta_t^2 = 0$. The lemma is proved. □

Corollary. *The rank of process $\hat{x}_t$ is at most m.*

This follows from the existence of a complete system of M-functionals containing m elements.

Theorem 9. *If condition* (A) *is fulfilled, then there exists an m-dimensional Wiener process $w(t)$ with respect to the current of σ-algebras $\mathscr{N}_t$, a vector function $a(x)$, and an operator function $B(x)$ such that for $t < \hat{\zeta}$*

$$\hat{x}_t - \hat{x}_0 = \int_0^t a(x_s)\,ds + \int_0^t B(x_s)\,dw(s). \tag{47}$$

Proof. We proceed from equalities (46). Let M-functionals β_t^i be constructed by means of orthogonalization of functionals $\hat{\alpha}_t^i$:

$$\beta_t^1 = \hat{\alpha}_t^1, \qquad \beta_t^k = \hat{\alpha}_t^k - \sum_{j=1}^{k-1} \int_0^t \frac{\partial \hat{\alpha}^k}{\partial \beta^j}(x_s)\, d\beta_s^j, \qquad k = 2, \ldots, m,$$

and $\tilde{b}_k(x)$ be determined by the inequalities

$$\langle \beta^k, \beta^k \rangle_t = \int_0^t \tilde{b}_k(x_s)\, ds \quad \text{and} \quad E_k = \{x: \tilde{b}_k(x) > 0\}.$$

Finally, let $\tilde{w}_1(t), \ldots, \tilde{w}_m(t)$ be one-dimensional processes independent of β_t^j and also mutually independent.

Set

$$w^k(t) = \int_0^t (1 - \chi_{E_k}(x_s))\, d\tilde{w}_k(s) + \int_0^t \chi_{E_k}(x_s) \frac{1}{\sqrt{\tilde{b}_k(x_s)}}\, d\beta_s^k.$$

It is easy to see that $w^k(t)$ are martingales and, moreover,

$$\langle w^k, w^k \rangle_t = \int_0^t (1 - \chi_{E_k}(x_s))^2\, ds + \int_0^t \chi_{E_k}(x_s) \frac{1}{\tilde{b}(x_s)}\, d\langle \beta^k, \beta^k \rangle_s = t.$$

Since $\tilde{w}_k$, $\tilde{w}_j$, β^k, and β^j are pairwise orthogonal we have $\langle w^k, w^j \rangle_t = 0$ for $k \neq j$. Therefore in view of Corollary 2 to the theorem 1 in Section 1 the m-dimensional process $w(t)$ with cooordinates $w^1(t), \ldots, w^m(t)$ is an m-dimensional Wiener process.

Note, furthermore, that

$$\beta_t^k = \int_0^t \sqrt{\tilde{b}_k(x_s)}\, dw^k(s).$$

Consequently,

$$\hat{\alpha}_t^k = \int_0^t \sqrt{\tilde{b}_k(x_s)}\, dw^k(s) + \sum_{j=1}^{k-1} \int_0^t \frac{\partial \hat{\alpha}^k}{\partial \beta^j}(x_s) \sqrt{\tilde{b}_j(x_s)}\, dw^j(s).$$

Thus equality (47) is valid, provided the matrix of the operator B in a natural basis of $\mathscr{R}^m$ is of the form $B(x) = \|\sigma_{ij}(x)\|$, where $\sigma_{ij}(x) = (\partial \hat{\alpha}_i / \partial \beta^j)(x) \sqrt{\tilde{b}_j(x)}$ for $i \leq j$ and $\sigma_{ij} = 0$ for $i > j$.

The theorem is proved. □

Hence under the conditions of Theorem 9, the process coincides locally in certain coordinates with a solution of a stochastic differential equation, i.e., is a (local) diffusion process. The question arises, when can a given process be (locally) transformed into a diffusion process by means of a random time substitution and a one-to-one continuous mapping of the space. To answer this question we introduce a new useful concept.

Let x_t be a Markov process in the domain $G \subset \mathcal{R}^m$ satisfying the conditions stipulated in the beginning of this section. Denote by $\mathcal{D}$ the set of continuous bounded functions f defined on G such that there exist a $\tilde{W}$-functional γ_t and an M-functional α_t satisfying

$$f(x_t) - f(x_0) = \gamma_t + \alpha_t. \tag{48}$$

As was shown above, representation (48) is valid for $f \in \mathcal{D}_{\mathsf{A}}$, provided

$$\gamma_t = \int_0^t \mathsf{A} f(x_s)\, ds,$$

so that $\mathcal{D} \supset \mathcal{D}_{\mathsf{A}}$. Let the process y_t be obtained from x_t by means of the random time substitution $y_t = x_{\tau_t}$. Then we have for $f \in \mathcal{D}$

$$f(y_t) - f(y_0) = f(x_{\tau_t}) - f(x_0) = \gamma_{\tau_t} + \alpha_{\tau_t}.$$

However, as follows from Lemmas 9 and 10, γ_{τ_t} and α_{τ_t} are correspondingly a $\tilde{W}$-functional and an M-functional on the process y_t. Therefore under a random time substitution the class $\mathcal{D}$ is transformed into itself.

Theorem 10. *If x_t is a Markov process such that in domain G coordinates $f_1, \dots, f_m$ exist which belong to $\mathcal{D}$, then a random time substitution τ_t exists such that the process $\hat{x}_t = (f_1(x_{\tau_t}), \dots, f_m(x_{\tau_t}))$ will be a diffusion process in the domain G_1 which is the image of G under the mapping $x \to (f_1(x), \dots, f_m(x))$.*

Proof. We may assume without loss of generality that $f_i(x) = x^i$, where x^i is the ith coordinate of point x.

Let

$$x_t^i - x_0^i = \gamma_t^i + \alpha_t^i.$$

Carrying out the random substitution of time τ_t, where τ_t is the solution of equation $\delta_{\tau_t} = t$, and δ_t is a W-functional with respect to which γ_t^i and $\langle \alpha^i, \alpha^i \rangle_t$ are absolutely continuous for all i. If $\gamma_t^i = \gamma_t^{i+} - \gamma_t^{i-}$ where $\gamma_t^{i\pm}$ are W-functionals, then one may choose for δ_t

$$\delta_t = t + \sum_{i=1}^m (\gamma_t^{i+} + \gamma_t^{i-} + \langle \alpha^i, \alpha^i \rangle_t).$$

Let

$$\hat{x}_t^i = x_{\tau_t}^i, \qquad \hat{\gamma}_t^i = \gamma_{\tau_t}^i, \qquad \hat{\alpha}_t^i = \alpha_{\tau_t}^i.$$

Then

$$\hat{x}_t^i - \hat{x}_0^i = \hat{\gamma}_t^i + \hat{\alpha}_t^i$$

and $\hat{\gamma}_t^i$ and $\langle \hat{\alpha}^i, \hat{\alpha}^i \rangle_t$ are absolutely continuous in t. Hence

$$\hat{x}_t^i - \hat{x}_0^i = \int_0^t a^i(\hat{x}_s)\, ds + \hat{\alpha}_t^i, \tag{49}$$

where $a^i(x)$ is a Borel function. Utilizing this representation one can, in the same manner as in Theorem 9, verify the existence of an m-dimensional Wiener process $w(t)$ such that equation (47) is satisfied for some $a(x)$ and $B(x)$.

The theorem is proved. □

Continuous processes in $\mathscr{R}^1$. The fact that $\mathscr{R}^1$ is an ordered space in which the order is consistent with the topology of the space allows us to study the structure of continuous Markov processes in $\mathscr{R}^1$ under much less restrictive assumptions. The strong Markov property is the only condition imposed on a continuous process x_t defined on a given interval $\Delta \subset \mathscr{R}^1$.

Let x and y belong to Δ. We say that point y *is accessible from point* x if

$$\mathsf{P}_x\{\tau_y < \infty\} > 0,$$

where τ_y is the moment at which the process hits point y for the first time (this is a Markov time). Let Δ_x be the set of y which is accessible from point x. It is easy to verify that Δ_x is an interval (open, half-open, or closed). This follows from the fact that on its route from point x to point y the process passes through all the points situated between x and y. If $z \in \Delta_x$, then $\Delta_z \subset \Delta_x$.

A point x is called a *regular* point if

1) x is an interior point of Δ_x,
2) $x_1 < x < x_2$, $x_1, x_2 \in \Delta_x$, exist such that $x \in \Delta_{x_1}$ and $x \in \Delta_{x_2}$.

If condition 2) is fulfilled, then each point on the interval $[x_1, x_2]$ is accessible from any other point. Indeed, accessibility of x from x_1 implies accessibility of x from all points of the interval $[x_1, x]$. In the same manner x is accessible from all points of the interval $[x, x_2]$, i.e., x is accessible from any point in $[x_1, x_2]$ and any point in $[x_1, x_2]$ is accessible from x.

Let

$$\alpha_x = \inf\,[y\colon y \in \Delta_x,\ \mathsf{P}_y\{\tau_x < \infty\} > 0],$$

$$\beta_x = \sup\,[y\colon y \in \Delta_x,\ \mathsf{P}_y\{\tau_x < \infty\} > 0].$$

If a point x is regular, then the interval (α_x, β_x) is nonempty and contains x. This interval is called *the interval of regularity of a process containing point* x.

In this subsection we shall be concerned mainly with the investigation of a process in the interval of regularity. As a preliminary we shall discuss the possible nature of nonregularity.

If x is a nonregular point, then at least one of the following conditions holds:

(I) $\forall y < x \quad \mathsf{P}_x\{\tau_y < \infty\} = 0$,
(II) $\forall y > x \quad \mathsf{P}_x\{\tau_y < \infty\} = 0$,

(III) $\forall y<x \quad \mathsf{P}_y\{\tau_x<\infty\}=0$,
(IV) $\forall y>x \quad \mathsf{P}_y\{\tau_x<\infty\}=0$.

In the case when (I) and (II) are satisfied for x, this point is an absorbing point.

Assume that (I) is satisfied, but not (II). In this case if τ^ε is the time of the first exit out of the neighborhood $(x-\varepsilon, x+\varepsilon)$, then $\tau^\varepsilon=\tau_{x+\varepsilon}$ and by the definition of a characteristic operator of a process we have

$$\mathfrak{A}f(x)=\lim_{\varepsilon\downarrow 0}\frac{f(x+\varepsilon)-f(x)}{\mathsf{E}_x\tau^\varepsilon}. \tag{50}$$

If (II) is satisfied but not (I), then the characteristic operator is evaluated by the formula

$$\mathfrak{A}f(x)=\lim_{\varepsilon\downarrow 0}\frac{f(x-\varepsilon)-f(x)}{\mathsf{E}_x\tau^\varepsilon}. \tag{51}$$

Observe that in both formulas the denominator can be written as an increment of a monotonic function constructed from the process.

Let δ be such that $\mathsf{E}_x\tau_{x+\delta}<\infty$. For $y\in[x, x+\delta]$ set

$$g(y)=\mathsf{E}_y\tau_{x+\delta}.$$

Since for $y_1<y_2$

$$\tau_{x+\delta}=\tau_{y_2}+\theta_{\tau_{y_2}}\tau_{x+\delta}$$

with probability $\mathsf{P}_{y_1}=1$, taking the expectation on both sides of the equality and taking the strong Markov property of the process into account we obtain

$$\begin{aligned}\mathsf{E}_{y_1}\tau_{x+\delta}&=\mathsf{E}_{y_1}\tau_{y_2}+\mathsf{E}_{y_1}\mathsf{E}(\theta_{\tau_{y_2}}\tau_{x+\delta}\,|\,\mathcal{N}_{\tau_{y_2}})\\&=\mathsf{E}_{y_1}\tau_{y_2}+\mathsf{E}_{y_1}\mathsf{E}_{x(\tau_{y_2})}\tau_{x+\delta}=\mathsf{E}_{y_1}\tau_{y_2}+\mathsf{E}_{y_2}\tau_{x+\delta},\end{aligned}$$

or

$$g(y_1)=\mathsf{E}_{y_1}\tau_{y_2}+g(y_2).$$

This implies the equality

$$\mathsf{E}_x\tau^\varepsilon=g(x)-g(x+\varepsilon).$$

An analogous equality is valid in the case of formula (51) by putting $g(y)=\mathsf{E}_y\tau_{x-\delta}$:

$$\mathsf{E}_x\tau^\varepsilon=g(x)-g(x-\varepsilon).$$

Thus in the first case we have

$$\mathfrak{A}f(x)=\lim_{\varepsilon\downarrow 0}\frac{f(x+\varepsilon)-f(x)}{g(x)-g(x+\varepsilon)}$$

and in the second

$$\mathfrak{A}f(x)=\lim_{\varepsilon\downarrow 0}\frac{f(x-\varepsilon)-f(x)}{g(x)-g(x-\varepsilon)}.$$

Hence an evaluation of a characteristic operator at nonregular points at which one of the conditions (I) or (II) is satisfied reduces to an evaluation of a one-sided derivative of a monotone function.

Assume now that both (I) and (II) are not satisfied. If (III) is satisfied but not (IV), then x is a lower bound of the interval of regularity accessible from this interval.

If, however, (IV) is fulfilled but not (III), then x is the upper bound of the interval of regularity. The behavior of the processes at the boundary of the interval of regularity is discussed below.

Finally, let (III) and (IV) be fulfilled, but not (I) and (II). Assume, furthermore, that x is not an absorbing point. Then for all $t>0$

$$\mathsf{P}_x\{x_t\in(-\infty,x)\}\quad\text{and}\quad \mathsf{P}_x\{x_t\in(x,\infty)\}$$

do not depend on t: the process must after time $t=0$ leave the point x, hit one of the sets $(-\infty, x)$ or (x, ∞), and will never depart from these sets since point x is inaccessible.

Let

$$p=\mathsf{P}_x\{\forall t>0\quad x_t>x\},\qquad q=\mathsf{P}_x\{\forall t>0\quad x_t<x\}.$$

Introduce for some δ the functions

$$g_1(x)=\frac{1}{p}\mathsf{E}_x\tau^\delta\chi_{(x,\infty)}(x_{\tau^\delta}),\qquad g_2(x)=\frac{1}{q}\mathsf{E}_x\tau^\delta\chi_{(-\infty,x)}(x_{\tau^\delta}).$$

Function $g_1(x)$ decreases on the interval $(x, x+\delta)$ while $g_2(x)$ increases on the interval $[x-\delta, x]$. The characteristic operator at point x is of the form

$$\mathfrak{A}f(x)=\lim_{\varepsilon\downarrow 0,\,\varepsilon_1\downarrow 0}\frac{p[f(x+\varepsilon)-f(x)]+q[f(x-\varepsilon_1)-f(x)]}{p[g_1(x)-g_1(x+\varepsilon)]+q[g_2(x)-g_2(x-\varepsilon_1)]}.\tag{52}$$

Utilizing conditions (I)–(IV) one can classify the points as follows: a point is inaccessible from the left [right] if condition (III) [(IV)] is fulfilled; the point is impassable to the left [right] if condition (I) [(II)] is fulfilled.

A point which is inaccessible from the left and right is called simply *inaccessible.*

Consider the possible locations of nonregular points relative to the interval of regularity:

1) such a point may be a boundary of the interval of regularity;
2) it may be a limit point for the boundaries of intervals of regularity;
3) it may be an interior point of the set of nonregular points.

If a nonregular point is a right-hand end-point of the interval of regularity or the limit of an increasing sequence of such points it is then impassable to the left. If a nonregular point is a left-hand end-point of the interval of regularity or a limit of a decreasing sequence of such points, then it is impassable to the right. Assume that G is an interval of nonregular points not containing absorbing points. Let E be the set of x such that Δ_x contains point x as an interior point, $x \in G$. If $x_1 \in E$ and $x_2 \in E$, then $x_2 \notin \Delta_{x_1}$ since otherwise there would be a point z between x_2 and x_1 such that x_2 would be accessible from z and z would be accessible from x_2 and hence the interval between z and x_2 would have consisted of regular points. The set E consists, however, of isolated points only and hence $G-E$ consists of at most a countable number of intervals.

We now consider a process on one such interval G_1. Let $x \in G_1$; then x is either a left-hand or right-hand end-point of the interval Δ_x. In the first case the point x is called a *left* point, and in the second case a *right* point. If x is a left point and $y \in \Delta_x$, then y is also a left point. This follows from the fact that if $x<z<y$ and z is accessible from y and y is accessible from x, then (z, y) contains regular points. Analogously if x is a right point, then Δ_x consists of right points. Furthermore, if x is a left point then the right-hand end-point $\bar{x}$ of the interval Δ_x is also a left point since otherwise $\Delta_x \cap \Delta_{\bar{x}}$ would be nonempty and the points in the intersection would be simultaneously both left and right. In the same manner the left end-point of the interval Δ_x cannot be a left point, provided x is a right point. The same arguments show that for any set B of left points the $\sup B$ will also be a left point, while for a set C of right points the $\inf C$ will be a right point. Hence any left point is to the right of any right point and one of the following possibilities holds: a) all points in G_1 are left points; b) all points in G_1 are right points; c) there exists a point $\bar{x} \in G_1$ such that the points in $G_1 \cap (-\infty, \bar{x})$ are right and those in $(\bar{x}, \infty) \cap G_1$ are left.

Now consider the behavior of a process on an interval consisting solely of left points.

Let G_1 be such an interval. Denote by E_1 the set of inaccessible points in G_1. If x is such a point, then since Δ_x is nonempty one can find $\delta > 0$ such that in the interval $(x, x+\delta)$ there will be no inaccessible points. Hence E_1 is at most countable and $G_1 - E_1$ represents a sum of intervals each one of which contains no inaccessible points. Let U be such an interval. Since for all $t>0$, $\mathsf{P}_x\{x_t > x\} = 1$, the process until its exit out of U is a monotone process. If the right-hand end-point of U is inaccessible, then the process will never depart U; it may be considered on U and is a monotone nondecreasing process with probability 1. A general form of such processes is described in the following theorem.

Theorem 11. *Let U be an interval in $\mathcal{R}^1$ on which a continuous strong Markov nondecreasing process is defined. If this interval contains no absorbing points, then a continuous strictly increasing function $\lambda(t)$ defined on $\mathcal{R}^1$ exists, taking on values in U such that for all $x \in U$*

$$\mathsf{P}_x\{x_t = \lambda(t+s_x)\} = 1,$$

where s_x is the solution of the equation $\lambda(s_x) = x$.

Proof. Let x and $y \in U$, $x<y$. Since there are no inaccessible points in U, y is accessible from x. Indeed if $\bar{x}=\inf [x: \mathsf{P}_x\{\tau_y<\infty\}>0]$, then $\bar{x}$ must coincide with the left-hand end-point of U since if $\bar{x}$ is accessible from $z<\bar{x}$ so is y. Furthermore, since for $z \in (x, y)$ the time τ of the first exit out of (x, y) coincides with τ_y (with probability $\mathsf{P}_z=1$) and

$$\mathsf{P}_z\{\tau_y<a\} \geqslant \mathsf{P}_x\{\tau_y<a\}=\lambda>0,$$

we have

$$\mathsf{P}_z\{x_a<y\} \leqslant 1-\lambda, \qquad \mathsf{P}_z\{x_{ka}<y\} \leqslant (1-\lambda)^k, \qquad \mathsf{P}_z\{\tau_y>ka\} \leqslant (1-\lambda)^k$$

for all $z \in (x, y)$. Therefore $\mathsf{E}_z\tau_y^m<\infty$ for all $m>0$.

We show that Var $\tau_y=0$, i.e., that τ_y is a nonrandom quantity. Consider the random process τ_z on $[x, y]$, where $z \in [x, y]$. Clearly, τ_z is nondecreasing. Since x_t is a nondecreasing continuous function without intervals on which it is constant, τ_z is also a process continuous in z (since τ_z is the inverse function for x_t: $x_{\tau_z}=z$). Finally, note that τ_z is a process with independent increments. If $\mathfrak{F}_z$ is the σ-algebra generated by variables τ_{z_1}, $z_1 \leqslant z$, then since $\mathfrak{F}_z \subset \mathcal{N}_{\tau_z}$ we have for $z_1<z_2$

$$\begin{aligned}
\mathsf{P}\{\tau_{z_2}-\tau_{z_1}<\alpha \mid \mathfrak{F}_{z_1}\} &= \mathsf{E}(\mathsf{P}\{\tau_{z_2}-\tau_{z_1}<\alpha \mid \mathcal{N}_{\tau_{z_1}}\} \mid \mathfrak{F}_{z_1}) \\
&= \mathsf{E}(\mathsf{P}\{\theta_{\tau_{z_1}}\tau_{z_2}<\alpha \mid \mathcal{N}_{\tau_{z_1}}\} \mid \mathfrak{F}_{z_1}) \\
&= \mathsf{E}(\mathsf{P}_{z_1}\{\tau_{z_2}<\alpha\} \mid \mathfrak{F}_{z_1})=\mathsf{P}_{z_1}\{\tau_{z_2}<\alpha\},
\end{aligned}$$

i.e., the distribution of $\tau_{z_2}-\tau_{z_1}$ does not depend on $\mathfrak{F}_{z_1}$. Since the process τ_z is continuous it must be Gaussian and relation $\tau_z>0$ implies that Var $\tau_z=0$.

For $x<y$ let

$$\Phi(x, y)=\mathsf{E}_x\tau_y.$$

Then

$$\mathsf{P}_x\{\tau_y=\Phi(x, y)\}=1.$$

It follows from the equality

$$\Phi(x, y)=\mathsf{E}_x[\tau_z+\theta_{\tau_z}\tau_y]=\Phi(x, z)+\Phi(z, y),$$

valid for $x<z<y$, that there exists a function $\Phi(x)$ such that

$$\Phi(x, y)=\Phi(y)-\Phi(x)$$

and, moreover, $\Phi(x)$ is a continuous strictly monotone function on U. (One can, for instance, choose for $\Phi(x)$ a function which is equal to $\Phi(z, x)$ for $x>z$ and to

$\Phi(x, z)$ for $x < z$, where z is a fixed point in U.) Let $\lambda(t)$ be the inverse function for Φ: $\lambda(\Phi(x)) = x$ for $x \in U$. Since

$$\mathsf{P}_x\{\tau_y = \Phi(y) - \Phi(x)\} = 1,$$

it follows that

$$\mathsf{P}_x\{y = \lambda(\tau_y + \Phi(x))\} = 1, \quad \mathsf{P}_x\{x_{\tau_y} = \lambda(\tau_y + \Delta(x))\} = 1.$$

Utilizing the continuity of x_t, τ_x, and λ we obtain

$$\mathsf{P}_x\{x_{\tau_y} = \lambda(\tau_y + \Phi(x)), \, y > x\} = 1.$$

Substituting an arbitrary $t > 0$ in place of τ_y we complete the proof of the theorem. □

Consider now a Markov process on the interval (α, β) consisting of regular points and such that α and β are accessible from the inside of the interval and Δ_α and Δ_β possess a nonvoid intersection with (α, β). In this case α is accessible from β and β is accessible from α. Indeed, since for any $x \in [\alpha, \beta]$ one can find a neighborhood such that all the points of this neighborhood are accessible one from the other, the interval $[\alpha, \beta]$ can be covered by a finite number of these neighborhoods. Therefore $\alpha = x_0 < x_1 < \cdots < x_a = \beta$ exists such that x_k is accessible from x_{k-1} and from x_{k+1}.

Now let ζ be the moment at which the process hits the boundary of $[\alpha, \beta]$ for the first time. Then in view of inequalities $\zeta \leq \tau_\alpha$, $\zeta \leq \tau_\beta$ we can write

$$\mathsf{P}_x\{\zeta < t\} \geq \max\,[\mathsf{P}_x\{\tau_\alpha < t\}, \mathsf{P}_x\{\tau_\beta < t\}] \geq \max\,[\mathsf{P}_\beta\{\tau_\alpha < t\}, \mathsf{P}_\alpha\{\tau_\beta < t\}],$$

since $\mathsf{P}_x\{\tau_\alpha < t\}$ decreases and $\mathsf{P}_x\{\tau_\beta < t\}$ increases. Hence $\mathsf{E}_x\zeta^m$ is bounded for all $m > 0$.

The following lemma will be useful.

Lemma 12. *For any* $\varepsilon > 0$

$$\lim_{t \downarrow 0} \sup_{\alpha \leq x \leq \beta} \mathsf{P}_x\{\sup_{s \leq t} |x_s - x| > \varepsilon\} = 0;$$

in particular, the process is uniformly stochastically continuous on the interval consisting of regular points.

Proof. Let $x \in [z_1, z_2]$ and $|z_2 - z_1| < \varepsilon/2$. Then

$$\begin{aligned}
&\mathsf{P}_x\{\sup_{s \leq t} |x_t - x| > \varepsilon\} \\
&\leq \mathsf{P}_x\left\{\tau_{z_1} < t, \sup_{\tau_{z_1} \leq s \leq \tau_{z_1} + t} |x_s - z_1| > \frac{\varepsilon}{2}\right\} \\
&\quad + \mathsf{P}_x\left\{\tau_{z_2} < t, \sup_{\tau_{z_2} \leq s \leq \tau_{z_2} + t} |x_s - z_2| > \frac{\varepsilon}{2}\right\} \\
&\leq \mathsf{P}_{z_2}\left\{\sup_{s \leq t} |x_s - z_1| > \frac{\varepsilon}{2}\right\} + \mathsf{P}_{z_2}\left\{\sup_{s \leq t} |x_s - z_2| > \frac{\varepsilon}{2}\right\}.
\end{aligned}$$

Therefore if $\alpha = z_0 < z_1 < \cdots < z_n = \beta$ and $z_k - z_{k-1} < \varepsilon/2$, we have

$$\sup_{\alpha \le x \le \beta} \mathsf{P}_x\{\sup_{s \le t} |x_s - x| > \varepsilon\} \le 2 \sup_{k \le n} \mathsf{P}_{z_k}\left\{\sup_{s \le t} |x_s - z_k| > \frac{\varepsilon}{2}\right\}. \tag{53}$$

It follows from the continuity with probability 1 of the process that for all z

$$\lim_{t \downarrow 0} \mathsf{P}_z\left\{\sup_{s \le t} |x_s - z| > \frac{\varepsilon}{2}\right\} = 0. \tag{54}$$

Relations (53) and (54) imply the assertion of the lemma. □

Some general idea concerning the behavior of the process in a neighborhood of a regular point can be obtained from the following theorem.

Theorem 12. *If x is a regular point, then for any $\delta > 0$*

$$\mathsf{P}_x\{\sup_{t \le \delta} x_t > x\} = 1, \qquad \mathsf{P}_x\{\inf_{t \le \delta} x_t < x\} = 1.$$

Proof. Both assertions are proved in the same manner; thus only the first one will be proved.

Denote by Γ_δ the event $\{\sup_{t \le \delta} x_t > x\}$, $\Gamma = \bigcap_\delta \Gamma_\delta$. Clearly, Γ_δ is $\mathscr{N}_\delta$-measurable, Γ_δ is monotonically decreasing with δ, and $\mathsf{P}_x\{\Gamma_\delta\} \to \mathsf{P}_x\{\Gamma\}$ as $\delta \downarrow 0$. Therefore to prove the theorem it is sufficient to show that $\mathsf{P}_x\{\Gamma\} = 1$. Since Γ is $\mathscr{N}_{0+}$-measurable in view of Lemma 2 in Volume II, Chapter II, Section 6, $\mathsf{P}_x\{\Gamma\}$ may take only the two values 0 and 1. Assume that $\mathsf{P}_x\{\Gamma\} = 0$. Then $\mathsf{P}_x\{\bar{\Gamma}\} = 1$, where $\bar{\Gamma}$ is the complement to Γ. If even $\bar{\Gamma}$ occurs it means that for some δ the event $\bar{\Gamma}_\delta$ occurred, i.e., $\{\sup_{s \le \delta} x_s \le x\}$. Denote by η the time of the first exit from the set $(-\infty, x]$; η is positive on the set $\bar{\Gamma}$ and is a Markov moment, being the time of the first exit out of a closed set (cf. Volume II, Chapter II, Section 5, p. 128). Since $x_\eta = x$ we have $\mathsf{P}_{x_\eta}(\bar{\Gamma}) = 1$. Hence

$$\mathsf{P}_x\{\sup_{s \le \eta + \theta_\eta \eta} x_s \le x\} = \mathsf{P}_x\{\bar{\Gamma} \cap \theta_\eta \bar{\Gamma}\} = \mathsf{E}_x \chi_{\bar{\Gamma}} \mathsf{P}\{\theta_\eta \bar{\Gamma} | \mathscr{N}_\eta\} = \mathsf{E}_x \chi_{\bar{\Gamma}} \mathsf{P}_{x_\eta}(\bar{\Gamma}) = 1$$

(here χ_A is the indicator of the set A). Therefore

$$\mathsf{P}_x\{\eta + \theta_\eta \eta \le \eta\} = 1,$$

which contradicts the fact that η is positive on $\bar{\Gamma}$. This contradiction implies that $\mathsf{P}_x(\Gamma) = 1$. The theorem is proved. □

Corollary. *If x is a regular point, then for all $t > 0$*

$$\lim_{y \to x} \mathsf{P}_x\{\tau_y < t\} = 1. \tag{55}$$

For example for $y > x$

$$\mathsf{P}_x\{\tau_y < t\} \leqslant \mathsf{P}_x\{\sup_{s \leqslant t} x_s \geqslant y\}$$

and formula (55) follows from Theorem 12.

We introduce the function

$$m(x) = \mathsf{P}_x\{x_\zeta = \beta\}. \tag{56}$$

Let $x_1 < x_2$. Then

$$\begin{aligned} m(x_1) &= \mathsf{P}_{x_1}\{x_\zeta = \beta\} = \mathsf{P}_{x_1}\{\tau_{x_2} < \zeta, x_\zeta = \beta\} \\ &= \mathsf{E}\chi_{\{\tau_{x_2} < \zeta\}}\theta_{\tau_{x_2}}\chi_{\{x_\zeta = \beta\}} = \mathsf{E}\chi_{\{\tau_{x_2} < \zeta\}}\mathsf{E}_{x_{\tau_{x_2}}}\chi_{\{x_\zeta = \beta\}} \\ &= \mathsf{E}\chi_{\{\tau_{x_2} < \zeta\}}\mathsf{P}_{x_2}\{x_\zeta = \beta\} = m(x_2)\mathsf{P}_{x_1}\{\tau_{x_2} < \zeta\}. \end{aligned}$$

Since $\mathsf{P}_x\{\tau_{x_2} < \zeta\} \leqslant 1$, we have $m(x_1) \leqslant m(x_2)$. The function $m(x)$ is nondecreasing on $[\alpha, \beta]$. Moreover, in view of equality (55) and the positiveness of ζ

$$\lim_{x_2 \downarrow x_1} \mathsf{P}_{x_1}\{\tau_{x_2} < \zeta\} = 1, \quad \text{i.e.,} \quad \lim_{x_2 \downarrow x_1} m(x_2) = m(x_1).$$

Therefore $m(x)$ is continuous on the right. Analogously one verifies that the function

$$1 - m(x) = \mathsf{P}_x\{x_\zeta = \alpha\}$$

is continuous on the left. Thus $m(x)$ is an increasing continuous function and, moreover, as it is easy to see that $m(\alpha) = 0$ and $m(\beta) = 1$.

We shall show that $m(x)$ is strictly increasing. Assume that for $x_1 < x_2$ $m(x_1) = m(x_2)$ and x_1 is the smallest number z such that $m(z) = m(x_2)$. Then

$$\mathsf{P}_{x_1}\{\tau_{x_2} < \zeta\} = 1.$$

This implies that with probability $\mathsf{P}_{x_1} = 1$ the process will reach the point x_2 before reaching point α. Therefore $x_1 > \alpha$ and x_1 is the smallest number from which x_2 is reached before α. In view of Theorem 12 the process may with probability $\mathsf{P}_{x_1} = 1$ reach the interval $[\alpha, x_1)$ during an arbitrarily small time period (i.e., before reaching point x_2) and also from the points of interval (α, x_1) it may—with positive probability—reach α before reaching x_2 (from the definition of point x_1). We arrived at a contradiction which shows that $m(x)$ is strictly increasing.

Theorem 13. *The function $m(x)$ defined by equality* (56) *maps the interval* $[\alpha, \beta]$ *into the interval* $[0, 1]$ *in a continuous, monotone, and one-to-one manner; the functional $\alpha_t = m(x_t) - m(x_0)$ for $t < \zeta$ and $\alpha_t = m(x_{\zeta-0}) - m(x_0)$ for $t \geqslant \zeta$ is an M-functional on the process.*

Proof. It is only required to show that α_t is an M-functional.

Since

$$\theta_h\alpha_t = \begin{cases} m(x_{t+h}) - m(x_h), & t+h \leq \zeta, \\ m(x_\zeta) - m(x_h), & h < \zeta,\ t+h \geq \zeta, \\ 0 , & h \geq \zeta, \end{cases}$$

it follows that $\theta_h\alpha_t = \alpha_{t+h} - \alpha_t$ and hence α_t is an additive functional.

Furthermore α_t is clearly continuous with probability 1.

Observe that

$$m(x) = \mathsf{P}_x\{x_\zeta = \beta\}m(\beta) + \mathsf{P}_x\{x_\zeta = \alpha\}m(\alpha) = \mathsf{E}_x m(x_\zeta).$$

Therefore for any Markov time $\tau \leq \zeta$

$$\mathsf{E}_x m(x_\tau) = \mathsf{E}_x \mathsf{E}_{x_\tau} m(x_\zeta) = \mathsf{E}_x \theta_\tau m(x_\zeta) = \mathsf{E}_x m(x_\zeta) = m(x).$$

In particular,

$$\mathsf{E}_x m(x_{t\wedge\zeta}) = m(x), \qquad \mathsf{E}_x \alpha_t = 0.$$

Next, let

$$\begin{aligned} f(t, x) &= \mathsf{E}_x[m(x_{t\wedge\zeta}) - m(x)]^2 \\ &= \mathsf{E}_x m^2(x_{t\wedge\zeta}) - m^2(x) + 2m(x)[m(x) - \mathsf{E}_x m(x_{t\wedge\zeta})] \\ &= \mathsf{E}_x m^2(x_{t\wedge\zeta}) - m^2(x). \end{aligned}$$

It follows from the continuity of $m^2(x)$ and the uniform stochastic continuity of the process (Lemma 12) that

$$\lim_{t\downarrow 0} \sup_{\alpha \leq x \leq \beta} f(t, x) = 0.$$

Hence in view of Theorem 3 in Volume II, Chapter II, Section 6, there exists a W-functional $\langle\alpha, \alpha\rangle_t$ such that $\mathsf{E}_x\alpha_t^2 = \mathsf{E}_x\langle\alpha, \alpha\rangle_t$. □

Remark. The relation

$$\mathsf{E}_x m(x_\tau) = m(x)$$

which was verified for any Markov time $\tau \leq \zeta$ implies that $m(x)$ is a harmonic function.

Clearly in order to study the Markov process x_t it is sufficient to investigate the process $m(x_t)$, i.e., a process on $[0, 1]$ for which $m(x) = x$.

From now on, we shall assume that $m(x) = x$ and that the interval $[\alpha, \beta]$ coincides with $[0, 1]$.

We introduce the function

$$n(x) = \mathsf{E}_x\zeta.$$

Let $0 \leqslant a < x < b \leqslant 1$. Denoting by τ the moment of the first exit out of (a, b), we have

$$\zeta = \tau + \theta_\tau \zeta,$$
$$\mathsf{E}_x\zeta = \mathsf{E}_x\tau + \mathsf{E}_x \mathsf{E}_{x_\tau}\zeta,$$
$$n(x) = \mathsf{E}_x\tau + \mathsf{E}_x n(x_\tau).$$

To compute $\mathsf{E}_x n(x_\tau)$ we note that $\mathsf{E}_x x_\tau = x$ (since $m(x) = x$). Hence

$$x = a\mathsf{P}_x\{x_\tau = a\} + b\mathsf{P}_x\{x_\tau = b\}.$$

Moreover, since

$$\mathsf{P}_x\{x_\tau = a\} + \mathsf{P}_x\{x_\tau = b\} = 1,$$

we have

$$\mathsf{P}_x\{x_\tau = a\} = \frac{b-x}{b-a}, \qquad \mathsf{P}_x\{x_\tau = b\} = \frac{x-a}{b-a}.$$

Thus

$$n(x) = \mathsf{E}_x\tau + n(a)\frac{b-x}{b-a} + n(b)\frac{x-a}{b-a}.$$

Set $b - x = t(b-a)$, then $x = ta + (1-t)b$ and hence

$$n(ta + (1-t)b) - tn(a) - (1-t)n(b) = \mathsf{E}_x\tau > 0.$$

This inequality is valid for all $a, b \in [0, 1]$, $0 < t < 1$. This implies that $n(x)$ is a strictly convex (upward) function. Therefore there exists a decreasing derivative $n'(x)$.

Consider now the characteristic operator of the process at the interior points of the interval $[0, 1]$. Let τ be the time of the first exit out of the interval $(x - \varepsilon_1, x + \varepsilon_2)$. Then

$$\mathsf{E}_x f(x_\tau) = f(x - \varepsilon_1)\frac{\varepsilon_2}{\varepsilon_1 + \varepsilon_2} + f(x + \varepsilon_2)\frac{\varepsilon_1}{\varepsilon_1 + \varepsilon_2},$$

(57) $$\mathsf{E}_x\tau = n(x) - n(x - \varepsilon_1)\frac{\varepsilon_2}{\varepsilon_1 + \varepsilon_2} - n(x + \varepsilon_2)\frac{\varepsilon_1}{\varepsilon_1 + \varepsilon_2},$$

$$\mathfrak{A}f(x) = \lim_{\varepsilon_1 \downarrow 0, \varepsilon_2 \downarrow 0} \frac{(1/\varepsilon_2)[f(x + \varepsilon_2) - f(x)] - (1/\varepsilon_1)[f(x) - f(x - \varepsilon_1)]}{(1/\varepsilon_2)[n(x) - n(x + \varepsilon_2)] - (1/\varepsilon_1)[n(x - \varepsilon_1) - n(x)]}.$$

The last formula is too complicated for computing a characteristic operator. However, for a class of functions which is sufficient for defining a generating operator, this formula may substantially be simplified.

Theorem 14. *Assume that $f(x)$ is absolutely continuous and a continuous function $g(t)$ exists such that the relation*

$$f'(x)=f'(0)+\int_0^x g(t)\,dn'(t) \tag{58}$$

is fulfilled. Then for all $x\in(0,1)$

$$\mathfrak{A}f(x)=-g(x).$$

Proof. We have

$$\begin{aligned}\frac{1}{\varepsilon}[f(x+\varepsilon)-f(x)]&=\frac{1}{\varepsilon}\int_x^{x+\varepsilon} f'(u)\,du\\ &=\frac{1}{\varepsilon}\int_x^{x+\varepsilon}\left[f'(x)+\int_x^u g(t)\,dn'(t)\right]du\\ &=f'(x)+\int_x^{x+\varepsilon}\frac{x+\varepsilon-t}{\varepsilon}g(t)\,dn'(t).\end{aligned}$$

Utilizing this representation (ε may also be a negative number) we obtain

$$\frac{1}{\varepsilon_2}[f(x+\varepsilon_2)-f(x)]-\frac{1}{\varepsilon_1}[f(x)-f(x-\varepsilon_1)]=\int_{x-\varepsilon_1}^{x+\varepsilon_2}\Delta_{\varepsilon_1,\varepsilon_2}(x,t)g(t)\,dn'(t),$$

where

$$\Delta_{\varepsilon_1,\varepsilon_2}(x,t)=\begin{cases}\dfrac{t-x+\varepsilon_1}{\varepsilon_1} & \text{for } x-\varepsilon_1\leqslant t\leqslant x\\[2ex] \dfrac{x+\varepsilon_2-t}{\varepsilon_2} & \text{for } x\leqslant t\leqslant x+\varepsilon_2.\end{cases}$$

Substituting $g(t)=1$ we have

$$\frac{1}{\varepsilon_2}[n(x+\varepsilon_2)-n(x)]-\frac{1}{\varepsilon_1}[n(x)-n(x-\varepsilon_1)]=\int_{x-\varepsilon_1}^{x+\varepsilon_2}\Delta_{\varepsilon_1,\varepsilon_2}(x,t)\,dn'(t).$$

Since $\Delta_{\varepsilon_1,\varepsilon_2}(x,t)$ is nonnegative, $n'(t)$ is strictly decreasing, and $g(t)$ is continuous, we verify using the mean value theorem that

$$\lim_{\varepsilon_1\downarrow 0,\,\varepsilon_2\downarrow 0}\frac{\int_{x-\varepsilon_1}^{x+\varepsilon_2}\Delta_{\varepsilon_1,\varepsilon_2}(x,t)g(t)\,dn'(t)}{\int_{x-\varepsilon_1}^{x+\varepsilon_2}\Delta_{\varepsilon_1,\varepsilon_2}(x,t)\,dn'(t)}=g(x).$$

The theorem is proved. □

Remark 1. The function $g(t)$ satisfying relation (58) is naturally denoted by $df'(t)/dn'(t)$. Thus, the characteristic operator of process x_t, $\mathfrak{A}$, is defined on absolutely continuous functions f, for which $df'(x)/dn'(x)$ exists and is continuous; moreover,

$$\mathfrak{A}f(x) = -\frac{df'(x)}{dn'(x)}. \tag{59}$$

Remark 2. The generating operator A of a process x_t which is cut off at the moment of its exit out of $(0, 1)$ is defined for all absolutely continuous functions f such that $df'(x)/dn'(x)$ is continuous on $[0, 1]$ and $f(0) = f(1) = 0$; moreover, $\mathsf{A}f$ coincides with $\mathfrak{A}f$.

Indeed, taking into account the relation between characteristic and generating operators for processes on a compact set (cf. Volume II, Chapter II, Section 5, Theorem 1) we need only verify that on functions belonging to $\mathscr{C}^0_{[0,1]}$ which vanish at points 0 and 1, operator $\mathfrak{A}$ is given by formula (59) provided such an operator is well defined. Since in this case $\mathfrak{A}$ and A coincide on $\mathscr{C}^0_{[0,1]}$ the closure of A implies the closure of $\mathfrak{A}$. It is easy to see that the differentiation operator appearing on the right-hand side of (59) is also closed. This implies that operators A and $\mathfrak{A}$ coincide everywhere and that $\mathsf{A}f$ coincides with the left-hand side of (59).

Finally, consider the behavior of the process on the interval of regularity. Let this interval be (α, β); here the points α and β are nonregular.

There are four possible cases: 1) the boundary point α of the interval (α, β) is accessible from inside the interval, any point $x \in (\alpha, \beta)$ is accessible from α (such a boundary point is called a *regular boundary*); 2) α is accessible from inside the interval but the points of the interval are inaccessible from it (such a point will be called a *capturing boundary*); 3) α is inaccessible from inside the interval but the points belonging to the interval are accessible from it [such a point will be called an *emitting* (*releasing*) *boundary*]; 4) α is inaccessible from the inside and all the points of the interval are also inaccessible from it (such a point is called a *natural boundary*).

If α is a regular boundary, then, being nevertheless a nonregular point, it is either inaccessible from the left or impassable to the left. In the last case α is a reflecting boundary for the interval (α, β). It is easy to find a form of a generating operator for a process with two reflecting boundaries.

Theorem 15. *If $\hat{x}_t$ is a process on $[0, 1]$ with $m(x) = x$, 0 and 1 are reflecting boundaries of the interval of regularity $(0, 1)$, then $\mathsf{A}f(x)$ is defined for $0 < x < 1$ by the left-hand side of (59) and*

$$\mathsf{A}f(0) = \lim_{\varepsilon \to 0} \frac{f(\varepsilon) - f(0)}{\mathsf{E}_0 \tau_\varepsilon}, \qquad \mathsf{A}f(1) = \lim_{\varepsilon \to 0} \frac{f(1-\varepsilon) - f(1)}{\mathsf{E}_1 \tau_{1-\varepsilon}},$$

and $\mathscr{D}_\mathsf{A}$ coincides with the set of functions for which $\mathsf{A}f$ is continuous.

Proof. The proof of the theorem is immediate if we evaluate the characteristic operator at points 0 and 1 and utilize Remark 2 above and Theorem 1 in Volume II, Chapter II, Section 5. □

If the boundaries of the interval are accessible while the interior points are inaccessible from the boundary, then it is natural to consider processes which are cut off after reaching the boundary. Their characteristic operators are described in Remark 2. If the boundaries are not reached the process always remains inside the interval of regularity. Since an interval is a locally compact space and a characteristic operator can be defined locally at each point using formula (59), the generating operator of the process will be defined with the aid of Theorem 1 in Volume II, Chapter II, Section 5. Still it will be necessary to utilize a whole collection of functions $m(x)$ and $n(x)$. We shall now show how this can be avoided.

Lemma 13. *There exists a strictly increasing continuous harmonic function $M(x)$ on (α, β). Any other continuous harmonic function $g(x)$ on (α, β) is of the form*

$$g(x)=aM(x)+b,$$

where a and b are constants.

Proof. Let $\alpha_n \downarrow \alpha$, $\beta_n \uparrow \beta$, $\alpha_1<\beta_1$. Denote by ζ_n the time of the first exit out of (α_n, β_n) and let

$$\gamma_n(x)=\mathsf{P}_x\{x_{\zeta_n}=\beta_n\}.$$

As was shown above, $\gamma_n(x)$ is a harmonic function in (α_n, β_n). Therefore we have for $m<n$ and $x\in(\alpha_m, \beta_m)$

$$\mathsf{E}_x\gamma_n(x_{\zeta_m})=\gamma_n(x).$$

However,

$$\begin{aligned}\gamma_n(x)=\mathsf{E}_x\gamma_n(x_{\zeta_m})&=\gamma_n(\alpha_m)\mathsf{P}_x\{x_{\zeta_m}=\alpha_m\}+\gamma_n(\beta_m)\mathsf{P}_x\{x_{\zeta_m}=\beta_m\}\\&=\gamma_n(\alpha_m)(1-\gamma_m(x))+\gamma_n(\beta_m)\gamma_m(x)\\&=\gamma(\alpha_m)+[\gamma_n(\beta_m)-\gamma_n(\alpha_m)]\gamma_m(x),\end{aligned}$$

i.e.,

$$\gamma_m(x)=\frac{\gamma_n(x)-\gamma_n(\alpha_m)}{\gamma_n(\beta_m)-\gamma_n(\alpha_m)}. \tag{60}$$

Set

$$g_n(x)=\frac{\gamma_n(x)-\gamma_n(\alpha_1)}{\gamma_n(\beta_1)-\gamma_n(\alpha_1)}, \qquad x\in(\alpha_n, \beta_n).$$

It is easy to verify that $g_n(x)=g_m(x)$ for $x\in(\alpha_m, \beta_m)$ and $m<n$. It is only necessary to utilize (60) for this purpose. Thus a function $M(x)$ exists, which coincides with $g_n(x)$ on each of the intervals (α_n, β_n). This function possesses the required properties.

Let $g(x)$ be an arbitrary locally bounded harmonic function. Then for $\alpha_m<x<\beta_m$

$$g(x)=g(\alpha_m)(1-\gamma_m(x))+g(\beta_m)\gamma_m(x),$$

i.e., g is a linear function of a constant and $\gamma_m(x)$. The same applies to $M(x)$ on (α_m, β_m). Hence constants a_m and b_m exist for each m such that

$$g(x)=a_mM(x)+b_m, \quad x\in(\alpha_m, \beta_m).$$

Since $M(\alpha_1)=0$, $M(\beta_1)=1$, it follows that $b_m=g(\alpha_1)$, $a_m=g(\beta_1)-g(\alpha_1)$, i.e., a_m and b_m do not depend on m.

The lemma is proved. □

Once again instead of the initial process x_t, it is possible to consider the process $\hat{x}_t=M(x_t)$ for which $M(x)=x$. The process $\hat{x}_t$ is defined on an (infinite or finite) interval. Therefore it is natural to assume from the beginning that for the initial process, $M(x)$ coincides with x.

Lemma 14. *A convex (upward) function $N(x)$ exists such that for $\alpha<x-\varepsilon_1<x<x+\varepsilon_2<\beta$*

$$\mathsf{E}_x\tau=N(x)-\frac{\varepsilon_2}{\varepsilon_1+\varepsilon_2}N(x-\varepsilon_1)-\frac{\varepsilon_1}{\varepsilon_1+\varepsilon_2}N(x+\varepsilon_2),$$

where τ is the time of the first exit outside the interval $(x-\varepsilon_1, x+\varepsilon_2)$.

Proof. Let α_k, β_k, and ζ_k be as in Lemma 13 and

$$n_k(x)=\mathsf{E}_x\zeta_k.$$

On (α_k, β_k) set

$$S_k(x)=n_k(x)-\frac{\beta_1-x}{\beta_1-\alpha_1}n_k(\alpha_1)-\frac{(x-\alpha_1)}{\beta_1-\alpha_1}n_k(\beta_1).$$

It is easy to verify that for $\alpha_k<x-\varepsilon_1<x<x+\varepsilon_2<\beta_k$

$$\mathsf{E}_x\tau=S_k(x)-\frac{\varepsilon_2}{\varepsilon_1+\varepsilon_2}S_k(x-\varepsilon_1)-\frac{\varepsilon_1}{\varepsilon_1+\varepsilon_2}S_k(x+\varepsilon)=S_k(x)-\mathsf{E}_xS_k(x_\tau),$$

where τ is as in the statement of the lemma. Therefore we have on (α_k, β_k) for $m>k$

$$S_m(x)-S_k(x)=\mathsf{E}_x[S_m(x_\tau)-S_k(x_\tau)],$$

i.e., $S_m(x)-S_k(x)$ is a harmonic function. Since $S_m(\alpha_1)=S_k(\alpha_1)=0$, $S_m(\beta_1)=S_k(\beta_1)=0$, it easily follows from Lemma 13 that $S_m(x)=S_k(x)$ on (α_k, β_k). Setting $N(x)=S_k(x)$ for $\alpha_k<x<\beta_k$, $k=1, 2, \ldots$, we arrive at the required function. □

Now with the aid of function $N(x)$ the characteristic operator $\mathfrak{A}$ can be defined at once for all the points of the interval (α, β) by the equation

(61)
$$\mathfrak{A}f(x)=-\frac{df'(x)}{dN'(x)}.$$

Finally, functions $M(x)$ and $N(x)$ permit us to characterize the boundary points α and β.

We present a more detailed classification of an inaccessible boundary. An inaccessible boundary α is called *attracting* if for any $\varepsilon>0$ there exists $\delta>0$ such that

(62)
$$\mathsf{P}_x\{\lim_{t\to\infty} x_t=\alpha\}>1-\varepsilon$$

for all $x\in(\alpha, \alpha+\delta)$. It is called *repelling* if for any $x>\alpha$ and $x_1<x$

$$\mathsf{P}_{x_1}\{\tau_x<\infty\}=1.$$

Theorem 16. *A boundary α is inaccessible if $N(\alpha+0)=-\infty$; moreover, in case* 1) *$M(\alpha+0)>-\infty$ and the boundary is attracting, while in case* 2) *$M(\alpha+0)=-\infty$ and the boundary is repelling.*

Proof. We may assume without loss of generality that $M(x)=x$. Then $x_t-\alpha$ is a nonnegative martingale since

$$\mathsf{E}(x_{t+h}-\alpha|\mathcal{N}_t)=\mathsf{E}_{x_t}(x_h-\alpha)=x_t-\alpha.$$

Consequently, in view of Theorem 1 in Volume I, Chapter II, Section 2, there exists for all x with probability $\mathsf{P}_x=1$ the limit

$$\lim_{t\to\infty} x_t=x_\infty.$$

Note that this limit cannot be an interior point in the interval (α, β) since from any interval with end-points which are interior points of (α, β) the process departs within a finite time interval. Therefore $x_\infty=\alpha$ or $x_\infty=\beta$.

If $\beta=\infty$, then

$$\mathsf{P}_x\{x_\infty=\alpha\}=1$$

since $\mathsf{E}_x x_\infty\leq \mathsf{E}_x x_t=x$.

If, however, $\beta < \infty$, then x_t is a bounded martingale and

$$\mathsf{E}_x x_\infty = x.$$

Consequently,

$$\mathsf{P}_x\{x_\infty = \alpha\} = \frac{\beta - x}{\beta - \alpha}, \qquad \mathsf{P}_x\{x_\infty = \beta\} = \frac{x - \alpha}{\beta - \alpha}.$$

The form of this probability implies the existence of a $\delta > 0$ such that (62) is satisfied.

We show that

$$\mathsf{P}_x\{x_t > \alpha\} = 1$$

for all t, i.e., that α is an inaccessible boundary. Assume that for some t

$$\mathsf{P}_x\{x_t = \alpha\} = \delta > 0.$$

Then for all $\varepsilon < x - \alpha$

$$\mathsf{P}_x\{\tau_{\alpha+\varepsilon} < t\} > \delta$$

and hence for $\bar{x} \in (\alpha + \varepsilon, x)$

$$\mathsf{P}_{\bar{x}}\{\tau_{\alpha+\varepsilon} < t\} > \delta.$$

Denote by τ the time of the first exit out of the interval $(\alpha + \varepsilon, x)$. Then $\tau \leqslant \tau_{\alpha+\varepsilon}$ and hence $\mathsf{P}_{\bar{x}}\{\tau < t\} > \delta$ for all $\bar{x} \in (\alpha + \varepsilon, x)$. However,

$$\mathsf{P}_{\bar{x}}\{x_t \notin (\alpha + \varepsilon, x)\} \geqslant \mathsf{P}_{\bar{x}}\{\tau > t\} > \delta.$$

Utilizing the argument presented on pp. 128–129 in Volume II, Chapter II, Section 5, we verify that

$$\mathsf{E}_{\bar{x}}\tau \leqslant \sum_{k=0}^{\infty} \mathsf{P}_{\bar{x}}\{\tau > kt\} \leqslant t \sum_{k=0}^{\infty} (1 - \delta)^k = \frac{t}{\delta}.$$

Hence,

$$\mathsf{E}_{\bar{x}}\tau = N(\bar{x}) - \frac{x - \bar{x}}{x - \alpha - \varepsilon} N(\alpha + \varepsilon) - \frac{\bar{x} - \alpha - \varepsilon}{x - \alpha - \varepsilon} N(x) \leqslant \frac{t}{\delta}.$$

The last inequality contradicts the condition $N(\alpha + 0) = -\infty$ since it is valid for all $\varepsilon > 0$. Assertion 1) is thus proved.

2) Now let $\alpha < x_1 < x$. We show that the probability $q(x_1, x)$ to reach x from x_1 before α equals 1.

If $\alpha < x_2 < x_1 < x$, then denoting by τ the moment of the first exit out of (x_2, x) we have

$$\mathsf{P}_{x_1}\{x_\tau = x\} = \frac{M(x_1) - M(x_2)}{M(x) - M(x_2)}.$$

However,

$$q(x_1, x) \geq \mathsf{P}_{x_1}\{x_\tau = x\}$$

for any $x_2 \in (\alpha, x_1)$. Approaching the limit as $x_2 \downarrow \alpha$ we verify that $q(x_1, x) = 1$.

Now we show that the boundary α is inaccessible. Let A be the event that the process reaches α before reaching x, B_k be the event that the process will intersect the interval (x_1, x) $2k$ times, and let τ_k be the time of the $2k$th intersection of the interval (x_1, x). Then τ_k is a Markov time and

$$\mathsf{P}_{x_1}\{x_{\tau_k} = x_1\} = 1.$$

Let V be the event that the point α will be reached from the point x_1. The point α can be reached after the interval (x_1, x) has been intersected $2k$ times, $k = 0, 1, 2, \ldots$. Hence noting that $\mathsf{P}_{x_1}\{A\} = 0$ we have

$$\begin{aligned} \mathsf{P}_{x_1}\{V\} &= \mathsf{P}_{x_1}\{A\} + \mathsf{P}_{x_1}\{B_1 \cap \theta_{\tau_1} A\} + \cdots + \mathsf{P}_{x_1}\{B_k \cap \theta_{\tau_k} A\} + \cdots \\ &= \mathsf{P}_{x_1}\{A\}\left(1 + \sum_{k=1}^{\infty} \mathsf{P}_{x_1}\{B_k\}\right) = 0. \end{aligned}$$

The theorem is proved. □

Remark. If $M(\alpha+0) > -\infty$ and $N(\alpha+0) > -\infty$, then the boundary α is accessible. The condition $M(\alpha+0) = -\infty$ necessarily implies that $N(\alpha+0) = -\infty$ since $N(M^{-1}(x))$ is a convex (upward) function which vanishes at two points (M^{-1} is the inverse for M).

We shall now study the behavior of the process at the interval of regular points assuming that both boundaries are repelling. It follows from Theorem 16 that in this case $M(\alpha+0) = -\infty$ and $M(\beta-0) = +\infty$. Therefore it may be assumed without loss of generality that $M(x) = x$ and (α, β) coincides with $(-\infty, +\infty)$. It was shown in the proof of Theorem 16 that $\mathsf{P}_x\{\tau_y < \infty\} = 1$. Theorem 12 implies that for all $\delta > 0$

$$\lim_{x \to y} \mathsf{P}_x\{\tau_y > \delta\} = 0.$$

Utilizing the relation

$$|\mathsf{E}_x f(x_t) - \mathsf{E}_y f(x_t)| \leq 2\|f\| \mathsf{P}_x\{\tau_y > \delta\} + \sup_{s \leq \delta} |\mathsf{E}_y f(x_{t-s}) - \mathsf{E}_y f(x_t)|,$$

we verify that $\mathsf{E}_x f(x_t)$ is a continuous function provided f is continuous since the second summand tends to 0 for $\delta > 0$ in view of the continuity of x_t. Thus x_t is a stochastically continuous Feller process.

We shall obtain the conditions under which $\mathsf{E}_x\tau_y$ is finite.

Lemma 15. *If*

$$\lim_{a\to-\infty}\frac{1}{a}N(a)=\gamma_1<+\infty \tag{63}$$

exists, then $\mathsf{E}_x\tau_y<\infty$ *for all* $x<y$; *if*

$$\lim_{b\to+\infty}\frac{1}{b}N(b)=\gamma_2>-\infty$$

exists, then $\mathsf{E}_x\tau_y<\infty$ *for all* $x>y$.

Proof. We shall establish, for instance, the first assertion of the lemma. Let $a<x<y$ and $\tau_{[a,y]}$ be the time of the first exit out of (a, y). Then in view of Lemma 14

$$\mathsf{E}_x\tau_{[a,y]}=N(x)-\frac{y-x}{y-a}N(a)-\frac{x-a}{y-a}N(y).$$

Clearly, $\tau_{[a,y]}\uparrow\tau_y$ as $a\to-\infty$. Therefore

$$\mathsf{E}_x\tau_y=N(x)-N(y)+(y-x)\lim_{a\to-\infty}\frac{N(a)}{a}. \tag{64}$$

The lemma is proved. □

Remark 1. If the limit (63) exists then the function $N'(x)$ is bounded for $x\to-\infty$ and $\gamma_1=\lim_{x\to-\infty}N'(x)$. Analogously, if γ_2 is finite, then $\gamma_2=\lim_{x\to+\infty}N'(x)$.

Remark 2. If $N'(-\infty)$ is finite, then the formula

$$\mathsf{E}_x\tau_y=\int_x^y[N'(-\infty)-N'(z)]\,dz \tag{65}$$

is valid for $\mathsf{E}_x\tau_y$ with $x<y$ (this formula is a consequence of (64)).

In particular, if

$$\int_{-\infty}^x[N'(-\infty)-N'(z)]\,dz<\infty,$$

then the variable $\mathsf{E}_x\tau_y$ is bounded for $x\in(-\infty, y]$. An analogous assertion is valid for $\mathsf{E}_x\tau_y$ if $x>y$, provided $N'(+\infty)>-\infty$.

Remark 3. Let $\alpha = -\infty$ be an emitting boundary; since for some $t>0$ and y

$$\mathsf{P}_\alpha\{\tau_y<t\}>0,$$

it follows that for $x \in (\alpha, z)$

$$\mathsf{P}_x\{\tau_y<t\} \geqslant \mathsf{P}_\alpha\{\tau_y<t\}.$$

In the same manner as in the proof of Theorem 11 we also obtain that

$$\sup_{x<z} \mathsf{E}_x\tau_y<\infty.$$

Therefore $\alpha = \pm\infty$ for an emitting boundary, the integral $\int_\alpha^y [N^1(z)-N^1(\alpha)]\, dz$ being finite.

If a repelling boundary is not emitting it is then a *natural* boundary.

Lemma 16. *Let the boundary* α $(\alpha = \pm\infty)$ *be natural. Then*

$$\lim_{x\to\alpha} \mathsf{P}_x\{\tau_a<t\}=0. \tag{66}$$

Proof. Let $\alpha=-\infty$. Clearly, $\mathsf{P}_x\{\tau_a<t\}$ monotonically decreases as $x\to-\infty$. If, however, $\inf_{x\leqslant a} \mathsf{P}_x\{\tau_a<t\} \geqslant \delta>0$ holds, then

$$\sup_{x\leqslant a} \mathsf{P}_x\{\tau_a>kt\} = \sup_{x\leqslant a} \mathsf{E}_x\chi_{\{\tau_a>(k-1)t\}} \mathsf{P}\{\theta_{(k-1)t}\tau_a>t \mid \mathcal{N}_{(k-1)t}\}$$

$$\leqslant (1-\delta) \sup_{x\leqslant a} \mathsf{P}_x\{\tau_a>(k-1)t\} \leqslant (1-\delta)^k$$

and the inequality

$$\sup_{x\leqslant a} \mathsf{E}_x\tau_a \leqslant \sum_{k=1}^{\infty} kt(1-\delta)^{k-1} = \frac{t}{\delta^2}$$

would be valid; this, however, contradicts the fact that $-\infty$ is a natural boundary.

The lemma is proved. □

Lemma 17. *For any continuous bounded function* $f(x)$ *for which the limits* $f(-\infty)=\lim_{x\to-\infty} f(x)$, $f(+\infty)=\lim_{x\to+\infty} f(x)$ *exist, the relation*

$$\lim_{t\to0} \|T_t f - f\| = 0$$

is satisfied.

Proof. Since for $a<b$

$$\sup_{a\leqslant x\leqslant b} |T_t f(x)-f(x)| \leqslant \sup_{a\leqslant x\leqslant b} \mathsf{P}_x\{|x_t-x|>\varepsilon\} + \sup_{\substack{a\leqslant x_1\leqslant b\\ |x_2-x_1|\leqslant\varepsilon}} |f(x_1)-f(x_2)|,$$

it follows from Lemma 12 that

$$\lim_{t\to 0} \sup_{a\leqslant x\leqslant b} |T_t f(x) - f(x)| = 0.$$

Therefore to prove the lemma it is sufficient to show that the quantity

$$\overline{\lim_{t\to 0}}\, [\sup_{x\leqslant a} |T_t f(x) - f(x)| + \sup_{x\geqslant b} |T_t f(x) - f(x)|]$$

may become arbitrarily small by a proper choice of a and b. Consider, for example, the first term to the right of the lim sign. Choosing $a_1 > a$ we have

$$\begin{aligned}\sup_{x\leqslant a} |T_t f(x) - f(x)| &= \sup_{x\leqslant a} |f(x) - f(-\infty)| + \sup_{x\leqslant a} |T_t f(x) - f(-\infty)| \\ &= \sup_{x\leqslant a} |f(x) - f(-\infty)| + \sup_{x\leqslant a} [\mathsf{E}_x |f(x_t) - f(-\infty)| \chi_{\{\tau_{a_1} < t\}} \\ &\quad + \mathsf{E}_x |f(x_t) - f(-\infty)| \chi_{\{\tau_{a_1} \geqslant t\}}] \\ &\leqslant 2 \sup_{x\leqslant a_1} |f(x) - f(-\infty)| + 2\|f\| \mathsf{P}_a\{\tau_{a_1} < t\}.\end{aligned}$$

Since $\lim_{t\to 0} \mathsf{P}_a\{\tau_{a_1} < t\} = 0$ it follows that

$$\overline{\lim_{t\to 0}} \sup_{x\leqslant a} |T_t f(x) - f(x)| \leqslant 2 \sup_{x\leqslant a_1} |f(x) - f(-\infty)|.$$

This implies the validity of the assertion of the lemma. □

Lemma 18. *Denote by $\hat{C}_{(-\infty,\infty)}$ the set of continuous functions $f(x)$ for which there exist the limits $f(-\infty)$, $f(+\infty)$, and $f(\alpha) = 0$, provided $\alpha = \pm\infty$ is a natural boundary. Then $T_t f \in \hat{C}_{(-\infty,\infty)}$ for $f \in \hat{C}_{(-\infty,\infty)}$.*

Proof. The continuity of $T_t f$ for continuous bounded functions f has already been verified. Consider the limiting behavior of $T_t f$ at the point $-\infty$.

1) Let $-\infty$ be a natural boundary. Then (66) is fulfilled. Hence for $x < a$ we have

$$|T_t f(x)| \leqslant \mathsf{E}_x |f(x_t)| \chi_{\{\tau_a \geqslant t\}} + \mathsf{E}_x |f(x_t)| \chi_{\{\tau_a < t\}} \leqslant \sup_{y\leqslant a} |f(y)| + \|f\| \mathsf{P}_x\{\tau_a < t\}$$

and

$$\overline{\lim_{x\to -\infty}} |T_t f(x)| \leqslant \sup_{y\leqslant a} |f(y)|,$$

$$\overline{\lim_{x\to -\infty}} |T_t f(x)| \leqslant \lim_{a\to -\infty} \sup_{y\leqslant a} |f(y)| = 0.$$

2) Now let $-\infty$ be an emitting boundary. Then for $x<a$

$$|T_tf(x)-T_tf(a)|=|\mathsf{E}_xf(x_t)\chi_{\{\tau_a\le\delta\}}+\mathsf{E}_xf(x_t)\chi_{\{\tau_a>\delta\}}-T_tf(a)|$$
$$\le 2\|f\|\mathsf{P}_x\{\tau_a>\delta\}+\int_0^\delta \mathsf{P}_x\{\tau_a\in ds\}|T_{t-s}f(a)-T_sf(a)|$$
$$\le 2\|f\|\frac{1}{\delta}\mathsf{E}_x\tau_a+\sup_{s\le\delta}\|T_sf-f\|.$$

Utilizing formula (65) we obtain

$$\sup_{x\le a}|T_tf(x)-T_tf(a)|\le 2\|f\|\frac{1}{\delta}\int_{-\infty}^a[N'(-\infty)-N'(z)]\,dz+\sup_{s\le\delta}\|T_sf-f\|,$$

whence

$$\overline{\lim_{a\to-\infty}}\sup_{x\le a}|T_tf(x)-T_tf(a)|\le\sup_{s\le\delta}\|T_sf-f\|.$$

It now remains only to utilize Lemma 17 and Cauchy's criterion for the existence of $\lim_{x\to-\infty}T_tf(x)$. The lemma is proved. □

If α ($\alpha=\pm\infty$) is an emitting boundary we define $T_tf(\alpha)=\lim_{x\to\alpha}T_tf(x)$ for all $f\in\hat{C}_{(-\infty,\infty)}$. Then

$$T_tf(\alpha)=\int P(t,\alpha,dy)f(y),$$

where $P(t,\alpha,\cdot)$ is a probability measure. This measure may be considered as a transition probability from the boundary point. Therefore emitting boundaries may be adjoined to the phase space of the process. After this is done the phase space becomes either compact if both boundaries are emitting, or locally compact if there is a natural boundary among the boundaries. On this extended space the process will be a Feller stochastically continuous and regular one (in the noncompact case). This follows from Lemmas 17 and 18.

Consider the characteristic operator at the emitting boundary points adjoined to the phase space. For instance, let $-\infty$ be such a point. Then

$$\mathfrak{A}f(-\infty)=\lim_{a\to\infty}\frac{f(a)-f(-\infty)}{\mathsf{E}_{-\infty}\tau_a}=\lim_{a\to-\infty}\frac{f(a)-f(-\infty)}{-\int_{-\infty}^a(a-t)\,dN'(t)}.\tag{67}$$

If

$$f(z)=\int_{-\infty}^z(z-t)\varphi(t)\,dN'(t),$$

where $\varphi\in\hat{C}_{(-\infty,\infty)}$, then, as follows from (59) and (67), $\mathfrak{A}f(x)=-\varphi(x)$ for all $x\in[-\infty,\infty)$.

We now determine the generating operator of the process.

Theorem 17. *Let the points of the interval* $(-\infty, \infty)$ *be regular for the process* x_t, $M(x)=x$, *and the boundary points be repelling. Then the generating operator* A *of the process is defined by equality*

$$\mathsf{A}f(x) = -\frac{df'(x)}{dN'(x)} \tag{68}$$

for all f belonging to $\hat{C}_{(-\infty,\infty)}$ *such that the right-hand side of* (68) *is defined and belongs to* $\hat{C}_{(-\infty,\infty)}$.

Proof. We shall assume that the emitting boundaries are adjoined to the phase space as indicated above. Since the process obtained in this manner will be regular it is sufficient to verify that operator (68) is a generating operator of a regular process (cf. Volume II, Chapter II, Section 4). By Theorem 2 in Volume II, Chapter II, Section 4, it is required for this purpose to establish that the equation

$$\lambda f(x) + \frac{df'(x)}{dN'(x)} = g(x) \tag{69}$$

for $\lambda > 0$ possesses a solution on an everywhere dense set of functions $g \in \hat{C}_{(-\infty,\infty)}$ (it is easy to verify from the form of operator A that it is defined on a certain dense set and satisfies the maximum principle). Consider the following three cases.

1. *Both boundaries are natural.* Choose a finite function $\varphi \in \hat{C}_{(-\infty,\infty)}$ such that the function f satisfying relation $df'/dN' = \varphi$ also belongs to $\hat{C}_{(-\infty,\infty)}$. Then

$$\begin{aligned} f(x) &= f(0) + f'(0)x + \textstyle\int_0^x (x-z)\varphi(z)\,dN'(z) \\ &= f(0) + [f'(0) + \textstyle\int_0^x \varphi(z)\,dN'(z)]x - \textstyle\int_0^x z\varphi(z)\,dN'(z). \end{aligned}$$

Hence $f \in \hat{C}_{(-\infty,\infty)}$, provided

$$f(x) = x\textstyle\int_{-\infty}^x \varphi(z)\,dN'(z) - \textstyle\int_{-\infty}^x z\varphi(z)\,dN'(z), \tag{70}$$

where the function φ is such that

$$\textstyle\int \varphi(z)\,dN'(z) = \textstyle\int z\varphi(z)\,dN'(z) = 0. \tag{71}$$

Therefore f is finite, provided only φ is such, and hence the function g satisfying (69) is also finite. Substituting (70) into (69) we obtain

$$\lambda[x\textstyle\int_{-\infty}^x \varphi(z)\,dN'(z) - \textstyle\int_{-\infty}^x z\varphi(z)\,dN'(z)] + \varphi(x) = g(x). \tag{72}$$

Consider the set of functions g representable in the form (72) is φ is nonvanishing only on $[a, b]$ and satisfies (71). Then g is nonvanishing only on $[a, b]$. Assume that $l(dx)$ is a sign-alternating measure on $[a, b]$ such that $\int g(x)l(dx) = 0$ for all g of the form (72). Integrating (72) and interchanging the order of integration we obtain

$$(73)\qquad \int_a^b \lambda\left[\int_z^b x l(dx) - z\int_z^b l(dx)\right]\varphi(z)\,dN'(z) + \int_a^b \varphi(z) l(dz) = 0.$$

This relation implies that the measure $l(dz)$ is absolutely continuous with respect to $dN'(z)$. Define $\rho(z) = l(dz)/dN'(z)$. We then obtain from (73) that

$$\int_a^b \left\{\lambda\left[\int_z^b x\rho(x)\,dN'(x) - z\int_z^b \rho(x)\,dN'(x)\right] + \rho(z)\right\}\varphi(z)\,dN'(z) = 0.$$

Since φ is an arbitrary function satisfying (71) it follows that

$$\lambda\int_z^b (x-z)\rho(x)\,dN'(x) + \rho(z) = \gamma + \delta z,$$

where γ and δ are constants.

This equality yields the relationship

$$(74)\qquad \lambda\rho(z) + \frac{d\rho'(z)}{dN'(z)} = 0.$$

Equation (74) possesses two solutions, $\rho_1(t)$ and $\rho_2(t)$, satisfying equations

$$(75)\quad \rho_1(x) = -\lambda\int_{-\infty}^z (z-t)\rho_1(t)\,dN'(t), \qquad \rho_2(t) = \lambda\int_z^\infty (z-t)\rho_2(t)\,dN'(t),$$

respectively, and also the following conditions: $\rho_i(z) > 0$, $\rho_i(z)$ are convex downward ($i = 1, 2$); $\rho_1(z)$ increases while $\rho_2(z)$ decreases. Functions $\rho_i(z)$ ($i = 1, 2$) may be defined using the equalities

$$\rho_1(z) = \lim_{a\to-\infty} \frac{\rho(a, z)}{\rho(a, 0)}, \qquad \rho_2(z) = \lim_{a\to+\infty} \frac{\rho(a, z)}{\rho(a, 0)},$$

where $\rho(a, z)$ is the solution of the integral equation

$$\rho(a, z) = 1 - \lambda\int_a^z (z-t)\rho(a, t)\,dN'(t).$$

(The existence and uniqueness of a solution for this equation can be established using the method of successive approximations.)

It is easy to verify that an arbitrary solution of equation (74) on a finite interval is representable in the form of a linear combination of functions $\rho_i(z)$. Hence the set of finite functions g representable by formula (72) under the restrictions (71) is dense in the set of functions g satisfying

$$\int g(x)\rho_i(x)\,dN'(x) = 0, \qquad i = 1, 2.$$

We show that this set of functions is dense in $\hat{C}_{(-\infty,\infty)}$. Observe that

$$-\lambda \int_{-\infty}^{0} \rho_1(x)\, dN'(x) \leqslant -\lambda \int_{-\infty}^{0} (1-t)\rho_1(t)\, dN'(t) \leqslant \rho_1(1) < \infty,$$
$$-\lambda \int_0^{\infty} \rho_1(x)\, dN'(x) \geqslant -\lambda \int_0^{\infty} \rho_1'(0) x\, dN'(x) = +\infty$$

since $+\infty$ is a natural boundary. Analogously,

$$-\int_0^{\infty} \rho_2(x)\, dN'(x) < \infty, \qquad -\int_{-\infty}^{0} \rho_2(x)\, dN'(x) = +\infty.$$

Therefore there exist finite functions $h_i(\varepsilon, x)$ satisfying the conditions

$$\|h_i(\varepsilon, \cdot\,)\| \leqslant 1,$$
$$\int h_i(\varepsilon, x)\rho_i(x)\, dN'(x) = \frac{1}{\varepsilon},$$
$$\left|\int h_i(\varepsilon, x)\rho_j(x)\, dN'(x)\right| < \varepsilon, \qquad i, j = 1, 2, \quad i \neq j.$$

For any finite $g \in \hat{C}_{(-\infty,\infty)}$ one can choose c_1 and c_2 such that

$$\int [g(x) + c_1 h_1(\varepsilon, x) + c_2 h_2(\varepsilon, x)]\rho_i(x)\, dN'(x) = 0, \qquad i = 1, 2.$$

Under this choice of c_i

$$c_i \sim -\varepsilon \int g(x)\rho_i(x)\, dN'(x)$$

and hence $\|c_1 h_1(\varepsilon, \cdot\,) + c_2 h_2(\varepsilon, \cdot\,)\| \to 0$ as $\varepsilon \to 0$. For the case 1 the theorem is proved.

2) *One boundary is natural and the other emitting.* (For example, let $-\infty$ be the emitting boundary.) In this case we shall consider the function $\varphi \in \hat{C}_{(-\infty,\infty)}$ to be finite if it equals zero on $[b, \infty)$ for some b. If $df'/dN' = \varphi$ is a finite function, then

$$f(x) = f(-\infty) + x \int_{-\infty}^{x} \varphi(z)\, dN'(z) - \int_{-\infty}^{x} z\varphi(z)\, dN'(z).$$

This function belongs to $\hat{C}_{(-\infty,\infty)}$, provided

$$\int \varphi(z)\, dN'(z) = 0, \qquad f(-\infty) - \int z\varphi(z)\, dN'(z) = 0.$$

Thus

$$f(x) = x \int_{-\infty}^{x} \varphi(z)\, dN'(z) + \int_x^{\infty} z\varphi(z)\, dN'(z), \qquad \int_{-\infty}^{\infty} \varphi(z)\, dN'(z) = 0,$$

and $f(x)$ is a finite function. Equation (69) can now be rewritten in the form

$$\lambda\left[x \int_{-\infty}^{x} \varphi(z)\, dN'(z) + \int_x^{\infty} z\varphi(z)\, dN'(z)\right] + \varphi(x) = g(x). \tag{76}$$

We show that the set of functions g representable in the form (76) is dense in $\hat{C}_{(-\infty,\infty)}$ provided $\int \varphi(z)\,dN'(z)=0$. The arguments analogous to those presented in the proof of case 1 show that any sign-alternating measure $l(dx)$ such that $\int_{-\infty}^{b} l(dz)g(z)=0$ for all g of the form (76), where $\varphi(t)=0$ for $t\geq b$, is of the form $l(dx)=\rho(x)\,dN'(x)$, where $\rho(x)$ satisfies equation

(77) $$\lambda\left[\int_z^b x\rho(x)\,dN'(x)+z\int_{-\infty}^{z}\rho(x)\,dN'(x)\right]+\rho(z)=c$$

and c is a constant. This equation yields

$$\rho'(z)+\lambda\int_{-\infty}^{z}\rho(x)\,dN'(x)=0;$$

if $\rho(x)>0$, then $\rho'(z)$ increases. Taking into account the uniqueness of the solution of (77) for a given c we verify that a function $\rho(z)$ exists satisfying the conditions: $\rho(z)>0$, $\rho(z)$ increases and is convex downward for any b and c in (77); and for some γ the function $\gamma\rho(z)$ is the unique solution of equation (77) Consequently, it is sufficient to show that the finite functions $g(x)$ such that $\int g(x)\rho(x)\,dN'(x)=0$ are dense in $\hat{C}_{(-\infty,\infty)}$. If $h(\varepsilon,x)$ is a finite function such that $\|h(\varepsilon,\cdot)\|\leq 1$ and $\int h(\varepsilon,x)\rho(x)\,dN'(x)=1/\varepsilon$ (the existence of this function follows from the fact that $\int_0^\infty \rho(x)\,dN'(x)=+\infty$), then the function

$$g_\varepsilon(x)=g(x)-\varepsilon h(\varepsilon,x)\int g(z)\rho(z)\,dN'(z)$$

satisfies the condition $\int g_\varepsilon(x)\rho(x)\,dN'(x)=0$ and $\|g_\varepsilon-g\|\to 0$ as $\varepsilon\to 0$.

3) *Both boundaries are emitting*. Then $\int|z|\,dN'(z)$ is finite. Let $df'/dN'=\varphi$. We have

$$f(x)=c+x\int_{-\infty}^{x}\varphi(z)\,dN'(z)+\int_x^\infty z\varphi(z)\,dN'(z),\qquad \int\varphi(z)\,dN'(z)=0,$$

where c is a constant. Equation (69) can now be rewritten in the form

$$\lambda\int\max\,[x,z]\varphi(z)\,dN'(z)+\lambda c+\varphi(x)=g(x).$$

From this equation φ and c must be determined. This integral equation possesses a solution for all c and, moreover, the solution depends linearly on c. Therefore a unique c exists such that $\int\varphi(z)\,dN'(z)=0$. Theorem 17 is thus completely proved. □

Remarks

Chapter I

§ 1. It was mentioned in Volume I that many important results in the theory of martingales, its applications, and its formalization as a separate branch in the theory of random processes are due to J. L. Doob. The first systematic account of the theory of martingales is given in Doob [1]. J. L. Doob also discovered the decomposition of a submartingale into a sum of an increasing process and a martingale in the case of discrete time. The proof of the existence of such a decomposition for continuous times turned out to be complicated and was given by P. Meyer [1]; in this book one also finds references to the original papers. We have utilized a simpler idea suggested by K. M. Rao [1]. The theorem on regular submartingales is also due to P. Meyer [1]. Local martingales were introduced in the paper by K. Itô and S. Watanabe [1]. D. L. Fisk [1] introduced quasi-martingales into the literature. Necessary and sufficient conditions for a process to be a quasi-martingale were obtained by K. M. Rao [2]. The theory of square integrable martingales was developed in the works of P. Meyer [1], [2] and H. Kunita and S. Watanabe [1].

§ 2. K. Itô [1], [2], [3] introduced and studied stochastic integration of random functions. Integration over square integrable martingales with an absolutely continuous characteristic was considered by J. L. Doob [1]. Further improvements and development of stochastic integrals was given in the works of P. Meyer [2] and C. Doléans-Dade [1].

§ 3. K. Itô [3] established the formula for the stochastic differential of a function of processes possessing a stochastic differential for the case of stochastic integration over a Wiener measure. A generalization to the case of integrals over arbitrary continuous martingales was given in papers by H. Kunita and S. Watanabe [1] and A. V. Skorohod [6]. A generalization of Itô's formula for the case of integration over discontinuous martingales was presented in the paper by I. I. Gihman and A. Ya. Dorogovcev [1], and for the case of integration over a Wiener process and a Poisson measure in the papers by H. Kunita and S. Watanabe [1], A. V. Skorohod [7], and P. Meyer [2]. In H. Kunita and S. Watanabe [1] and in A. V. Skorohod [7] processes which are functionals on a fixed Markov process were considered. There is no such restriction in P. Meyer [2]; however, the isolation of the martingale part of a stochastics differential was given only for this case as well. A multiplicative decomposition of a supermartingale was introduced by K. Itô and S. Watanabe [1]. A more general result was obtained by P. Meyer (cf. also C. Doléans-Dade [1]).

Chapter II

§ 1. Stochastic line integrals are discussed in I. I. Gihman and A. V. Skorohod [1].

§ 2. The term "stochastic differential equations" was introduced by S. N. Bernstein [1] for a certain finite-difference scheme devised to obtain a sequence of Markov chains which in the limit becomes a Markov process of diffusion type (1). A stochastic differential equation for determination of a trajectory of random processes was introduced by I. I. Gihman [1], [3] and in a different form by K. Itô [3], [4]. A subsequent development of Itô's stochastic differential equations was presented in the works of I. V. Girsanov [2], A. V. Skorohod [3], and I. I. Gihman and A. V. Skorohod [1]. K. Itô and M. Nisio [1] suggested the study of stochastic equations with an unrestricted lag. The derivation of A. N. Kolmogorov's equations via stochastic differential equations was given by I. I. Gihman [1], [3] in the diffusion case and by A. V. Skorohod [3] for the discontinuous process. In this book stochastic differential equations in a Hilbert space are not studied. V. V. Baklan [1] and Yu. L. Daleckiĭ [1] initiated these investigations.

§ 3. A. Ya. Khinchin [1] was first to consider convergence of sequences of Markov chains to a continuous time Markov process. The theorem on weak compactness of measures associated with processes constructed from sums of independent random variables is due to Yu. V. Prokhorov [1]. General theorems on convergence of distributions of functionals on processes constructed from sums of independent random variables were studied by Yu. V. Prokhorov [1] and A. V. Skorohod [1] and theorems on convergence of Markov chains to Markov processes by A. V. Skorohod [2]. Theorems on convergence of distributions of functionals on sequences of series of random variables with an arbitrary dependence on the past presented in this text are based on I. I. Gihman [6]. The derivation of limit theorems for stochastic differential equations was stimulated by N. M. Krylov and N. N. Bogolyubov [1]. General limit theorems were obtained in I. I. Gihman [2], [6]. Limit theorems for stochastic equations with a small parameter are the subject of numerous investigations; among those the papers by R. Z. Hasminskiĭ [4], and I. I. Gihman [2] may be mentioned.

Chapter III

§ 1. I. V. Girsanov initiated the study of Itô's processes. One-dimensional Itô processes were studied by M. P. Ershov [1]; he proved the uniqueness theorem on representation of the process (Theorem 2). Theorem 9 and its corollary follow from K. Itô [4]. The proof on the possibility of representing a wide-sense diffusion process in the form of a solution of stochastic differential equation is contained in J. L. Doob [1], Chapter VI, Section 3. Theorem 12 was proved by A. A. Novikov [1].

§ 2. I. V. Girsanov was the first to consider equations of a diffusion type. Theorem 2 easily follows from his results. Theorem 3 is a generalization to the multi-dimensional case of results obtained by M. P. Ershov [2] and R. Sh. Liptser and A. N. Shiryayev [1]. The existence of a solution of the equation in the one-dimensional case was established in K. Itô and M. Nisio [1]. Lemma 3 is due to I. V. Girsanov [2]. The representation of Itô processes as processes of a diffusion type was discovered by A. N. Shiryayev [1] and M. P. Ershov [1].

§ 3. The absolute continuity of measures associated with diffusion processes with identical diffusion has been established by Yu. V. Prokhorov [1], I. V. Girsanov [1], and A. V. Skorohod [3]. Lemma 2 is a minor modification of a lemma due to D. W. Stroock and S. R. S. Varadhan [1]. A. D. Ventzel [1] studied the absolute continuity of transition probabilities of one-dimensional continuous Markov processes. The proof of the existence of a solution utilizing weak compactness of measures was first suggested by A. V. Skorohod [5] for an equation with continuous coefficients. An analogous idea for the proof of the existence of a solution was used by K. Itô and M. Nisio [1] and by D. W. Stroock and S. R. S. Varadhan [1]. A. V. Skorohod [5], I. V. Girsanov [2], and S. Watanabe and T. Yamada [1] studied the uniqueness of a solution for equations with coefficients which do not satisfy the Lipschitz condition. The weak existence and weak uniqueness in the homogeneous case was established by H. Tanaka [1] and N. V. Krylov [1], [2] under the condition of continuity of coefficients. The most general conditions for the weak existence and uniqueness (when the diffusion operator is continuous and nondegenerate) were obtained by D. W. Stroock and S. R. S. Varadhan [1]. The connection between diffusion processes and various problems related to differential equations was studied by M. Kac [1] and R. Z. Has'minskiĭ [1]. General equations involving characteristic operators for Markov processes are presented in E. B. Dynkin [1], Chapter 13. Solutions of boundary problems with the aid of Markov processes are presented in papers by R. Z. Has'minskiĭ [2], [3] and M. I. Freidlin [1], [2]. E. B. Dynkin [3], A. V. Skorohod [4], and A. D. Ventzel [2] studied representations of additive functionals in terms of stochastic integrals.

§ 4. The first five subsections contain the results of A. V. Skorohod [6]. A description of one-dimensional continuous Markov processes was given by W. Feller [1], [2], [3] and E. B. Dynkin [2] and [1], Chapters 15–17.

Bibliography

1. Baklan, V. V.:
[1] Variational differential equations and Markov processes in Hilbert space. Soviet Math. **5**(6), 1553–1557 (1964).
2. Bernstein, S. N.:
[1] Principes de la théorie des équations differentielles stochastiques. Trudy Steklov Fiz–Mat. Institute **5**, 95–125 (1934).
3. Chantladze, T. L.:
[1] Stochastic differential equations in Hilbert space. Commun. Akad. Nauk Gruz. SSR **33**(3), 529–539 (1964) [In Russian]
4. Daleckiĭ, Ju. L.:
[1] Differential equations with functional derivatives and stochastic equations for generalized random processes. Soviet Math. **7**, 220–223 (1966).
5. Doléans-Dade, C.:
[1] Quelques applications de la formule de changement de variables pour les semimartingales. Z. Wahrscheinlichkeitstheorie und verw. Geb. **16**, 181–194 (1970).
6. Doleans-Dade, C., Meyer, P. S.:
[1] Intégrales stochastiques par rapport aux martingales locales. Séminaire de Probabilités IV. Berlin–Heidelberg–New York: Springer-Verlag 1970.
7. Doob, J. L.:
(1) Stochastic Processes. New York: Wiley 1953.
8. Dynkin, E. B.:
[1] Markov Processes. New York: Academic Press 1965 [Translated from Russian by J. Fabius].
[2] One-dimensional continuous strong Markov process. Theory Prob. and Applic. **4**, 1–52 (1959).
[3] Additive functionals of a Wiener process determined by stochastic integrals. Theory Prob. and Applic. **5**, 402–411 (1960).
9. Ershov, M. P.:
[1] Representations of Itô processes. Theory Prob. and Applic. **17**, 165–169 (1972).
[2] On the absolute continuity of measures corresponding to diffusion type processes. Ibid., 169–174 (1972).
10. Feller, W.:
[1] Diffusion processes in one dimension. Trans. Amer. Math. Soc. **77**, 1–31 (1954).
[2] The general diffusion operator and positivity preserving semi-groups in one dimension. Ann. Math. **60**, 427–436 (1954).
[3] On second order differential operators. Ann. Math. **61**, 90–105 (1955).
11. Fisk, D. L.:
[1] Quasi-martingales. Trans. Amer. Math. Soc. **120**, 369–389 (1965).
12. Freidlin, M. I.:
[1] On the stochastic equations of Itô and singular elliptic equations. Izv. Akad. Nauk SSSR, Ser. Mat. **26**(5), 653–676 (1962) [In Russian].
[2] A note on the generalized solution of the Dirichlet problem. Theory Prob. and Applic. **10**, 161–164 (1965).
13. Friedman, A.:
[1] Partial Differential Equations of Parabolic Type, Englewood Cliffs, N.J.: Prentice-Hall 1964.

14. Gihman, I. I.:
[1] On a method of formation of random processes. Dokl. Akad. Nauk SSR **58**, 961–964 (1947) [In Russian].
[2] On differential equation with random functions. Ukr. Mat. Zhurn. **2**(3), 45–69 (1950) [In Russian].
[3] On the theory of differential equations for random processes. Part I. Ukr. Mat. Zhurn. **2**(4), 37–63 (1950); Part II. Idem. **3**(3), 317–339 (1951) [In Russian].
[4] Differential equations with random functions. Winter School in Prob. Theory, Užgorod 1964, Kiev, 41–86 (1964) [In Russian].
[5] On a weak compactness of a set of measures associated with solutions of stochastic differential equations. Mat. fizika, Mežved. Sb., Kiev **7**, 49–65 (1970) [In Russian].
[6] Limit theorems for sequences of series of random variables. Theory of Random Processes, Mežved. Sb., Kiev **2** (1973).
15. Gihman, I. I., Dorogovcev, A. Ya.:
[1] On stability of solutions of stochastic differential equations. Ukr. Mat. Zhurn. **17**(6), 3–21 (1965).
16. Gihman, I. I., Skorohod, A. V.:
[1] Stochastic Differential Equations (Translated by K. Wickwire). Berlin–Heidelberg–New York: Springer-Verlag 1972.
17. Girsanov, I. V.:
[1] On transforming a certain class of stochastic processes by absolutely continuous substitution of measures. Theory Prob. and Applic. **5**, 285–301 (1960).
[2] On Itô's stochastic integral equation. Soviet Mathematics **2**(3), 506–509 (1961).
[3] An example of non-uniqueness of the solution of the stochastic equation of K. Itô. Theory Prob. and Applic. **7**, 325–330 (1962).
18. Has'minskiĭ, R. Z.:
[1] Probability distribution for functional of trajectories of random processes of diffusion type. Dokl. Akad. Nauk SSSR **104**, 22–25 (1955) [In Russian].
[2] A probability approach to boundary problems for elliptic and parabolic equations. Theory Prob. and Applic. **2**, 414–425 (1957).
[3] Diffusion processes and elliptic differential equations degenerating at the boundary of the domain. Theory Prob. and Applic. **3**, 400–413 (1958).
[4] A limit theorem for the solutions of differential equations with random right-hand sides. Theory Prob. and Applic. **11**, 390–406 (1966).
19. Itô, K.:
[1] Stochastic integral. Proc. Japanese Acad. Tokyo **20**, 519–524 (1944).
[2] On a stochastic integral equation. Proc. Japanese Acad. Tokyo **22**, 32–35 (1946).
[3] On a formula concerning stochastic differentials. Nagoya Math. J. **3**, 55–65 (1951).
[4] On stochastic differential equations. Mem. Amer. Math. Soc. **4**, 1–51 (1951).
[5] Multiple Wiener integral. J. Math. Soc. Japan **3**, 157–169 (1951).
20. Itô, K., Watanabe, S.:
[1] Transformation of Markov processes by additive functionals. Ann. Inst. Fourier **15**, 13–30 (1965).
21. Itô, K., McKean, H. P., Jr.:
[1] Diffusion Processes and Their Sample Paths, New York: Academic Press 1965.
22. Itô, K., Nisio, M.:
[1] Stationary solutions of stochastic differential equations. J Math. Kyoto Univ. **4**, 1–75 (1964).
23. Kac, M.:
[1] On some connections between probability theory and differential and integral equations. Proc. 2nd Berkeley Sympos. on Math. Statist. and Probab., Berkeley, 189–215 (1951).
24. Khinchin, A. Ya.:
[1] Asymptotische Gesetze der Wahrscheinlichkeitsrechnung. New York: Chelsea 1948 [Published originally by Springer-Verlag: Berlin 1933].
25. Krylov, N. V.:
[1] On quasi-diffusional processes. Theory Prob. and Applic. **11**, 373–389 (1966).
[2] Itô's stochastic integral equations. Theory Prob. and Applic. **14**, 330–336 (1969).
[3] An inequality in the theory of stochastic integrals. Theory Prob. and Applic. **16**, 438–448 (1971).
26. Krylov, N. M., Bogolyubov, N. N.:
[1] On Fokker–Planck equations derived in perturbation theory by a method based on spectral properties of perturbation Hamiltonian. Zapiski Kaf. Mat. Fiz. Nauk, Akad. Nauk UkrSSR **4**, 5–158 (1939) [In Ukranian].

27. Kunita, H., Watanabe, S.:
[1] On square integrable martingales. Nagoya Math. J. **30**, 209–246 (1967).
28. Liptser, R. Sh., Shiryayev, A. N.:
[1] On absolute continuity of measures associated with processes of diffusion type with respect to the Wiener measure. Izv. A. N. USSR, Ser. Mat. **36**(4), 847–889 (1972) [In Russian].
29. Maruyama, G.:
[1] Continuous Markov processes and stochastic equations. Rend. Circ. Math. Palermo **4**, 1–43 (1955).
30. McKean, H. P., Jr.:
[1] Stochastic Integrals. New York: Academic Press 1969.
31. Meyer, P. A.:
[1] Probability and Potentials. Waltham, Mass.: Blaisdell 1966.
[2] Intégrales stochastiques. Séminaire de Probabilités. Berlin: Springer-Verlag 1977.
32. Novikov, A. A.:
[1] On an identity for stochastic integrals. Theory Prob. and Applic. **17**, 717–720 (1972).
33. Prokhorov, Yu. V.:
[1] Convergence of random processes and limit theorem in probability theory. Theory Prob. and Applic. **1**, 157–216 (1956).
34. Rao, K. M.:
[1] On decomposition theorems of Meyer. Math. Scand. **24**, 66–78 (1969).
[2] Quasi-martingales. Math. Scand. **24**, 79–92 (1969).
35. Schryayev, A. N.:
[1] Nowye resultaty teorii upravlyaemyh sluchajnyh prozessov. Trudy 4. Prashskoy Konferenzii po Teorii Informazii 1965, Praga 1967, pp. 131–203.
36. Skorohod, A. V.:
[1] Limit theorems for stochastic processes with independent increments. Theory Prob. and Applic. **2**, 122–142 (1957).
[2] Limit theorems for Markov processes. Theory Prob. and Applic. **3**, 202–246 (1958).
[3] On differentiability of measures corresponding to random processes, II. Markov processes. Theory Prob. and Applic. **5**, 40–44 (1960).
[4] Additive functionals of a process of Brownian motions. Theory Prob. and Applic. **6**, 396–404 (1961).
[5] Studies in the Theory of Random Processes [Translated from Russian by Scripta Technica]. Reading, Mass: Addison-Wesley 1965.
[6] On the local structure of continuous Markov processes. Theory Prob. and Applic. **11**, 336–372 (1966).
[7] Homogeneous Markov processes without discontinuities of the second kind. Theory Prob. and Applic. **12**, 222–240 (1967).
37. Stratonovich, R. L.:
[1] Uslovnye Markovskie Prozessy i ih Primenenie v Teorii Optimalnogo Upravleniya. Moscow. Izdatelstvo MGU, 1966.
38. Stroock, D. W., Varadhan, S. R. S.:
[1] Diffusion processes with continuous coefficients, I, II. Comm. Pure Appl. Math. **12**, 345–400, 479–530 (1969).
39. Tanaka, H.:
[1] Existence of diffusions with continuous coefficients. Memb. Fac. Sci. Kyushu Univ., Ser A **18**, 89–103 (1964).
40. Ventzel, A. D.:
[1] On the absolute continuity of transition probabilities of an one-dimensional diffusion process. Theory Prob. and Applic. **6**, 404–410 (1961).
[2] On continuous additive functionals of a multidimensional Wiener process. Soviet Math. **3**(1), 264–266 (1962).
41. Watanabe, S.:
[1] On stochastic differential equations for multidimensional diffusion processes with boundary conditions. J. Math. Kyoto Univ. **11**, 169–180 (1971).
42. Watanabe, S., Yamada, T.:
[1] On the uniqueness of solutions of stochastic differential equations. J. Math. Kyoto Univ. **11**, 553–563 (1971).

Appendix: Corrections to Volumes I and II

Volume I

Pp. 340–341. The proof of the boundedness of $m(z)$ presented on pp. 340–341 is erroneous. Below we present a corrected proof.

Since $m_n(z)\uparrow m(z)$ and $m_n(z)$ is continuous, $m(z)$ is continuous from below. Moreover, by virtue of Minkowski's inequality,

$$\begin{aligned}[m(z_1+z_2)]^{1/k} &= [\int |(x, z_1+z_2)|^k \mu(dx)]^{1/k} \\ &\leqslant [\int |(x, z_1)|^k \mu(dx)]^{1/k} + [\int |(x, z_2)|^k \mu(dx)]^{1/k} \\ &\leqslant [m(z_1)]^{1/k} + [m(z_2)]^{1/k}.\end{aligned}$$

Thus $[m(z)]^{1/k}$ is a semiadditive function continuous from below and hence in view of the theorem by I. M. Gelfand (see, e.g., L. V. Kantorovich and G. P. Akilov, *Functional Analysis in Normed Spaces*, Moscow, Fizmatgiz, 1959, p. 233) there exists a constant M such that $[m(z)]^{1/k} \leqslant M|z|$.

P. 408. Omit the last line on this page.

Pp. 525–531. In Section 1 of Chapter VIII two theorems on measurable linear functions and their operators are presented. As stated therein the theorems are not correct. We now present the correct statements and proofs.

A function $l(x)$ is called a measurable linear functional with respect to measure μ on a Hilbert space $(X, \mathfrak{B})$ if $l(x)$ is the limit in measure μ of a sequence of continuous linear functionals $l_n(x)$.

Theorem 1. *In order that a $\mathfrak{B}$-measurable function $l(x)$ be a measurable linear functional with respect to measure μ it is necessary and sufficient that a symmetric convex compact set K exist such that the following conditions be satisfied:*

1) *if $\mathscr{D}$ is a linear hull of K, then $\mu(\mathscr{D})=1$;*
2) *$l(x)$ is linear on $\mathscr{D}$;*
3) *$l(x)$ is continuous on K.*

Proof. Necessity. If $l(x)$ is a measurable linear functional then a sequence of continuous functionals $l_n(x)$ exists such that

$$l(x)=\mu\text{-}\lim_{n\to\infty} l_n(x)$$

and

$$\mu\left(\left\{x: |l_n(x)-l_{n+1}(x)|>\frac{1}{n^2}\right\}\right)\leqslant\frac{1}{n^2}.$$

Set $\mathscr{G}_k=\bigcap_{n=k}^{\infty}\{x: |l_n(x)-l_{n+1}(x)|\leqslant 1/n^2\}$. Clearly, $\mathscr{G}_k$ is a symmetric convex closed set and

$$\mu(\mathscr{G}_k)=1-\sum_{n\geqslant k}\mu\left(\mathscr{X}-\left\{x: |l_n(x)-l_{n+1}(x)|\leqslant\frac{1}{n^2}\right)\geqslant 1-\sum_{n\geqslant k}^{\infty}\frac{1}{n^2}.$$

Since $l_n(x)$ is uniformly convergent to $l(x)$ on each one of the sets $\mathscr{G}_k$, it is convergent to $l(x)$ on the linear hull $\bar{\mathscr{G}}_k$ of the set $\mathscr{G}_k$ and hence $l(x)$ is linear on $\bar{\mathscr{G}}_k$. Let F_k be a symmetric convex compact such that $F_k\subset\mathscr{G}_k$, $\mu(\mathscr{G}_k-F_k)<1/k$, and $\mathscr{D}_k$ be the linear hull of F_k. Choose a sequence $\rho_k\downarrow 0$ such that

$$\sum\rho_k\left(\sup_{x\in F_k}|x|+\sup_n\sup_{x\in F_k}|l_n(x)|\right)<\infty$$

and let

$$K=\{x: x=\sum\rho_k x_k,\ x_k\in F_k\}.$$

It is easy to see that: a) K is a symmetric convex compact set; b) the linear hull $\mathscr{D}$ of the set K contains all the sets $\mathscr{D}_k$ and hence $\mu(\mathscr{D})=1$; c) $l_n(x)$ converges to $l(x)$ uniformly on K and therefore $l(x)$ is linear on $\mathscr{D}$. The necessity of the theorem's conditions is thus verified.

Sufficiency. Denote $K_n=\{x: (1/n)x\in K\}$. Then $\mathscr{D}=\bigcup_n K_n$ and hence for any $\varepsilon>0$ there exists n such that $\mu(K_n)>1-\varepsilon$. Clearly K_n is a symmetric convex compact set. We show that for any $\delta>0$ a continuous linear functional $\varphi(x)$ exists such that $|\varphi(x)-l(x)|<\delta$ for $x\in K_n$. Set

$$S_1=\left\{x: l(x)\geqslant\frac{\delta}{2}\right\}\cap K_n,\qquad S_2=\left\{x: l(x)\leqslant-\frac{\delta}{2}\right\}.$$

In view of conditions 2) and 3), S_1 and S_2 are convex compacts symmetrical with respect to the origin 0. Therefore there exists a hyperplane which passes through the origin and separates these sets. Let this hyperplane be represented by $\{x: (a,x)=0\}=L$. Denote by $\varphi(x)$ the functional

$$\varphi(x)=l(x_0)\frac{(a,x)}{(a,x_0)},$$

where $x_0 \in S_1$ is a point such that (a, x_0) is maximal. $\varphi(x)$ is a continuous functional. Furthermore,

$$|l(x)-\varphi(x)| = \left|l(x)-l\left(x_o\frac{(a, x)}{(a, x_0)}\right)\right| = \left|l\left(x - x_0\frac{(a, x)}{(a, x_0)}\right)\right|.$$

However,

$$\frac{1}{2}\left(x - x_0\frac{(a, x)}{(a, x_0)}\right) \in K_n \cap L \subset K_n \setminus S_1 \setminus S_2,$$

since $|(a, x)/(a, x_0)| \leq 1$. Hence

$$\left|l\left(\frac{1}{2}\left(x - x_0\frac{(a, x)}{(a, x_0)}\right)\right)\right| \leq \frac{\delta}{2}, \qquad |l(x)-\varphi(x)| < \delta. \quad \square$$

Remark. If $l(x)$ is a μ-measurable linear functional, then there exists an orthonormal basis $\{e_k\}$ in $\mathscr{D}$ such that $l(x)$ is the limit in measure μ of the sequence $l(P_n x)$. Indeed, let $l(x) = \lim_{n\to\infty} (a_n, x)$ in measure μ. There exists an everywhere positive function $\rho(x)$ such that

$$\lim_{n\to\infty} \int [(a_n, x) - l(x)]^2 \rho(x)\mu(dx) = 0, \qquad \int |x|^2 \rho(x)\mu(dx) < \infty.$$

Let A be a symmetric operator satisfying

$$(Az, z) = \int (z, x)^2 \rho(x)\mu(dx).$$

As follows from the lemma in Volume I, Chapter V, Section 5, A is a symmetric kernel operator. Denote its eigenvectors by $\{e_k\}$ and the corresponding eigenvalues by λ_k. Then

$$\begin{aligned}\int [(a_m, x) - (a_n, x)]^2 \rho(x)\mu(dx) &= (A(a_n - a_m), a_n - a_m) \\ &= \sum_{k=1}^{\infty} \lambda_k [(a_n, e_k) - (a_m, e_k)]^2.\end{aligned}$$

Therefore the limits $\lim_{n\to\infty} (a_n, e_k) = \alpha_k$ exist such that $\sum_{k=1}^{\infty} \alpha_k^2 \lambda_k < \infty$, $l(x) = \sum_{k=1}^{\infty} \alpha_k (x, e_k)$, where the series is convergent in measure μ (cf. Volume I, Chapter V, Section 6).

A measurable function $A(x)$ defined on $\mathscr{X}$ with values in $\mathscr{X}$ is called a *measurable linear operator* if a sequence of linear operators A_n exists such that $A_n x$ is weakly convergent to $A(x)$ in measure μ, i.e., for all $y \in \mathscr{X}$ the numerical sequence $(A_n x, y)$ converges in measure μ to $(A(x), y)$.

Theorem 2. *In order that $A(x)$ be a measurable linear operator it is necessary and sufficient that there exist a symmetric compact set K such that the following conditions are satisfied:*

1) *if $\mathscr{D}$ is a linear hull of K, then $\mu(\mathscr{D})=1$;*
2) *$A(x)$ is linear on $\mathscr{D}$;*
3) *$A(x)$ is continuous on K.*

Proof. The necessity of the theorem's conditions is established in exactly the same manner as in Theorem 1. We now prove their sufficiency. Let $\mathscr{L}_n$ be a monotone sequence of finite-dimensional subspaces such that $\bigcup \mathscr{L}_n$ is dense in $\mathscr{X}$ and let P_n be the projection operator on $\mathscr{L}_n$. Since for all x for which $A(x)$ is defined $P_nA(x)\to A(x)$ as $n\to\infty$ it is sufficient to show that the operator $P_nA(x)$ for each n is a limit of a sequence of operators $A_m^{(n)}(x)$ convergent in measure μ. However, in that case $P_nA_m^{(n)}x$ also converges in measure μ to $P_nA(x)$. In order that the latter be fulfilled it is sufficient that $(A_m^{(n)}x, e_k)$ converge in measure μ to $(P_nA(x), e_k)$, where $\{e_k\}$ is a basis such that its intercepts are bases in $\mathscr{L}_n$. Since $(P_nA(x), e_k)=(A(x), e_k)$ is a μ-measurable linear functional, in view of Theorem 1 one can find a sequence of vectors $a_k^{(m)}$ such that

$$(x, a_k^{(m)})\to(A(x), e_k)$$

as $m\to\infty$ in measure μ. However, then

$$\sum_{k=1}^{\infty}(x, a_k^{(m)})P_ne_k\to P_nA(x)$$

also in measure μ (only a finite number of summands are nonzero in the sum on the left of the last relationship). Thus the required sequence of operator $A_m^{(n)}$ is defined by the equation

$$(A_m^{(n)}x, e_k)=\begin{cases}(x, a_k^{(m)}), & e_k\in\mathscr{L}_n\\ 0, & e_k\notin\mathscr{L}_n,\end{cases}$$

and the sufficiency of the conditions of Theorem 2 is verified. □

Volume II

P. 321, line 1. Replace "so that" with "since."

Subject Index